Multiple Cholecystokinin Receptors
in the CNS

Multiple Cholecystokinin Receptors in the CNS

Edited by

C. T. DOURISH, S. J. COOPER,
S. D. IVERSEN, and L. L. IVERSEN

Oxford   New York   Tokyo
OXFORD UNIVERSITY PRESS
1992

Oxford University Press, Walton Street, Oxford OX2 6DP
Oxford   New York   Toronto
Delhi   Bombay   Calcutta   Madras   Karachi
Petaling Jaya   Singapore   Hong Kong   Tokyo
Nairobi   Dar es Salaam   Cape Town
Melbourne   Auckland
and associated companies in
Berlin   Ibadan

Oxford is a trade mark of Oxford University Press

Published in the United States
by Oxford University Press, New York

A catalogue record for this book is available from the British Library

Library of Congress Cataloging in Publication Data
Multiple cholecystokinin receptors in the CNS / edited by C. T. Dourish . . . [et al.].
Based on a symposium held at Neuroscience Research Centre in
Terlings Park, Essex, U.K.
Includes bibliographical references.
1. Cholecystokinin — Receptors — Congresses.   2. Cholecystokinin —
Antagonists — Congresses.   3. Central nervous system — Physiology —
Congresses.   4. Cholecystokinin — Physiological effect — Congresses.
[DNLM: 1. Central Nervous System Diseases — drug therapy —
congresses.   2. Receptors, Cholecystokinin — physiology — congresses.
WL 300 M961]
QP364.7.M84   1992   616.8′061 — dc20   91-37284

ISBN 0 19 857756 7

Typeset by Colset Private Ltd., Singapore
Printed in Great Britain by
Biddles Ltd, Guildford and King's Lynn

Preface

L.L. Iversen

Cholecystokinin (CCK) epitomizes the typical 'gut–brain peptide'. First discovered because of its role in the gastrointestinal system, it has only been recognized as a 'neuropeptide' quite recently. The existence of CCK in brain was first described by Vanderhaeghen and colleagues in the mid-1970s, and the first international symposium on CCK in the nervous system was organized in Brussels in 1984 (Vanderhaeghen and Crawley 1985).

Since then a great deal of information about the neuronal localizations of this peptide has been obtained by immunohistochemical studies and latterly by *in situ* hybridization studies. It has proved to be one of the most abundant of the brain peptides, present in large numbers of the principal neurons in the cerebral cortex and other brain regions.

Another important development during the past decade has been the recognition that CCK receptors in the brain are predominantly of a different subtype (CCK_B) from those found in the alimentary tract (CCK_A). The discovery in Merck Sharp & Dohme Research Laboratories of a series of non-peptide antagonists with A and B selectivity derived from the natural product asperlicin represents a major theme of the present volume. The availability for the first time of potent CCK antagonist drugs led to an upsurge of research activity aiming to understand the functional importance of CCK in both the central nervous system (CNS) and the periphery. We have supplied research samples of the Merck antagonists to many hundreds of investigators, and this, together with our own in-house research, has yielded several hundred publications in the scientific literature (Fig. 1).

This volume summarizes the proceedings of a symposium held to review this area of neuroscience research. It is the third such symposium held at the Neuroscience Research Centre in Terlings Park, Essex, UK. The earlier meetings also resulted in publications by Oxford University Press (Iversen and Goodman 1986; Stahl *et al*. 1987). We were fortunate in having many of the leading scientists involved in research on CCK present at this CCK Symposium and contributing to this volume. The latest information on the anatomical distribution of CCK in CNS and its pharmacology and possible functions is reviewed by these experts. New non-peptide CCK antagonists are also described by colleagues from the Parke-Davis and Eli Lilly Research Laboratories.

There are four broad areas of CNS function in which CCK has been implicated. These form the main themes of later sections of this volume, and may

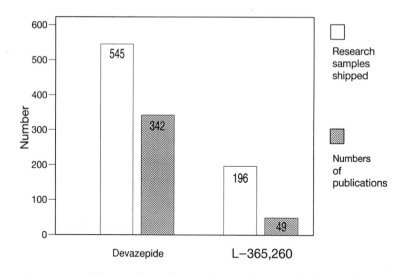

Fig. 1. Summary of the number of research samples supplied to academic investigators and publications in the scientific literature on devazepide and L-365,260.

in the future represent areas in which CCK antagonists could find therapeutic applications. They are the role of CCK in anxiety/panic disorders, its role in appetite control mechanisms, the interaction of CCK with opiate analgesic mechanisms, and the interaction of CCK with dopaminergic mechanisms.

It is still too early to know whether CCK antagonists will find important therapeutic niches in the treatment of CNS disorders in man. These compounds will undoubtedly continue to represent important research tools for neuropeptide research for many years to come.

References

Iversen, L. L. and Goodman, E. (ed.) (1986). *Fast and slow chemical signalling in the nervous system.* Oxford University Press, Oxford.

Stahl, S. M., Iversen, S. D., and Goodman, E. (ed.) (1987). *Cognitive neurochemistry.* Oxford University Press, Oxford.

Vanderhaeghen, J. J. and Crawley, J. N. (ed.) (1985). Neuronal Cholecystokinin. *Ann. NY Acad. Sci.* **448**.

Contents

Contributors

Mark C. Austin
Unit on Behavioral Neuropharmacology, Clinical Neuroscience Branch,
National Institute of Mental Health, Bethesda, MD 20892, USA.

Joanne Bacon
AFRC Institute of Animal Physiology and Genetics Research, Babraham,
Cambridge, CB2 4AT, UK.

B.A. Baldwin
AFRC Institute of Animal Physiology and Genetics Research, Babraham,
Cambridge CB2 4AT, UK.

G. Ballejo
Department of Pharmacology, FMRP, Campus USP, Ribeirao Preto, SP,
BR 14049, Brazil.

R. Bandopadhyay
Department of Biochemistry, Charing Cross and Westminster Medical
School, Fulham Palace Road, London W6 8RF, UK.

M.C. Beinfeld
Department of Pharmacology, St Louis University School of Medicine,
St Louis, MO 63104, USA.

J. de Belleroche
Department of Biochemistry, Charing Cross and Westminster Medical
School, Fulham Palace Road, London W6 8RF, UK.

J.J. Benoliel
INSERM U 288, Neurobiologie Cellulaire et Fonctionnelle, Faculté de
Médecine Pitié-Salpêtrière, 91, Boulevard de l'Hôpital, 75013 Paris, France.

Florence Bergeron
INSERM U 266, CNRS UA 498, Facultê de Pharmacie, 4 Avenue de
l'Observatoire, Paris VI, France.

John E. Blundell
Department of Psychology, University of Leeds, Leeds LS2 9JT, UK.

P. Boden
Parke-Davis Research Unit, Addenbrooke's Hospital Site, Hills Road, Cambridge CB2 2QB, UK.

S. Bourgoin
INSERM U 288, Neurobiologie Cellulaire et Fonctionnelle, Faculté de Médecine Pitié-Salpêtrière, 91, Boulevard de l'Hôpital, 75013 Paris, France.

Susan Boyce
Merck Sharp & Dohme Research Laboratories, Neuroscience Research Centre, Terlings Park, Eastwick Road, Harlow, Essex CM20 2QR, UK.

Jacques Bradwejn
McGill University, St Mary's Hospital Centre, Montreal, Quebec, Canada.

R. F. Brown
Lilly Research Laboratories, Eli Lilly & Company, Indianapolis, IN 46285, USA.

F. Cesselin
INSERM U 288, Neurobiologie Cellulaire et Fonctionnelle, Faculté de Médecine Pitié-Salpêtrière, 91, Boulevard de l'Hôpital, 75013 Paris, France.

Robin W. Clarke
Department of Physiology and Environmental Science, University of Nottingham School of Agriculture, Sutton Bonington, Loughborough, Leics LE12 5RD, UK.

Peter G. Clifton
Laboratory of Experimental Psychology, University of Sussex, Brighton BN1 9QG, UK.

Steven J. Cooper
School of Psychology, The University of Birmingham, Edgbaston, Birmingham B15 2TT, UK.

Pierre-Jean Corringer
INSERM U 266, CNRS UA 498, Faculté de Pharmacie, 4 Avenue de l'Observatoire, Paris VI, France.

Jacqueline N. Crawley
Unit on Behavioral Neuropharmacology, Clinical Neuroscience Branch, National Institute of Mental Health, Bethesda, MD 20892, USA.

Valérie Daugé
INSERM U 266, CNRS UA 498, Faculté de Pharmacie, 4 Avenue de
l'Observatoire, Paris VI, France.

J.C. De Aguiar
Department of Pharmacology, FMRP, Campus USP, Ribeirao Preto, SP,
BR 14049, Brazil.

Muriel Derrien
INSERM U 266, CNRS UA 498, Faculté de Pharmacie, 4 Avenue de
l'Observatoire, Paris VI, France.

Anthony H. Dickenson
Department of Pharmacology, University College London, London
WC1E 6BT, UK.

G.J. Dockray
MRC Secretory Control Research Group, Department of Physiology,
University of Liverpool, Liverpool, UK.

Colin T. Dourish
Wyeth Research (UK) Ltd, Huntercombe Lane South, Taplow,
Maidenhead, Berkshire SL6 0PH, UK.

Christiane Durieux
INSERM U 266, CNRS UA 498, Faculté de Pharmacie, 4 Avenue de
l'Observatoire, Paris VI, France.

Piers C. Emson
AFRC Institute of Animal Physiology and Genetics Research, Babraham,
Cambridge CB2 4AT, UK.

John R. Evers
Unit on Behavioral Neuropharmacology, Clinical Neuroscience Branch,
National Institute of Mental Health, Bethesda, MD 20892, USA.

Susan M. Fiske
Unit on Behavioral Neuropharmacology, Clinical Neuroscience Branch,
National Institute of Mental Health, Bethesda, MD 20892, USA.

Timothy W. Ford
Department of Physiology and Environmental Science, University of
Nottingham School of Agriculture, Sutton Bonington, Loughborough, Leics
LE12 5RD, UK.

John Francis
School of Psychology, The University of Birmingham, Edgbaston,
Birmingham B15 2TT, UK.

Dale Frankenfield
Merck Sharp & Dohme Research Laboratories, West Point, PA 19486,
USA.

Roger M. Freidinger
Merck Sharp & Dohme Research Laboratories, West Point, PA 19486,
USA.

P. Frey
Sandoz Research Institute, CH-3001, Bern, Switzerland.

J. Gibbs
Department of Psychiatry, Cornell University Medical College, E.W.
Bourne Behavioral Research Laboratory, New York Hospital–Cornell
Medical Center, White Plains, NY 10605, USA.

M. Goldstein
Department of Psychiatry, New York University Medical Center,
New York, NY 10016, USA.

F.G. Graeff
Laboratory of Psychobiology, FFCLRP, Campus USP, Ribeirao Preto, SP,
BR 14049, Brazil.

F.S. Guimarães
Department of Pharmacology, FMRP, Campus USP, Ribeirao Preto, SP,
BR 14049, Brazil.

Mitsuko Hamamura
AFRC Institute of Animal Physiology and Genetics Research, Babraham,
Cambridge, CB2 4AT, UK.

M. Hamon
INSERM U 288, Neurobiologie Cellulaire et Fonctionnelle, Faculté de
Médecine Pitié-Salpêtrière, 91, Boulevard de l'Hôpital, 75013 Paris, France.

J.S. Han
Department of Physiology, Beijing Medical University, 38 Xue Yuan Road,
Beijing 100083, China.

Richard J. Hargreaves
Merck Sharp & Dohme Research Laboratories, Neuroscience Research
Centre, Terlings Park, Eastwick Road, Harlow, Essex CM20 2QR, UK.

John Harris
Department of Physiology and Environmental Science, University of
Nottingham School of Agriculture, Sutton Bonington, Loughborough, Leics
LE12 5RD, UK.

Jaanus Harro
Psychopharmacology Laboratory, Tartu University, 34 Veski Street, 202400
Tartu, Estonia.

C.A. Hendrie
Department of Psychology, University of Leeds, Leeds LS2 9JT, UK.

D.R. Hill
Parke-Davis Research Unit, Addenbrooke's Hospital Site, Hills Road,
Cambridge CB2 2QB, UK.

J.P. Hodgkiss
Department of Pharmacology, University of Edinburgh, Edinburgh
EH8 9JZ, UK.

T. Hökfelt
Department of Histology and Neurobiology, Karolinska Institutet, S-10401
Stockholm, Sweden.

J.J. Howbert
Lilly Research Laboratories, Eli Lilly & Company, Indianapolis, IN 46285,
USA.

J. Hughes
Parke-Davis Research Unit, Addenbrooke's Hospital Site, Hills Road,
Cambridge CB2 2QB, UK.

P.H. Hutson
Merck Sharp & Dohme Research Laboratories, Neuroscience Research
Centre, Terlings Park, Eastwick Road, Harlow, Essex CM20 2QR, UK.

L.L. Iversen
Merck Sharp & Dohme Research Laboratories, Neuroscience Research
Centre, Terlings Park, Eastwick Road, Harlow, Essex CM20 2QR, UK.

Susan D. Iversen

Merck Sharp & Dohme Research Laboratories, Neuroscience Research Centre, Terlings Park, Eastwick Road, Harlow, Essex CM20 2QR, UK.

Robert Jackson

Merck Sharp & Dohme Research Laboratories, West Point, PA 19486, USA.

G. J. van Kamp

Department of Clinical Chemistry, Free University Hospital, De Boelelaan 1117, 1081 HV Amsterdam, The Netherlands.

David E. Kellstein

The Procter & Gamble Company, Health and Personal Care Technology, Miami Valley Laboratories, Cincinnati, OH 45239, USA.

J. S. Kelly

Department of Pharmacology, University of Edinburgh, Edinburgh EH8 9JZ, UK.

J. A. Kemp

Merck Sharp & Dohme Research Laboratories, Neuroscience Research Centre, Terlings Park, Harlow, Essex CM20 2QR, UK.

J. N. C. Kew

Merck Sharp & Dohme Research Laboratories, Neuroscience Research Centre, Terlings Park, Harlow, Essex CM20 2QR, UK.

S. J. Kitchener

Merck Sharp & Dohme Research Laboratories, Neuroscience Research Centre, Terlings Park, Eastwick Road, Harlow, Essex CM20 2QR, UK.

Hiroshi Kiyama

MRC Group, Department of Neuroendocrinology, AFRC Institute of Animal Physiology and Genetics Research, Babraham, Cambridge CB2 4AT, UK.

Hilton Klein

Merck Sharp & Dohme Research Laboratories, West Point, PA 19486, USA.

Diana Koszycki

McGill University, St Mary's Hospital Centre, Montreal, Quebec, Canada.

M.A. Kuiper
Department of Neurology, Free University Hospital, De Boelelaan 1117, 1081 HV Amsterdam, The Netherlands.

Aavo Lang
Psychopharmacology Laboratory, Tartu University, 34 Veski Street, 202400 Tartu, Estonia.

Gareth Leng
AFRC Institute of Animal Physiology and Genetics Research, Babraham, Cambridge CB2 4AT, UK.

Rodger A. Liddle
Department of Medicine, PO Box 3083, Duke University Medical Center, Durham, NC 27710, USA.

Juinn H. Lin
Merck Sharp & Dohme Research Laboratories, West Point, PA 19486, USA.

A. Lindén
Department of Psychiatry and Clinical Research Centre, Karolinska Institutet, S-141 86 Huddinge, Sweden.

K.L. Lobb
Lilly Research Laboratories, Eli Lilly & Company, Indianapolis, IN 46285, USA.

David S. Magnuson
Department of Pharmacology, University College London, London WC1E 6BT, UK.

G.R. Marshall
Merck Sharp & Dohme Research Laboratories, Neuroscience Research Centre, Terlings Park, Harlow, Essex CM20 2QR, UK.

N.R. Mason
Lilly Research Laboratories, Eli Lilly & Company, Indianapolis, IN 46285, USA.

A. Mauborgne
INSERM U 288, Neurobiologie Cellulaire et Fonctionnelle, Faculté de Médecine Pitié-Salpêtrière, 91, Boulevard de l'Hôpital, 75013 Paris, France.

David J. Mayer
Department of Physiology, Medical College of Virginia, Virginia
Commonwealth University, Richmond, VA 23298, USA.

Gwendolyn McCormick
Merck Sharp & Dohme Research Laboratories, West Point, PA 19486,
USA.

E. McGowan
AFRC Institute of Animal Physiology and Genetics Research, Babraham,
Cambridge CB2 4AT, UK.

Paul R. McHugh
Department of Psychiatry and Behavioral Sciences, Johns Hopkins
University School of Medicine, Baltimore, MD 21205, USA.

L. G. Mendelsohn
Lilly Research Laboratories, Eli Lilly & Company, Indianapolis, IN 46285,
USA.

Tamara Montgomery
Merck Sharp & Dohme Research Laboratories, West Point, PA 19486,
USA.

Timothy H. Moran
Department of Psychiatry and Behavioral Sciences, Johns Hopkins
University School of Medicine, Baltimore, MD 21205, USA.

Kathy Murray
Charles River Laboratories, 251 Ballardvale Street, Wilmington, MA 01887,
USA.

D. A. Neel
Lilly Research Laboratories, Eli Lilly & Company, Indianapolis, IN 46285,
USA.

J. C. Neill
Department of Psychology, University of Leeds, Leeds LS2 9JT, UK.

Michael F. O'Neill
Laboratorios Almirall, Cardoner 68–74, 08024 Barcelona, Spain.

Mina Patel
Department of Gastrointestinal Pharmacology, Glaxo Group Research
Limited, Ware, Herts SG12 0DP, UK.

R. Pinnock
Parke-Davis Research Unit, Addenbrooke's Hospital Site, Hills Road,
Cambridge CB2 2QB, UK.

Anu Pôld
Psychopharmacology Laboratory, Tartu University, 34 Veski Street, 202400
Tartu, Estonia.

Walter Pouch
Merck Sharp & Dohme Research Laboratories, West Point, PA 19486,
USA.

J.K. Reel
Lilly Research Laboratories, Eli Lilly & Company, Indianapolis, IN 46285,
USA.

Jens F. Rehfeld
Department of Clinical Biochemistry, State University Hospital
(Rigshospitalet), DK-2100, Copenhagen, Denmark.

P.H. Robinson
Institute of Psychiatry, De Crespigny Park, Denmark Hill, London
SE5 8AF, UK.

Raquel E. Rodriguez
Department of Biochemistry, Faculty of Medicine, University of
Salamanca, 37007 Salamanca, Spain.

Peter J. Rogers
Psychobiology Section, Consumer Sciences Department, AFRC Institute of
Food Research, Shinfield, Reading RG2 9AT, UK.

Bernard P. Roques
INSERM U 266, CNRS UA 498, Faculté de Pharmacie, 4 Avenue de
l'Observatoire, Paris VI, France.

Mariano Ruiz-Gayo
INSERM U 266, CNRS UA 498, Faculté de Pharmacie, 4 Avenue de
l'Observatoire, Paris VI, France.

Nadia M.J. Rupniak
Merck Sharp & Dohme Research Laboratories, Neuroscience Research
Centre, Terlings Park, Eastwick Road, Harlow, Essex, CM20 2QR, UK.

A.S. Russo
Laboratory of Psychobiology, FMRP FFCLRP, Campus USP, Ribeirao Preto, SP, BR 14049, Brazil.

Maria P. Sacristan
Department of Biochemistry, Faculty of Medicine, University of Salamanca, 37007 Salamanca, Spain.

M. Schalling
Department of Histology and Neurobiology, Karolinska Institutet, S-104 01 Stockholm, Sweden.

S.N. Schiffmann
Laboratory of Neuropathology and Neuropeptide Research, Brugmann and Erasme Academic Hospitals, Université Libre de Bruxelles, Campus Anderlecht, CP 601 B-1070, Brussels, Belgium.

K. Seroogy
Department of Anatomy and Neurobiology, Albert B. Chandler Medical Center, University of Kentucky, Lexington, KY 40536–0084, USA.

L. Singh
Parke-Davis Research Unit, Addenbrooke's Hospital Site, Hills Road, Cambridge CB2 2QB, UK.

A.D. Smith
University Department of Pharmacology, South Parks Road, Oxford OX1 3QT.

G.P. Smith
Department of Psychiatry, Cornell University Medical College, E.W. Bourne Behavioral Research Laboratory, New York Hospital–Cornell Medical Center, White Plains, NY 10605, USA.

P. Södersten
Department of Psychiatry and Clinical Research Centre, Karolinska Institutet, S-141 86 Huddinge, Sweden.

Keith Soper
Merck Sharp & Dohme Research Laboratories, West Point, PA 19486, USA.

Colin F. Spraggs
Department of Gastrointestinal Pharmacology, Glaxo Group Research Limited, Ware, Herts SG12 0DP, UK.

A. Jon Stoessl

Department of Clinical Neurological Sciences, University Hospital and Robarts Research Institute, University of Western Ontario, London, Ontario N6A 5A5, Canada.

Gillian Sturman

Division of Physiology and Pharmacology, Polytechnic of East London, Romford Road, London E15 4LZ, UK.

Ann F. Sullivan

Department of Pharmacology, University College London, London WC1E 6BT, UK.

N. Suman-Chauhan

Parke-Davis Research Unit, Addenbrooke's Hospital Site, Hills Road, Cambridge CB2 2QB, UK.

E. Szczutkowski

Department of Clinical Neurological Sciences, University Hospital and Robarts Research Institute, University of Western Ontario, London, Ontario N6A 5A5, Canada.

S. Totterdell

University Department of Pharmacology, South Parks Road, Oxford OX1 3QT, UK.

Spencer Tye

Merck Sharp & Dohme Research Laboratories, Neuroscience Research Centre, Terlings Park, Eastwick Road, Harlow, Essex CM20 2QR, UK.

J.-J. Vanderhaeghen

Laboratory of Neuropathology and Neuropeptide Research, Brugmann and Erasme Academic Hospitals, Université Libre de Bruxelles, Campus Anderlecht, CP 601 B-1070, Brussels, Belgium.

Eero Vasar

Psychopharmacology Laboratory, Tartu University, 34 Veski Street, 202400 Tartu, Estonia.

J. Walsh

Center for Ulcer Research and Education (CURE), Los Angeles, CA 90073, USA.

E. Ch. Wolters

Department of Neurology, Free University Hospital, De Boelelaan 1117, 1081 HV Amsterdam, The Netherlands.

G. N. Woodruff

Parke-Davis Research Unit, Addenbrooke's Hospital Site, Hills Road, Cambridge CB2 2QB, UK.

Part I

CCK neurons and receptors in the CNS

1. CCK neurons and receptors in the CNS: introduction

G. J. Dockray

Studies of the function and dysfunction of CNS systems that use CCK as a transmitter have now entered an especially exciting phase. In particular, the development in recent years of orally active antagonists for CCK receptors has opened the way for experimental studies that were hitherto impossible. The idea that CCK has a transmitter function in the CNS has developed over the last 15 years on the basis of evidence from many different disciplines. Even so, the availability of antagonists has been of crucial importance because for the first time it has become possible to formulate and test specific hypotheses of the functions of CCK in the intact nervous system.

The familiar sulphated CCK octapeptide has been recognized for some years as the most abundant of the CCK group of peptides in the CNS; while many other forms of CCK have also been identified in brain, their existence can be attributed to variations in (or incomplete) post-translational processing of the CCK8 precursor first characterized by cDNA sequencing (see Dockray 1988). The combined application of peptide immunohistochemistry and localization of mRNA by *in situ* hybridization has made it possible to define with some confidence the distribution of CCK in the CNS, including the identification of its patterns of coexistence with other transmitter substances. However, whether or not CCK is the only endogenous ligand for central CCK receptors remains an open question. In this context it is interesting, for example, that there is now evidence to indicate that in pig and rat brain there may be a second member of the gastrin/CCK group which has a closer resemblance to the amphibian peptide caerulein than to CCK (Varro 1989). The possible representation in mammals of peptides first found in lower species is also of interest in view of two recent studies on submammalian members of the family (Fig. 1.1). The characteristic feature that distinguishes mammalian CCK and gastrin is usually considered to be a sulphated tyrosine residue at position 7 (CCK) or 6 (gastrin) counting from the C-terminus. In chicken gastrin, there is a sulphated tyrosine in the characteristic position of CCK followed by a proline residue; the latter appears to shift the orientation of the tyrosine side-chain relative to the C-terminus and to lower affinity at CCK_A type receptors. This peptide therefore appears at first sight to be a CCK but its biological properties are clearly those of a gastrin (Dimaline *et al.* 1986; Dimaline and Lee 1990). In the case

CCK	Arg–Asp–Tyr–Met–Gly–Trp–Met–Asp–Phe–NH$_2$
Gastrin	Glu–Glu–Ala–Tyr–Gly–Trp–Met–Asp–Phe–NH$_2$
Caerulein	Gln–Asp–Tyr–Thr–Gly–Trp–Met–Asp–Phe–NH$_2$
Chicken gastrin	His–Phe–Tyr–Pro–Asp–Trp–Met–Asp–Phe–NH$_2$
Cionin	Asn–Tyr–Tyr–Gly–Trp–Met–Asp–Phe–NH$_2$

Fig. 1.1. Amino acid sequences of the major representatives of the CCK family (C-terminal nonapeptide amides only are shown). In each case the tyrosine residue is known to exist in the sulphated form; caerulein has CCK-like biological actions but chicken gastrin, which is CCK-like in the primary amino acide sequence (i.e. sulphated tyrosine at position 7 from the C-terminus) has a gastrin-like pattern of actions. In chemical terms, cionin is both a gastrin and a CCK.

of cionin, from the neural gland of *Ciona intestinalis*, there are sulphated tyrosines in both positions, so that this peptide is both a gastrin and a CCK (Johnsen and Rehfeld 1990). Thus it appears that across a range of species a number of different peptide structures have evolved that together challenge present ideas of the factors determining the affinity of the natural ligands for different types of gastrin/CCK receptor. Among other things, it will be of obvious interest to see if peptides similar to those mentioned above occur in mammals.

The idea that different types of receptor distinguish between CCK and gastrin first developed from bioassay work done 20 years ago. Autoradiographic studies have since clearly indicated that within the CNS there is a major binding site (the B-receptor in the terminology of Moran *et al.* (1986)) that differentiates poorly between gastrin and CCK, and a less abundant site that has the properties of the peripheral CCK receptor (the A-receptor) (Table 1.1). Identification of the chemical differences between these sites has been brought nearer by a recent report of the molecular cloning of a putative CCK receptor from

Table 1.1. CCK receptor types

Type	Characteristic tissue	Agonist selectivity	Antagonist
A(h)	Pancreas Gall bladder	CCK $\ggg$ gastrin	Devazepide
A(l)	Gall bladder	CCK $\ggg$ gastrin	Devazepide JMV 180
B	CNS neurons	CCK $>$ gastrin	L-365,260
G	Parietal cells, antral muscle	CCK $=$ gastrin	L-365,260

The major receptors for CCK and their characteristics. See text for further information on the high (h) and low (l) affinity CCK$_A$ receptors.

rat striatum. McVittie *et al.* (1990) used the polymerase chain reaction employing multiple degenerate primers for conserved transmembrane domains of
G-protein-coupled receptors to identify a 2.2 kb clone encoding a novel 326
residue protein with seven predicted membrane-spanning domains. When this
clone was expressed in *Xenopus* oocytes, both CCK and gastrin evoked Ca^{2+}
efflux, which is compatible with stimulation of a B-type receptor. Given that
different gastrin/CCK receptors are likely to be structurally related to each
other, it should now be possible to characterize other members of the group.

The prospects for further development of antagonists for the gastrin/CCK
group of peptides are encouraging. It is notable, for example, that several
chemically quite different types of useful gastrin/CCK antagonist have now
been produced. These include, among the A-selective antagonists, devazepide,
which is a benzodiazepine, and the proglumide-related compounds such as
CR 1409 (lorglumide) derived from glutaramic acid, and similarly, among the
B-selective antagonists, benzodiazepines (L-365,260) as well as structures
initially derived from CCK itself such as PD 134308 (Freidinger 1989; Hughes
et al. 1990; see also Chapters 2 and 5 of this volume). The characterization of
functionally important receptor types inevitably moves in parallel with the
development of new antagonists. Two illustrations of the ways in which present
ideas of gastrin/CCK receptor classification might be refined are worth men-
tioning. First, it is difficult to distinguish between CCK_B and gastrin receptors
using the presently available range of compounds, and indeed until recently it
was even difficult to say whether or not they were separate receptor types at all.
But recent studies by Howbert *et al.* (1990; see also Chapter 3 of this volume)
now suggest that it might be possible to develop antagonists that distinguish between these receptors, indicating that they are indeed separate entities. Second,
in the pancreas, it is well established that there is a pronounced biphasic
amylase response to CCK — stimulatory effects are associated with activation of
a high affinity receptor and inhibitory effects with a low affinity site. One compound that distinguishes between these two sites is the CCK-8 analogue
JMV 180 (Stark *et al.* 1989; Matozaki *et al.* 1990). This compound is interesting
because it appears to be capable of stimulating a maximal secretory response
and yet does not promote hydrolysis of phosphatidyl inositol. It has been suggested that JMV 180 acts as an agonist at the high affinity site and as an
antagonist at the low affinity site. The precise mechanisms involved are a matter
of debate: there may be separate high and low affinity CCK_A sites, a single
receptor site that exists in different states, or an abundance of spare binding
sites of a single receptor at which JMV 180 acts as a partial agonist. The resolution of this issue is of interest in its own right, but since it is generally thought
that pancreatic and brain CCK_A sites are very similar, if not identical, it also
becomes important to examine the biological properties of compounds like
JMV 180 in the CNS.

Rapid progress in science often follows the introduction of new experimental approaches. It is also the case that while new approaches answer some

questions, other questions are brought into sharper focus and hitherto unsuspected questions are raised. Fifteen years ago, when the peptides of the gastrin/CCK family were first found in brain, considerations of their possible messenger roles were not much more than idle speculation. The results now coming from studies with the new generation of CCK antagonists mark a major turning point in providing convincing evidence for the idea that CCK functions in several different CNS systems, including those involved in control of food intake, analgesia, and anxiety. Interpretation of the data obtained using these antagonists depends, of course, on a knowledge of the identity and distribution of both CCK and its receptors in the CNS. These issues and the strategies for developing non-peptide antagonists are the subject of chapters in the first section of this volume. Together they set the scene in defining the old questions which are now nearing a solution, and in providing a basis for formulating the new problems that face us in the future.

References

Dimaline, R., and Lee, C. M. (1990). Biological properties of chicken gastrin: a member of the gastrin/CCK family with novel structure–activity relationships. *Am. J. Physiol.* **259**, 9882–88.

Dimaline, R., Young, J., and Gregory, H. (1986). Isolation from chicken antrum and primary amino acid sequence of a novel 36-residue peptide of the gastrin/CCK family. *FEBS Lett.* **205**, 318–22.

Dockray, G. J. (1988). Regulatory peptides and the neuroendocrinology of gut–brain relations. *Q.J. Exp. Physiol.* **73**, 703–27.

Freidinger, R. M. (1989). Cholecystokinin and gastrin antagonists. *Med. Res. Rev.* **9**, 271–90.

Howbert, J. J., Lobb, K. L., Brown, R. F., Reel, J. K., Mendelsohn, L. G., Mason, N. R., Mahoney, D. F., and Bruns, R. F. (1990). Evidence that gastrin receptors on AR4-2J cells are distinct from CCK-B receptors using novel pyrazolidinone CCK antagonists. *Soc. Neurosci. Abstr.* **16**, 82.

Hughes, J., Boden, P., Costall, B., Domeney, A., Kelly, E., Horwell, D. C., Hunter, J. C., Pinnock, R. D., and Woodruff, G. N. (1990). Development of a class of selective cholecystokinin type B receptor antagonists having potent anxiolytic activity. *Proc. Natl Acad. Sci. USA* **87**, 6728–32.

Johnsen, A. H., and Rehfeld, J. F. (1990). Cionin: a disulfotyrosyl hybrid of cholecystokinin and gastrin from the neural ganglion of the protochordate *Ciona intestinalis*. *J. Biol. Chem.* **265**, 3054–8.

McVittie, L. D., Monsma, F. J., Jr., Gerfen, C. R., Burch, R. M., Sibley, D. R., and Mahan, L. C. (1990). Molecular cloning of a CCK/gastrin responsive G-protein coupled receptor. *Neurosci. Abstr.* **16**, 82.

Matozaki, T., Göke, B., Tsunoda, Y., Rodriguez, M., Martinez, J., and Williams, J. A. (1990). Two functionally distinct cholecystokinin receptors show different modes of actions on Ca^{2+} mobilization and phospholipid hydrolysis in isolated rat pancreatic acini. Studies using a new cholecystokinin analog, JMV-180. *J. Biol. Chem.* **265**, 6247–54.

Moran, T. H., Robinson, P. H., Goldrich, M. S., and McHugh, P. R. (1986). Two brain cholecystokinin receptors: Implications for behavioral actions. *Brain Res.* **362**, 175–9.

Stark, H. A., Sharp, C. M., Sutliff, V. E., Martinez, J., Jensen, R. T., and Gardner, J. D. (1989). CCK-JMV-180: A peptide that distinguishes high-affinity cholecystokinin receptors from low-affinity cholecystokinin receptors. *Biochem. Biophys. Acta* **1010**, 145–50.

Varro, A. (1989). Identification and characterization of a novel CCK-like immunoreactive peptide in pig CNS. *Neurochem. Int.* **14**, 505–10.

2. Synthesis of non-peptide CCK antagonists

Roger M. Freidinger

Introduction

The current list of peptide hormones and neurotransmitters presents many potential opportunities for the development of new therapeutic agents. Useful drugs might be obtained from compounds which either mimic or block the effects of the native peptides. To date, however, few peptide-based drugs have been obtained, primarily because of limitations inherent in the peptide structure itself. These deficiencies include rapid degradation by proteases leading to short duration, poor oral absorption, and lack of specificity.

One approach to overcoming the limitation of peptide structures is to discover non-peptide ligands for the receptors of interest (Freidinger 1989*a*). Non-peptides present a much broader range of structural possibilities, and there are many non-peptide drugs with appropriate properties as therapeutic agents. This chapter will review the current state of development of non-peptide CCK receptor antagonists.

The antagonists discussed in this review will be compared primarily using the [^{125}I]CCK binding assay in rat pancreas (Innis and Snyder 1980) and guinea pig brain tissue (Saito *et al.* 1981; Chang *et al.* 1983). Gastrin receptor binding is closely comparable with brain CCK binding affinity (guinea pig gastric glands) (Praissman *et al.* 1983) and is included in certain instances. Selected functional assay data are included to confirm the antagonist character of the compounds.

Early non-peptide CCK antagonists

The first CCK antagonists (Fig. 2.1) were weakly potent (IC$_{50}$s of 100–500 μM) and non-selective for CCK versus gastrin receptors. The initial report of a CCK receptor antagonist was that of dibutyryl cyclic guanosine monophosphate (Bt$_2$cGMP) (Peikin *et al.* 1979). Certain amino acid derivatives have also been reported to have weak CCK and/or gastrin antagonist properties. Researchers at Rotta Laboratories reported that benzoyl-DL-glutamic acid dipropylamide (proglumide) could inhibit gastrin-stimulated gastric acid secretion (Rovati *et al.* 1967). Further studies confirmed that

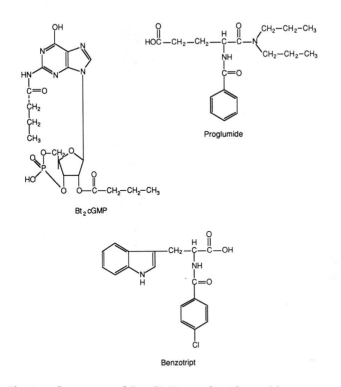

Fig. 2.1. Structures of Bt₂cGMP, proglumide, and benzotript.

proglumide and *p*-chlorobenzoyl-L-tryptophan (benzotript) inhibited binding of radio-labelled gastrin to gastrin receptors (Vidal *et al.* 1980) and that the antagonism was competitive (Rovati 1976). Proglumide and benzotript were tested for their ability to antagonize the effects of CCK (Hahne *et al.* 1981). Both were found to be competitive antagonists of comparable potency ($IC_{50} = 500 \, \mu M$). The potency for inhibition of [^{125}I]CCK binding and inhibition of CCK-stimulated amylase secretion in pancreatic acini was comparable. Both these antagonists are readily synthesized compounds. As a result, prior to the development of more potent and selective CCK and gastrin antagonists, proglumide in particular was used extensively in studies of CCK (Maton *et al.* 1986).

Analogues of proglumide

An important advance in analogues of proglumide was reported by workers at Rotta Laboratories (Makovec *et al.* 1985). They investigated the effects of substituents on the benzoyl group, the variation of the propyl groups, and the length of the side-chain of proglumide (Table 2.1) (Makovec *et al.* 1986*a*). On

Table 2.1.   Structure–activity relationships of proglumide analogs

Compound	n	R_1	R_2	R_3	IC$_{50}$ mg/ml[a]
Proglumide	2	H	n-Pr	n-Pr	340.00
1	2	4-CL	n-Pr	n-Pr	34.40
2	2	3-CL	n-Pr	n-Pr	500.00
3	2	3,4-di-CL	n-Pr	n-Pr	4.20
4	2	3,4-di-CL	n-Bu	n-Bu	0.33
5	2	3,4-di-CL	n-Pentyl	n-Pentyl	0.06
6	2	3,4-di-CL	n-Hexyl	n-Hexyl	0.93
7	2	3,4-di-Me	n-Pentyl	n-Pentyl	0.53
8	2	3,4-di-CL	n-Pentyl	H	25.70
9	2	3,4-di-Me	Hexamethylene		275.20
10	1	3,4-di-CL	n-Pentyl	n-Pentyl	0.83

[a]Inhibition of CCK-8 (5ng/ml)-stimulated contractions of guinea pig gall bladder *in vitro*.

the aromatic ring of the benzoyl group, both steric and electronic effects as well as regiochemistry were explored using a wide variety of substituents. The most preferred substitution pattern was 3,4-dichloro (compare compounds 1, 2, and 3) which resulted in a potency increase *in vitro* of 80-fold compared with proglumide. The alkyl amide groups were also found to play an important role. Comparison of straight-chain lengths of one to seven carbons and branched alkyls in secondary and tertiary amides and cyclic tertiary amides (e.g. compounds 4–9) showed optimal potency for the di-*n*-pentylamide. Aspartic acid analogues were found to be less potent than glutamic derivatives. The optimal antagonist (compound 5) from these studies resulted from combination of the individual best substituents (Jensen *et al.* 1986). This compound, which is referred to as C R-1409 or lorglumide, is about 5600 times more potent than proglumide. This higher potency has been confirmed *in vivo* (Makovec *et al.* 1987), and the antagonist was shown to be competitive by Schild analysis (Makovec *et al.* 1985).

Interestingly, the more potent proglumide analogues also have improved selectivity relative to proglumide. Receptor binding studies have demonstrated that compounds 5 and 7 are much more selective ligands for CCK$_A$ compared with CCK$_B$ receptors than were the earlier weaker antagonists (Table 2.2)

Table 2.2. [^{125}I] CCK-8 binding inhibition comparison of selected antagonists and CCK

Compound	IC$_{50}$, rat pancreatic acini (μM)	IC$_{50}$, mouse cortex (μM)
CCK-8	0.00051	0.0031
CCK-4	5000	0.032
Proglumide	6300	11 000
Benzotript	82	40
5	0.13	30
7	0.93	5600
11	0.33	9.1

Reproduced with permission from Freidinger (1989b).

(Makovec *et al.* 1986b; Niederau *et al.* 1986). These analogues, together with the benzodiazepine CCK antagonists to be discussed, helped to confirm the existence of at least two subclasses of CCK receptors (see Chapter 5 of this volume). The interaction of lorglumide with CCK$_A$ receptors was also shown to be stereospecific. The enantiomer derived from D-glutamic acid showed a 50-fold greater affinity than the L-enantiomer for pancreatic acini receptors (Makovec *et al.* 1987; Makovec *et al.* 1988). An analogue of lorglumide in which oxygen replaces one methylene (compound 11 (Fig. 2.2) (C R-1505, loxiglumide), has properties similar to those of lorglumide and is currently undergoing clinical evaluation (Table 2.2) (Setnikar *et al.* 1987a; Setnikar *et al.* 1987b).

The detailed pharmacological properties of lorglumide and loxiglumide have been reported (Makovec *et al.* 1987; Setnikar *et al.* 1987a). Both have been confirmed to be moderately potent selective CCK$_A$ receptor antagonists *in vitro* and *in vivo*. Both are devoid of agonist activity. A report on the pharmacokinetics of loxiglumide in humans shows it to have nearly 100 per cent oral bioavailability and good duration of blood levels of drug (Setnikar *et al.* 1988). These compounds are synthesized in three steps from DL-glutamic acid (Makovec *et al.* 1986a). The more potent proglumide analogues are significant

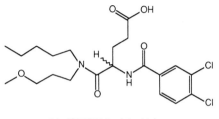

11 (CR1505, Loxiglumide)

Fig. 2.2. Structure of compound 11 (CR1505, loxiglumide).

because they represent non-peptide ligands for a peptide receptor, and they appear to have suitable properties to be drug molecules.

Asperlicin and analogues

In 1985, Chang *et al.* reported a novel natural product CCK antagonist which proved to be the lead for the Merck CCK programme. Using a screening assay based on radio-ligand binding to rat pancreas and guinea pig brain CCK receptors, they discovered a fermentation product from *Aspergillus alliaceus* which was given the name asperlicin (compound 12) (Chang *et al.* 1985; Goetz *et al.* 1985). Structurally, asperlicin can be viewed as a 1,4-benzodiazepine with a large 3-substituent derived from tryptophan and leucine and a quinazolone fused to the 1,2-positions (Liesch *et al.* 1985). At the time of its discovery, asperlicin was the most potent non-peptide CCK antagonist known ($IC_{50} = 1\mu M$, pancreas binding). Asperlicin was shown to be competitive and selective for CCK receptors with no agonist activity. It was also one of the first antagonists to show selectivity for CCK receptors from different tissues, thereby confirming the existence of the CCK_A and CCK_B receptor subtypes. Its binding to brain CCK receptors is 100 times weaker than its binding to pancreas receptors, and it is also a poor gastrin antagonist (Chang *et al.* 1985). Many analogues of asperlicin were prepared by chemical synthesis (Bock *et al.* 1986) and biosynthesis (Houck *et al.* 1988), and some of these had improved properties. For example, reduction of the 7a–8 imine linkage gave a sevenfold potency gain (compound 13), and acylation of this antagonist with succinic anhydride produced compound 14 (Fig. 2.3) which has much improved water solubility and retains the potency and selectivity of asperlicin (Bock *et al.* 1986). The desired oral activity was not achieved with these compounds, however.

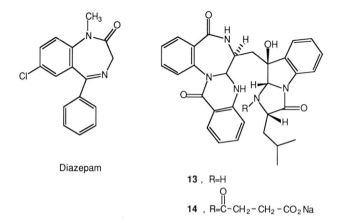

Diazepam

13 , R=H

14 , R=$\overset{\overset{\textstyle O}{\|}}{C}$-CH$_2$-CH$_2$-CO$_2$Na

Fig. 2.3. Structures of diazepam and compounds 13 and 14.

Benzodiazepine analogues

CCK$_A$ selective antagonists

An alternative to the asperlicin derivatization approach to improved antagonists involved attempts to design simpler totally synthetic CCK antagonists based on the asperlicin lead. Evans and colleagues reasoned that a structure combining the elements of diazepam (Fig. 2.3) with D-tryptophan might mimic asperlicin and have CCK receptor affinity (Evans *et al.* 1986). Such a structure and its relationship to asperlicin is outlined in Fig. 2.4. Compounds such as 15 are readily synthesized from an aminobenzophenone and a D-tryptophan ester, for example, and 15 proved to have a CCK$_A$ receptor affinity identical with that of asperlicin with good selectivity. In accord with the design hypothesis, the binding of structures like 15 to the CCK receptor was found to be stereospecific with the S-enantiomers having lower affinity. Furthermore, this new class of antagonists displays oral activity (Evans *et al.* 1987).

Efforts to optimize the CCK$_A$ antagonist activity of these benzodiazepines also proved very successful. Early structure activity studies showed that potency could be increased several-fold with *N*-1-methyl and 2'-fluoro substituents, and *N*-1-carboxymethyl increased water solubility (Table 2.3) (Evans *et al.* 1987). Subsequently, it was found that the nature of the group linking the benzodiazepine with the 3-position aromatic group as well as the point of attachment to the aromatic group were critical factors (Table 2.4) (Evans *et al.* 1986, 1988). When the 3-position methylene linking group in compound 15 is replaced with a carboxamide and the indole is attached at its 2-position rather than the 3-position, the new analogue 20 is obtained. The increase in CCK$_A$ receptor binding affinity with these changes is substantial, about 500-fold. The 3-indolyl derivative 19, however, shows almost no increase in binding. Reduction of the amide carbonyl to methylene, as in the case of compound 21, also gives an analogue less potent than 20, thus illustrating the key role of the

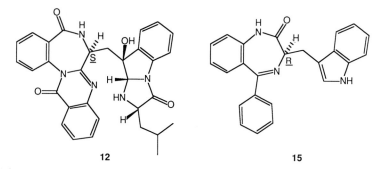

12   15

Fig. 2.4. Asperlicin and a diazepam–D-tryptophan hybrid. (Reproduced with permission from Freidinger 1989*b*.)

Table 2.3. Receptor binding data for 3-indolylmethyl-5-phenyl-1, 4-benzodiazepines

Compound	R	X	IC$_{50}$ (μM)		
			Rat pancreas[a]	Guinea pig brain[a]	Guinea pig gastric gland[b]
12 (asperlicin)	—	—	1.4	>100	>100
15	H	H	1.2	50	50
16	H	F	0.5	80	40
17	CH$_3$	F	0.3	10	13
18	CH$_2$COOH	F	0.3	23	5

[a]Inhibition of ^{125}I-CCK binding.
[b]Inhibition of ^{125}I-gastrin binding.
Reproduced with permission from Freidinger (1989b).

amide. *N*-1 methylation of 20 and resolution provided the enantiomers 22 and 23. This comparison points out that methylation further enhances potency and that the 3-position stereochemistry is important, just as for the original lead 15.

Compound 23 is the optimal CCK antagonist in this series with a binding constant for the CCK$_A$ receptor comparable to that of the native peptide CCK (Chang and Lotti 1986; Evans *et al.* 1986, 1988). This compound, which is known as devazepide (also L-364,718 or MK-329), is a competitive antagonist with high selectivity versus CCK$_B$, gastrin, and other peptide and neurotransmitter receptors (Chang and Lotti 1986). Devazepide has also been shown to be highly potent by several routes of administration, including oral, in a variety of functional assays, such as gastric emptying and gall bladder contraction, in a number of different species, and no agonist activity has been observed (Lotti *et al.* 1987; Pendleton *et al.* 1987). It has been demonstrated that devazepide crosses the blood–brain barrier efficiently (Pullen and Hodgson 1987). This antagonist is the most potent known CCK$_A$ antagonist of any structural type, being at least 100 times more potent than loxiglumide. Devazepide was chosen for clinical trials to evaluate its therapeutic potential in humans.

Table 2.4. Receptor binding data for 3-amino-5-phenyl-1,4-benzodiazepine derivatives

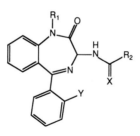

| | | | | | IC$_{50}$ (μM) | | |
| | | | | | Rat | Guinea pig | Guinea pig |
Compound	R$_1$	X	Y	R$_2$	Stereo	pancreas[a]	brain[a]	gastric gland[b]
19	H	O	H	3-indolyl	RS	1.1	8.4	18
20	H	O	H	2-indolyl	RS	0.0047	8	4
21	H	H,H	F	2-indolyl	RS	0.087	100	>10
22	CH$_3$	O	H	2-indolyl	R	0.0083	3.7	1.4
23	CH$_3$	O	H	2-indolyl	S	0.00008	0.27	0.17
27	CH$_3$	O	F	p-Cl-phenyl	S	0.002	2.9	3.2
28	CH$_3$	O	F	m-Br-phenyl	S	0.0035	3.5	0.75
29	CH$_2$COOH	O	H	2-indolyl	RS	0.0014	6.0	0.65

[a]Inhibition of ^{125}I-CCK binding.
[b]Inhibition of ^{125}I-gastrin binding.
Reproduced with permission from Freidinger (1989b).

The 3-aminobenzodiazepine derivatives such as devazepide are readily prepared by total synthesis (Fig. 2.5) (Bock *et al.* 1987). The unsubstituted 1-methyl-5-phenyl-1,4-benzodiazepine (24) is first prepared in four steps from aminobenzophenone (25). This intermediate is then nitrosated, followed by reduction to give the parent racemic 3-aminobenzodiazepine (26). An elegant resolution–racemization procedure converts most of the racemic amine to the desired s-enantiomer (Reider *et al.* 1987). Acylation with an indole-2-carboxylic acid derivative provides devazepide in high yield.

A number of structurally related CCK$_A$ antagonists with nanomolar-level potency have been found (Table 2.4). For example, indole can be replaced with *p*- or *m*-substituted phenyl (e.g. compounds 27 and 28). Groups providing increased water solubility such as an *N*-1-carboxymethyl can be incorporated as in compound 29 (Evans *et al.* 1988). Other excellent antagonists within the 1,4-benzodiazepine class include triazolo compounds such as 30 (Fig. 2.6), which is very potent *in vitro* (IC$_{50}$=0.2 nM, pancreas binding) and *in vivo* (ED$_{50}$ = 0.04 mg/kg mouse gastric emptying, p.o.), but has somewhat lower selectivity than devazepide (Bock *et al.* 1988).

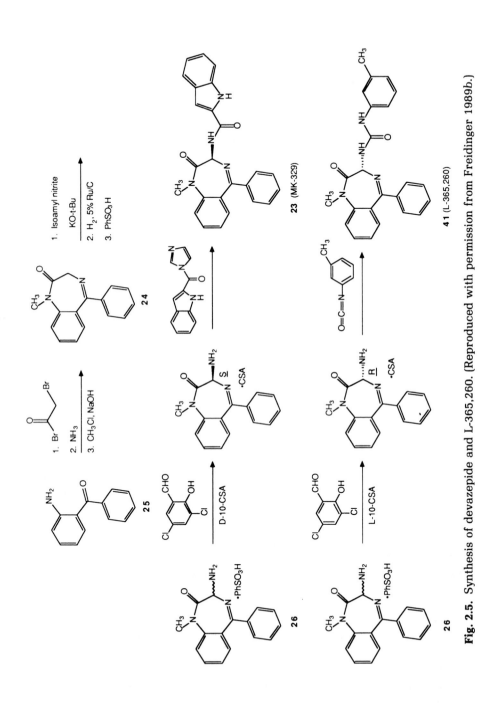

Fig. 2.5. Synthesis of devazepide and L-365,260. (Reproduced with permission from Freidinger 1989b.)

In addition to the 3-substituted benzodiazepines, certain 2-substituted structures were also found to be CCK_A selective antagonists (Fig. 2.6). One of these compounds is tifluadom (31), which had previously been shown to be a potent opioid agonist (Romer *et al.* 1982). Tifluadom has moderate and selective affinity for the CCK_A versus the CCK_B receptor (IC_{50} values of 47 nM and $> 100 \mu$M respectively), but it has about 30-fold greater affinity for the opioid receptor (Chang *et al.* 1986). The S-(−)-tifluadom enantiomer has higher affinity than the R-(+)-enantiomer (the sign of rotation refers to toluene) for both CCK and opioid receptors (Petrillo *et al.* 1985; Chang *et al.* 1986). A number of tifluadom analogues were synthesized in an effort to separate the CCK and opioid activities (Bock *et al.* 1990). While some potent analogues such as the indole-2-carbonyl derivative (32) were obtained, the desired change in receptor selectivity was not achieved. Importantly, devazepide and related structures have a low affinity for opioid receptors (Chang and Lotti 1986).

Another related structure class in which CCK_A antagonist activity has been found is the 3-aminobenzolactams (Parsons *et al.* 1989) (Fig. 2.6). The most potent analogue (compound 33: $IC_{50} = 3$ nM, pancreas binding) is competitive and highly selective for CCK_A versus CCK_B receptors. Oral activity was also found. Lactam ring size was investigated and it was found that potency increases in the order $6 < 8 < 7$. Stereospecificity was demonstrated, and a molecular modelling study showed good correspondence between structural features of the more potent R-enantiomer and devazepide. The authors concluded that the *tert*-butyloxycarboxymethyl group of compound 33 and the 5-phenyl of devazepide may share the same space on the receptor.

CCK_B/gastrin-selective antagonists

All the antagonists discussed so far have shown selectivity for the CCK_A receptor subtype. None of these compounds has been shown to bind selectively to the CCK_B or gastrin type receptors. In 1989, 3-substituted benzodiazepine analogues were described which are potent and are the first non-peptide ligands to bind selectively to CCK_B and gastrin receptors rather than to CCK_A receptors (Bock *et al.* 1989). These antagonists were designed from the earlier CCK_A-selective benzodiazepines such as compound 25 by combining individual structural modifications which led to loss of CCK_A selectivity. An example of this type of antagonist is compound 35 in which both the urea linkage and the large *N*-1 substituent are crucial for the indicated potency and selectivity (Table 2.5). Compound 35 is the most potent known antagonist with this type of selectivity.

An alternative method for achieving CCK_B/gastrin selectivity was found to be resolution of 1-methyl-3-arylurea benzodiazepines (Bock *et al.* 1989). Compounds such as 36 and 39, which are racemic, are non-selective (Table 2.5). Separation into the S- and R-enantiomers revealed selective antagonists. The S-enantiomers such as 37 and 40 bind selectively to the CCK_A receptor subtype, while 38 and 41 are a new type of CCK_B/gastrin-selective ligand.

Table 2.5. Receptor binding stereospecificity of selected benzodiazepine derivatives

Compound	R_1	Y	R_2	Stereo	Rat pancreas[a]	Guinea pig brain[a]	Guinea pig gastric gland[b]
						IC$_{50}$ (μM)	
27	CH_3	F	p-Cl-phenyl	S	0.002	2.9	3.2
34	CH_3	F	p-Cl-phenyl	R	0.049	11	7.6
35	(pyrrolidine-CH_2C=O)	H	(HN-C$_6H_4$-Cl)	RS	0.52	0.0003	0.0005

No.							
36	CH$_3$	H	HN— (4-Cl phenyl)	RS	0.051	0.023	0.022
37	CH$_3$	H	HN— (4-Cl phenyl)	S	0.026	0.41	0.54
38	CH$_3$	H	HN— (4-Cl phenyl)	R	1.1	0.005	0.012
39	CH$_3$	H	HN— (3-CH$_3$ phenyl)	RS	0.008	0.007	0.003
40	CH$_3$	H	HN— (3-CH$_3$ phenyl)	S	0.003	0.151	0.13
41	CH$_3$	H	HN— (3-CH$_3$ phenyl)	R	0.28	0.002	0.001

[a] Inhibition of [^{125}I]-CCK binding.
[b] Inhibition of [^{125}I]-gastrin binding.
Reproduced with permission from Freidinger (1989b).

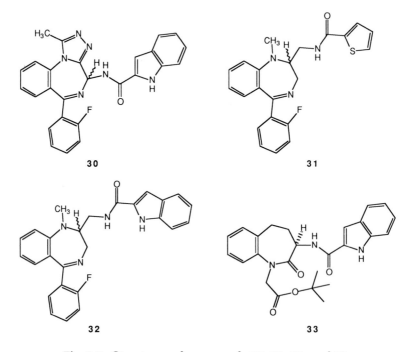

Fig. 2.6. Structures of compounds 30, 31, 32, and 33.

Investigation of a number of substituents on the phenyl showed *m*-methyl to be among the most potent, and, considering all properties, compound 41 (L-365,260) was the most preferred antagonist (Bock *et al.* 1989; Lotti and Chang 1989). Competitive binding inhibition was demonstrated for guinea pig gastrin and brain CCK receptors, and no agonist activity was observed. Upon oral administration, L-365,260 potently antagonized gastrin-stimulated acid secretion in several animal species (mouse, $ED_{50} = 30\ \mu g/kg$) with good duration of action. Good selectivity was demonstrated with respect to 13 other receptors (Lotti and Chang 1989). Interestingly, none of these antagonists demonstrated significant separation of CCK_B and gastrin receptor binding. L-365,260 is readily available through modification of the synthetic route to devazepide (Fig. 2.2).

Structural hybrid antagonists

Since the structures of the amino acid analogues such as lorglumide and the benzodiazepines such as devazepide were reported, hybrid structures combining elements of both types of antagonist have been synthesized. This work was based on molecular modelling studies which suggested which structural features of these compounds might correspond to receptor binding

Table 2.6. Receptor binding data for hybrid glutamic-acid-based CCK antagonists

Compound	Stereo	R_1	R_2	IC$_{50}$ (μM) Rat pancreas	IC$_{50}$ (μM) Guinea pig Brain
42[a]	DL	(2-indolyl)	n-Pentyl	0.008	0.23
43[a]	D	(3-methoxyphenyl-HN)	n-Pentyl	0.005	0.71
44[a]	L	(3-methoxyphenyl-HN)	n-Pentyl	0.5	9.7
45 (A-65186)[b]	D	(3-quinoline)	n-Pentyl	0.005	3.6
5 (Lorglumide)[a]	DL	(3,4-dichlorophenyl)	n-Pentyl	0.02	2.2

[a]Freidinger et al. (1990).
[b]Kerwin et al. (1989).
Reproduced with permission from Freidinger (1989b).

elements. The indole 2-carboxamide of devazepide was matched to the 3, 4-dichlorobenzoylamide of lorglumide. Furthermore, the two aromatic rings of the 5-phenylbenzodiazepine of the former antagonist were matched with the two n-pentyl chains of the latter. An experimental test of this analysis would be to substitute the 2-indolyl for the 3, 4-dichlorophenyl in lorglumide.

This compound was synthesized at both Merck (Freidinger et al. 1990) and Abbott (Kerwin et al. 1989), and, in support of the modelling study, its CCK$_A$ receptor affinity proved to be greater than that of lorglumide (Table 2.6). Kerwin et al. found that a 3-quinoline group also provided a potent antagonist, and that D-enantiomers were more potent that L-enantiomers. Freidinger et al. also prepared 3-methoxy-phenylurea analogues in an attempt to introduce CCK$_B$ receptor selectivity into the glutamic acid analogue series. The

Table 2.7.   CCK antagonist activity of benzotript hybrid analogues

Ar	Enantiomer	IC$_{50}$ (μM)		Inhibition of amylase release
		Pancreas	Cortex	
Cl— (4-chlorophenyl)	R	330	> 10 000	—
Cl— (4-chlorophenyl)	S	540	—	—
indol-2-yl	R	51 ± 13(5)	8000 ± 850(3)	670(2)
indol-2-yl	S	660	—	—
quinolin-3-yl	R	23.4 ± 5.4(6)	15,000	42(2)
quinolin-3-yl	S	1130 ± 270(3)	6300	10 000 inhib. 68%

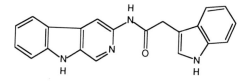

Fig. 2.7. Structure of compound 46.

resultant compounds 43 and 44 were still CCK$_A$ selective, however, with the D-enantiomer 43 being as potent as compounds 42 and 45.

The Abbott group has also synthesized hybrid analogues of benzotript (Table 2.7) which incorporate features from devazepide and lorglumide (Kerwin *et al.* 1990). These compounds provide yet another way of obtaining CCK$_A$-selective antagonists with nanomolar potency.

Finally, Evans (1989) has combined another benzodiazepine receptor ligand, the β-carboline, with an indole-containing acid to give a further novel class of CCK antagonist. Compounds such as 46 are reasonably potent and highly selective for the CCK$_A$ receptor subtype (Fig. 2.7).

Peptoid CCK antagonists

Very recently, researchers at Parke-Davis have reported the second structural class of non-peptide CCK$_B$ selective antagonists (Hughes *et al.* 1990; see also Chapter 5 of this volume). In contrast with the other CCK antagonist structures which have already been described, these compounds were designed starting from the CCK-4 tetrapeptide. The preferred antagonist of this type (PD 134308 or CI988) has a potency comparable with that of L-365,260 (IC$_{50}$ = 1.7 nM, mouse cortex binding) and higher selectivity (IC$_{50}$ = 2717 nM, rat pancreas binding). Furthermore, PD134308 is orally active, exhibits activity in anxiolytic assays, and has water solubility (Fig. 2.8).

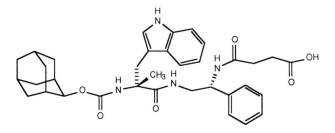

Fig. 2.8. Structure of PD134308.

Conclusion

The last five years have seen an impressive output of potent, selective, and orally active non-peptide antagonists of CCK. These developments in medicinal chemistry and pharmacology are being followed closely by a surge of biological studies using these antagonists as tools for the study of the physiological roles of CCK and gastrin (Hughes *et al.* 1989; see also later sections of this volume). The next few years should see further refinement of selective antagonists, and some answers will emerge on whether a therapeutic utility in humans will be discovered. In addition, pharmaceutical researchers will be attempting to take advantage of what has been learned with CCK for the design of non-peptide antagonists of other peptides. It will be exciting to be involved in these efforts.

Acknowledgments

It is a pleasure to acknowledge the contributions of my many colleagues at Merck who developed asperlicin, MK-329, and L-365,260. They are listed as authors of the papers cited in appropriate sections of this chapter. I also thank Ms Jean Kaysen for skilful help in preparing the manuscript.

References

Bock, M. G., DiPardo, R. M., Rittle, K. E., Evans, B. E., Freidinger, R. M., Veber, D. F., Chang, R. S. L., Chen T., Keegan M. E., and Lotti, V. J. (1986). Cholecystokinin antagonists. Synthesis of asperlicin analogues with improved potency and water solubility. *J. Med. Chem.* **29**, 1941–5.

Bock, M. G., DiPardo, R. M., Evans, B. E., Rittle, K. E., Veber, D. F., Freidinger, R. M., Hirshfield, J., and Springer, J. P. (1987). Synthesis and resolution of 3-amino-1,3-dihydro-5-phenyl-2H-1,4-benzodiazepin-2-ones. *J. Org. Chem.* **52**, 3232–9.

Bock, M. G., DiPardo, R. M., Evans, B. E., Rittle, K. E., Veber, D. F., Freidinger, R. M., Chang, R. S. L., and Lotti, V. J. (1988). Cholecystokinin antagonists. Synthesis and biological evaluation of 4-substituted 4H-[1,2,4] triazolo [4,3-a] [1,4] benzodiazepines. *J. Med. Chem.* **31**, 176–81.

Bock, M. G., DiPardo, R. M., Evans, B. E., Rittle, K. E., Whitter, W. L., Veber, D. F., Anderson, P. S., and Freidinger, R. M. (1989). Benzodiazepine gastrin and brain cholecystokinin receptor ligands: L-365,260. *J. Med. Chem.* **32**, 13–16.

Bock, M. G., DiPardo, R. M., Evans, B. E., Rittle, K. E., Whitter, W. L., Veber, D. F., Freidinger, R. M., Chang, R. S. L., Chen, T. B., and Lotti, V. J. (1990). Cholecystokinin-A receptor ligands based on the κ-opioid agonist tifluadom. *J. Med. Chem.* **33**, 450–3.

Chang, R. S. L., and Lotti, V. J. (1986). Biochemical and pharmacological characterization of an extremely potent and selective non-peptide cholecystokinin antagonist. *Proc. Natl Acad. Sci. USA* **83**, 4923–6.

Chang, R. S. L., Lotti, V. J., Martin, G. E., and Chen, T. B. (1983). Increase in ^{125}I-cholecystokinin receptor binding following chronic haloperidol treatment, intracisternal 6-hydroxydopamine or ventral tegmental lesions. *Life Sci.* **32**, 871–8.

Chang, R. S. L., Lotti, V.J., Monaghan, R. L., Birnbaum, J., Stapley, E. O., Goetz, M. A., Albers-Schonberg, G., Patchett, A. A., Liesch, J. M., Hensens, O. D., and Springer, J. P. (1985). A potent non-peptide cholecystokinin antagonist selective for peripheral tissue isolated from *Aspergillus alliaceus*. Science **230**, 177-9.

Chang, R. S. L., Lotti, V. J., Chen, T. B., and Keegan, M. E. (1986). Tifluadom, a kappa opiate agonist, acts as a peripheral cholecystokinin receptor antagonist. *Neurosci. Lett.* **72**, 211-14.

Evans, B. E. (1989). Recent developments in cholecystokinin antagonist research. *Drugs Fut.* **14**, 971-9.

Evans, B. E., Bock, M. G., Rittle, K. E., DiPardo, R. M., Whitter, W. L., Veber, D. F., Anderson, P. S., and Freidinger, R. M. (1986). Design of potent, orally effective, non-peptidal antagonists of the peptide hormone cholecystokinin. *Proc. Natl Acad. Sci USA* **83**, 4918-22.

Evans, B. E., Rittle, K. E., Bock, M. G., DiPardo, R. M., Freidinger, R. M., Whitter, W. L., Gould, N. P., Lundell, G. F., Homnick, C. F., Veber, D. F., Anderson, P. S., Chang, R. S. L., Lotti, V. J., Cerino, D. J., Chen, T. B., Kling, P. J., Kunkel, K. A., Springer, J. P., and Hirshfield, J. (1987). Design of non-peptidal ligands for a peptide receptor: cholecystokinin antagonists. *J. Med. Chem.* **30**, 1229-39.

Evans, B. E., Rittle, K. E., Bock, M. G., DiPardo, R. M., Freidinger, R. M., Whitter, W. L., Lundell, G. F., Veber, D. F., Anderson, P. S.,Chang, R. S. L., Lotti, V. J., Cerino, D. J., Chen, T. B., Kling, P. J., Kunkel, K. A., Springer, J. P., and Hirshfield, J. (1988). Methods for drug discovery: development of potent, selective, orally effective cholecystokinin antagonists. *J. Med. Chem.* **31**, 2235-46.

Freidinger, R. M. (1989*a*). Non-peptide ligands for peptide receptors. *Trends Pharm. Sci.* **10**, 270-4.

Freidinger, R. M. (1989*b*). Cholecystokinin and gastrin antagonists. *Med. Res. Rev.* **9**, 271-90.

Freidinger, R. M., Whitter, W. L., Gould, N. P., Holloway, M. K., Chang, R. S. L., and Lotti, V. J. (1990). Novel glutamic acid-derived cholecystokinin receptor ligands. *J. Med. Chem.* **33**, 591-5.

Goetz, M. A., Lopez, M., Monaghan, R. L., Chang, R. S. L., Lotti, V. J., and Chen, T. B. (1985). Asperlicin, a novel non-peptidal cholecystokinin antagonist from Aspergillus alliaceus. Fermentation, isolation and biological properties. *J. Antibiot.* **38**, 1633-37.

Hahne, W. F., Jensen, R. T., Lemp, G. F., and Gardner, J. D. (1981). Proglumide and benzotript: members of a different class of cholecystokinin receptor antagonists. *Proc. Natl Acad. Sci. USA* **78**, 6304-8.

Houck, D. R., Ondeyka, J., Zink, D. L., Inamine, E., Goetz, M. A., and Hensens, O. D. (1988). On the biosynthesis of asperlicin and the directed biosynthesis of analogs in Aspergillus alliaceus. *J. Antibiot.* **41**, 882-91.

Hughes, J., Dockray, G. J., and Woodruff, G. N. (1989). *The neuropeptide cholecystokinin.* Ellis Horwood, Chichester.

Hughes, J., Boden, P., Costall, B., Domeney, A., Kelly, E., Horwell, D. C., Hunter, J. C., Pinnock, R. D., and Woodruff, G. N. (1990). Development of a class of selective cholecystokinin type B receptor antagonists having potent anxiolytic activity. *Proc. Natl Acad. Sci. USA* **87**, 6728-32.

Innis, R. B., and Snyder, S. H. (1980). Distinct cholecystokinin receptors in brain and pancreas. *Proc. Natl Acad. Sci. USA* **77**, 6917-21.

Jensen, R. T., Zhou, Z.-C., Murphy, R. B., Jones, S. W., Setnikar, I., Rovati, L. A., and Gardner, J. D. (1986). Structural features of various proglumide-related cholecystokinin receptor antagonists. *Am. J. Physiol.* **251**, G839–46.

Kerwin, J. F., Nadzan, A. M., Kopecka, H., Lin, C. W., Miller, T., Witte, D., and Burt, S. (1989). Hybrid cholecystokinin (CCK) antagonists: new implications in the design and modification of CCK antagonists. *J. Med. Chem.* **32**, 739–42.

Kerwin, J. F., Wagenaar, F., Kopecka, H., Lin, C. W., Miller, T., Witte, D., and Nadzan, A. M. (1990). CCK antagonists: investigations of the relationship between benzotript and glutamic acid-based antagonists. *Peptides, Chemistry, Structure and Biology* (eds J. E. Rivier and G. R. Marshall) pp. 149–51. ESCOM, Leiden.

Liesch, J. M., Hensens, O. D., Springer, J. P., Chang, R. S. L., and Lotti, V. J. (1985). Asperlicin, a novel non-peptidal cholecystokinin antagonist from *Aspergillus alliaceus*. Structure elucidation. *J. Antibiot.* **38**, 1638–41.

Lotti, V. J. and Chang, R. S. L. (1989). A new potent and selective non-peptide gastrin antagonist and brain cholecystokinin receptor (CCK-B) ligand: L-365,260. *Eur. J. Pharmacol.* **162**, 273–80.

Lotti, V. J., Pendleton, R. G., Gould, R. J., Hanson, H. M., Chang, R. S. L., and Clineschmidt, B. V. (1987). *In vivo* pharmacology of L-364,718, a new potent non-peptide peripheral cholecystokinin antagonist. *J. Pharmacol. Exp. Ther.* **241**, 103–9.

Makovec, F., Chiste, R., Bani, M., Pacini, M. A., Setnikar, I., and Rovati, L. A. (1985). New glutaramic acid derivatives with potent competitive and specific cholecystokinin antagonist activity. *Arzneimittelforsch.* **35**(II), 1048–51.

Makovec, F., Chiste, R., Bani, M., Revel, L., Setnikar, I., and Rovati, A. L. (1986*a*). New glutamic and aspartic derivatives with potent CCK-antagonistic activity. *Eur. J. Med. Chem.* **21**, 9–20.

Makovec, F., Bani, M., Chiste, R., Revel, L., Rovati, L. C., and Rovati, L. A. (1986*b*). Differentiation of central and peripheral cholecystokinin receptors by new glutaramic acid derivatives with cholecystokinin antagonistic activity. *Arzneimittelforsch.* **36**(I), 98–102.

Makovec, F., Bani, M., Cereda, R., Chiste, R., Pacini, M. A., Revel, L., Rovati, L. A., Rovati, L. C., and Setnikar, I. (1987). Pharmacological properties of lorglumide as a member of a new class of cholecystokinin antagonists. *Arzneimittelforsch.* **37**(II), 1265–8.

Makovec, F., Chiste, R., Rovati, L. C., and Setnikar, I. (1988). Stereospecific cholecystokinin antagonistic activity of lorglumide (CR 1409). *Gastroenterology* **94** (2), A279.

Maton, P. N., Jensen, R. T., and Gardner, J. D. (1986). Cholecystokinin antagonists. *Horm. Metabol. Res.* **18**, 2–9.

Niederau, C., Niederau, M., Williams, J. A., and Grendell, J. H. (1986). New proglumide analogue CCK receptor antagonists: very potent and selective for peripheral tissues. *Am. J. Physiol.* **251**, G856–60.

Parsons, W. H., Patchett, A. A., Holloway, M. K., Smith, G. M., Davidson, J. L., Lotti, V. J., and Chang R. S. L. (1989). Cholecystokinin antagonists. Synthesis and biological evaluation of 3-substituted benzolactams. *J. Med. Chem.* **32**, 1681–5.

Peikin, S. R., Costenbader, C. L., and Gardner, J. D. (1979). Actions of derivatives of cyclic nucleotides on dispersed acini from guinea pig pancreas. *J. Biol. Chem.* **254**, 5321–7.

Pendleton, R. G., Bendesky, R. J., Schaffer, L., Nolan, T. E., Gould, R. J., and Clineschmidt, B. V. (1987). Roles of endogenous cholecystokinin in biliary, pancreatic and gastric function: studies with L-364,718, a specific cholecystokinin receptor antagonist. *J. Pharmacol. Exp. Ther.* **241**, 110–16.

Petrillo, P., Amato, M., and Tavani, A. (1985). The interaction of the two isomers of the opioid benzodiazepine tifluadom with μ-,δ, and k-binding sites and their analgesic and intestinal effects in rats. *Neuropeptides* **5**, 403-6.

Praissman, M., Walden, M. E., and Pellechia, C. (1983). Identification and characterization of a specific receptor for cholecystokinin on isolated fundic glands from guinea pig gastric mucosa using a biologically active ^{125}I-CCK-8 probe. *J. Recept. Res.* **3**, 647-65.

Pullen, R. G. L. and Hodgson, O. J. (1987). Penetration of diazepam and the nonpeptide CCK antagonist, L-364,718, into rat brain. *J. Pharm. Pharmacol.* **39**, 863-4.

Reider, P. J., Davis, P., Hughes, D. L., and Grabowski, E. J. J. (1987). Crystallization-induced asymmetric transformation: stereospecific synthesis of a potent peripheral CCK antagonist. *J. Org. Chem.* **52**, 955-7.

Romer, D., Buscher, H. H., Hill, R. C., Mauer, R., Petcher, T. J., Zeugner, H., Benson, W., Finner, E., Milkowski, W., and Thies, P. W. (1982). An opioid benzodiazepine. *Nature* **298**, 759-60.

Rovati, A. L. (1976). Inhibition of gastric secretion by anti-gastric H_2-blocking drugs. *Scand. J. Gastroenterol.* **11**(Supp. 42), 113-18.

Rovati, A. L., Casula, P. L., and DeRe, G. (1967). Attivita antisecretiva di alcuni nuovi composti privi di attivita anticolenergica. *Minerva Med.* **58**, 3651-3.

Saito, A., Goldfine, I. D., and Williams, J. A. (1981). Characterization of receptors for cholecystokinin and related peptides in mouse cerebral cortex. *J. Neurochem.* **37**, 483-90.

Setnikar, I., Bani, M., Cereda, R., Chiste, R., Makovec, F., Pacini, M.A., Revel, L., Rovati, L.C., and Rovati, L.A. (1987a). Pharmacological characterization of a new potent and specific nonpolypeptidic cholecystokinin antagonist. *Arzneimittelforsch.* **37**(I), 703-7.

Setnikar, I., Bani, M., Cereda, R., Chiste, R., Makovec, F., Pacini, M. A., and Revel, L. (1987b). Anticholecystokinin activities of loxiglumide. *Arzneimittelforsch.* **37**(II), 1168-71.

Setnikar, I., Chiste, R., Makovec, F., Rovati, L. C., and Warrington, S. J. (1988)., Pharmacokinetics of loxiglumide after single intravenous or oral doses in man. *Arzeimittelforsch.* **38**(I), 716-20.

Vidal, Y., Plana, R. R., Cifarelli, A., and Bizzari, D. (1980). Effects of antigastrin drugs on the interaction of ^{125}I-human gastrin with rat gastric mucosa membranes. *Hepatogastroenterology* **27**, 41-9.

3. A novel series of non-peptide CCK and gastrin antagonists: medicinal chemistry and electrophysiological demonstration of antagonism

*J. J. Howbert, K. L. Lobb, R. F. Brown, J. K. Reel,
D. A. Neel, N. R. Mason, L. G. Mendelsohn, J. P. Hodgkiss,
and J. S. Kelly*

The neuropeptides are a rapidly growing class of mediators which have important roles in a great diversity of physiological processes, particularly in the CNS and the gastrointestinal tract. Historically, the tools available for studying these processes have been almost exclusively the peptides themselves and related peptide agonists and antagonists. This has proved a considerable hindrance to research because of the pharmacokinetic shortcomings of many peptides, which include poor oral absorption, protease lability, and poor blood–brain barrier penetration. As a result, attention has recently turned to the development of non-peptide ligands for neuropeptide receptors. One of the most generally successful approaches has been the identification of non-peptide lead structures through broad screening with radio-ligand binding assays. Initial leads are then optimized with respect to potency, selectivity, and other desired properties via systematic manipulation of the structure. This method was previously used to develop non-peptide CCK and gastrin antagonists, using initial screening at a peripheral (CCK_A) type CCK receptor (Chang *et al.* 1985; Evans *et al.* 1986).

We have applied such a sequence of broad screening and structure optimization at the brain CCK (CCK_B) receptor. Structure–activity work on early leads led to identification of compound 1, containing a 1-(phenylaminocarbonyl)-4, 5-diphenyl-3-pyrazolidinone structure, as promising for further modification. Several domains within this structure were extensively varied, as detailed below. We have also shown that one compound from the series, LY262691, functions as an antagonist of neuronal excitation induced by CCK at a reported CCK_B type receptor (Boden and Hill 1988; Hill *et al.* 1988; Hill and Boden 1989).

Compounds were synthesized in two steps from appropriately substituted α-phenyl cinnamate esters. These were converted by condensation with hydrazine in refluxing ethanol to the corresponding 4,5-diphenyl-3-pyrazolidinones,

(Carpino 1958). The pyrazolidinones were subsequently reacted with the appropriate phenylisocyanate or phenylisothiocyanate, typically in tetrahydrofuran at room temperature, followed by purification via column chromatography or recrystallization, to provide target compounds 1–29. In the case of the unsubstituted pyrazolidinone used in compounds 1–16, *trans* stereochemistry of the phenyl rings was established by X-ray crystallography. All other pyrazolidinones are presumed but not proved to be *trans*, except that used to prepare compound 23. All compounds were prepared and tested in racemic form.

The affinities of the compounds were assessed at three different receptors for CCK and gastrin, using modifications of published procedures. Binding to the CCK_B receptor was defined according to the method of Chang and Lotti (1986), using mouse brain membranes and 20 pM [^{125}I]-CCK-8 sulphated as radio-ligand. CCK_A receptor binding to rat pancreas was as described by Chang *et al.* (1986) using [^{3}H]-devazepide at a concentration of 0.4–0.6 nM. The method used for gastrin binding to guinea pig stomach mucosal membranes was similar to that of Takeuchi *et al.* (1979) using 20 pM [^{125}I]-gastrin I. Use of bovine serum albumin (BSA) was avoided in all binding assays, as it tended to bind the compounds of the series and make them appear inactive. IC_{50} values of displacement curves were determined using seven concentrations of compound and calculated using the ALLFIT program (De Lean *et al.* 1978).

For electrophysiological studies, mice of either sex, 3–6 months old, were killed by decapitation and the brain placed in artificial cerebrospinal fluid (aCSF) of the following composition (mM): NaCl, 134; KCl, 5; KH_2PO_4, 1.25; $MgSO_4 \cdot 7H_2O$, 2; $NaHCO_3$, 16; D-glucose, 10; $CaCl_2$, 2.5. Coronal slices, 400–500 μM thick, containing the ventromedial nucleus of the hypothalamus were placed on nylon netting at the gas–liquid interface of a slice chamber, exposed to a well-humidified stream of gas, and superfused continuously at a rate of 0.3–1.0 mL/min with aCSF at 35–37 °C which was pregassed with 95 per cent O_2, 5 per cent CO_2. CCK-8S (Bachem) was used at a standard concentration of 800 nM, since this was found to give a readily detectable and reversible response. The peptide was dissolved in sterile distilled water at 1 mM, and a few drops of dilute NH_4OH were added to aid solubilization. LY262691 (compound 4) was dissolved in dimethylsulphoxide (DMSO) and the required volume of this stock solution was dissolved in aCSF just prior to application to the slice; the appropriate controls were subsequently performed. Intracellular recording electrodes were made using 1.2 mm o.d. glass capillary tubing (Clark Electromedical Instruments); when filled with 1 M potassium acetate they had d.c. resistances of 80–100 MΩ. Active and passive membrane properties were examined by applying depolarizing and hyperpolarizing current pulses across the membrane using a bridge amplifier (M701, WP Instruments). Data were only considered for analysis from those cells which generated action potentials with amplitudes greater than 60 mV measured from the resting potential. The data reported were obtained from neurons which provided stable impalements for periods up to 6 h.

Structure–activity relationships (SAR) in receptor binding

Numerous alterations in the substituents on each of the three phenyl rings in structure 1 were explored. On the ring attached to the aminocarbonyl nitrogen (compounds 2–14) (Table 3.1), such changes alone improved potency at the CCK_B receptor by over 65-fold (1.2 μM for compound 1 to 18 nM for compound 14). High potency was attained with certain 4-monosubstituted (compounds 3, 4, 6) or 3,4-disubstituted (compounds 12–14) systems, but not when the same substituents were present at the 3- or 2-position alone (cf. compounds 9, 11 with 3, 12). Most compounds in this group had limited affinity at the CCK_A receptor ($\geq 10 \mu$M), with the exception of certain 3,4-disubstituted patterns. The 3-CF_3, 4-Cl pattern seemed almost unique in this respect, providing CCK_A potencies of 300 nM or better (compound 12; also 16, see below). Affinities at the gastrin receptor within this set were intermediate, ranging from five-fold to 20-fold less than at CCK_B. Less discrimination was seen with 3,4-disubstituted compounds (CCK_B-to-gastrin ratios of 5–7) than with 4-monosubstitution (ratios of 10–20), parallelling the situation at the CCK_A receptor.

The trends in substituent SAR on the two phenyl rings attached directly to the pyrazolidinone (compounds 17–29) were very similar to one another (Table 3.2). Placement of a chloro group at the 2- and/or 3-positions led to modest improvements in potency at the CCK_B and gastrin receptors. Methoxy substituents at the same positions, however, were uniformly less potent. Substitution of any kind at the 4-position of either ring resulted in dramatic loss of affinity. The 3-cyano group on compound 18 afforded a CCK_B gastrin ratio of nearly 35, the best yet in this series. Compounds 22 and 23 provided a crucial test of the geometrical requirements for these two aryl groups. *Trans* compound 22 proved to be at least 10-fold more potent in CCK_B binding than its *cis* congener 23.

In addition to needing certain well-defined patterns of substituents for maximal potency, there appeared to be a general requirement for all three of the phenyl rings found in lead 1 *per se*. Replacement of each of the rings with a variety of heteroaromatic, alkyl, and aralkyl groups led to generally reduced affinities (not shown). Only certain fused aromatic systems (e.g. naphthyl, quinolinyl) gave potencies comparable with the parent all-phenyl compounds (not shown).

Virtually all compounds involving modifications of the embedded acylsemicarbazide had mediocre potency at all receptors, including those in which N^2 or N^3 was methylated, or N^3 was deleted or replaced by O, S, or CH_2 (not shown). The only exception was compound 16, where, as for 12, the 3-CF_3, 4-Cl pattern was associated with good potency at CCK_A (Table 3.1). Unlike

Table 3.1. Binding to CCK and gastrin receptors

No.	R	CCK brain IC$_{50}$ (μM)	CCK pancreas		Gastrin IC$_{50}$ (μM)	Brain-to-pancreas ratio	Brain-to-gastrin ratio
			Percentage inhibition at 10 μM	IC$_{50}$ (μM)			
X = O							
1	2,3-diCl	1.15[b]	53				
2	H	5.2	10				
3	4-CF$_3$	0.044[a]		10.6	0.42	240	10
4	4-Br (**LY262691**)	0.031[c]		11.6[b]	0.49[a]	370	16
5	4-Me	0.21	14				
6	4-i-Pr	0.025[b]	47[a]		0.26	~450	10
7	4-OMe	0.34	20				
8	4-OCH$_2$Ph	0.054[a]	34		1.1	~360	20
9	3-CF$_3$	0.33[b]	52				
10	3-Me	1.6	21		50% at 10		
11	2-CF$_3$	1.2	17				
12	3-CF$_3$,4-Cl	0.022[a]		0.30[a]	0.15	14	6.8
13	3,4-diCl	0.042		1.0[a]	0.21	24	5.0
14	3,4-(CH$_2$)$_4$	0.018[a]		1.2	0.081	67	4.5
X = S							
15	4-CF$_3$	2.2	37				
16	3-CF$_3$,4-Cl	0.61[a]		0.051[a]	3.0	0.084	4.9

[a] Average of two determinations.
[b] Average of three determinations.
[c] Average of five determinations.

Table 3.2. Binding to CCK and gastrin receptors

R³—[benzene] ... N–H ... N ... R¹ ; R²—[benzene] (pyrrolidinone–urea structure, with substituents R¹, R², R³)

No.	R²	R³	R¹	CCK brain IC$_{50}$ (µM)	CCK pancreas		Gastrin IC$_{50}$ (µM)	Brain-to-pancreas ratio	Brain-to-gastrin ratio
					Percentage inhibition at 10 µM	IC$_{50}$ (µM)			
R¹ = CF₃									
17	2-Cl	H		0.007[a]	47		0.18[a]	~1610	26
18	3-CN	H		0.020[a]	35		0.67[a]	~930	34
19	3-OMe	H		0.072	42		1.4	~190	19
R¹ = Br									
20	2-Cl	H		0.020[c]	38		0.36	~820	18
21	2-OMe	H		0.21[a]	53		1.0[a]	~59	4.8
22	2,3-diCl	H (trans)		0.031[a]	81[a]		0.23	~76	7.4
23	2,3-diCl	H (cis)		0.40	64		1.7[a]	~14	4.3
24	4-NO₂	H		1.2	64				
25	H	2-Cl		0.015[b]	23	8.6	0.22	570	15
26	H	3-OMe		0.068[b]	59		0.69	~490	10
27	H	3-Cl		0.011[b]			0.21	~630	19
28	H	4-Cl		0.16	69[a]		0.86	~28	5.4
29	2-Cl	2-Cl		0.006[b]		7.9	0.033[a]	1320	5.5
Devazepide				0.050[c]		0.0005[d]	0.30[a]	0.01	6.0
L-365,260				0.008[a]		>1[a]	0.003[a]	>125	0.4

[a] Average of two determinations.
[b] Average of three determinations.
[c] Average of four determinations.
[d] Average of seven determinations.

12, compound 16 bound poorly to CCK_B, so that it was actually *c.* 12-fold selective for CCK_A. This presumably stemmed from the presence within the molecule of a thiocarbonyl moiety (X = S), which produced a loss of potency at CCK_B for all substitution patterns, relative to the parent oxocarbonyl structures (X = O).

Electrophysiology

One compound, LY262691 (compound 4), was fully investigated for antagonism of the CCK-8S-mediated excitation of neurons in the ventromedial nucleus of the hypothalamus (Boden and Hill 1988). Intracellular recordings from two neurons are shown in Figs 3.1 and 3.2. At a concentration of $1\,\mu M$, LY262691 caused a marked reduction in the excitatory action of CCK-8S

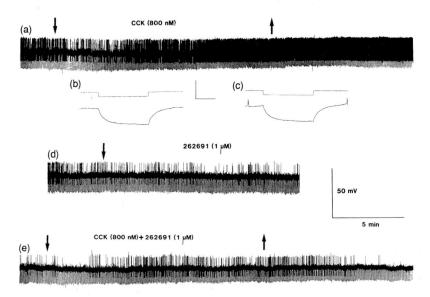

Fig. 3.1. Application of CCK (800 nM) for 15 min (indicated by arrows) caused a slowly developing increase in the spontaneous discharge frequency of a ventromedial hypothalamic neuron (**a**). The effect of CCK was reversible and the discharge rate fell to control levels after about 25 min (start of record (**d**)). During CCK-evoked excitation in (**a**) there was no measurable change in input resistance of the cell, as shown in records obtained before (**b**) and after (**c**) exposure to CCK. Each trace is the average of eight consecutive records and shows the change in membrane potential (lower records) following injection of a current of 0.1 nA (upper). The calibration bar in the inset represents 20 mV, 10 ms. (**d**) Application of LY262691 ($1\,\mu M$) caused a slight reduction in the spontaneous discharge rate. (**e**) A second challenge with CCK in the presence of LY262691 resulted in an excitation which was 46 per cent less than that in (**a**), as determined by the peak discharge rate, and which returned to control discharge levels within 5 min of wash-out. The records in (**d**) and (**e**) are continuous.

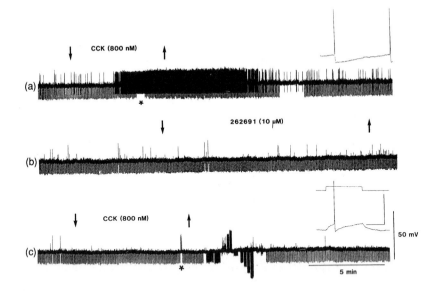

Fig. 3.2. (a) CCK (800 nM) reversibly increased the discharge frequency of a spontaneously active ventromedial hypothalamic neuron (exposure indicated by arrows). The inset to the right shows spontaneous action potentials at a faster time base, recorded at the time marked with an asterisk. **(b)** LY262691 at 10 μM was without effect on resting potential or input resistance of the cell. **(c)** A second application of CCK 14 min after change-over to compound-free aCSF elicited no response, suggesting that sufficient LY262691 was present to antagonize the excitatory actions of CCK. It was established that the membrane was still capable of discharging action potentials; the inset to the right of **(c)** shows a single action potential elicited with a current of 0.1 nA (upper trace), recorded at the time marked with an asterisk. The calibration bar in the lower inset represents 50 mV, 50 ms.

(Fig. 3.1). The antagonism was dose-dependent, with little or no antagonism seen at concentrations below about 400 nM. No evidence of agonist activity was detected when LY262691 was used at concentrations as high as 10 μM; indeed, after exposure to this concentration of LY262691 and subsequent wash-out for 14 min the action of CCK-8S remained fully blocked, although the neuron was still in good condition and capable of firing action potentials (Fig. 3.2).

Other functional assays

Various members of the series have also been shown to be potent antagonists of the contraction of guinea pig ileum induced by either CCK-4 or CCK-8 (Lucaites *et al.* 1990), and of phosphoinositide turnover induced in AR4-2J rat acinar cells by gastrin (Howbert *et al.* 1990). None of the compounds showed any apparent agonist activity at any concentration tested.

Conclusions

Using a strategy of identifying and optimizing leads through screening at the CCK_B receptor, we have succeeded in developing non-peptide ligands which have the following properties:

(1) they have excellent affinities for CCK_B receptors (down to 6 nM);
(2) they generally maintain selectivity for this same receptor, relative to CCK_A and gastrin receptors (up to 1000- and 35-fold selectivity respectively);
(3) they are antagonists of several types of CCK- and gastrin-stimulated responses in both neuronal and peripheral tissues.

We have also found a compound (16) with good potency and moderate selectivity toward the CCK_A receptor. For all receptors, the best affinities remain associated with the 1-(phenylaminocarbonyl)-4,5-diphenyl-3-pyrazolidinone structure found in the original lead compound. Stringent requirements for the type and position of substitution on each phenyl ring were also evident.

Only two other series of potent non-peptide CCK_B antagonists have previously been described, exemplified by L-365,260 (Bock *et al.* 1989; see also Chapter 2 of this volume) (Table 3.2) and PD134308 (Hughes *et al.* 1990; see Chapter 5 of this volume). The present compounds are independent in genesis and structurally distinct from both these series, yet offer potencies and CCK_B–CCK_A selectivities that are quite comparable. A limitation in the L-365,260 series was the inability to distinguish CCK_B from gastrin receptors, such that affinity ratios for these two receptors were all near unity (range 0.3–3). Data available to date on the PD compounds indicate they are also inherently non-selective between these receptors. In contrast, the CCK_B/gastrin affinity ratios of the present series vary from 4 to 35. While still greater selectivities are desirable, these do represent a significant advance, and in fact appear to be the best yet reported for any ligands, peptide or non-peptide (Howbert *et al.* 1990). Because of their potency and profile of selectivity, these compounds are expected to be valuable tools in the further exploration of the physiological and pathophysiological roles of CCK and gastrin in the CNS and gastrointestinal tract. They may be particularly useful in the debate over whether the CCK_B and gastrin receptors are truly distinct entities.

The structural relationship of this new non-peptide series to other CCK_B agonists and antagonists is unclear. Attempts to correlate certain domains with the side-chains of the peptide agonist CCK-4 by structural modification of the non-peptide lead were inconclusive. Some features of the present series are reminiscent of the non-peptide antagonists related to L-365,260 (e.g. the *N*-phenylurea fragment), but the correspondence does not seem to hold if the effects of substituent patterns are examined (e.g. compare the poor potency of compound 10 with that of L-365,260, which likewise contains a 3-methyl group). Molecular modelling work is currently underway to improve the definition of domain mapping among the various CCK and gastrin receptor ligands.

The electrophysiological study demonstrated that a compound from the series, LY262691 (4) which has high affinity for brain CCK (CCK$_B$) receptors, could effectively antagonize the potent excitatory actions of CCK-8S on neurons of the ventromedial nucleus of the hypothalamus. No evidence was obtained for even partial agonist activity with this compound. The clear-cut reduction in the spontaneous discharge rate seen with LY262691 may reflect the antagonism of a tonic release of endogenous CCK, although non-specific actions of the compound cannot be excluded. It is clear, however, that further physiological studies with LY262691 and other compounds in the series are warranted.

References

Bock, M. G., DiPardo, R. M., Evans, B. E., Rittle, K. E., Whitter, W. L., Veber, D. F., Anderson, P. S., and Freidinger, R. M. (1989). Benzodiazepine gastrin and brain cholecystokinin receptor ligands: L-365,260. *J. Med. Chem.* **32**, 13–16.

Boden, P., and Hill, R. G. (1988). Effects of cholecystokinin and related peptides on neuronal activity in the ventromedial nucleus of the rat hypothalamus. *Br. J. Pharmacol.* **94**, 246–52.

Carpino, L. A. (1958). A new synthesis of unsaturated acids. II. α, β-Olefinic acids. *J. Am. Chem. Soc.* **80**, 601–4.

Chang, R. S. L. and Lotti, V. J. (1986). Biochemical and pharmacological characterization of an extremely potent and selective non-peptide cholecystokinin antagonist. *Proc. Natl Acad. Sci. USA* **83**, 4923–6.

Chang, R. S. L., Lotti, V. J., Monaghan, R. L., Birnbaum, J., Stapley, E. O., Goetz, M. A., Albers-Schonberg, G., Patchett, A. A., Liesch, J. M., Hensens, O. D. and Springer, J. P.(1985). A potent non-peptide cholecystokinin antagonist selective for peripheral tissues isolated from *Aspergillus alliaceus*. *Science* **230**, 177–9.

Chang, R. S. L., Lotti, V. J., Chen, T. B., and Kunkel, K. A. (1986). Characterization of the binding of [^{3}H]-($\pm$)-L-364,718: A new potent, non-peptide cholecystokinin antagonist radioligand selective for peripheral receptors. *Mol. Pharmacol.* **30**, 212–17.

De Lean, A., Munson, P. J., and Rodbard, D. (1978). Simultaneous analysis of families of sigmoidal curves application to bioassay radioligand assay and physiological dose response curves. *Am. J. Physiol.* **235**, E97–102.

Evans, B. E., Bock, M. G., Rittle, K. E., DiPardo, R. M., Whitter, W. L., Veber, D. F., Anderson, P. S., and Friedinger, R. M. (1986). Design of potent, orally effective, non-peptidal antagonists of the peptide hormone cholecystokinin. *Proc. Natl Acad. Sci. USA* **83**, 4918–22.

Hill, R. G. and Boden, P. (1989). Electrophysiological actions of CCK in the mammalian central nervous system. In *The neuropeptide cholecystokinin (CCK)* (ed. J. Hughes, G. Dockray, and G. Woodruff), pp. 186–94. Wiley, New York.

Hill, R. G., Boden, P. R., Hall, M. D., Hewson, G., Hunter, J. C., and Hughes, J. (1988). Characterization of the actions of cholecystokinin at central nervous system receptors by the use of selective antagonists. In *Neurology and neurobiology*, Vol. 47, *Cholecystokinin antagonists* (ed. R. Y. Wang and R. Schoenfeld), pp. 149–64. Alan R. Liss, New York.

Howbert, J. J., Lobb, K. L., Brown, R. F., Reel, J. K., Mendelsohn, L. G., Mason, N. R., Mahoney, D. F., and Bruns, R. F. (1990). Evidence that gastrin receptors on AR4-2J cells are distinct from CCK-B receptors, using novel pyrazolidinone CCK antagonists. *Soc. Neurosci. Abstr.* **16**, 82.

Hughes, J., Boden, P., Costall, B., Domeney, A., Kelly, E., Horwell, D. C., Hunter, J. C., Pinnock, R. D., and Woodruff, G. N. (1990). Development of a class of selective cholecystokinin type B receptor antagonists having potent anxiolytic activity. *Proc. Natl Acad. Sci. USA* **87**, 6728-32.

Lucaites, V. L., Mendelsohn, L. G., Mason, N. R., Cohen, M. L., Lobb, K. L., Brown, R. F., Reel, J. K., and Howbert, J. J. (1990). A novel pyrazolidinone CCK antagonist: studies of contraction in guinea-pig ileum. *Soc. Neurosci Abstr.* **16**, 82.

Takeuchi, K., Speir, G. R., and Johnson, L. R. (1979). Mucosal gastrin receptor. I. Assay standardization and fulfillment of receptor criteria. *Am. J. Physiol.* **237**, E284-94.

4. Distribution of brain neuronal CCK. An *in situ* hybridization study

J.-J. Vanderhaeghen and S. N. Schiffmann

CCK-like immunoreactivity was discovered in the vertebrate central nervous system in the mid 1970s (Vanderhaeghen *et al.* 1975). It was shown later that most of the immunoreactivity corresponds to the carboxyl terminal octapeptide of CCK (Dockray 1976) in its sulphated biologically active form (Robberecht *et al.* 1978). Several immunohistochemical studies have mapped the distribution of CCK octapeptide (CCK-8) in rat brain (Innis *et al.* 1979; Larsson and Rehfeld 1979; Vanderhaeghen *et al.* 1980, 1981; Vanderhaeghen 1985). Preprocholecystokinin (preproCCK) has been cloned in the rat (Deschenes *et al.* 1984) as well as in man (Takahashi *et al.* 1985), and is identical in brain and gut. Localization of CCK mRNA has been conducted using Northern blot analysis (Burgunder and Young 1988; Voigt and Uhl 1988; Iadarola *et al.* 1989). Recently, two extensive mapping studies of CCK mRNA using *in situ* hybridization histochemistry (ISHH) (Ingram *et al.* 1989; Schiffmann and Vanderhaeghen 1991) have confirmed the immunohistochemical studies. In addition, some selected regions such as the thalamus and the mesencephalon have also been reported to contain CCK mRNA (Siegel and Young 1985; Burgunder and Young 1988, 1989*a*, *b*; Savasta *et al.* 1988, 1989, 1990; Voigt and Uhl 1988; Seroogy *et al.* 1989). In addition, studies on human brain material have recently been performed (Palacios *et al.* 1989; Schalling *et al.* 1989; Savasta *et al.* 1990). CCK mRNA is also present in large motoneurons of the rat spinal cord (Schiffmann and Vanderhaeghen 1991). Identification of CCK mRNA has changed our understanding of the brain distribution of CCK. Indeed, immunohistochemistry, particularly on human tissue, favours visualization of terminals, whereas ISHH favours visualization of neuronal perikarya. In this chapter we summarize our experience of the use of ISHH for mapping CCK perikarya in the rat brain.

Rat brains were frozen in 2-methylbutan (Merck) and cooled on dry ice. Coronal sections 4 μm thick were mounted on poly-L-lysine-coated slides and stored at −20 °C. Sections were fixed in 4 per cent paraformaldehyde, delipidated with ethanol, and air-dried before being incubated with ^{35}S-labelled probes for hybridization. After several treatments (for details see Schiffmann and Vanderhaeghen 1991) sections were covered with hyperfilm beta-max (Amersham 40 day) and then dipped in Ilford K5 emulsion for 4 weeks. Film and emulsion were then developed. The intensity of neuronal labelling was rated as heavy, moderate, or light. Two CCK probes were used, the first (ACCK) complementary to nucleotide 300-343 (N-terminal extended CCK-8) and the second (BCCK) complementary to nucleotide 249-293 (N-terminal part of preproCCK) of rat CCK CDNA (Deschenes *et al.* 1984).

The specificity of the results was assessed by the recognition of the appropriate sized band (800 bp) (Deschenes *et al.* 1984) on Northern blot analysis of rat total cortical RNA performed under the same conditions of stringency and using the same probe as used for ISHH, and also by the identical pattern of distribution obtained with two probes, ACCK and BCCK, complementary to different regions of preproCCK. Moreover, the signal was abolished both by RNAse pretreatment and by hybridization with labelled probe and an excess of cold probe. Although the background was low, it could not be excluded that isolated silver grains in the neuropil represent a true hybridization to CCK mRNA.

The two CCK probes produced similar labelling of cells which were widely distributed throughout many rat brain areas. These cells were always identified as neurons. Figure 4.1 shows the macroscopic distribution of the CCK-mRNA-containing cells and Tables 4.1–4.3 summarize the distribution, relative density, and intensity of these labelled cells. Background hybridization over white matter was low and similar to that observed in some adjacent sections hybridized with the human tyrosine hydroxylase probe.

Telencephalon

Olfactory bulb and anterior olfactory nuclei

Many CCK-mRNA-containing cells were observed in the external plexiform layer, particularly in its external segment. These were probably tufted cells and were moderately labelled. Mitral cells were unlabelled. Abundant hybridized perikarya were present in all subdivisions of the anterior olfactory nucleus (Fig. 4.1(a)) in its whole rostrocaudal extent. Most were moderately labelled, but some were heavily labelled.

Cerebral cortex

The cerebral cortex contained many hybridized perikarya. Their distribution in two bands (Figs 4.1 and 4.2(a)) was similar in all cortical areas (Fig. 4.1). The superficial band of labelling corresponded to layers II and III, whilst the deeper band represented a labelling of cells in layers V and VI. The density of labelled cells was high in layers II–III, and very high in layers V–VI (Fig. 4.2(a)). Most of these cells were lightly labelled, although a low density of neurons was heavily labelled, especially in the superficial layers. In the deep layers, some lightly labelled cells had a pyramidal shape (Fig. 4.2(c)). In addition, occasional moderately to heavily labelled neurons were observed in layer I (Fig. 4.2(b)). In the piriform cortex, a uniform light labelling of layer II as well as scattered

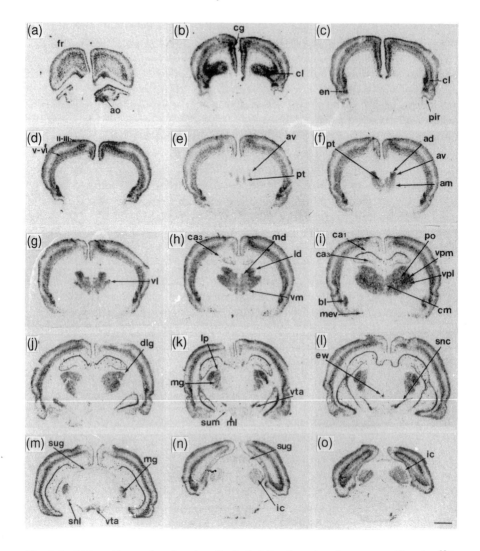

Fig. 4.1. Autoradiographs showing the hybridization signal generated by the ^{35}S-labelled ACCK probe–CCK mRNA hybrid in several areas of rat brain. The distribution of CCK mRNA is presented in selected coronal sections, rostral to caudal ((a)–(o)). Abbreviations: ad, anterodorsal thalamic nucleus; am, anteromedial thalamic nucleus; ao, anterior olfactory nucleus; av, anteroventral thalamic nucleus; bl, basolateral amygdaloid nucleus; CA1 and CA3, first and third subfields of Ammon's horn of the hippocampus; cg, cingulate cortex; cl, claustrum; cm, centromedial thalamic nucleus; dlg, doral lateral geniculate nucleus; en, endopiriform nucleus; ew, Edinger–Westphal nucleus; fr, frontal cortex; ic, inferior colliculus; ld, laterodorsal thalamic nucleus; lp, lateroposterior thalamic nucleus; md, mediodorsal thalamic nucleus; mev, ventral medial amygdaloid nucleus; ml, lateral part of the medial mammillary nucleus; mg, medial geniculate nucleus; pir, piri-

Table 4.1. Distribution, density, and intensity of labelling of CCK-mRNA-containing cells in the rat telencephalon

Area		CCK mRNA	
Olfactory bulb		+ +	M
Anterior olfactory n.		+ +	M
		+/−	H
Cerebral cortex	Layer I	+/−	H
	Layer II–III	+ + +	L–M
		+/−	H
	Layer IV	−	
	Layer V–VI	+ + +	L–M
		+/−	H
Piriform cortex		+ +	L–M
		+/−	H
Amygdaloid nuclei	Cortical	+	H
	Basolateral	+ + +	L
		+	H
	Basomedial	+	H
	Central	+/−	H
	Medial Ventral	+ + +	L
	Lateral	+	H
Subiculum		+ +	L–M
		+	H
Hippocampus	Dentate gyrus	+	H
	Stratum lacunosum	+	H
	Stratum radiatum	+	H
	Stratum pyramidale	+ + +	L
		+/−	H
	Stratum oriens	+	H
Septum	Lateral	+/−	L
	Medial	−	
Olfactory tubercle		−	
Caudate-putamen		−	
Nucleus accumbens		−	
Globus pallidus		−	
Substantia innominata		−	
Bed nucleus stria terminalis		−	
Claustrum		+ + +	M
		+	H
Endopiriform nucleus		+ + +	M
		+	H

Cell density: −, no labelled cells; +/−, very few; +, occasional; + +, frequent; + + +, abundant.
Labelled cell intensity: H, heavy; M, moderate; L, light.

form cortex; po, posterior thalamic nuclear group; pt, paratenial thalamic nucleus; snc, substantia nigra pars compacta; snl, substantia nigra pars lateralis; sug, superficial grey of the superior colliculus; sum, supramammillary nucleus; vl, ventrolateral thalamic nucleus; vm, ventromedial thalamic nucleus; vpl, lateral ventroposterior thalamic nucleus; vpm, medial ventroposterior thalamic nucleus; vta, ventral tegmental area. (Scale bar, 2 μm.)

Table 4.2. Distribution, density, and intensity of labelling of CCK-mRNA-containing cells in the rat diencephalon

Area		CCK mRNA	
Hypothalamus	Posterior area	+/−	M
	Suprachiasmatic nucleus	−	
	Periventricular nucleus	+/−	L
	Paraventricular nucleus	+	L
	Supra-optic nucleus	+	L
	Arcuate nucleus	−	
	Dorsomedial nucleus	+	L
	Supramammillary nucleus	+	L
		+/−	M
	Medial mammillary nucleus		
	lateral part	+/−	L
	posterior part	+/−	L
Zona incerta			
Thalamus	Habenula	−	
	Anteromedial nucleus	+++	M
	Anteroventral nucleus	+++	M
	Anterodorsal nucleus	+++	M
	Paraventricular nucleus	−	
	Reticular nucleus	−	
	Centromedial nucleus	++	L
	Reuniens nucleus	++	L
	Rhomboid nucleus	++	L
	Paracentral nucleus	++	L
	Paratenial nucleus	++	L
	Centrolateral nucleus	++	L
	Parafascicular nucleus	−	
	Laterodorsal nucleus	+++	M
	Posterior nuclear group	+++	M
	Ventrolateral nucleus	+++	M
	Ventromedial nucleus	+++	M
	Ventral posterolateral nucleus	+++	M
	Ventral posteromedial nucleus	+++	M
	Laterodorsal geniculate nucleus	+++	M
	Lateroventral geniculate nucleus	−	
	Lateroposterior nucleus	+++	M
	Medial geniculate nucleus	+++	M

For explanation of symbols see footnote to Table 4.1.

Table 4.3. Distribution, density, and intensity of labelling of CCK-mRNA-containing cells in the rat brain stem, spinal cord, and spinal ganglia

Area	CCK mRNA	
Brain stem		
Mesencephalon		
Ventral tegmental area	+ +	H
Interfascicularis nucleus	+	L
Substantia nigra	+ + +	H
Linearis rostralis	+ +	M
Central grey	+	M
Edinger–Westphal nucleus	+ +	H
Superior colliculus	+	L
Inferior colliculus	+ + +	L
Pons and cerebellum		
Locus coeruleus	−	
Pontine reticular nucleus	+/−	M
Parabrachial nucleus	+	M
Raphe nuclei	+/−	L–M
Medulla oblongata		
Inferior olive	−	
Reticular formation	+/−	M
Paragigantocellular nucleus	+	M
Spinal trigeminal nucleus	+ +	M
Solitary tract nucleus	−	
Spinal cord		
Layer II–III	+ +	M
Layer IX	+ + +	L–M
Layer X	+	M
Dorsal root ganglia	−	

For explanation of symbols see footnote to Table 4.1.

heavily labelled neurons in layers II and III were observed. No hybridization-positive cells were present in the subcortical white matter.

Claustrum, endopiriform nucleus, and amygdala

A very high density of moderately labelled cells was observed in both the claustrum and the endopiriform nucleus (Figs 4.1(b)–(l)). Scattered heavily labelled cells were also present in these nuclei. A low density of heavily labelled perikarya was found in all nuclei (lateral, medial, central, cortical, basomedial, and basolateral) of the amygdaloid complex (Fig. 4.1(i)). In addition, numerous lightly labelled cells were also present in the basolateral and ventromedial amygdaloïd nuclei.

Hippocampus

The fine macroscopically identified labelling in Ammon's horn (Figs 4.1(h)–(m)) corresponded to a light labelling of all pyramidal cells of the stratum

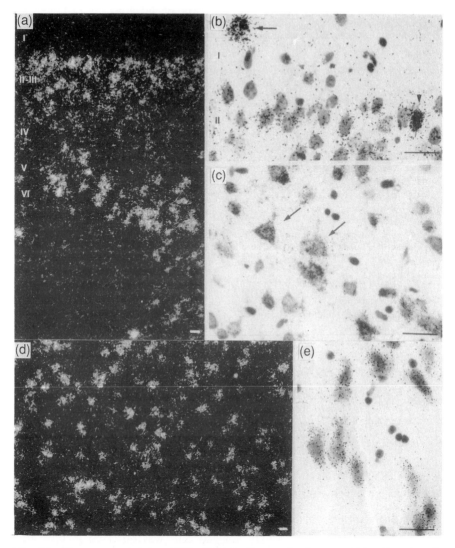

Fig. 4.2. Neurons containing CCK mRNA in the cerebral cortex ((a)–(c)) and in the thalamus ((d), (e)). In the interior cingulate cortex, a high density of labelled cells were present in layers II–III and layers V–VI(a). Heavily labelled neurons were found in layer I (arrow) and II (arrowhead) where they were intermingled with lightly labelled neurons (b). In the deep layers lightly labelled cells had a pyramidal shape (arrows) (c). A high density (d) of lightly to moderately labelled neurons (e) was observed in the ventrolateral thalamic nucleus ((d), (e)). (Bright-field ((b), (c), (e)) and dark-field ((a), (d)) illumination; scale bar, 25 μm.)

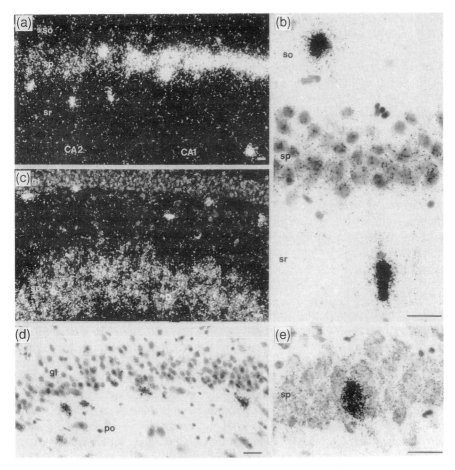

Fig. 4.3. Neurons containing CCK mRNA in the hippocampal formation. Heavily labelled neurons were observed in the strata oriens (so) (b), pyramidale (sp) (e) and radiatum (sr) ((a), (b)) of Ammon's horn, and in the polymorph zone (po) of the dentate gyrus ((c), (d)). The pyramidal layer of Ammon's horn was lightly labelled ((a)-(c), (e)) with a higher labelling in the CA1 subfield than in the CA2 subfield (a), whilst the granular layer (gl) of the dentate gyrus was totally devoid of silver grains (c). (Bright-field ((b), (d), (e)) and dark-field ((a), (c)) illumination; scale bar, 25 μm.)

pyramidale (Figs 4.3(a)–(e)). This labelling was greater in the CA1 subfield of Ammon's horn and decreased abruptly at the CA1–CA2 junction (Fig. 4.3(a)). Some heavily labelled neurons were also present in the stratum pyramidale (Figs 4.3(b), (c), (e)). Heavily labelled neurons were observed in the strata oriens, radiatum, and lacunosum (Figs 4.3(a), (b)) in the different subfields of Ammon's horn. In the area dentata, heavily labelled perikarya were present in the polymorph zone of the hilus and more prominently along the hilar border of the granular layer (Figs 4.3(c), (d)). The granular layer was completely devoid of labelling (Fig. 4.3(c)). In the subiculum, heavily labelled perikarya were present in all layers, including the pyramidal cell layer, together with a high density of lightly to moderately labelled neurons.

The heavily labelled CCK-mRNA-containing neurons (Figs 4.3(b), (e)) found in the hippocampal formation, including Ammon's horn, the subiculum, and the area dentata, exhibited the heaviest labelling of all neurons detected in the rat CNS.

Caudate-putamen

No labelled perikarya were observed in the caudate-putamen or in the nucleus accumbens or in neighbouring structures such as the olfactory tubercle, the globus pallidus, the ventral pallidum, the substantia innominata, or the bed nucleus of the stria terminalis.

Diencephalon

Thalamus

A very high density of neurons were labelled throughout the thalamus (Figs 4.2(d), (e)). Microscopic examination revealed that the vast majority of thalamic nuclei contained labelled cell bodies. In these nuclei a high to very high density of neurons contained CCK mRNA. As shown in Table 4.1 and Fig. 4.1, the anteromedial, anteroventral, anterodorsal, mediodorsal, centro-lateral, centromedial, reuniens, rhomboid, paratenial, laterodorsal, ventro-lateral, ventromedial, ventral posterolateral, ventral posteromedial, medial geniculate, lateroposterior, and laterodorsal geniculate nuclei, and the poster-ior nuclear group contained either lightly or moderately labelled neurons. In contrast, the habenula, paraventricular, reticular, lateroventral geniculate and parafascicular nuclei were devoid of CCK-mRNA-containing cells.

Hypothalamus

A moderate density of lightly labelled neurons was found in both the para-ventricular and the supramamillary nuclei (Fig. 4.1(k)). Scattered lightly labelled neurons were observed in the lateral and posterior parts of the medial mammillary nucleus. Scattered lightly labelled cells were also observed in the supra-optic, dorsomedial, suprachiasmatic, and periventricular nuclei, and moderately labelled neurons were present in the posterior hypothalamic area. No labelled cells were observed in the arcuate or medioventral nuclei.

Brainstem

Mesencephalon

Numerous moderately to heavily labelled neurons were present in the substantia nigra pars compacta (A9) (Figs 4.1(k)–(m)) and pars lateralis (Figs 4.1(l), (m)), as well as in the ventral tegmental area (A10) (Figs 4.1(k)–(m)) and the rostral and caudal linear nuclei. The intensity of labelling (Fig. 4.1(m)) and the density of labelled cells were slightly lower in a segment of substantia nigra pars compacta just medial to the substantia nigra pars lateralis. The presence of lightly labelled neurons in the interfascicular nucleus, scattered moderately labelled neurons in the central grey, and numerous heavily to moderately labelled neurons in the Edinger-Westphal nucleus (Fig. 4.1(l)) was also detected.

A low density of lightly labelled cells was present in the superficial grey layer of the superior colliculus (Figs 4.1(m), (n)), and, more caudally, a high density of lightly to moderately labelled perikarya was present in the inferior colliculus (Figs 4.1(n), (o)).

Metencephalon and myelencephalon

No labelled cells were observed in the cerebellar cortex or the deep nuclei of the cerebellum. Scattered moderately labelled cells were present in the pontine and bulbar reticular formation and in the parabrachial nucleus. Some large neurons of the spinal trigeminal nucleus were also lightly to moderately labelled, while no labelled cells were observed in the olivary complex, the area postrema, or the nucleus of the solitary tract.

Spinal cord

Scattered moderately labelled neurons were observed in layer X around the central canal and a moderate density of moderately labelled neurons was present in layers II–III of the posterior horn. In the anterior horn numerous large neurons (25–50 μm in diameter) were moderately labelled in layer IX. These neurons have the same morphology as motoneurons. In layer IX some unlabelled motoneurons were intermingled with the labelled ones. This pattern was similar at the cervical thoracic and lumbar/sacral levels (Schiffmann and Vanderhaeghen 1991).

Spinal ganglia

No labelled cells were observed in the dorsal root ganglia even when hybridization was performed with a mix of the ACCK and BCCK probes and sections were exposed for as long as 8 weeks. This confirms the previous results of Seroogy *et al.* (1990) that the CCK-like immunoreactivity detected there using immunohistochemistry does not represent genuine CCK (Schiffmann *et al.* 1990).

The specificity of the results was assessed by the recognition of the appropriate sized band (800 bp) (Deschenes *et al.* 1984) on Northern blot analysis of rat total cortical RNA performed under the same conditions of stringency and using the same probe as used for ISHH, and also by the identical pattern of distribution obtained with two probes, ACCK and BCCK, complementary to different regions of preproCCK. Indeed, a selective dendritic transport of mRNA has been demonstrated in hippocampal neurons (Davis *et al.* 1987), and labelled proximal dendrites were observed in the present study as reported by Bloch *et al.* (1990).

Comparison of the present distribution of CCK-mRNA-containing cells with the immunocytochemical distribution of CCK-LI (Innis *et al.* 1979; Hökfelt *et al.* 1980, 1988; Vanderhaeghen *et al.* 1980, 1981; Vanderhaeghen 1985) raises several questions which could be addressed to four selected brain areas which have different patterns of distribution.

Regions with similar content and distribution of CCK mRNA and CCK-LI

The group of regions with similar content and distribution of CCK mRNA and CCK-LI includes the regions where, in earlier histochemical studies, CCK-LI-containing cell bodies were detectable only if colchicine was injected prior to sacrifice (Innis *et al.* 1979; Hökfelt *et al.* 1980, 1988; Vanderhaeghen *et al.* 1980, 1981; Maciewicz, *et al.* 1984; Vanderhaeghen 1985; Seroogy *et al.* 1989). The following areas fulfil these conditions: the olfactory bulb, the cortical, basomedial, lateral, and central amygdalöïd nuclei, the dorsomedial, suprachiasmatic, and periventricular nuclei, and the posterior area of the hypothalamus, the ventral tegmental area, the substantia nigra, the linearis rostralis, the Edinger–Westphal nucleus, the central grey, the reticular formation including the gigantocellular nucleus, the parabrachial nucleus, and the raphé nuclei. In some regions such as the hippocampal formation, some subregions or layers have a similar content and distribution of CCK and mRNA and CCK-LI; nevertheless, these areas will be discussed below. The spinal trigeminal nucleus contained some lightly labelled neurons. Some CCK-LI-containing neurons were also present in this nucleus, but it has been reported that this CCK-LI probably represents calcitonin gene-related peptide (CGRP) (Hökfelt *et al.* 1988). The presence of CCK mRNA in spinal trigeminal nucleus neurons suggests that the two peptides CCK and CGRP are expressed in this nucleus.

Regions with a high content of CCK mRNA and a low content or absence of CCK-LI

The regions with a high content of CCK mRNA and a low content or absence of CCK-LI are the medial mammillary nucleus pars lateralis and posterior, the inferior colliculus, the superficial grey layer of the superior colliculus, and most of the thalamic nuclei. The thalamus contains a high density of CCK-mRNA-containing cells, as reported in other ISHH studies using different CCK probes (Siegel and Young 1985; Burgunder and Young 1988, 1989*b*; Savasta *et al.* 1988; Voigt and Uhl 1988; Ingram *et al.* 1989; Seroogy *et al.* 1989) and confirmed by Northern blot analysis (Burgunder and Young 1988; Voigt and Uhl

1988; Iadarola *et al.* 1989). These CCK-mRNA-containing thalamic neurons are essentially neurons projecting to the cortex or the striatum as demonstrated by combined ISHH and retrograde tracing (Burgunder and Young 1988). They were detected as early as the 17th day of gestation in medial geniculate nucleus, and their maturation is achieved in the entire thalamus at 1 month (Burgunder and Young 1989*b*). However, immunohistochemical reports generally do not mention the presence in the thalamus of CCK-LI-containing cell bodies (Innis *et al.* 1979; Vanderhaeghen *et al.* 1980, 1981; Vanderhaeghen 1985; Hökfelt *et al.* 1988), although CCK-LI-containing fibres are detected there. Nevertheless, some studies note the presence of scattered CCK-LI-containing neurons in the thalamus, particularly in the rat ventroposterior nucleus (Mantyh and Hunt 1984) and the cat intralaminar nuclei (Wahle and Albus 1985). Radio-immunoassay studies (Rehfeld 1978; Larsson and Rehfeld 1979; Barden *et al.* 1981; Beinfeld *et al.* 1981; Emson *et al.* 1982; Marley *et al.* 1984; Vanderhaeghen 1985) have given varying results, ranging from no CCK-LI detectable to 25 per cent of the cortical CCK-LI content depending on the species studied and also, probably, on the antisera used.

Similar data exist for the medial mammillary nucleus pars lateralis and posterior, the inferior colliculus, and the superficial grey layer of the superior colliculus in which no CCK-LI-containing cell bodies were present, although CCK-LI-containing fibres and a small amount of radio-immunoassayable CCK-LI were detected.

At the present time, explanations for these discrepancies can only be speculative. The existence of a differential splicing of the CCK primary transcript resulting in a different peptide, or the existence of a highly homologous mRNA encoded by an unknown gene are improbable since a single mRNA species similar to the one present in other brain areas is detected in the thalamus by Northern blot analysis (Burgunder and Young 1988; Voigt and Uhl 1988; Iadarola *et al.* 1989). In addition, whereas no CCK-LI has been detected in the pig cerebellum, a CCK mRNA identical with the cortical CCK mRNA is present there, and *in vitro* translation of cerebellar mRNA results in the production of radio-immunoassayable CCK-LI (Gubler *et al.* 1987). A differential processing of the CCK peptide precursor resulting in peptide species unrecognizable by the available CCK antisera or differences in rates of synthesis or turnover could be more convincing possibilities. This latter hypothesis is proposed for substance P and somatostatin in the dorsal root ganglion which contains a higher content of substance P and somatostatin mRNAs per cell than is found in other CNS areas, although the peptide content itself is relatively low (Boehmer *et al.* 1989). Finally, whereas similar sensitivities of ISHH and immunohistochemistry have been reported for the detection of somatostatin-containing neurons (Fitzpatrick-McElligott *et al.* 1988), ISHH can be more sensitive than immunohistochemistry in detecting CCK-expressing cell bodies, as has also been proposed for neurons containing substance P and neurokinin B (Warden and Young 1988). Moreover, the combination of a low level of synthesis in individual neurons, assessed by their light to moderate ISHH labelling, with a possible rapid transport from the cell body to the distal fibres, i.e. in the cortical terminals of CCK-mRNA-containing thalamic neurons, could render these perikarya below the immunohistochemical level of detection.

Regions with a moderate to high content of CCK mRNA and CCK-LI but with a higher density of cells containing CCK mRNA than of cells containing CCK-LI

Contrary to the conclusions of Ingram *et al.* (1989), who stated that in most areas ISHH confirms immunohistochemistry, numerous brain areas exhibit a far higher density of CCK-mRNA-containing neurons than of CCK-LI-containing cells. This is true for layers II–III and V–VI of the cerebral cortex, the stratum pyramidale of Ammon's horn of the hippocampus, the claustrum, the endopiriform nucleus, the anterior olfactory nucleus, and the medial ventral and basolateral amygdaloid nuclei. All these areas are rich in CCK-LI as detected by radio-immunoassay (Rehfeld 1978; Larsson and Rehfeld 1979; Barden *et al.* 1981; Beinfeld *et al.* 1981; Emson *et al.* 1982; Marley *et al.* 1984; Vanderhaeghen 1985) and rich in CCK-LI-containing fibres. The difference between the densities of CCK-mRNA-containing cells and of CCK-LI-containing neurons (Kiyama *et al.* 1984) is less, although also present, in the supramammillary nucleus.

The distribution and the morphological characteristics of the cortical CCK-LI-containing neurons have been extensively reported in several species (Innis *et al.* 1979; Vanderhaeghen *et al.* 1980, 1981; Demeulemeester *et al.* 1985; Seroogy *et al.* 1985; Vanderhaeghen 1985; Lotstra and Vanderhaeghen 1987*b*). They are mostly found in layers II and III, but also in the deeper layers, and they represent 0.6 per cent of the neuronal population in the cat visual cortex, reaching a proportion of 2.2 per cent in layers II–III (Demeulemeester *et al.* 1985). They are mostly bipolar or bitufted, but multipolar and triangular or pyramidal-like cells are also observed, particularly in the deeper layers (Seroogy *et al.* 1985). Combined immunohistochemistry and retrograde tracing show that few detected cortical CCK-LI-containing neurons have long projections (Seroogy *et al.* 1985). Experimental lesions have demonstrated that various areas of the cerebral cortex could contain neuronal cell bodies at the origin of the CCK-LI-containing terminals in the striatum (Meyer *et al.* 1982, 1988; Gilles, *et al.* 1983; Meyer and Protopapas 1985). Moreover, CCK-LI-containing fibres are detected in subcortical tracts after colchicine pretreatment or mechanical damage (Vanderhaeghen *et al.* 1981; Seroogy *et al.* 1985). In the present ISHH study, the pyramidal shape of some CCK-mRNA-containing neurons and the very high density of CCK-mRNA-containing neurons in the deep cortical layers suggest that cortical pyramidal neurons are labelled. All these arguments lead us to propose that cortical pyramidal neurons express CCK, and that CCK could be a major component of the cortical projections, as has been suggested for the thalamostriatal and thalamocortical projection neurons (Burgunder and Young 1988). The expression of neuropeptide in pyramidal neurons has already been reported for neurotensin in the human developing subiculum (Lotstra *et al.* 1989).

Several arguments exist which suggest that CCK-LI-containing neurons in the claustrum, endopiriform nucleus, and amygdaloid nuclei also have long projections. Indeed, studies combining lesions and radio-immunoassay have demonstrated that, as well as the cerebral cortex, these nuclei could also contain neuronal perikarya at the origin of the CCK-LI-containing terminals in the striatum (Meyer *et al.* 1982; Gilles *et al.* 1983). Moreover, the striosomal distribution of CCK-LI-containing terminals in the human caudate nucleus has suggested that they originate in the amygdala (Schiffmann *et al.* 1989).

In the hippocampal formation, the distribution and the density of CCK-mRNA-containing cells in the strata oriens, radiatum, lacunosum, and the polymorph zone of the hilus are in agreement with the previous immunohistochemical data (Innis *et al.* 1979; Vanderhaeghen *et al.* 1980, 1981; Gall 1984; Harris *et al.* 1985; Vanderhaeghen 1985; Lotstra and Vanderhaeghen 1987*a*; Hökfelt *et al.* 1988). The stratum pyramidale contains some heavily labelled CCK-mRNA-containing neurons, most probably corresponding to the CCK-LI-containing neurons described in this layer (Innis *et al.* 1979; Vanderhaeghen *et al.* 1980, 1981; Gall 1984; Harris *et al.* 1985; Vanderhaeghen 1985; Lotstra and Vanderhaeghen 1987*a*; Hökfelt *et al.* 1988). However, virtually all the pyramidal cells of the stratum pyramidale are lightly labelled. The stratum pyramidale, like the granular layers of the dentate gyrus and the cerebellum, has a high cellular density. It has been noted that, in ISHH studies, these areas show 'a diffuse accumulation of silver grains which might result from weak unspecific binding of the probe' (Julien *et al.* 1987). However, such an unspecific labelling of the stratum pyramidale appears unlikely since, using our CCK probes, the granular layer of the dentate gyrus on the same tissue section was totally devoid of silver grains. The presence of specific labelling which could represent a CCK mRNA content suggests once again that, since pyramidal neurons have long projections, they could express CCK which should be, as in the cerebral cortex, cotransported with glutamate to the target structures. The pyramidal cells of Ammon's horn project mainly to the septal nuclei and to intrinsic areas of the hippocampal formation such as the subiculum or the other subfields of Ammon's horn. A slightly higher CCK mRNA content is observed in the CA1 pyramidal cells compared with the CA2, CA3, and CA4 pyramidal cells. A similar difference is observed for other neurochemical markers such as the M1 and M2 muscarinic receptors (Joyce *et al.* 1989). Although the significance of this pattern has yet to be established, it could be linked with differences in input and/or output pathways of the subfields of Ammon's horn.

The discrepancies between the density of CCK-LI- and CCK-mRNA-containing neurons in these brain areas could be explained in several ways as discussed above. However, the greater sensitivity of ISHH compared with immunohistochemistry, particularly for the detection of putative neurons with long projections which express neuropeptide, is the most convincing hypothesis. Nevertheless, the fact that in these brain areas, including the stratum pyramidale of Ammon's horn of the hippocampus, the density of heavily labelled neurons in ISHH is similar to the density of the detected CCK-LI-containing neurons, described as interneurons, favours this hypothesis.

Regions with a low content of CCK mRNA and a moderate to high content of CCK-LI

Some regions have the surprising characteristic of having a low content of CCK mRNA and a high content of CCK-LI as demonstrated by both immunohistochemistry (Innis *et al.* 1979; Vanderhaeghen *et al.* 1980, 1981; Vanderhaeghen 1985; Hökfelt *et al.* 1988) and radio-immunoassay (Rehfeld 1978; Larsson and Rehfeld 1979; Barden *et al.* 1981; Beinfeld *et al.* 1981; Marley *et al.* 1984; Vanderhaeghen 1985). These areas are the paraventricular and supra-optic nuclei of the hypothalamus and the septal nuclei. Colchicine pretreatment was required to detect CCK-LI-containing cell bodies in these areas, but this is also the case for numerous brain areas with moderate to high CCK mRNA contents. Voigt and Uhl (1988), who have recently reported similar observations, suggest that a technique-related loss of mRNA in these particular regions is most unlikely since detection of other mRNA species in adjacent sections is successful. Using a similar protocol to ours on post-fixed sections, Ingram *et al.* (1989) failed to detect hypothalamic CCK mRNA. This confirms the low level of CCK mRNA in hypothalamic nuclei. This could reflect a low level of functional activity and/or of CCK synthesis or a higher stability of the peptide. Furthermore, as in the study by Ingram *et al.* (1989), no CCK mRNA was detected in the bed nucleus of the stria terminalis and the solitary tract nucleus. Thus the identity of the immunohistochemically detected peptide in these nuclei is questionable.

In conclusion, ISHH has permitted the localization of CCK mRNA in the rat CNS and the identification of new putative CCK-containing neuronal pathways undetectable by immunocytochemistry. The significance of the presence of CCK mRNA in the absence of detectable CCK-LI in several brain areas, or in certain neuronal types such as cortical and hippocampal pyramidal neurons or thalamocortical projecting neurons, remains to be established and merits further study.

Acknowledgements

We wish to thank J.-L. Conreur for photography, P. Halleux and R. Menu for fine technical assistance and M. De Baerdemaeker for secretarial work. This work was supported by grants from Fonds de la Recherche Scientifique Médicale (3.4523–86 and 3.4574.90), Fondation Médicale Reine Elisabeth

(Belgium) (Neurobiology 86–92), and Loterie Nationale Belge (1987). S. N. Schiffmann is a Research Assistant of the Fonds National de la Recherche Scientifique (Belgium).

References

Barden, N., Merand, Y., Rouleau, D., Moore, S., Dockray, G. J., and Dupont, A. (1981). Regional distribution of somatostatin and cholecystokinin-like immunoreactivities in rat and bovine brain. *Peptides* 2, 299–302.

Beinfeld, M. C., Meyer, D. K., Eskay, R. L., Jensen, R. T., and Brownstein, M. J. (1981). The distribution of cholecystokinin immunoreactivity in the central nervous system of the rat as determined by radioimmunoassay. *Brain Res.* 212, 51–7.

Bloch, B., Guitteny, A.F., Normand, E., and Chouam S. (1990). Presence of neuropeptide messenger mRNAs in neuronal processes. *Neurosci. Lett.* 109, 259–64.

Boehmer, C. G., Norman, J., Catton, M., Fine, L. G., and Mantyh, P. W. (1989). High level of substance P, somatostatin and alpha-tubulin are expressed by rat and rabbit dorsal root ganglia neurons. *Peptides* 10, 1179–94.

Burgunder, J. M. and Young, W. S., III (1988). The distribution of thalamic projection neurons containing cholecystokinin messenger RNA, using *in situ* hybridization histochemistry and retrograde labeling. *Mol. Brain Res.* 4, 179–89.

Burgunder, J. M. and Young, W. S., III (1989*a*). Neurons containing cholecystokinin mRNA in the mammillary region: ontogeny and adult distribution in the rat. *Cell. Mol. Neurobiol.* 9, 281–94.

Burgunder, J. M. and Young, W. S., III (1989*b*). Ontogeny of cholecystokinin gene expression in the rat thalamus. A hybridization histochemical study. *Dev. Brain Res.* 46, 221–32.

Davis, L., Banker, G. A., and Steward, O. (1987). Selective dendritic transport of RNA in hippocampal neurons in culture. *Nature* 330, 477–9.

Demeulemeester, H., Vandesande, F., and Orban, G. A. (1985). Immunocystochemical localization of somatostatin and cholecystokinin in the cat visual cortex. *Brain Res.* 332, 361–4.

Deschenes, R. J., Lorenz, L. J., Haun, R. S., Boos, B. A., Collier, K. J., and Dixon, J. E. (1984). Cloning and sequence analysis of a cDNA encoding rat preprocholecystokinin. *Proc. Natl Acad. Sci. USA* 81, 726–30.

Dockray, G. J. (1976). Immunochemical evidence of cholecystokinin-like peptides in brain. *Nature* 274, 711–13.

Emson, P. C., Rehfeld, J. F., and Rossor, M. N. (1982). Distribution of cholecystokinin-like peptides in the human brain. *J. Neurochem.* 38, 1177–9.

Fitzpatrick-McElligott, S., Card, J. P., Lewis, M. E., and Baldino, F. Jr. (1988). Neuronal localization of prosomatostatin mRNA in the rat brain with *in situ* hybridization histochemistry. *J. Comp. Neurol.* 273, 558–72.

Gall, C. (1984). The distribution of cholecystokinin-like immunoreactivity in the hippocampal formation of the guinea pig: localization in the mossy fibres. *Brain Res.* 306, 73–83.

Gilles, C., Lotstra, F., and Vanderhaeghen, J.-J. (1983). CCK nerve terminals in the rat striatum and limbic areas originate partly in the brain stem and partly in telencephalic structures. *Life Sci.* 32, 1683–90.

Gubler, U., Chua, A., Young, D., Fan, Z. W., and Eng, J. (1987). Cholecystokinin mRNA in the pig cerebellum. *J. Biol. Chem.* 262, 15242–5.

Harris, K. M., Marshall, P. E., and Landis, D. M. D. (1985). Ultrastructural study of cholecystokinin-immunoreactive cells and processes in area CA1 of the rat hippocampus. *J. Comp. Neurol.* **233**, 147–58.

Hökfelt, T., Herrera-Marschitz, M., Seroogy, K., Gong, J., Staines, W. A., Holets, V., Schalling, M., Ungerstedt, U., Post, C., Rehfeld, J. F., Frey, P., Fischer, J., Dockray, G., Hamaoka, T., Walsh, J. H., and Goldstein, M. (1988). Immunohistochemical studies on cholecystokinin (CCK)-immunoreactive neurons in the rat using sequence specific antisera and with special reference to the caudate nucleus and primary sensory neurons. *J. Chem. Neuroanat.* **1**, 11–52.

Hökfelt, T., Skirboll, J. F., Goldstein, M., Markey, K., and Dann, O. (1980). A subpopulation of mesencephalic dopamine neurons projecting to limbic areas contains a cholecystokinin-like peptide: evidence from immunohistochemistry combined with retrograde tracing. *Neuroscience* **5**, 2093–124.

Iadarola, M. J., Naranjo, J. R., Duchemin, A. M., and Quach, T. T. (1989). Expression of cholecystokinin and enkephalin mRNAs in discrete brain regions. *Peptides* **10**, 687–92.

Ingram, S. M., Krause, R. G., II, Baldino, F., Jr, Skeen, L. C., and Lewis, M. E. (1989). Neuronal localization of cholecystokinin mRNA in the rat brain by using *in situ* hybridization histochemistry. *J. Comp. Neurol.* **287**, 260–72.

Innis, R. B., Correa, F. M. A., Uhl, G. R., Schneider, B., and Snyder, S. H. (1979). Cholecystokinin octapeptide-like immunoreactivity: histochemical localization in the rat brain. *Proc. Natl Acad. Sci. USA* **76**, 521–5.

Joyce, J. N., Gibbs, R. B., Cotman, C. W., and Marshall, J. F. (1989). Regulation of muscarinic receptors in hippocampus following cholinergic denervation and reinnervation by septal and striatal transplants. *J. Neurosci.* **9**, 2776–91.

Julien, J. F., Legay, F., Dumas, S., Tappaz, M., and Mallet, J. (1987). Molecular cloning, expression and *in situ* hybridization of rat brain glutamic acid decarboxylase messenger RNA. *Neurosci. Lett.* **73**, 173–80.

Kiyama, H., Shiosaka, S., Tateishi, K., Hashimura, E., Hamakoa, T., and Tohyama, M. (1984). Cholecystokinin-8-like immunoreactive neuron pathway from the supramammillary region to the ventral tegmental area of Gudden of the rat. *Brain Res.* **304**, 397–400.

Larsson, L. J. and Rehfeld, J. F. (1979). Localization and molecular heterogeneity of cholecystokinin in the central and peripheral nervous system. *Brain Res.* **165**, 201–18.

Lotstra, F. and Vanderhaeghen, J.-J. (1987a). Distribution of immunoreactive cholecystokinin in the human hippocampus. *Peptides* **8**, 911–20.

Lotstra, F. and Vanderhaeghen, J.-J. (1987b). High concentration of cholecystokinin neurons in the newborn human entorhinal cortex. *Neurosci. Lett.* **80**, 191–6.

Lotstra, F., Mailleux, P., Schiffmann, S. N., Vierendeels, G., and Vanderhaeghen, J.-J. (1989). Neurotensin containing neurones in the human hippocampus of the adult and during development. *Neurochem. Int.* **14**, 143–51.

Maciewicz, P. W., Phipps, B. S., Grenier, J., and Poletti, C. E. (1984). Edinger-Westphal nucleus: cholecystokinin immunocytochemistry and projections to spinal cord and trigeminal nucleus in the cat. *Brain Res.* **299**, 139–45.

Mantyh, P. W. and Hunt, S. P. (1984). Neuropeptides are present at all levels in visceral and taste pathways: from periphery to sensory cortex. *Brain Res.* **299**, 297–311.

Marley, P. D., Rehfeld, J. F., and Emson, P. C. (1984). Distribution and chromatographic characterisation of gastrin and cholecystokinin in the rat central nervous system. *J. Neurochem.* **42**, 1523–35.

Meyer, D. K. and Protopapas, Z. (1985). Studies on cholecystokinin-containing neuronal pathways in the rat cerebral cortex and striatum. *Ann. NY Acad. Sci.* **448**, 133–43.

Meyer, D. K., Beinfeld, M. C., Oertel, W. H., and Brownstein, M. J. (1982). Origin of the cholecystokinin-containing fibers in the rat caudatoputamen. *Science* **215**, 187–8.

Meyer, D. K., Schultheiss, K., and Hardung, M. (1988). Bilateral ablation of frontal cortex reduces concentration of cholecystokinin-like immunoreactivity in rat dorsolateral striatum. *Brain Res.* **452**, 113–17.

Palacios, J. M., Savasta, M., and Mengod, G. (1989). Does cholecystokinin colocalize with dopamine in the human substantia nigra? *Brain Res.* **488**, 369–75.

Rehfeld, J. F. (1978). Immunochemical studies on cholecystokinin. II. Distribution and molecular heterogeneity in the central nervous system and small intestine of man and hog. *J. Biol. Chem.* **253**, 4022–30.

Robberecht, P., Deschodt-Lanckman, M., and Vanderhaeghen, J.-J. (1978). Demonstration of biological activity of brain gastrin-like peptidic material in the human: its relationship with the COOH-terminal octapeptide of cholecystokinin. *Proc. Natl Acad. Sci. USA* **75**, 524–8.

Savasta, M., Palacios, J. M., and Mengod, G. (1988). Regional localization of the mRNA coding for the neuropeptide cholecystokinin in the rat brain studied by *in situ* hybridization. *Neurosci. Lett.* **93**, 132–8.

Savasta, M., Ruberte, E., Palacios, J. M., and Mengod, G. (1989). The colocalization of cholecystokinin and tyrosine hydroxylase mRNAs in mesencephalic neurons in the rat brain examined by *in situ* hybridization. *Neuroscience* **29**, 363–9.

Savasta, M., Palacios, J. M., and Mengod, G. (1990). Regional distribution of the messenger RNA coding for the neuropeptide cholecystokinin in the human brain examined by *in situ* hybridization. *Mol. Brain Res.* **7**, 91–104.

Schalling, M., Friberg, K., Bird, E., Goldstein, M., Schiffmann, S. N., Mailleux, P., Vanderhaeghen, J.-J., and Hökfelt T. (1989). Presence of cholecystokinin mRNA in dopamine cells in the ventral mesencephalon of a human with schizophrenia. *Acta Physiol. Scand.* **137**, 467–8.

Schiffmann, S. N., and Vanderhaeghen, J.-J. (1991). Distribution of cells containing mRNA encoding for cholecystokinin in the rat central nervous system. *J. Comp. Neurol.* **304**, 219–33.

Schiffmann, S. N., Mailleux, P., Przedborski, S., Halleux, P., Lotstra, F., and Vanderhaeghen, J.-J. (1989). Cholecystokinin distribution in the human striatum and related subcortical structures. *Neurochem. Int.* **14**, 167–73.

Seroogy, K., Fallon, J. H., Loughlin, S. E., and Leslie, F. M. (1985). Few cortical cholecystokinin immunoreactive neurons have long projections. *Exp. Brain Res.* **59**, 533–42.

Seroogy, K., Schalling, M., Brene, S., Dagerlind, A., Chai, S. Y., Hökfelt, T., Persson, H., Brownstein, M., Huan, R., Dixon, J. E., Filer, D., Schlessinger, D., and Goldstein, M. (1989). Cholecystokinin and tyrosine hydroxylase messenger RNAs in neurons of rat mesencephalon: peptide/monoamine coexistence studies using *in situ* hybridization combined with immunocytochemistry. *Exp. Brain Res.* **74**, 149–62.

Seroogy, K., Mohapatra, N. K., Lund, P. K., Rethelyi, M., McGehee, D. S., and Perl, E. R. (1990). Species-specific expression of cholecystokinin messenger RNA in rodent dorsal root ganglia. *Mol. Brain Res.* **7**, 171–6.

Siegel, R. E. and Young, W. S., III (1985). Detection of preprocholecystokinin and preproenkephalin A mRNAs in rat brain using complementary RNA probes. *Neuropeptides* **6**, 573–80.

Takahashi, Y., Kato, K., Hayashizaki, Y., Wakabayashi, T., Othsuka, E., Matsuki, S., Ikehara, M., and Matsubara, K. (1985). Molecular cloning of the human cholecystokinin gene by use of a synthetic probe containing deoxynosine. *Proc. Natl Acad. Sci. USA* **82**, 1931–5.

Vanderhaeghen, J.-J. (1985). Neuronal cholecystokinin. GABA and Neuropeptides in the CNS, Part I. In *Handbook of chemical neuroanatomy* (ed. A. Björklund and T. Hökfelt), Vol. 4, pp. 406–35. Elsevier, Amsterdam.

Vanderhaeghen, J.-J., Signeau, J. C., and Gepts, W. (1975). New peptide in the vertebrate CNS reacting with antigastrin antibodies. *Nature* **257**, 604–5.

Vanderhaeghen, J.-J., Lotstra, F., De Mey, J., and Gilles C. (1980). Immunohistochemical localization of cholecystokinin- and gastrin-like peptides in the brain and hypophysis of the rat. *Proc. Natl Acad. Sci. USA* **77**, 1190–4.

Vanderhaeghen, J.-J., Lotstra, F., Vierendeels, G., Gilles, C., Deschepper, C., and Verbanck, P. (1981). Cholecystokinins in the central nervous system and neurohypophysis. *Peptides* **2** (suppl. 2), 81–8.

Voigt, M. M. and Uhl, G. R. (1988). Preprocholecystokinin mRNA in the rat brain: regional expression includes thalamus. *Mol. Brain Res.* **4**, 247–53.

Wahle, P. and Albus, K. (1985). Cholecystokinin octapeptide immunoreactive material in neurons of the intralaminar nuclei of the cat's thalamus. *Brain Res.* **327**, 348–53.

Warden, M. K. and Young, W. S., III (1988). Distribution of cells containing mRNAs encoding substance P and neurokinin B in the rat central nervous system. *J. Comp. Neurol.* **272**, 90–113.

5. Detection of CCK receptor subtypes in mammalian brain using highly selective non-peptide antagonists

D. R. Hill, L. Singh, P. Boden, R. Pinnock, G. N. Woodruff, and J. Hughes

Introduction

The neuropeptide CCK occurs in the mammalian CNS in a variety of molecular forms, but it is the sulphated octapeptide (CCK-8S) that predominates (Vanderhaeghen *et al.* 1975; Dockray 1976; Rehfeld 1978; Larsson and Rehfeld 1979). Since its discovery in brain, the effects of CCK-8S and of related peptide fragments such as CCK-4 and pentagastrin have been studied extensively in a wide variety of systems. These studies have led to the implication of CCK in a variety of physiological systems. For instance, CCK has been suggested as an endogenous satiety factor and as such may play an important role in the regulation of feeding (Gibbs *et al.* 1973; Morley *et al.* 1982; Crawley and Kiss 1985; Dourish *et al.* 1989*a, b*; Chapter 17 of this volume). Other studies have indicated that the peptide may be intimately involved in mediating the transmission of painful stimuli or modulating the endogenous opioid system (Jurna and Zetler 1981; Faris *et al.* 1983; Watkins *et al.* 1984; see also Pittaway and Hill 1987; Baber *et al.* 1989; Chapters 43–45 of this volume). This suggestion is supported by the presence of CCK immunoreactivity and also by the presence of mRNA in brain regions such as the periaqueductal grey, the thalamus, and the spinal cord which are known to be of importance in the central processing of sensory information (Skirboll *et al.* 1983; Innis and Aghajanian 1986; Savasta *et al.* 1988; Chapter 4 of this volume).

Another system that has received considerable attention is the dopamine pathway arising in the mid-brain and projecting to the nucleus accumbens — the mesolimbic dopamine pathway. In this pathway CCK appears to be colocalized with dopamine, at least in a proportion of the neurons (Hökfelt *et al.* 1980*a, b*; Chapter 32 of this volume). This finding raised the possibility that the physiological roles of dopamine and CCK are closely related, and many laboratories have studied the effects of CCK on dopamine function in this pathway (Crawley *et al.* 1986; Voigt *et al.* 1986; Brodie and Dunwiddie 1987; Freeman and Chiodo 1988; Artaud *et al.* 1989; Barrett *et al.* 1989; Crawley 1989; Debonnel *et al.* 1990).

A major factor that has hampered progress in understanding the function of CCK in the brain has been the lack of selective antagonists with high affinity and selectivity for CCK receptors. However, this situation has changed with the recent development of non-peptide antagonists for the two subtypes of CCK receptor: the CCK_A and CCK_B receptors (Chang and Lotti 1986; Evans *et al.* 1986; Bock *et al.* 1989; Lotti and Chang 1989; Hughes *et al.* 1990; Horwell *et al.* 1991).

CCK receptor subtypes

Radio-ligand binding studies using tissue homogenates provided the first evidence that CCK receptors were heterogeneous. Innis and Snyder (1980) demonstrated that the unsulphated form of CCK-8 (CCK-8US) together with a number of peptide fragments including pentagastrin and CCK-4, the C-terminal tetrapeptide of CCK, displayed considerable selectivity towards the receptor site in brain labelled by [^{125}I]Bolton-Hunter CCK as opposed to pancreatic tissue. These studies, which were the first to indicate that central CCK receptors have less stringent structural requirements for binding than do pancreatic sites, led to the use of the terms 'peripheral type' and 'brain type' to differentiate the two classes of receptor.

Although no evidence for CCK receptor heterogeneity within the brain was obtained in these early radio-ligand binding studies, subsequent investigation suggested that CCK receptors resembling those in the periphery were present within discrete regions of the CNS. For instance, Hommer and colleagues (Hommer and Skirboll 1983; Hommer *et al.* 1985, 1986) examined the effects of CCK-8S, CCK-8US and CCK-4, given intravenously, on the activity of dopamine neurons within the substantia nigra. These investigators were able to define two separate effects of CCK that appeared to be mediated by separate receptors. CCK-8S, but not CCK-8US or CCK-4, caused excitation of the mesecephalic dopamine neurons, whilst all three peptides were able to potentiate the inhibitory actions of dopamine itself. In view of the fact that the pharmacological profile of these two responses paralleled data from binding studies and as previous investigations by the same group had shown that a significant part of the CCK response remained after removing all afferent input, it was concluded that peripheral type CCK receptors might be present on mid-brain dopamine neurons.

Behavioural studies in the nucleus accumbens also suggested that peripheral type CCK receptors may be present in this tissue in addition to the brain type receptor. Crawley and coworkers (Crawley *et al.* 1986; Chapter 35 of this volume) demonstrated that central injections of CCK-8S potentiated dopamine-induced hypolocomotion but that this effect was not mimicked by CCK-8US or CCK-4, again suggesting the involvement of a CCK receptor with a stringent requirement for sulphation analogous to the peripheral CCK receptor.

More recently, Vickroy *et al.* (1988) and Marshall *et al.* (1990) have shown that the facilitatory effects of CCK-8S on evoked dopamine release from slices of the nucleus accumbens *in vitro* are not mimicked by CCK-8US and can be antagonized by low (nanomolar) concentrations of devazepide, which suggests that they are mediated via a CCK_A receptor.

Following the first measurements of CCK receptor binding in the brain, numerous laboratories have described the autoradiographic localization of CCK receptor sites within the CNS of rodent and primates, including man, using a variety of radio-ligands such as [^{125}I]CCK-33 (Zarbin *et al.* 1983), [^{125}I]CCK-8 (Dietl *et al.* 1987; Dietl and Palacios 1989), [^{3}H]CCK-8 (Van Dijk *et al.* 1984), and [^{3}H]BOC[Nle$^{28, 31}$]CCK$_{27-33}$ (Pélaprat *et al.* 1987), and the more selective CCK_B radio-ligands [^{3}H]pentagastrin (Gaudreau *et al.* 1983) and [^{3}H]CCK-4 (Durieux *et al.* 1988). There is general agreement between laboratories concerning the overall distribution of CCK receptor sites throughout the brain, with high concentrations being present in the cerebral cortex especially the limbic regions, the basal ganglia, the nucleus accumbens, and the olfactory tubercle. Intermediate concentrations are found in the hypothalamus and hippocampus, and lower levels still in the medulla and spinal cord. None the less, even within regions of low overall density, discrete nuclei with high concentrations of CCK receptors are present. The cerebellum remains enigmatic, with some species including man having a high density of receptor in this tissue whilst the rat is devoid of cerebellar CCK receptors (Mantyh and Mantyh 1985).

Despite the availability of abundant information on the localization of CCK receptors, relatively little data on the presence of CCK receptor subtypes in the brain was available until the work of Moran *et al.* (1986). These authors provided the first direct autoradiographical evidence for the presence in brain of CCK receptors resembling those in the periphery. They described binding sites in a number of brain regions which bound [^{125}I]CCK-33 but had relatively low affinity for CCK-8US, CCK-4, and gastrin. The terms 'type A' and 'type B' CCK receptors were suggested as alternatives to 'peripheral type' and 'brain type' in order to classify CCK receptors, thereby resolving the problem of referring to the receptor type in terms of an anatomical location.

Type A receptors were found in the area postrema, the nucleus tractus solitarius, the interpeduncular nucleus, and the posterior hypothalamus, but no evidence was found for their existence in other areas such as the nucleus accumbens or the substantia nigra.

Non-peptide CCK receptor antagonists

The possibility that CCK is involved in many physiological processes and that potent CCK antagonists may have potential as therapeutic agents has lead to the development of several non-peptide antagonists (see Chapter 2 of this volume). Broadly speaking, these compounds can be divided into three main categories:

(1) derivatives of amino acids: CR1409 (lorglumide);
(2) derivatives of benzodiazepines: devazepide (L-364,718 or MK-329) and L-365,260;
(3) non-peptide analogues of CCK: CI-988 (PD134308).

CR1409 was perhaps the first non-peptide antagonist that combined both high affinity with good selectivity for the CCK_A receptor (Makovec *et al.* 1986) (Table 5.1). However, the pre-eminence of this compound was soon eclipsed by the development of the benzodiazepine derivative devazepide (Chang and Lotti 1986; Evans *et al.* 1986) which bound to CCK_A receptors with picomolar affinity, whilst acting at the CCK_B receptor present in the cerebral cortex of several species only at much higher concentrations. L-365,260, an analogue of devazepide, represented the first non-peptide CCK_B receptor antagonist (Bock *et al.* 1989; Lotti and Chang 1989), although this compound does not display the same picomolar affinity for the CCK_B receptor that is shown by devazepide for the CCK_A site.

The original lead compound for each of the compounds discussed so far was a non-peptide. Work in our laboratories has led to the design of small-molecule non-peptide analogues of the endogenous peptide and has resulted in the discovery of 'dipeptoids' that are potent CCK_B receptor antagonists (Hughes *et al.* 1990; Horwell *et al.* 1991) A representative of this series of compounds is CI988 which has high affinity and selectivity for the CCK_B receptor. Indeed, when measured under identical conditions CI988 surpasses L-365,260 in terms of both affinity for the CCK_B receptor and selectivity over the CCK_A receptor. Table 5.1 summarizes the affinity and selectivity of these and other reference compounds at the CCK_A and CCK_B receptor. Their structures are illustrated in Fig. 5.1.

Table 5.1. IC_{50} for the inhibition of specific [^{125}I] CCK-8 binding to mouse cortex (CCK_B) and rat pancreas (CCK_A) by CCK receptor agonists and antagonists

Compound	IC_{50} (nM)		Pancreas/cortex ratio
	Mouse cortex CCK_B	Rat pancreas CCK_A	
Agonist			
CCK-8S	0.3	0.1	0.3
Pentagastrin	0.8	600	750
CCK-8US	2.6	59	23
CCK-4	2.6	5330	2050
Antagonist			
PD134308	1.7	2717	1598
L-365,260	5.2	240	46
Devazepide	31	0.2	0.006
CR1409	500	7.1	0.01

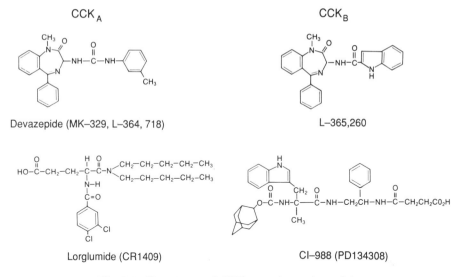

Fig. 5.1. Structures of CCK receptor antagonists.

Biochemical studies using non-peptide antagonists

The first autoradiographical studies showing the presence of CCK_A receptors in the brain relied on unsulphated peptide fragments to displace [^{125}I]CCK-33 selectively from the majority of CCK receptors in the brain (Moran *et al.* 1986). These studies were confirmed and extended using devazepide to displace [^{125}I]CCK-8 selectively from CCK_A receptors in sections of rat brain containing the area postrema (AP), the nucleus tractus solitarius (NTS), and the interpeduncular nucleus (IPN) (Hill *et al.* 1987). The tritiated form [^{3}H]devazepide was also utilized to label CCK_A receptors directly in autoradiographic and homogenate binding studies. The very high selectivity of devazepide for the CCK_A site permitted the clear delineation of the localization of the two subclasses of CCK receptor even in an anatomically discrete area such as the NTS. Thus, [^{125}I]CCK-8 binding to the medial but not the lateral NTS was sensitive to devazepide, and only those receptor sites in the medial NTS were labelled by [^{3}H]devazepide (Hill *et al.* 1987).

Subsequent studies using several different species, including the monkey, have revealed quite striking differences in the receptor types that are prevalent in particular areas of the CNS. One area of particular interest, given the role that CCK may play in modulating afferent sensory information, is the spinal cord. CCK receptors are localized to the dorsal horn of the spinal cord, and in particular to the substantia gelatinosa, in a wide variety of species (Hill *et al.* 1988*a*). In the rat these receptor sites have the properties of CCK_B receptors and bind CCK-8US and devazepide with relatively high and low affinities

Table 5.2. Inhibition of [^{125}I] CCK binding to sections of rodent and primate spinal cord

Compound	IC$_{50}$(nM)		
	Rat	Monkey	Human
CCK-8S	0.68	0.89	0.65
CCK-8US	14	245	117
Devazepide	71	0.19	0.28
L-365,260	13	1700	440

Data from Hill *et al.* (1988a) and Hill and Woodruff (1990)

respectively (Table 5.2). In the spinal cord of the non-human and human primate CCK receptor sites are localized in a comparable fashion to the rodent in laminae 2 and 3 of the substantia gelatinosa. Although sharing a similar localization, the receptor type is different as devazepide binds with very high affinity in both the monkey and human spinal cord, strongly suggesting that the CCK$_A$ receptor subtype predominates in these species (Table 5.2). Moreover, [^{3}H]devazepide shows comparable affinity for the CCK receptor in the spinal cord and the pancreas (Hill *et al.* 1988a).

Measurements of the affinity of the CCK$_B$ selective antagonist L-365,260 at CCK receptors confirm the finding that there are pronounced species differences in the receptor present in rodent and primate spinal cord. Thus L-365,260 showed higher affinity for CCK receptors in the rodent cord than in the primate cord (Table 5.2) (Hill and Woodruff 1990).

The finding that primate spinal cord contained a predominance of CCK$_A$ receptors prompted an investigation of their distribution in the higher centres of the cynomolgus monkey brain (Hill *et al.* 1990). Both [^{125}I]CCK-8 with devazepide and L-365,260 and [^{3}H]devazepide were utilized to determine the distribution of CCK$_A$ and CCK$_B$ receptors throughout the monkey brain. In addition to the dorsal vagal complex, where CCK$_A$ receptors were found in the rat, devazepide-sensitive CCK$_A$ receptor sites were localized to a number of hypothalamic nuclei including the supra-optic and paraventricular nuclei, the dorsomedial and infundibular nuclei, and the neurohypophysis. The mammillary bodies and supramammillary nuclei also contained CCK$_A$ receptor sites.

Perhaps the most interesting finding, given the close association of CCK with dopamine (Hökfelt *et al.* 1980a, b; Chapter 32 of this volume) was the presence of high concentrations of CCK$_A$ receptors in the substantia nigra and ventral tegmental regions of the cynomolgus monkey (Fig. 5.2). The localization of [^{3}H]devazepide binding as well as devazepide-sensitive [^{125}I]CCK-8 binding closely paralleled the distribution of the large cell bodies of the pars compacta, although the highest density of binding of CCK$_A$ receptors in the mesencephalon was found in an area corresponding to the ventral tegmental area.

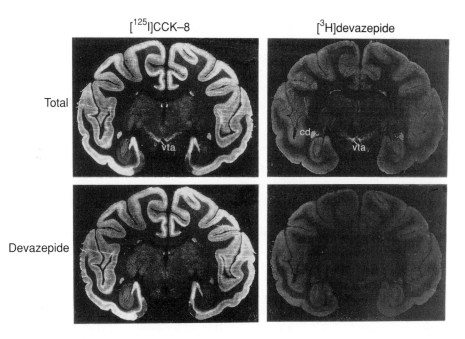

Fig. 5.2. Autoradiographs showing the presence of CCK$_A$ receptor binding sites in sections of cynomolgus monkey brain. Sections were incubated with either 50 pM [^{125}I]Bolton-Hunter CCK-8 or 200 pM [3H]devazepide in the absence and presence of 10 nM devazepide. The CCK$_A$ antagonist devazepide selectively inhibited [^{125}I]Bolton-Hunter CCK-8 binding to the zona compacta of the substantia nigra and the ventral tegmental area (vta) and also partially reduced binding to the caudate nucleus (cd). These same regions were selectively labelled by [^{3}H]devazepide, with specific binding to the ventral tegmental region being particularly clear. (Reproduced from Hill *et al.* 1990.)

The pattern of CCK$_A$ receptor binding in the mesencephalon strongly suggested the presence of CCK$_A$ receptors on dopamine cell bodies. This was further supported by the finding that [^{125}I]CCK-8 and [^{3}H]devazepide binding was lost following a unilateral lesion of the dopamine neurons using 1-methyl-4-phenyl-1,2,3,6-tetrahydropyrridine (MPTP). Furthermore, specific [^{3}H]devazepide binding was detected in sections of control human substantia nigra but was absent from tissue obtained from parkinsonian brains.

As outlined in the Introduction, there is some evidence suggesting the presence of CCK$_A$ receptors in the rat substantia nigra although these are not readily detectable using autoradiography. The failure to detect these receptors in the rat may be due to a number of factors. Clearly, one possibility is that they are present in the rodent in lower concentrations than in the primate and this prevents their detection.

Low densities of CCK$_A$ receptor binding sites also appear to be present in the caudate nucleus and ventral putamen of the monkey brain (Hill *et al.* 1990). This is perhaps not surprising as these sites probably represent binding sites on the terminals of the dopamine neurons having their origin in the pars compacta of the substantia nigra. Evidence for this premise again comes from experiments using tissue from unilateral MPTP lesioned monkeys. [³H]devazepide binding in the medial caudate was lost on the side ipsilateral to the lesion.

The CCK_A receptor subtype is clearly present in the CNS to a greater extent than was hitherto believed. Indeed, functional studies suggest that it may be present in several areas of the rat brain. None the less, the CCK_B receptor appears to be more ubiquitous and is present in many regions of the brain including limbic regions, and thus may have an important role in emotional behaviour. Until recently, L-365, 260, a close analogue of devazepide, was the only selective CCK_B receptor antagonist available. Binding studies show that this compound binds to the CCK_B receptor in brain with nanomolar affinity ($K_i = 2.0$ nM) (Lotti and Chang 1989), and it is active in producing a functional block of CCK excitation although it is somewhat less potent ($K_b = 33$ nM) (Kemp *et al.* 1989; see Chapter 7 of this volume).

The novel dipeptoid CI-988 represents a new class of CCK receptor antagonist with high affinity and selectivity for the CCK_B receptor. Indeed, the high selectivity of CI-988 (Table 5.1) lends itself to the detection of CCK receptor subtypes in the brain in biochemical and pharmacological models. This can be illustrated in autoradiographical experiments by selective inhibition of CCK_B receptor binding to sections of rat brain containing both subclasses of CCK receptor. In sections of rat brain taken approximately 3 mm posterior to bregma, [¹²⁵I]CCK-8 labels receptor binding sites in the cerebral cortex, the ventromedial and dorsomedial hypothalamus, the hippocampus, the amygdaloid complex, and the habenular nuclei. CCK-8S displaced uniformly from all areas of the brain section. CI-988 inhibited the binding of [¹²⁵I]CCK from CCK_B sites in the cortex, amygdala, hippocampus, and ventromedial hypothalamus, but was less effective at inhibiting binding in the dorsomedial hypothalamic nucleus and the habenular nucleus (Fig. 5.3). Binding in these regions was inhibited by devazepide, confirming that [¹²⁵I]CCK was binding to CCK_A sites.

The presence of CCK_A sites in the habenular nucleus is of interest as it is here that the neurons, whose terminal field lies in the interpeduncular nucleus and where CCK_A sites were first detected, have their origin. Indeed, electrolytic lesions of the habenular nucleus decreased CCK_A binding in the IPN (Hill *et al.* 1988*b*). Moreover, the existence of CCK_A receptors in the rat dorsomedial hypothalamic nucleus parallels the situation in the primate.

Electrophysiological studies using non-peptide antagonists

Until recently, the majority of electrophysiological studies in brain have relied heavily on agonists such as pentagastrin and CCK-8US to differentiate CCK

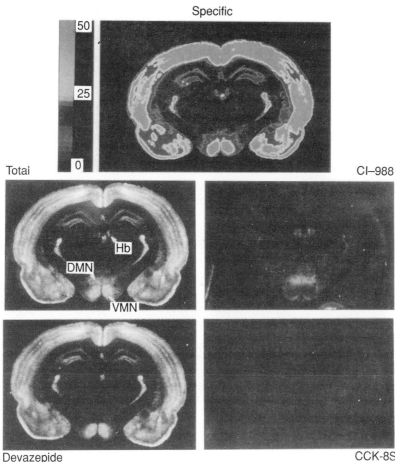

Fig. 5.3. Autoradiographs of Bolton-Hunter [^{125}I]CCK-8 binding to sections of rat brain showing selective displacement from CCK$_A$ and CCK$_B$ sites by devazepide and Cl-988. Sections were incubated with [^{125}I]CCK-8 (50 pM) in the absence and presence of devazepide (10 nM), PD134308 (300 nM) and CCK-8S (1 μM). Specific binding of [^{125}I]CCK to most regions of the section was inhibited by CI-988, whereas devazepide was relatively ineffective. However, binding to the dorso-medial hypothalamic nucleus and the habenular nucleus was sensitive to devazepide, indicating the presence of CCK$_A$ sites, and in these areas CI-988 produced little inhibition of binding. CCK-8S displaced uniformly across the section.

receptor subtypes. The availability of potent and selective antagonists makes it easier to identify functional CCK receptor subtypes clearly.

The ventromedial nucleus (VMN) of the rat maintained *in vitro* has been shown to be sensitive to a number of peptides including CCK (Kow and Pfaff 1986). Furthermore, there is strong immunocytochemical evidence for the presence of CCK in a projection from the parabrachial nucleus to the VMN (Zaborski *et al.* 1984) and the VMN also contains high densities of CCK receptor binding sites (Day *et al.* 1987). Consequently, the VMN brain slice preparation represents an ideal system for the study of CCK agonists and antagonists using both extracellular and intracellular recording techniques (Boden and Hill 1988).

Inclusion of CCK-8S in the medium superfusing the VMN slice preparation resulted in a concentration-dependent increase in the firing rate of neurons within the nucleus with an EC_{50} of between 30 and 100 nM (Boden and Hill 1988; Kemp *et al.* 1989; Chapter 7 of this volume) (Fig. 5.4). Both CCK-8US and pentagastrin also increased cell firing with similar potency to the sulphated peptide, suggesting that the response is mediated by a CCK_B receptor. CI-988 applied for 10 min produced a reduction in the response to CCK-8S (Fig. 5.4)

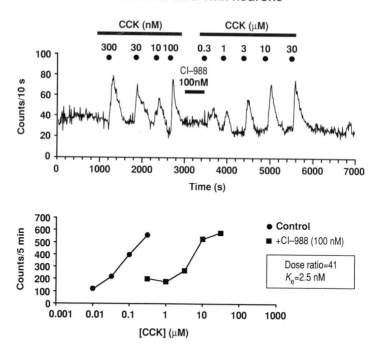

Fig. 5.4. Ratemeter recordings (upper panel) of spontaneous action potential firing with time recorded from a neuron in the ventromedial nucleus of the hypothalamus. Application of CCK-8S for 1 min produced a concentration-dependent increase in firing. Application of CI-988 for 10 min antagonized the response. The lower panel shows the same data plotted as concentration–response curves showing the parallel shift produced by CI-988. (Reproduced from Hughes *et al.* 1990.)

and pentagastrin (not shown). The dose–response curve to CCK-8S was progressively shifted to the right in a parallel fashion by increasing concentrations of CI-988. Moreover, Schild analysis indicated that the antagonism was competitive and an equilibrium constant of 2.5 nM was obtained for the data shown in Fig. 5.4.

The CCK_A receptor antagonist devazepide has also been shown to antagonize responses to CCK-8 in the VMN preparation (Boden and Hill 1988) but only at concentrations considerably greater than those needed to block CCK_A receptors (see below). In contrast, L-365,260 blocked the response to CCK-8 at lower concentrations, which is consistent with the presence of the CCK_B receptor type in the VMH (see Chapter 7 of this volume).

Recordings made posterior to the VMN in the intrafasicular nucleus, an area close to the IPN, revealed neurons that were also sensitive to CCK-8S but not

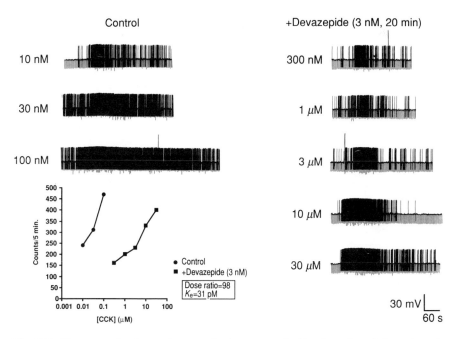

Fig. 5.5. Pen recorder trace from a single neuron in the dorsal raphé nucleus *in vitro* showing concentration-dependent excitation by CCK-8S and its antagonism by devazepide. Traces in the left-hand column show the effects of 1 min applications of CCK-8S (10–100 nM). Traces in the right-hand column were taken from the same neuron after incubation for 20 min with devazepide and show that the response to CCK-8S was considerably reduced by the antagonist. The inset shows the same data presented as a concentration–response curve in which the number of action potentials occurring in a 5 min period immediately following peptide application is plotted as a function of concentration. Devazepide produced a parallel shift to the right indicative of competitive antagonism. The equilibrium constant for this experiment was 31 pM. (Data from Boden *et al.* 1991.)

to CCK-4 or pentagastrin and responded with an excitation. As noted above, high concentrations of devazepide will antagonize the actions of CCK in the VMH, but responses in the intrafasicular nucleus were sensitive to nanomolar concentrations of the CCK_A receptor antagonist (Hill and Boden 1989).

Another area of the brain in which the effects of CCK have been studied electrophysiologically is the dorsal raphé nucleus (Boden *et al.* 1991). CCK-like immunoreactivity is present in the nucleus (Vanderhaeghen *et al.* 1980; Van Der Koy *et al.* 1981), and one of the major inputs to it originates in the NTS where CCK_A receptors exist and where CCK has been shown to inhibit spontaneously active neurons (Morin *et al.* 1983). The dorsal raphé also has major efferent connections to the IPN (Steinbusch and Niewenhuys 1983), a region known to possess CCK_A binding sites (Hill *et al.* 1987), and to the ventromedial hypothalamus.

Two distinct neuronal populations were identified in the dorsal raphé nucleus, but only those that responded to the application of 5-HT also responded to CCK, and most of these were excited. Application of CCK-8S to neurons resulted in a concentration-dependent increase in cell firing (Fig. 5.5). However, pentagastrin did not mimic the action of the sulphated octapeptide, sug-

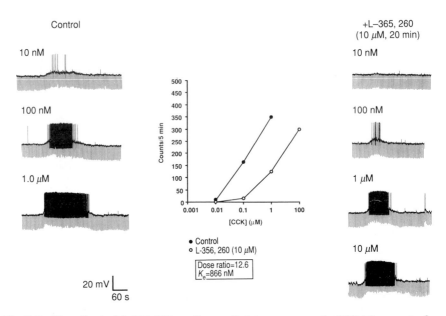

Fig. 5.6. The effect of L-365,260 on the excitatory response to CCK-8S on a single neuron in the dorsal raphé nucleus. Control traces are shown in the left-hand column and recordings made after a 20 min exposure to L-365,260 (10 μM) are shown on the right. The concentration–response curve to CCK-8S was shifted to the right by L-365,260 and an equilibrium constant of 866 nM was obtained for the antagonist.

gesting that the receptor mediating this response was not the same as in the VMN. In keeping with this, low concentrations of devazepide produced a rapid antagonism of the response. Devazepide shifted the CCK dose–response curve to the right in a parallel fashion and a mean equilibrium constant of 0.13 nM was calculated for the antagonist (Fig. 5.5). L-365,260 also produced a block of the CCK response but was considerably weaker than devazepide, having a mean equilibrium constant of 725 nM (Fig. 5.6).

The potent antagonism of this excitatory response by devazepide and the lack of effect of pentagastrin clearly indicate the presence of a CCK_A receptor in the dorsal raphé nucleus located on 5-hydroxytryptamine (5-HT) neurons. This receptor population may play a role in mediating CCK–5-HT interactions in relation to satiety (see Chapter 23 of this volume).

Behavioural effects of non-peptide CCK_B receptor antagonists

Since the discovery of CCK in the brain 15 years ago, considerable effort has been expended in designing experiments to elucidate its physiological role in the brain. As with other neurotransmitter systems whose functions are more clearly understood than CCK, most information has come from studies using antagonists. CCK itself is widely distributed throughout the brain, especially the limbic system, and in some neurons it is colocalized with classical neurotransmitters such as γ-aminobutyrate (GABA) and 5-HT (Van der Koy *et al.* 1981; MacDonald *et al.* 1982; Hendry *et al.* 1984; Somogyi *et al.* 1984). Furthermore, CCK_B receptors are found in high concentrations in regions thought to be involved in the control of affective behaviour.

On the basis of these observations it was hypothesized that a CCK_B receptor antagonist may influence emotional behaviour, and so CI-988 was tested in several models that are predictive of anxiolytic activity in man. These were the mouse black–white box, the rat elevated plus maze, the rat social interaction test, and the marmoset human threat test (Hughes *et al.* 1990). In all four models CI-988 proved to be an extremely effective agent, producing anxiolysis over a wide dose range without any indication of the sedative effects characteristic of the benzodiazepines. Figure 5.7 illustrates the effect of CI-988 and two other CCK antagonists in one of the anxiolytic models – the rat elevated plus maze (Singh *et al.* 1991). In this model untreated animals spend most of their time in one or other of the enclosed arms of the plus-shaped maze rather than in the more aversive open arms from which the animal is potentially more likely to fall. Known anxiolytics such as diazepam and chlordiazepoxide increase the amount of time that the animal spends in the open arms. Intraperitoneal injection of CI-988 over the dose range 0.1–10 mg/kg increased the amount of time that animals spent on the open arm of the maze to a similar extent to that observed with chlordiazepoxide. The possibility that this anxiolytic behaviour was attributable to blockade of CCK_B receptors was supported by the finding that the CCK_B antagonist L-365,260, at similar doses

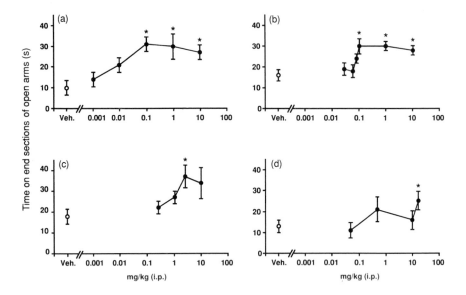

Fig. 5.7. The effect of (a) CI-988, (b) L-365,260, (c) chlordiazepoxide, and (d) devaz-epide in the rat elevated plus maze. Compounds were administered intraperi-toneally 40 min before the test. The time spent on the distal end sections of the open arms was determined during the 5 min observation period. Results are the mean time spent on the open arms (vertical bars represent the SEM) by 10 animals per group. Significant differences from vehicle, which were determined by Dunnett's *t* test after significant ANOVA, are indicated by asterisks.

to CI-988, also increased the time that the animals spent on the open arms of the plus maze. In contrast, the CCK_A antagonist devazepide showed signifi-cant effects only at doses in excess of 10 mg/kg which are probably sufficient to block CCK_B receptors (but see also Chapter 12 of this volume).

In order to study this phenomenon further we investigated the effects of CCK_B receptor agonists given by the intracerebroventricular (i.c.v.) route, again using the elevated plus maze as a model. To increase the sensivity for anxiogenic effects, the total time spent on the open arm was measured rather than just the time spent on the distal portion of the arm as in anxiolytic meas-urements. I.c.v. injection of low doses of caerulein and the CCK_B receptor selective agonist pentagastrin produced a significant decrease in the time spent on the open arm when compared with control indicating an anxiogenic response (Fig. 5.8) (see also Chapters 13 and 14 of this volume). Similarly, the convulsant pentylenetetrazol (PTZ) given systemically also produced an anxiogenic response. The effects of both caerulein and pentagastrin were dose-dependently blocked by peripheral administration of CI-988 given immediately after the i.c.v. injection of agonist (Fig. 5.8). The same dose of CI-988 did not block the anxiogenic response to PTZ.

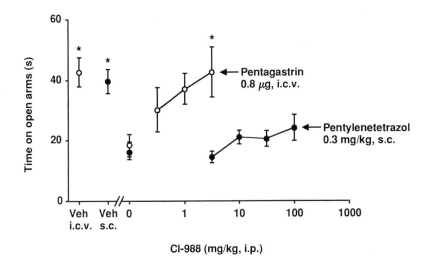

Fig. 5.8. The effect of CI-988 on the anxiogenic action of pentagastrin and PTZ in the rat elevated plus maze. Pentagastrin and PTZ were administered i.c.v. and subcutaneously (s.c.) respectively immediately before CI-988. After 15 min the animals were placed on the plus maze and the total time spent on the open arms was measured during the 5 min observation period. Results are the mean time spent on the open arms and the SEM of 10 animals per group. Both pentagastrin and PTZ produced a significant anxiogenic response. The effect of pentagastrin but not that of PTZ was reversed by CI-988. Significant differences from the group given pentagastrin or PTZ are shown by asterisks.

Summary and conclusions

Devazepide and CI-988 represent two non-peptide CCK receptor antagonists with high affinity and selectivity for the two classes of CCK receptor present in the brain. Both compounds antagonize the excitatory effects of CCK with equilibrium constants close to their binding affinities.

Our studies using these compounds show that the CNS contains both subtypes of CCK receptor and that functional responses characteristic of each receptor type can be identified in many areas of the brain. The distribution of CCK_A receptors is clearly more widespread than early studies suggested, although there remains a discrepancy between the measurement of a functional response and the presence of CCK_A binding sites in some areas.

Two major conclusions regarding the function of CCK in the brain can be made on the basis of the data presented above. First, the anxiolytic effects of two CCK_B receptor antagonists of quite different chemical classes together with the anxiogenic effect of the selective CCK_B receptor agonist pentagastrin provide evidence for the involvement of endogenous CCK in the control of anxiety. Additional evidence in support of this comes from recent studies showing the anxiogenic effects of CCK agonists in humans (De Montigny 1989; Bradwejn *et al.* 1990; Chapters 10 and 11 of this volume) and rodents (Daugé *et al.* 1989; Ravard and Dourish 1990; Harro *et al.* 1990; Chapters 12, 13, and 14 of this volume). Second, the need for relatively high doses of devazepide (compared with CI-988 or L-365,260) to produce an anxiolytic effect suggests the physiological involvement of CCK_B rather than CCK_A receptors in the process even though the anxiogenesis produced by direct injection of CCK into the nucleus accumbens can be blocked by devazepide (Daugé *et al.* 1989). Further work is clearly required to determine the functional role of CCK_A receptors in the central nervous system.

The finding that CCK_B antagonists are potential anxiolytics raises many questions, particularly with regard to the neural substrates that are involved and their mechanism of action. None the less compounds such as CI-988 and L-365,260 may be representative of a novel class of anxiolytic agent with an improved profile compared with benzodiazepines and 5-HT-related anxiolytics.

Irrespective of this, the availability of highly potent non-peptide antagonists with which to study CCK receptors will provide a strong stimulus for basic research in this area.

References

Artaud, F., Baruch, P., Stutzmann, M., Saffroy, M., Godeheu, G., Barbeito, L., Herve, D., Studler, J. M., Glowinski, J., and Cheramy, A. (1989). Cholecystokinin: co-release with dopamine from nigrostriatal neurons in cat. *Eur. J. Neurosci.* **1**, 162–72.

Baber, N. S., Dourish, C. T., and Hill, D. R. (1989). The role of CCK caerulein and CCK antagonists in nociception. *Pain* **39**, 307–28.

Barrett, R. W., Steffey, M. E., and Wolfram, A. W. (1989). Type A cholecystokinin binding sites in cow brain: characterization using (-)-[3H]-L364,718. *Mol. Pharmacol.* **36**, 285–90.

Bock, M. G., Dipardo, R. M., Evans, B. E., Rittle, K. E., Whitter, W. L., Veber, D. F., Anderson, P. S., and Freidinger, R. M. (1989). Benzodiazepine gastrin and cholecystokinin receptor ligands: L-365,260. *J. Med. Chem.* **32**, 13–16.

Boden, P. and Hill, R. G. (1988). Effects of cholecystokinin and related peptides on neuronal activity in the ventromedial nucleus of the rat hypothalamus. *Br. J. Pharmacol.* **94**, 246–52.

Boden, P., Woodruff, G. N., and Pinnock, R. D. (1991). Pharmacology of a cholecystokinin receptor on serotonin neurones in the dorsal raphe of the rat brain. *Br. J. Pharmacol.* **102**, 635–8.

Bradwejn, J., Koszycki, D., Shriqui, C., and Meterissian, G. (1990). Cholecystokinin induces panic attacks in patients with panic disorder. *Can. J. Psychiat.* **35**, 83–5.

Brodie, M. and Dunwiddie, T. V. (1987). Cholecystokinin potentiates dopamine inhibition of mesencephalic dopamine neurones in vitro. *Brain Res.* **425**, 106–14.

Chang, R. S. L., and Lotti, V. J. (1986). Biochemical and pharmacological characterization of an extremely potent and selective non-peptide cholecystokinin antagonist. *Proc. Natl Acad. Sci. USA* **83**, 4923–6.

Crawley, J. N. (1989). Microinjection of cholecystokinin into the rat ventral tegmental area potentiates dopamine-induced hypolocomotion. *Synapse* **3**, 346–55.

Crawley, J. N. and Kiss, F. (1985). Paraventricular nucleus lesions abolish the inhibition of feeding induced by systemic cholecystokinin. *Peptides* **6**, 927–35.

Crawley, J. N., Stivers, J. A., Hommer, D. W., Skirboll, L. R., and Paul, S. M. (1986). Antagonists of central and peripheral behavioral effects of cholecystokinin octapeptide. *J. Pharm. Exp. Ther.* **236**, 320–30.

Daugé, V., Steimes, P., Derrien, M., Beau, N., Roques, B. P., and Feger, J. (1989). CCK-8 effects on motivational and emotional states of rats involve CCK-A receptors of the postero-median part of the nucleus accumbens. *Pharmacol. Biochem. Behav.* **34**, 157–63.

Day, N. C., Hall, M. D., Clarke, C. D. and Hughes, J. (1987). High concentrations of cholecystokinin receptor binding sites in the ventromedial hypothalamic nucleus. *Neuropeptides* **8**, 1–18.

Debonnel, G., Gaudreau, P., Quirion, R., and DeMontigny, C. (1990). Effects of long term haloperidol treatment on the responsiveness of accumbens neurones to cholecystokinin and dopamine: electrophysiological and radioligand binding studies in the rat. *J. Neurosci.* **10**, 469–78.

Deitl, M. M., and Palacios, M. (1989). The distribution of cholecystokinin receptors in the vertebrate brain: species differences studied by receptor autoradiography. *J. Chem. Neuroanat.* **2**, 149–61.

Deitl, M. M., Probst, A., and Palacios, J. M. (1987). On the distribution of cholecystokinin binding sites in the human brain: an autoradiographic study. *Synapse* **1**, 169–83.

De Montigny, C. (1989). Cholecystokinin tetrapeptide induces panic-like attacks in healthy volunteers. *Arch. Gen. Psychiat.* **46**, 511–17.

Dockray, G. J. (1976). Immunochemical evidence of cholecystokinin-like peptides in brain. *Nature* **264**, 568–70.

Dourish, C. T., Rycroft, W. and Iversen, S. D. (1989*a*). Postponement of satiety by blockade of brain cholecystokinin (CCK-B) receptors. *Science* **245**, 1509–11.

Dourish, C. T., Ruckert, A. C., Tattersall, F. D., and Iversen, S. D. (1989*b*). Evidence that decreased feeding induced by systemic injection of cholecystokinin is mediated by CCK-A receptors. *Eur. J. Pharmacol.* **173**, 233–4.

Durieux, C., Pelaprat, D., Charpentier, B., Morgat, J.-L., and Roques, B. P. (1988). Characterization of [3H]-CCK-4 binding sites in mouse and rat brain. *Neuropeptides* **12**, 141–8.

Evans, B. E., Bock, M. G., Rittle, K. E., Dipardo, R. M., Whitter, W. L., Veber, D. F., Anderson, P. S., and Freidinger, R. M. (1986). Design of potent orally effective nonpeptidal antagonists of the peptide hormone cholecystokinin. *Proc. Natl Acad. Sci. USA* **83**, 4918–22.

Faris, P. L., Komisaruk, B. R., Watkins, L. R. and Mayer, D. J. (1983). Evidence for the neuropeptide cholecystokinin as an antagonist of opiate analgesia. *Science*, **219**, 310–12.

Freeman, A. S. and Chiodo, L. A. (1988). Electrophysiological effects of cholecystokinin octapeptide on identified rat nigrostriatal dopaminergic neurones. *Brain Res.* **439**, 266–74.

Gaudreau, P., Quirion, R., St-Pierre, S., and Pert, C. (1983). Characterization and visualization of cholecystokinin receptors in rat brain using [³H]pentagastrin. *Peptides* **4**, 755–62.

Gibbs, J., Young, R. C., and Smith, G. P. (1973). Cholecystokinin decreases food intake in rats. *J. Comp. Physiol. Psychol.* **84**, 488–95.

Harro, J., Pold, M. and Vasar, E. (1990). Anxiogenic-like action of caerulein, a CCK-8 receptor agonist in the mouse: influence of acute and subchronic diazepam treatment. *Naunyn-Schmeidebergs Arch. Pharmacol.* **341**, 62–7.

Hendry, S. H. C., Jones, E. G., DeFilipe, J., Schmechel, D., Brandon, C., and Emson, P. C. (1984). Neuropeptide-containing neurons of the cerebral cortex are also GABAergic. *Proc. Natl Acad. Sci. USA* **81**, 6526–30.

Hill, R. G. and Boden, P. (1989). Electrophysiological actions of CCK in the mammalian CNS. In *The neuropeptide cholecystokinin (CCK). Anatomy biochemistry, receptors and physiology* (ed. J. Hughes, G. Dockray, and G. N. Woodruff), pp. 186–94. Ellis Horwood Chichester.

Hill, D. R. and Woodruff, G. N. (1990). Differentiation of central cholecystokinin receptor binding sites using the non-peptide antagonists MK-329 and L-365,260. *Brain Res.* **526**, 276–83.

Hill, D. R., Campbell, N. J., Shaw, T. M., and Woodruff, G. N. (1987). Autoradiographic localization and biochemical characterization of peripheral type CCK receptors in rat CNS using highly selective non-peptide CCK antagonists. *J. Neurosci.* **7**, 2976–76.

Hill, D. R., Shaw, T. M., and Woodruff, G. N. (1988a). Binding sites for [125I]cholecystokinin in primate spinal cord are of the CCK-A subclass. *Neurosci. Lett.* **89**, 133–9.

Hill, D. R., Shaw, T. M., Dourish, C. T., and Woodruff, G. N. (1988b). CCK-A receptors in the rat interpeduncular nucleus: evidence for a presynaptic location. *Brain Res.* **454**, 101–5.

Hill, D. R., Shaw, T. M., Graham, W., and Woodruff, G. N. (1990). Autoradiographical detection of cholecystokinin (CCK-A) receptors in primate brain using [125I]-Bolton Hunter CCK8 and [3H]-MK-329. *J. Neurosci.* **10**, 1070–81.

Hökfelt, T., Rehfeld, J. F., Skirboll, L., Ivemark, B., Goldstein, M., and Markey, K. (1980a). Evidence for coexistence of dopamine and cholecystokinin in meso-limbic neurons. *Nature* **285**, 476–78.

Hökfelt, T., Skirboll, L. R., Rehfeld, J., Goldstein, M., Markey, K., and Dann, O. (1980b). A subpopulation of mesencephalic dopamine neurones projecting to limbic areas contains a cholecystokinin-like peptide: evidence from immunochemistry combined with retrograde tracing. *Neuroscience* **5**, 2093–124.

Hommer, D. W. and Skirboll, L. R. (1983). Cholecystokinin-like peptides potentiate apomorphine-induced inhibition of dopamine neurones. *Eur. J. Pharmacol.* **91**, 151–2.

Hommer, D. W., Palkowitz, M., Crawley, J. N., Paul, S. M., and Skirboll, L. R. (1985). Cholecystokinin-induced excitation in the substantia nigra: evidence for peripheral and central components. *J. Neurosci.* **5**, 1387–92.

Hommer, D. W., Stoner, G., Crawley, J. N., Paul, S. M., and Skirboll, L. R. (1986). Cholecystokinin–dopamine coexistence: electrophysiological evidence corresponding to cholecystokinin receptor subtypes. *J. Neurosci.* **6**, 3039–43.

Horwell, D. C., Hughes, J., Hunter, J. C., Pritchard, M. C., Richardson, R. S., Roberts, E., and Woodruff, G. N. (1991). Rationally designed 'dipeptoid' analogues of CCK. Alpha-methyl trytophan derivatives as highly selective and orally active gastrin and CCK-B antagonists with potent anxiolytic properties. *J. Med. Chem.* **34**, 404-14.

Hughes, J., Boden, P., Costall, B., Domeney, A., Kelly, E., Horwell, D. C., Hunter, J. C., Pinnock, R. D. and Woodruff, G. N. (1990). Development of a class of cholecystokinin type B receptor antagonists having potent anxiolytic activity. *Proc. Natl Acad. Sci. USA* **87**, 6728-32.

Innis, R. B. and Aghajanian, G. K. (1986). Cholecystokinin-containing and nociceptive neurons in the Edinger-Westphal nucleus. *Brain Res.* **363**, 230-8.

Innis, R. B. and Snyder, S. H. (1980). Distinct cholecystokinin receptors in brain and pancreas. *Proc. Natl Acad. Sci. USA* **77**, 6917-21.

Jurna, I. and Zetler, G. (1981). Antinociceptive effect of centrally administered caerulein and cholecystokinin (CCK). *Eur. J. Pharmacol.* **73**, 323-31.

Kemp, J. A., Marshall, G. and Woodruff, G. N. (1989). Antagonism of CCK-induced excitation of the rat ventromedial hypothalamic neurons by a new selective CCK-B receptor antagonist. *Br. J. Pharmacol.* **98** (Proc. Suppl), 630P.

Kow, L.-M. and Pfaff, D. W. (1986). CCK-8 stimulation of ventromedial hypothalamic nucleus in vitro: a feeding related event? *Peptides* **7**, 473-9.

Larsson, L.-I. and Rehfeld, J. F. (1979). Localization and molecular heterogeneity in the central and peripheral nervous system. *Brain Res.* **165**, 201-18.

Lotti, V. and Chang, R. S. L. (1989). A potent and selective non peptide gastrin antagonist and brain (CCK-B) receptor ligand: L-365,260. *Eur. J. Pharmacol.* **162**, 273-80.

MacDonald, J. K., Parnavelas, J. G., Karamanlidis, A. N., Rosenquist, G., and Brecha, N. (1982). The morphology and distribution of peptide containing neurons in the adult and developing visual cortex of the rat III. Cholecystokinin. *J. Neurocytol.* **11**, 881-95.

Makovec, F., Bani, M., Chiste, R., Revel, L., Rovati, L. C., and Setnikar. (1986). Different peripheral and central antagonistic activity of new glutaramic acid derivatives on satiety induced by cholecystokinin in rats. *Regul. Pept.* **16**, 281-90.

Mantyh, C. R. and Mantyh, P. W. (1985). Differential localization of cholecystokinin-8 binding sites in the rat vs the guinea pig brain. *Eur. J. Pharmacol.* **113**, 137-9.

Marshall, F. H., Barnes, S., Pinnock, R. D., and Hughes, J. (1990). Characterization of cholecystokinin octapeptide-stimulated dopamine release from rat nucleus accumbens *in vitro. Br. J. Pharmacol.* **99**, 845-8.

Moran, T. H., Robinson, P., Goldrich, M. S., and McHugh, P. (1986). Two brain cholecystokinin receptors: implications for behavioral actions. *Brain Res.* **362**, 175-9.

Morin, M. P., Demarchi, P., Champagnat, J., Vanderhaeghen, J.-J., Rossier, J., and Denauvit-Saubie, M. (1983). Inhibitory effects of cholecystokinin on neurons in the nucleus tractus solitarius. *Brain Res.* **265**, 333-8.

Morley, J. E., Levine, A. S., Kneip, J., and Grace, M. (1982). The effects of vagotomy on the satiety effects of neuropeptides and naloxone. *Life Sci.* **30**, 1943-7.

Pélaprat, D., Broer, Y., Studler, J. M., Pechanski, M., Tassin, J. P., Glowinski, J., Rostene, W., and Roques, B. P. (1987). Autoradiography of CCK receptors in the rat brain using $[^3H]BOC[Nle^{28}, Phe^{31}]$-$CCK_{27,33}$ and $[^{125}I]$Bolton Hunter CCK8. Functional significance of subregional distributions. *Neurochem. Int.* **10**, 395-508.

Pittaway, K. M. and Hill, R. G. (1987). Cholecystokinin and pain. *Pain Headache* **9**, 213-46.

Ravard, S. and Dourish, C. T. (1990). Cholecystokinin and anxiety. *Trends Pharmacol. Sci.* **11**, 271–3.

Rehfeld, J. F. (1978). Immunochemical studies on cholecystokinin II. Distribution and molecular heterogeneity in the central nervous system of man and hog. *J. Biol. Chem.* **253**, 4022–30.

Savasta, M., Palacios, J. and Mengod, G. (1988). Regional localization of the mRNA coding for the neuropeptide cholecystokinin in the rat brain studied by *in situ* hybridization. *Neurosci. Lett.* **93**, 132–8.

Singh, L., Lewis, A. S., Field, M. J., Woodruff, G. N., and Hughes, J. (1991). Evidence for the involvement of the central CCK-B receptor in anxiety. *Proc. Natl Acad. Sci. USA* **88**, 1130–3.

Skirboll, L., Hökfelt, T., Dockray, G., Rehfeld, J., Brownstein, M. and Cuello, A. (1983). Evidence for periaqueductal cholecystokinin-substance P neurons projecting to the spinal cord. *J. Neurosci.* **3**, 1151–7.

Somogyi, P., Hodgson, A. J., Smith, A. D., Nunzi, M. G., Gorio, A. and Wu, J.-Y. (1984). Different populations of GABAergic neurons in the visual cortex and hippocampus of cat contain somatostatin or cholecystokinin-immunoreactive material. *Neuroscience* **14**, 2590–603.

Steinbusch, H. W. M. and Nieuwenhuys, R. (1983). The raphe nuclei of the rat brainstem: A cytoarchitectonic and immunohistochemical study. In *Chemical neuroanatomy* (ed. P. C. Emson), pp. 131–207. Raven Press, New York.

Van Dijk, A., Richards, J. G., Treciak, A., Gillessen, D. and Mohler, H. (1984). Cholecystokinin receptors: Biochemical demonstration and autoradiographic localization in rat brain and pancreas using [^{3}H]cholecystokinin as radioligand. *J. Neurosci.* **4**, 1021–33.

Vanderhaeghen, J.-J. Signeau, J. C., and Gepts, W. (1975). New peptide in vertebrate CNS reacting with antigastrin antibodies. *Nature* **257**, 604–5.

Vanderhaeghen, J.-J., Lostra, F., Demay, J. and Gilles, C. (1980). Immunochisto-chemical localization of cholecystokinin and gastrin-like peptides in the brain and hypophysis of the rat. *Proc. Natl Acad. Sci. USA* **77**, 1190–4.

Van Der Koy, D., Hunt, S., Steinbusch, H. W. M., and Verhofstad, A. A. J. (1981). Separate populations of cholecystokinin and 5HT containing cells in the dorsal raphe and their contribution to its ascending projections. *Neurosci. Lett.* **26**, 25–31.

Vickroy, T. W., Bianchi, B. R., Kerwin, J. F., Kopecka, H., and Nadzan, A. M. (1988). Evidence that type A CCK receptors facilitate dopamine efflux in rat brain. *Eur. J. Pharmacol.* **152**, 371–2.

Voigt, M., Wang, R. Y., and Westfall, T. C. (1986). Cholecystokinin octapeptides alter the release of endogenous dopamine from the rat nucleus accumbens *in vitro*. *J. Pharm. Exp. Ther.* **237**, 147–53.

Watkins, L. R., Kinsheck, I. B., and Mayer, D. J. (1984). Potentiation of opiate analgesia and apparent reversal of morphine tolerance by proglumide. *Science* **224**, 395–6.

Zaborsky, L., Beinfield, M. C., Palkovits, M. and Heimer, L. (1984). Brainstem connection to the hypothalamus ventromedial nucleus in the rat: a CCK-containing pathway. *Brain Res.* **303**, 225–331.

Zarbin, M. A., Innis, R. B., Walmsley, K., Snyder, S. H., and Kuhar, M. J. (1983). Autoradiographic localization of cholecystokinin receptors in rodent brain. *J. Neurosci.* **3**, 877–906.

6. In vitro and in vivo characterization of CCK_B receptors and behavioural responses induced by mesolimbic CCK receptor stimulation with compounds designed to recognize CCK_A or CCK_B sites selectively

Bernard P. Roques, Christiane Durieux,
Mariano Ruiz-Gayo, Muriel Derrien, Florence Bergeron,
Pierre-Jean Corringer, and Valérie Daugé.

CCK-8, the sulphated octapeptide fragment of CCK, has been found in specific neuronal cells where it is concentrated in synaptic vesicles and released by a calcium-dependent mechanism to interact with a similar nanomolar affinity with at least two binding sites designated CCK_B (or central site) and CCK_A (or peripheral site) (Vanderhaeghen and Crawley 1985; Chapter 5 of this volume). The former site resembles the gastrin receptor and the latter appears to be similar to the pancreatic receptor. High concentrations of CCK-8 have been found in brain regions such as the limbic lobe (hippocampus, amygdala, and temporal cortex) and the mesocorticolimbic pathway (ventral tegmental area (VTA), nucleus accumbens, and frontal cortex), raising interesting questions as to the possible role of this peptide in psychiatric disorders including schizophrenia, the etiopathology of which has been related to dopamine (DA) overactivity (Chapters 36 and 37 of this volume). Schizophrenic patients with negative symptoms (poverty of speech and affective disinterest) are characterized by a reduction in hippocampal size associated with lower amounts of CCK-8 and CCK binding sites in the hippocampus and amygdala (Ferrier *et al.* 1985). Furthermore, reduced levels of CCK have been found in the cerebrospinal fluid of schizophrenics (Verbanck *et al.* 1983). Although there is some controversy about the reported antipsychotic action of the CCK analogue ceruletide (Chapter 36 of this volume), a direct CCK abnormality in the limbic system in which cognition, attention, and emotion are integrated was reported to be one of the possible biochemical bases of type II schizophrenia (Vanderhaeghen and Crawley 1985). On the other hand, CCK-8 is colocalized at least partly with DA in the mesolimbic pathway, and schizophrenia as well as severe depressive syndromes could be caused by a dysfunction in the inter-relationships between the limbic lobe (hippocampus, amygdala) and mesencephalic DA inputs to the nucleus accumbens. All these hypotheses require extensive investigation, but this has been hampered by the lack of selective agonists and antagonists for CCK_A and especially CCK_B binding sites which are capable of crossing the blood–brain barrier. Moreover, CCK_A binding sites appear by autoradiography to be sparsely distributed in few brain regions, whereas a high concentration of CCK_B receptors has been found in several regions including the limbic lobe and the mesolimbic pathway (Chapter 5 of this volume). In contrast, most of the reported behavioural responses induced by CCK-8 seem to involve stimulation of CCK_A receptors (Chapters 17 and 35 of this volume). In this chapter we report on the structural approach used to design potent and selective CCK_B and CCK_A agonists and antagonists, the *in vitro* and *in vivo* binding properties of a CCK_B agonist which fulfils all the required criteria to be used by the systemic route, and the role of mesolimbic CCK_A and CCK_B receptors in modulating the emotional states of rats.

Design of potent and selective CCK_B agonists with appropriate bioavailability

Investigations of the role of brain CCK_A and CCK_B receptors require highly potent and selective agonists and antagonists endowed with chemical stability, high resistance to peptidases, in the case of peptides, and lipophilicity, facilitating brain penetration following systemic administration. Replacing Met^{28} and Met^{31} by Nle residues in CCK-8 to avoid methionine oxidation led to BDNL, $Boc[Nle^{28,31}]CCK_{27-33}$, a peptide equipotent with CCK-8 in binding and pharmacological assays (Ruiz-Gayo *et al.* 1985). Cyclic CCK-8-derived peptides, such as BC 254, were designed by taking into account the preferential folded conformations of CCK-8 in aqueous solution. Thus BC 254 was greater than 500-fold more potent than its linear analogue at CCK_B receptors (Table 6.1) without any significant change for CCK_A binding sites. This suggested that the high affinity and selectivity for CCK_B sites could be related to structural changes induced in the C-terminal part of the peptide by the bulky cyclized N-terminal region. This is supported by nuclear magnetic resonance (NMR) studies showing that the C-terminal sequence has a folded conformation with a solvent-buried localization of the Asp^{32} NH evidenced only in the cyclic peptide (Roy *et al.* 1989).

A high selectivity for CCK_B sites was also obtained with BC 264 by modification of the peptide backbone, i.e. retroinversion of the Nle—Gly amide bond and N-methylation of the Nle^{31} residue (Table 6.1). Interestingly, preliminary NMR studies suggested that the cyclic BC 254 and the linear BC 264 peptides have similar C-terminal folding. Because of its higher hydrophobicity and considerable resistance to peptidases, BC 264 was preferred to the cyclic BC 254 derivative for *in vitro* and *in vivo* binding and for pharmacological studies

Table 6.1. Activity and selectivity of various synthetic CCK-8 analogues at CCK_A and CCK_B receptors

Compounds		Affinities Ki (nM) CCK binding sites		Potencies IC$_{50}$ (nM) amylase release	
		CCK_A	CCK_B	Agonist	Antagonist
[26]Asp —Tyr(SO$_3$H)—Met —Gly —Trp—Met —Asp—PheNH$_2$[33]	CCK8	0.30	0.64	0.13	—
Boc-γD.Glu—Tyr(SO$_3$H)—Nle —D.Lys—Trp—Nle —Asp—PheNH$_2$		301	1390	>1000	—
Boc-γD.Glu—Tyr(SO$_3$H)—Nle —D.Lys—Trp—Nle —Asp—PheNH$_2$	BC 254	0.56	2500	>1000	—
Boc —Tyr(SO$_3$H)—gNle—mGly—Trp—(Me)Nle—Asp—PheNH$_2$	BC 264	0.15	78	377	—
Boc —Tyr(SO$_3$H)—Nle —Gly —Trp—Nle —Asp—PheNH$_2$	BDNL	0.23	0.93	0.28	—
Boc —Tyr(SO$_3$H)—Nle —Gly —Trp—Phe —Asp—PheNH$_2$		3.7	220	3.4	—
Boc —Tyr(SO$_3$H)—Nle —Gly —Trp—Orn —Asp—PheNH$_2$		110	340	22	—
Boc —Tyr(SO$_3$H)—Nle —Gly —Trp—Orn(Z)—Asp—PheNH$_2$	IM 9/11	93	310	—	480
Boc —Trp—Orn —AspNH$_2$		1400	14 000	13 800	—
Boc —Trp—Orn(Z)—AspNH$_2$		4200	200	—	230

All the experiments were done with guinea pig tissues; [^{3}H]pCCK-8 was used in binding studies.

(Charpentier *et al.* 1988). Recently, the Tyr(SO$_3$H) amino acid was replaced by the non-hydrolysable Phe(ϕCH$_2$SO$_3$H) residue without changing the binding properties, therefore leading to CCK-8-derived peptides fulfilling all the required criteria for systemic administration in pharmacological studies.

CCK$_B$ versus CCK$_A$ receptor recognition: criteria for agonist versus antagonist properties

Introduction of Phe[31] in place of Nle[31] in BDNL led to an increase in CCK$_B$ selectivity, since Boc[Nle[28], Phe[31]]CCK$_{27-33}$ is about 60-fold more potent

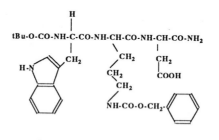

Antagonist CCK$_A$

PD 134 308

Adamantyl-O-CO-NH-C-CO-NH-CH$_2$-CH-NH-CO-CH$_2$

Antagonist CCK$_B$

A 71623

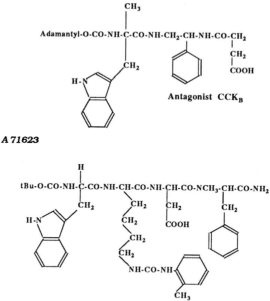

Agonist CCK$_A$

Fig. 6.1.

towards CCK_B than towards CCK_A receptors (Table 6.1). Introduction of an Orn^{31} residue led to a full agonist with an approximately 100-fold decrease in affinity for CCK_B and CCK_A receptors. Interestingly, when the amine function in the side-chain of Orn^{31} was protected by a benzyloxycarbonyl, the resulting peptide $Boc[Nle^{28}, Orn^{31}(Z)]CCK^{27-33}$ (IM 9/11) was a mixed antagonist acting at both CCK_B and CCK_A receptors ($K_A = 4.8 \times 10^{-7}$ M, $pA_2 = 6.32$ on CCK-8-induced pancreatic amylase release). The same antagonist potency at CCK_A binding sites was obtained by shortening the sequence of this peptide to $Boc-Trp-Orn(Z)-Asp-NH_2$ ($K_A = 2.3 \times 10^{-7}$, $pA_2 = 6.63$ on amylase release assay).

Short modified peptides related to the C-terminal part of CCK-8 can be of use in defining the structural characteristics necessary for the recognition of CCK_A or CCK_B receptors in their agonist or antagonist forms. Thus, modifications similar to those reported here (Fig. 6.1) were independently introduced into the sequence of CCK-4 leading either to a selective CCK_A agonist A 71623 (presence of Phe NH_2^{33}) or a CCK_B antagonist PD 134 308 (presence of a phenyl ring in position 31 and absence of Phe NH_2^{33}) (Hughes *et al.* 1990; Shiosaki *et al.* 1991).

Binding characteristics of the selective CCK_B agonist BC 264 in rodent brain

BC 264 has been radio-labelled by replacing the Boc group by the commercially available tritiated propionyl residue, and the subsequent radio-ligand [^{3}H]pBC 264 has been obtained in one step with a high specific activity

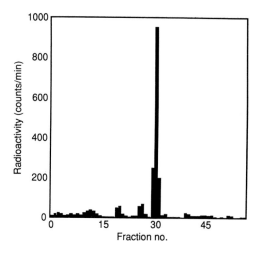

Fig. 6.2. Chromatogram of cerebral radioactivity 15 min after i.v. injection of [^{3}H]pBC 264 in mice.

(98 Ci/mmol). Studies of the bioavailability of [³H]pBC 264 have been carried out after intravenous (i.v.) administration in the mouse. The radioactivity present in the brain 15 min after injection was at least 1.6/10 000 of the total radioactivity injected, and 85 per cent of the recovered material corresponded to the intact molecule, illustrating its high *in vivo* resistance to peptidases (Fig. 6.2).

The binding of [³H]pBC 264 has been studied using guinea pig, rat, and mouse brains. The specific binding was high (80–90 per cent at the K_D concentration), leading to accurate measurements of the binding parameters. The Scatchard and Hill transformations of the saturation data indicated that [³H]pBC 264 interacted with the same high affinity with a single class of binding sites in the three species (K_D = 0.15–0.2 nM). The binding capacities measured with the selective CCK_B agonist [³H]pBC 264 were compared with those determined with [³H]pCCK-8 which has a similar affinity for CCK_A (1.02 nM) and CCK_B receptors (1.2 nM). The maximal number of binding sites for [³H]pBC 264 was significantly lower ($p < 0.05$) than the maximal number of binding sites determined using [³H]pCCK-8: 29.9 versus 40.6 fmol/mg protein in the mouse (Fig. 6.3), 17.2 versus 26.0 fmol/mg protein in the rat, and 22.3 versus 30.1 fmol/mg protein in the guinea pig. In contrast, in the cerebral cortex of rat and guinea pig, the B_{max} values were identical for both ligands. In the presence of 10 nM pBC 264, a concentration estimated to occupy 98.5 per cent of CCK_B sites, [³H]pCCK-8 interacted with a single class of binding sites (K_D = 3.8 nM, B_{max} = 11 fmol/mg protein) assumed to correspond to CCK_A

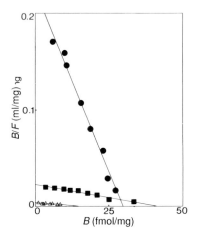

Fig. 6.3. Scatchard plots of specific binding to mouse brain membranes of [³H]pBC 264 (●) [³H]pCCK-8 (■), and [³H]pCCK-8 in the presence of 10 nM pBC 264(△). The data shown are from a single representative experiment with each point in triplicate.

Table 6.2. Inhibitory potencies of CCK analogues on the specific binding of 0.2 nM [³H]pBC 264 at CCK$_B$ sites in different species

	K_I (nM)		
	Guinea pig	Rat	Mouse
CCK 8	0.23 ± 0.03	1.01 ± 0.05	0.48 ± 0.09
pCCK 8	0.13 ± 0.01	0.55 ± 0.15	0.33 ± 0.04
pBC 264	0.13 ± 0.01	0.17 ± 0.03	0.17 ± 0.04
CCK-8US	6.0 ± 1.1	10.2 ± 3.1	4.6 ± 0.2
CCK-5	11.0 ± 24.2	19.7 ± 3.4	13.2 ± 2.4
CCK-4	49.9 ± 9.9	55.7 ± 10.2	16.4 ± 4.3
BC 254	0.69 ± 0.07	43.9 ± 6.7	23.3 ± 1.2
Devazepide	132.0 ± 18.3	549 ± 69	62.4 ± 8.9
L365,260	2.32 ± 0.32	10.9 ± 2.0	5.2 ± 0.7

Results are the mean ± SEM from three to four experiments performed in triplicate. Each determination was obtained from analysis of Hill plots using eight to ten determinations of analogues.

sites. The presence of this rather high level of CCK$_A$ receptors in the brain was unexpected, since so far autoradiographic studies have indicated the existence of peripheral type CCK receptors in only a few structures of rodent brain (see Chapter 5 of this volume). The result strongly suggests that CCK$_A$ sites are present in other brain regions but are either too diffusively distributed or have too low a concentration to be detected by autoradiography. This would be in agreement with the pharmacological responses observed after stimulation of CCK$_A$ receptors of the nucleus accumbens (Crawley 1988; Daugé *et al.* 1989). Some interesting differences have been evidenced in competition experiments using guinea pig, rat, and mouse brains (Table 6.2). Whereas pBC 264 showed the same very high affinity in the three species, the CCK$_B$ antagonist L-365,260 was found to be 18-fold, 30-fold, and 64-fold less potent in guinea pig, mouse, and rat respectively. Likewise, as already reported, the cyclic analogue BC 254 displayed a 33-fold and 63-fold lower affinity for mouse and rat than for guinea pig brain preparations. These differences suggest that CCK$_B$ receptor subtypes are present in different concentrations in each species (Durieux *et al.* 1988).

In vivo binding of [³H]pBC 264

The high affinity and selectivity of [³H]pBC 264 allowed its *in vivo* binding parameters to be characterized after intracerebroventricular (i.c.v.) administration in mouse. The maximum specific binding was reached 15 min after injection and high pressure liquid chromatography (HPLC) analysis showed that less than 15 per cent of the radio-ligand was metabolized. The specific binding was reversible and saturable: a Scatchard plot indicated that the dose of

[^{3}H]pBC 264 producing 50 per cent receptor occupancy was 25 pmol and that the maximum number of binding sites was 0.89 pmol/brain. In competition experiments, various CCK analogues showed the same order of potency as in *in vitro* experiments: specific binding of [^{3}H]pBC 264 was inhibited by 24 per cent, 38 per cent, and 46 per cent with CCK-5, CCK-4, and CCK-8US respectively. Surprisingly, CCK-8 was 200-fold less potent than pBC 264 (IC_{50} = 8500 pmol), but this is probably due to its short half-life *in vivo* as pCCK-8, which is protected from aminopeptidases, had a greater affinity (IC_{50} = 84 pmol). Accordingly, CCK-8 in the presence of the non-specific aminopeptidase inhibitor bestatin had a better receptor affinity (IC_{50} = 797 pmol).

Radio-labelled selective agonists are needed for the investigation of CCK receptor coupling mechanisms. [^{3}H]pBC 264 not only has a very high affinity and a high selectivity for CCK_B receptors, but also high specific binding, and therefore appears to be the ideal molecule to characterize CCK_B receptors. Moreover, the characterization of *in vivo* binding could allow pharmacological responses and CCK-8 receptor occupancy to be correlated.

Behavioural effects resulting from the stimulation of CCK_A and CCK_B receptors in rat nucleus accumbens

The role of CCK in the nucleus accumbens, a brain structure involved in adaptational processes, remains unclear.

In some electrophysiological studies an antagonism of DA by CCK-8 has been reported (Debonnel and De Montigny 1988), whereas potentiation of DA-induced locomotor activity by CCK-8 has been observed in behavioural studies (Crawley 1988). In addition, CCK-8 administration into the nucleus accumbens has been shown to attenuate the K^+-evoked release of DA, measured by push–pull cannula, whereas with slices superfused *in vitro* it was found to increase the basal levels of released neosynthesized [^{3}H]DA and either to decrease (Voigt *et al.* 1986) or increase (Vickroy and Bianchi 1989) the K^+-evoked release of [^{3}H]DA.

The pharmacological relevance of these apparently contradictory results was investigated by direct injection of selective CCK_A or CCK_B agonists into different parts of the nucleus accumbens. Very low doses (about 1 fmol) of CCK-8S injected into the postero-median nucleus accumbens were able to induce hypoexploration, measured in the four-hole box, and an increase in the emotional state, evaluated in the elevated plus maze, in rats exposed to a novel external stimuli (Daugé *et al.* 1989). Although few CCK_A receptors have been visualized in this part of the nucleus accumbens, these effects are likely to involve CCK_A receptors since they were suppressed by the selective CCK_A antagonist devazepide (100, 200 μg/kg, i.p.) (Chang and Lotti 1986) and were not observed after injection of 0.1–1000 fmol CCK-8US (Daugé *et al.* 1989) or 3 fmol–300 pmol of the selective CCK_B agonist BC 264 (Daugé *et al.* 1990).

The posteromedian nucleus accumbens receives CCK-8 terminals originating from the VTA, most of them also containing DA (Hökfelt *et al.* 1980). There-fore the possibility that the hypoexploration and increase in emotionality induced by CCK-8 were related to the mesolimbic DAergic pathway was investigated by injecting 6-OHDA into the nucleus accumbens to destroy mesencephalic DA inputs. The hypoexploration and increased emotionality induced by CCK-8 did not occur in the first, second, and third weeks in the 6-OHDA lesioned rats (Fig. 6.4). A supersensitivity to the effects of CCK-8 on nucleus accumbens neurons, eventually associated with an increase in the number of CCK binding sites, has been described in electrophysiological studies after 6-OHDA lesions of the DA mesolimbic pathway (Debonnel and De Montigny 1988). Furthermore, CCK-8 injected into the nucleus accumbens antagonizes the hyperlocomotion produced by apomorphine when rats are chronically treated with neuroleptics or after 6-OHDA lesions of the mesolimbic pathway (Weiss *et al.* 1989). In contrast with these probably postsynaptic related responses, our observation of a suppression of behavioural effects of CCK-8 produced by mesolimbic deafferentation may be due to the

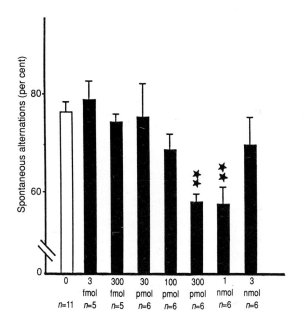

Fig. 6.4. Dopamine lesions of the mesolimbic system suppressed the hypoexplor-ation (number of hole visits) induced 15 min after CCK-8 (3 fmol) administration in the postero-median nucleus accumbens. Behaviour was observed in the first, second and third weeks after 6-OHDA injected into the nucleus accumbens (results of the second week were similar to those of the first week). Bars represent the SEM: ★; $p < 0.05$ ★★; $p < 0.01$ (Newman–Keuls test).

disappearance of a direct or indirect presynaptic action of CCK-8. Taking these
results into account, it can be speculated that the modulatory effects of CCK-8
in this part of the nucleus accumbens depend on DAergic activity, which in turn
is influenced by environmental factors.

BC 264 (selective CCK_B agonist) has previously been shown to be at least 50
times more potent than CCK-8 in stimulating the firing of rat hippocampal
neurons. Furthermore, stereotaxic injection of BC 264 or CCK-8 into the VTA
of rats resulted in potentiation of DA-induced hypolocomotion (Daugé et al.
1990). These two types of CCK-8 responses have previously been shown to
involve CCK_B receptors (Chapters 7 and 35 of this volume). Although the
anterior part of the nucleus accumbens is highly enriched in CCK_B receptors
(Pélaprat et al. 1987), the behavioural effects induced by stimulating this type
of receptor are still unclear. The only results suggesting an involvement of
CCK_B receptors in CCK-8 effects are those showing facilitation of the extinc-
tion of active avoidance responses and attenuation of the retention of passive
avoidance behaviour by both CCK-8 and CCK-8US (Fekete et al. 1984). As
shown in Fig. 6.5, BC 264 (0.3, 1 nmol) injected into the anterolateral part of

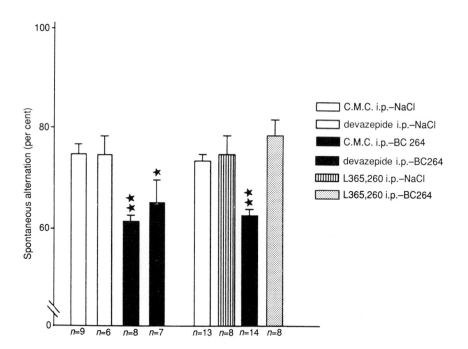

Fig. 6.5. Effect of various doses of BC 264 15 min after injection into the anterolateral part of the nucleus accumbens on the percentage of alternation of arm
choice measured in a Y-maze. Bars represent the SEM: ★, $p < 0.05$; ★★, $p < 0.01$
(Dunnett's t test).

the nucleus accumbens decreased the percentage of spontaneous alternations in the arms of the Y-maze but did not change the number of visits to the arms. This effect was suppressed by L-365,260 (CCK$_B$ antagonist, 5 mg/kg i.p.) (Lotti and Chang 1989) but not by devazepide (200 μg/kg i.p.) previously shown to suppress CCK$_A$-mediated responses (Fig. 6.6).

It is interesting to observe that stimulation of CCK$_B$ receptors in the nucleus accumbens (posteromedian, anterolateral) by BC 264 does not modify rat emotionality as measured in the elevated plus maze. Therefore in this structure CCK$_B$ receptor stimulation does not induce the anxiogenic-like effects expected from the reported anxiolytic effects of CCK$_B$ antagonists (Hughes *et al.* 1990; Ravard and Dourish 1990; Chapter 12 of this volume). At present, the type of receptor involved in these responses (CCK$_A$, CCK$_B$) and their localization (central and/or peripheral) is unknown. Finally, it is possible that

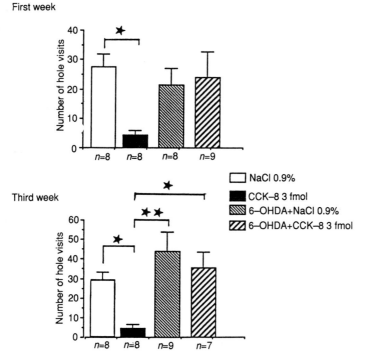

Fig. 6.6. Antagonism of the effect induced by BC 264 (300 pmol) by L-365,260 (5 mg/kg i.p.) or devazepide (200 μg/kg i.p.) injected 1 h before BC 264 CMC (= 0.5 per cent carboxymethylcellulose). Bars represent the SEM: ★ $p < 0.05$; ★★ $p < 0.01$ (Newman–Keuls test).

the nucleus accumbens does not represent an essential structure for the effects observed after peripheral administration of CCK_B antagonists.

The participation of CCK_B receptors in the transmission of information originating from the hippocampus, prefrontal cortex, and mesencephalon and their integration into the nucleus accumbens could explain the effects on alternation behaviour, a hypothesis which remains to be confirmed.

Conclusion

As has already been shown for the delta opioid peptides BUBU and BUBUC (Gacel *et al.* 1990), pseudopeptides with appropriate pharmacokinetic properties for systemic administration, such as BC 264 or Boc–Trp–Orn(Z)–AspNH$_2$ can be designed by taking into account the results of conformational studies, enzymatic investigations, and structure–activity relationships. This has been confirmed by the recent development of the pseudopeptide PD 134 308 which behaves as a systemically active CCK_B antagonist (see Chapter 5 of this volume).

Owing to its favourable physicochemical properties (lipophilicity but aqueous solubility, chemical and enzymatic stability) associated with a very high affinity and selectivity for CCK_B sites, [^{3}H]pBC 264 and its unlabelled counterpart appear to be the most appropriate agonists to date for studying the biochemical and pharmacological properties of CCK_B receptors. Thus the first clear demonstration of the involvement of CCK_B receptors in a behavioural response has been obtained after injection of BC 264 in the nucleus accumbens (anterior part). This effect (alternation perturbation) could be related to a modulation of the mesolimbic DAergic activity induced by CCK inputs. The proposed DA overactivity in schizophrenia, which could reflect a severe alteration of the prefrontal–hippocampus–accumbens CCK link, remains to be established.

Acknowledgements

We thank Dr A. Beaumont for a critical reading of the manuscript and C. Dupuis for typing it. This research was supported by a grant from Rhône Poulenc Rorer.

References

Chang, R. S. L. and Lotti, V. J. (1986). Biochemical and pharmacological characterization of an extremely potent and selective nonpeptide cholecystokinin antagonist. *Proc. Natl Acad. Sci. USA* **83**, 4923–6.

Charpentier, B., Durieux, C., Pelaprat, D., Dor, A., Reibaud, M., Blanchard, J. C., and Roques, B. P. (1988). Enzyme–resistant CCK analogs with high affinities for central receptors. *Peptides* **9**, 835–41.

Crawley, J. N. (1988). Modulation of mesolimbic dopaminergic behaviour by cholecystokinin. In *The mesocorticolimbic dopamine system* (ed. P. W. Kalivas and C. B. Nemeroff). *Ann. NY Acad. Sci.* **537**, 381-96.

Daugé, V., Steimes, P., Derrien, M., Beau, N., Roques, B. P., and Feger, J. (1989). CCK8 effects on motivational and emotional states of rats involve CCKA receptors of the postero-median part of the nucleus accumbens. *Pharmacol. Biochem. Behav.* **34**, 157-63.

Daugé, V., Böhme, G. A., Crawley, J. N., Durieux, C., Stutzmann, J. M., Féger, J., Blanchard, J. C., and Roques, B. P. (1990). Investigation of behavioral and electrophysiological responses induced by selective stimulation of CCKB receptors by using a new highly potent CCK analog: BC 264. *Synapse* **6**, 73-80.

Debonnel, G. and De Montigny, C. (1988). Increased neuronal responsiveness to cholecystokinin and dopamine induced by lesioning mesolimbic dopaminergic neurons: an electrophysiological study in the rat. *Synapse* **2**, 537-45.

Durieux, C., Pham, H., Charpentier, B., and Roques, B. P. (1988). Discrimination between CCK receptors of guinea-pig and rat brain by cyclic CCK8 analogues. *Biochem. Biophys. Res. Commun.* **154**, 1301-7.

Fekete, M., Lengyel, A., Hegedüs, B., Penke, B., Zarandy, M., Toth, G. K., and Telegdy, G. (1984). Further analysis of the effects of cholecystokinin octapeptides on avoidance behavior in rats. *Eur. J. Pharmacol.* **98**, 79-91.

Ferrier, I. N., Crow, T. J., Farmery, S. M., Roberts, G. W., Owen, F., Adrian, T. E., and Bloom, S. R. (1985). Reduced cholecystokinin levels in the limbic lobe in schizophrenia: A marker for pathology underlying the defect state. In *Neuronal cholecystokinin* (ed. J. J. Vanderhaeghen and J. N. Crawley). *Ann. NY Acad. Sci.* **448**, 495-506.

Gacel, G., Fellion, E., Baamonde, A., Daugé, V., and Roques, B. P. (1990). Synthesis, biochemical and pharmacological properties of BUBUC, a highly selective and systemically active agonist for in vivo studies of delta opioid receptors. *Peptides* **11**, 983-8.

Hökfelt, T., Skirboll, L., Rehfeld, J. F., Goldstein, M., Markey, K., and Dann, O. (1980). A subpopulation of mesencephalic dopamine neurons projecting to limbic areas contains a cholecystokinin-like peptide: evidence from immunohistochemistry combined with retrograde tracing. *Neuroscience* **5**, 2093-124.

Hughes, J., Boden, P., Costall, B., Domeney, A., Kelly, E., Horwell, D. C. Hunter, J. C., Pinnock, R. D., and Woodruff, G. N. (1990) Development of a class of selective cholecystokinin type B receptor antagonists having potent anxiolytic activity. *Proc. Natl Acad. Sci. USA* **87**, 6728-32.

Lotti, V. J. and Chang, R. S. L. (1989). A new potent and selective non-peptide gastrin antagonist and brain cholecystokinin receptor (CCKB) ligand: L365,260. *Eur. J. Pharmacol.* **162**, 273-80.

Pélaprat, D., Broer, Y., Studler, J. M., Peschanski, M., Tassin, J. P., Glowinski, J., Rostene W., and Roques, B. P. (1987). Autoradiography of CCK receptors in the rat brain using [^{3}H] Boc[Nle 28,31]CCK27-33 and [^{125}I]Bolton-Hunter CCK8. Functional significance of subregional distributions. *Neurochem. Int.* **10**, 495-498.

Ravard, S. and Dourish, C. T. (1990). Cholecystokinin and anxiety. *Trends Pharmacol. Sci.* **11**, 271-3.

Roy, P., Charpentier, B., Durieux, C., Dor, A., and Roques, B. P. (1989). Conformational behavior of cyclic CCK-related peptides determined by 400 MHz 1H NMR: relationships with affinity and selectivity for brain receptors. *Biopolymers* **28**, 69-79.

Ruiz-Gayo, M., Daugé, V., Menant, I., Begué, D., Gacel, G., and Roques, B. P. (1985).

Synthesis and biological activity of Boc(Nle28,Nle31)CCK27–33, a highly potent CCK8 analogue. *Peptides* **6**, 415–20.

Shiosaki, K., Lin, C. W., Asin, K., Kopecka, H., Craig, R., Wagenaar, F. L., Bianchi, B., Miller, T., Witte, D., Hodges, L., Montana, N. and Nadzan, A. M. (1991). Development of non-sulphated tetrapeptides as potent and selective CCK-A receptor agonists. In *Peptides 1990* (ed. E. Giralt and D. Andreu), pp. 710–11. *Proceedings of the 21st European Peptide Symposium.*

Vanderhaeghen, J. J. and Crawley, J. N. (eds) (1985). *Neuronal cholecystokinin. Ann. NY. Acad. Sci.* **448**.

Verbanck, P. M. P., Lotstra, F., Gilles, C., Linkowski, P., Mendlewicz, J., and Vanderhaeghen, J. J. (1983). Reduced cholecystokinin immunoreactivity in the cerebrospinal fluid of patients with psychiatric disorders. *Life Sci.* **34**, 67–72.

Vickroy, T. W. and Bianchi, B. R. (1989) Pharmacological and mechanistic studies of cholecystokinin-facilitated [^{3}H] dopamine efflux from rat nucleus accumbens. *Neuropeptides* **13**, 43–50.

Voigt, M., Wang, R. Y., and Westfall, T. C. (1986) Cholecystokinin octapeptides alter the release of endogenous dopamine from the rat nucleus accumbens *in vitro. J. Pharmacol. Exp. Ther.* **237**, 147–53.

Weiss, F., Ettenberg, A., and Koob, G. F. (1989). CCK-8 injected into the nucleus accumbens attenuates the supersensitive locomotor response to apomorphine in 6-OHDA and chronic neuroleptic treated rats. *Psychopharmacology* **93**, 409–15.

7. Agonist and antagonist pharmacology of CCK-receptor-mediated responses of rat ventromedial hypothalamic neurons

J. A. Kemp, G. R. Marshall, and J. N. C. Kew

The ventromedial hypothalamic (VMH) nucleus of the rat contains high levels of CCK immunoreactivity which is thought to arise from an afferent input from the parabrachial nucleus (Inagaki *et al.* 1984; Zaborsky *et al.* 1984). It also contains high levels of CCK_B receptor binding sites (Day *et al.* 1986), and CCK-8S is a potent excitant of VMH neurons (Kow and Pfaff 1986; Bowden and Hill 1988; Kemp *et al.* 1989). Thus the VMH slice has become a useful preparation with which to study the pharmacology of brain CCK_B receptors and has enabled comparisons to be made with results from receptor binding studies. Recently, several selective agonists and antagonists for CCK receptors have become available, and we have used these to examine in greater detail the pharmacology of the CCK-receptor-mediated excitatory responses of VMH neurons.

Extracellular recordings of single-cell action potential firing rates were made using conventional techniques on $350\,\mu m$ thick slices containing the VMH nucleus. These were placed in a small recording chamber (volume approximately 0.3 ml) where they were continuously perfused, at a rate of 2 ml/min, with warmed ($33\,°C$) oxygenated artificial cerebrospinal fluid (aCSF). Agonists were dissolved in the aCSF, from concentrated stock solutions, and were applied for periods of 1 min.

Agonist pharmacology

CCK-8S (1–1000 nM) produced concentration-dependent increases in the firing rate of VMH neurons with an EC_{50} of approximately 10 nM (Fig. 7.1). CCK-8US and CCK-4 also excited VMH neurons, being respectively only 5 and 18 times less potent than CCK-8S (Table 7.1). A-57696 (Lin *et al.* 1989) and BC 264 (Daugé *et al.* 1990), two peptide analogues reported to be selective for CCK_B receptors, were also potent excitants (Fig. 7.2) with equipotent molar ratios, compared with CCK-8S, of 1.7 and 2.5 respectively (Table 7.1).

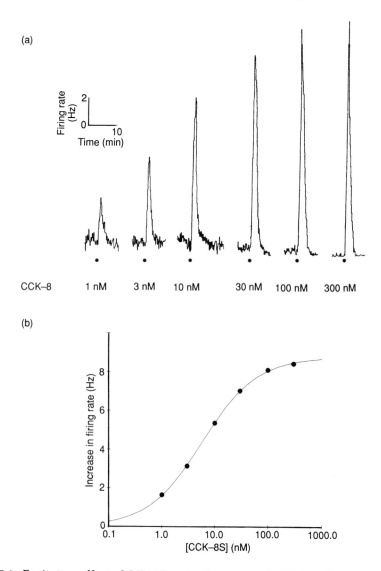

Fig. 7.1. Excitatory effect of CCK-8S on the firing rate of a VMH cell: (a) ratemeter recordings from a single neuron showing the concentration-dependent changes in firing rate produced by 1 min applications of CCK-8S (1–100 nM); (b) data from (a) plotted as a concentration response curve for CCK-8S. For this cell the EC_{50} for CCK-8S was 5 nM.

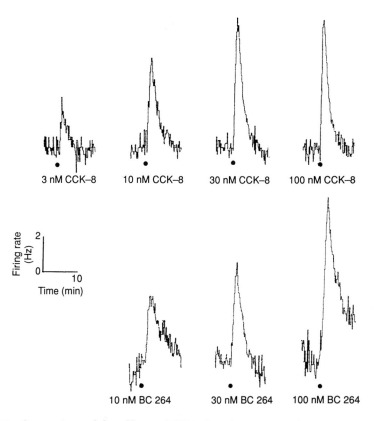

Fig. 7.2. Comparison of the effects of CCK-8S and BC 264 on the same neuron. For this cell BC 264 was only slightly less potent than CCK-8S. Combined data for all cells are shown in Table 7.1.

Table 7.1. Equipotent molar ratios for several CCK$_B$ receptor agonists compared with CCK-8S estimated from concentration–response curves from single cells as illustrated in Fig. 7.2

Compound	Equipotent molar ratios
CCK-8S	1.0
CCK-8US	5.2 ± 0.22 (n = 5)
CCK-4	18.1 ± 8.2 (n = 6)
A-57696	1.7 ± 0.37 (n = 7)
BC 264	2.5 ± 0.45 (n = 5)

Effects of pretreatment with pertussis toxin

The CCK receptor on rat pancreatic acinar cells (CCK_A type) has been shown
to be a G-protein-linked receptor, and evidence suggests that G_i proteins sensitive to both cholera toxin and pertussis toxin are involved in the functional
coupling of these receptors (Schnefel *et al.* 1990). In contrast, central CCK_B
receptors have not been shown to be linked to a second messenger system and
guanyl nucleotides have only a small influence on the affinity of agonists at
CCK_B receptors (Wennogle *et al.* 1988). Therefore we investigated the effect
of pertussis toxin treatment on the responsiveness of VMH neurons to CCK-8S.

Pertussis toxin (1.5 μg in 5 μl) or vehicle (50% glycerol, 50% phosphate
buffer) was injected into the third ventricle of rats under equithesin anaesthesia
(4 ml/kg, i.p.) 3–5 days before they were killed. There was no difference in the
response to CCK-8S of cells in slices from animals treated with pertussis
toxin or from control animals (Fig. 7.3). In contrast, the inhibitory effects of
baclofen (Fig. 7.3) and 5-hydroxytryptamine (5-HT) (Newberry and Priestley
1988) were completely absent in cells from animals treated with pertussis toxin.
In control slices baclofen (1 μM) always produced a complete inhibition of
cell firing rate. These results indicate that in cells where pertussis toxin produced an inhibition of certain G-protein-linked receptor mediated responses
(e.g. $GABA_B$ and $5\text{-}HT_{1A}$ receptors) it had no effect on responses to CCK-
receptor-mediated responses.

Antagonist pharmacology

L-365,260 (0.1 and 0.3 μM) antagonized the excitatory responses evoked
by CCK-8S. The antagonism developed rather slowly and reached a plateau

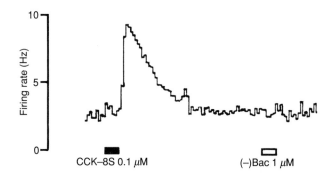

Fig. 7.3. Response of a VMH cell to CCK-8S (0.1 μM) and (—)baclofen (1 μM) in a
slice taken from an animal treated with pertussis toxin. The response to CCK-8S
is comparable with that seen in slices from control animals, whereas there is no
response to (—)baclofen at a concentration which caused 100 per cent inhibition
of the firing rate of cells in slices from control animals.

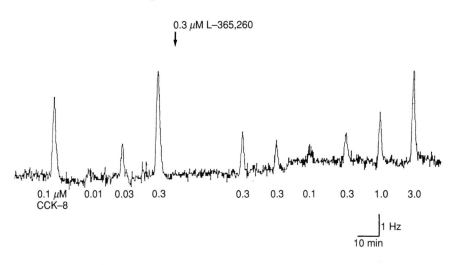

Fig. 7.4. Effect of the CCK$_B$-selective antagonist L-365,260 (0.3 μM) on the excitatory response to CCK-8S. Control responses to CCK-8S were obtained first and then L-365,260 was continuously applied from the arrow. The response to CCK-8S was inhibited in the presence of L-365,260, and this effect reached a plateau after approximately 1 h. After this time increasing the concentration of CCK-8S overcame the antagonism and resulted in a parallel shift to the right of the CCK-8S concentration-response curve.

after approximately 1 h of continuous perfusion (Fig. 7.4). If left to equilibrate for 1 h, L-365,260 produced parallel shifts to the right of the CCK-8S concentration–response curve which resulted in an estimated pA$_2$ of 7.48 ± 0.11 ($n = 6$).

Devazepide (0.3 and 1.0 μM) also produced parallel shifts to the right of the CCK-8S concentration–response curve, which led to an estimated pA$_2$ of 6.6 ± 0.17 ($n = 8$) for its antagonism of CCK-8S-induced excitation of VMH cells.

Conclusions

The absolute and relative potencies of CCK-8S, CCK-8US, CCK-4, A-57696, and BC 264 as excitants of rat VMH neurons are consistent with an action mediated by CCK$_B$ receptors. A-57696 has been reported to be a high affinity partial agonist at CCK$_B$/gastrin receptors and a much lower affinity antagonist at CCK$_A$ receptors. Its high agonist potency in the VMH clearly suggests an action at CCK$_B$ receptors. The partial agonist nature of agonists on VMH neurons is difficult to assess because of receptor desensitization produced by high agonist concentrations, and in many cases A-57696 and CCK-4 gave maximal responses as large as that to CCK-8S. However, on some cells both compounds appeared to give maximal responses which were slightly less (about 80 per cent) than that produced by CCK-8S. The relative potencies of CCK-8S, CCK-8US, and CCK-4 are in close agreement with their abilities to inhibit [^{3}H]pCCK-8 binding to rat cortical membranes (Daugé *et al.* 1990), although in this study BC 264 was found to be more potent than CCK-8S, particularly in electrophysiological experiments in hippocampal slices.

The lack of effect of pertussis toxin on CCK-8S-induced excitations of VMH neurons suggests that pertussis-toxin-sensitive G proteins (possibly G_i and G_o) are not coupled to the CCK receptor on VMH neurons. This is further evidence to indicate that this receptor is different from that found on pancreatic acinar cells (Schnefel *et al.* 1990). However, the nature of the G proteins or second messenger system linked to CCK_B receptors in brain still needs to be resolved.

The potencies of devazepide and L-365,260 as antagonists of CCK-8S responses in the VMH are in good agreement with their affinities for CCK_B receptors. Thus devazepide is 1000 times more potent as an antagonist of CCK_A receptors (Chang and Lotti 1986). The parallel shifts to the right of the CCK-8S concentration-response curve indicate a competitive mechanism of action, although at the fairly low concentrations used long incubation times appeared necessary in order for the antagonists to reach equilibrium.

These results for selective agonists and antagonists clearly indicate that CCK_B receptors mediate the excitation of VMH neurons and are in good agreement with autoradiographical studies which show a high density of CCK_B receptor binding sites in this brain region.

References

Boden, P. and Hill, R. G. (1988). Effects of cholecystokinin and related peptides on the neuronal activity in the ventromedial nucleus of the rat hypothalamus. *Br. J. Pharmacol.* **94**, 246–52.

Chang, R. S. L. and Lotti, V. J. (1986). Biochemical and pharmacological characterization of an extremely potent and selective non-peptide cholecystokinin antagonist. *Proc. Natl Acad. Sci. USA* **83**, 4923–6.

Daugé, V., Bohme, G. A., Crawley, J. N., Durieux, C., Stutzmann, J. M., Feger, J., Blanchard, J. C., and Roques, B. P. (1990). Investigation of behavioural and electrophysiological responses induced by selective stimulation of CCKB receptors using a new highly potent CCK analog, BC 264. *Synapse* **6**, 73–80.

Day, N. C., Hall, M. D., Clark, C. R., and Hughes, J. (1986). High concentrations of cholecystokinin receptor binding sites in the ventromedial hypothalamic nucleus. *Neuropeptides* **8**, 1–18.

Inagaki, S., Shiotani, Y., Yamano, M., Shiosaka, S., Takagi, H., Tateishi, K., Hashimura, E., Hamaoka, T., and Tohyama, M. (1984). Distribution, origin and fine structures of cholecystokinin-8-like immunoreactive terminals in the nucleus ventromedialis hypothalami of the rat. *J. Neurosci.* **4**, 1289–99.

Kemp, J. A., Marshall, G. R. and Woodruff, G. N. (1989). Antagonism of CCK-induced excitation of rat ventromedial hypothalamic neurones by a new selective CCK-B receptor antagonist. *Br. J. Pharmacol.* **98**, 630P.

Kow, L.-M. and Pfaff, D. W. (1986). CCK-8 stimulation of ventromedial hypothalamic neurons *in vitro*: A feeding-relevant event? *Peptides* **7**, 473.

Lin, C. W., Holladay, M. W., Barrett, R. W., Wolfram, C. A. W., Miller, T. R., Witte, D., Kerwin, J. F., Wagenaar, F., and Nadzan, A. M. (1989). Distinct requirements for activation at CCK-A and CCK-B/gastrin receptors: Studies with a C-terminal hydrazide analogue of cholecystokinin tetrapeptide (30–33). *Mol. Pharmacol.* **36**, 881.

Newberry, N. R. and Priestley, T. (1988). A 5-HT$_1$-like receptor mediates a pertussis toxin-sensitive inhibition of rat ventromedial hypothalamic neurones *in vitro*. *Br. J. Pharmacol.* **95**, 6–8.

Schnefel, S., Profrock, A., Hinsch, K.-D. and Schulz, I. (1990). Cholecystokinin activates G$_i$1-, G$_i$2-, G$_i$3- and several G$_s$-proteins in rat pancreatic acinar cells. *J. Biochem.* **269**, 483–8.

Wennogle, L., Wysowskyj, H., Steel, D. J., and Petrack, B. (1988). Regulation of central cholecystokinin recognition sites by guanyl nucleotides. *J. Neurochem.* **50**, 954–9.

Zaborszky, L., Beinfeld, M. C., Palkovits, M. and Heimer, L. (1984). Brainstem projection to the hypothalamic ventromedial nucleus in the rat: A CCK-containing long ascending pathway. *Brain Res.* **303**, 225–31.

8. Studies of gastrin/CCK receptors using functional isolated preparations

Mina Patel and Colin F. Spraggs

CCK is widely distributed in the CNS, and attention at this symposium has focused on the potential central activities of this family of peptides. In addition, gastrin (G-17) and CCK are important physiological mediators of digestive function (Walsh 1987; Dockray and Gregory 1990). These observations have suggested the possibility of a number of therapeutic indications for selective agonists and antagonists at CCK receptors for use in the treatment of both CNS diseases (anxiety and schizophrenia) and gastrointestinal diseases (peptic ulceration, mucosal hyperplasia and tumour growth, motility, and digestive disorders).

Studies using gastrin/CCK peptide agonists have suggested that the peptides mediate their actions through three receptor subtypes (CCK_A, CCK_B, and gastrin receptor) (Jensen 1989). However, complementary studies using non-peptidic antagonists have failed to distinguish CCK_B from gastrin receptors and have suggested that these are the same receptor subtype (Freidinger 1989). Binding studies have shown that receptor heterogeneity occurs in the CNS (Moran *et al.* 1986; Hill and Woodruff 1989), but further characterization has been hampered by lack of robust functional assay systems with which to study central neuronal function and so classify receptors *in vitro*.

In addition to central actions, gastrin and CCK have important physiological roles in the gastrointestinal tract, acting to stimulate acid secretion (Dockray and Gregory 1990) and modulate gastrointestinal motility (Walsh 1987). Isolated preparations from the gastrointestinal tract are readily accessible, and in this chapter we describe the application of such preparations in the investigation of gastrin/CCK receptor subtypes using selective gastrin/CCK peptide agonists and the non-peptidic antagonists devazepide and L-365,260.

Acid secretion was measured from rat isolated gastric mucosa (RGM) prepared as described by Reeves and Stables (1985). Secretory responses to pentagastrin (standard agonist) and other peptides were obtained by cumulative addition. Contractile responses were measured on strips of guinea pig fundus (GPF) with mucosa removed and the longitudinal muscle–myenteric plexus preparation of guinea pig ileum (GPI) (Yau *et al.* 1974). The contractile effects of CCK-8 and CCK-4 (standard agonists), and of other peptides, were obtained by sequential addition using a contact time of 1 min and a cycle time of 5 min. All preparations were maintained in Krebs–Henseleit solution gassed with 95 per cent O_2, 5 per cent CO_2 at 37 °C.

Preliminary experiments established that concentration-response curves to the standard agonists could be reliably repeated in all three preparations, and therefore the effects of other agonists and of antagonists (following 30 min equilibration) could be compared with control responses in individual tissues.

Studies with agonists

Comparisons of agonists in these preparations were performed following pretreatment of tissues with the peptidase inhibitors bestatin (100 μM) and phosphoramidon (1 μM). These inhibitors did not modify concentration-response curves to pentagastrin in RGM or CCK-8 in GPI. However, they did improve the responsiveness of CCK-8 in GPF, resulting in a fivefold leftward displacement of the CCK-8 concentration–response curve and an increase in maximum response of 26 per cent.

The activities of gastrin/CCK peptides in these gastrointestinal preparations are summarized in Table 8.1 and Fig. 8.1. In RGM, the peptides stimulated acid secretion with a rank order of agonist potency consistent with involvement of gastrin receptors. The relative agonist potencies of the peptides were similar to those obtained in other gastrin-receptor-mediated systems including stimulation of aminopyrine uptake in canine isolated parietal cells (Soll *et al.* 1984) and contraction of guinea pig dispersed fundic smooth muscle cells (Bitar *et al.* 1982; Menozzi *et al.* 1989). In contrast with the present studies, CCK-8 and

Table 8.1. Agonist potencies of gastrin/CCK peptides in isolated preparations of rat gastric mucosa, guinea pig, and guinea pig ileum

Agonist	EC_{50} (nM)		
	RGM	GPF	GPI
CCK-8S	0.9 (0.2–5)*	2.5 (1.2–5.4)	1.4 (0.9–2.1)
CCK-8US	384 (34–4337)	NT	>2000
CCK-4	613 (280–1341)*	51 (4–712)*	>30 000
G-17S	4.3 (3.6–5.1)	4 (3–6)*	1170
G-17US	16 (5–34)	NT	>10 000
Pentagastrin	27 (15–49)	28 (48–203)*	>10 000

Values are expressed as the geometric mean with 95 per cent confidence limits ($n > 4$). Some peptides produced maxima that were clearly smaller than the standard agonist (pentagastrin in RGM and CCK-8 in GPF) and are denoted by an asterisk. In these cases the EC_{50} value was determined from individual 50 per cent responses for that peptide. In GPI, the peptides were compared by measuring their effects on the sustained portion of the response (see text). NT, not tested.

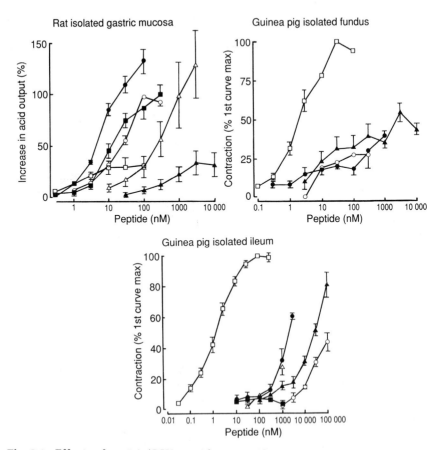

Fig. 8.1. Effects of gastrin/CCK peptides on acid secretion in rat isolated gastric mucosa and on contraction in guinea pig isolated fundus and ileum. Responses are expressed as percentages of the maximal response to pentagastrin in RGM and CCK-8 in GPF and GPI. Values are mean (vertical bars show SEM) (n = 4–6) for the following peptides: ●, G-17S; ■, G-17US; ○, pentagastrin; □, CCK-8S; △, CCK-8US; ▲, CCK-4.

CCK-4 were full agonists in these latter preparations, which may reflect differences in receptor number and efficiency of coupling of the different systems (Table 8.1). The relative agonist potencies of the peptides in RGM also differed from those reported for excitatory responses recorded from cells in hippocampal and hypothalamic brain slices (Boden and Hill 1988; Bohme *et al.* 1988) where pentagastrin, CCK-4, and non-sulphated CCK-8 were equipotent. This supports the contention that gastrin receptors may differ from central CCK_B receptors (Jensen 1989).

In GPI, at all concentrations tested, CCK-8 produced a biphasic contractile

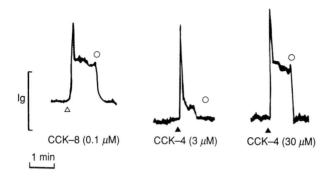

Fig. 8.2. Tracings of representative responses to CCK-8S (0.1 μM) and CCK-4 (3 μM and 30 μM) in GPI. Peptides were added at time points (△, ▲) and were removed from the bath following a 60 s contact time (○). CCK-8S produced a biphasic contraction with a secondary tonic phase that was maintained until wash-out. In contrast, CCK-4 (3 μM) produced only a 'spike' contraction which returned to basal levels before wash-out. Higher concentrations of CCK-4 (30 μM) produced a response that resembled that of CCK-8S.

response consisting of a transient increase in tension followed by a sustained contraction which was maintained until removal of the drug from the organ bath by washing (Fig. 8.2). For comparison of relative agonist potencies in GPI, the effect of peptides on the sustained portion of the response was measured. Under these conditions, CCK-8 was greater than 1000 times more potent than the other peptides, and this relative potency resembled the rank order of agonist potency described for gastrin/CCK peptides as stimulants of amylase release in rat pancreatic acini (Jensen *et al.* 1982). These rank orders are consistent with the view that the sustained contraction in GPI is mediated by CCK_A receptors.

A second type of contractile response was observed in GPI (Fig. 8.2). In contrast with the biphasic contractile response produced by CCK-8, CCK-4 at concentrations of less than 30 μM produced only a transient increase in tension which returned to resting levels within the drug contact time. The sustained contractile component was absent. At concentrations of 30 μM and above, CCK-4 produced a biphasic contraction identical with that observed for CCK-8. The different contractile responses in GPI suggested that more than one receptor may mediate contraction in this tissue. Although CCK-4 was approximately 30 000 times less potent than CCK-8 at producing a sustained contraction in GPI, it was only approximately 50 times less potent than CCK-8 at producing a transient contraction. This latter relative potency was not consistent with activity solely at CCK_A receptors (Jensen 1989). Thus in subsequent experiments the most predominant responses to the agonist was measured, i.e. the sustained contraction for CCK-8 and the transient contractile response for CCK-4.

CCK-8 was a potent agonist in GPF, whereas G-17, pentagastrin, and CCK-4 were less potent and produced maximal responses that were less than 50 per cent of the CCK-8 maximum (Fig. 8.1). The lower efficacy of these peptides did not allow classification of the receptor mediating contraction in GPF based on rank order of agonist potency.

Studies with antagonists

The effects of devazepide and L-365,260 are summarized in Table 8.2 and Fig. 8.3.

In RGM, devazepide was a weak competitive antagonist of pentagastrin-stimulated acid secretion with a pK_B estimate of 5.77. This affinity estimate agreed well with reported IC_{50} values for binding in guinea pig gastric glands, inhibition of G-17-stimulated acid secretion in mouse isolated whole stomach (Chang and Lotti 1986), and inhibition of G-17-induced contraction of guinea pig dispersed gastric smooth muscle cells (Huang *et al*. 1989). L-365,260 was a more potent competitive antagonist ($pA_2 = 7.62$) than devazepide. L-365,260 was approximately 50 times less potent in RGM than would have been predicted by its binding affinity at gastrin receptors (guinea pig gastric glands) and CCK_B receptors (guinea pig cortex) (Bock *et al*. 1989). However, recent functional studies of L-365,260 as an antagonist in rat hypothalamic brain slices (Kemp *et al*. 1989; see Chapter 7 of this volume) and guinea pig dispersed smooth muscle cells (Huang *et al*. 1989) yielded affinity estimates that agree closely with the present data in RGM.

In GPF, devazepide and L-365,260 behaved as competitive antagonists of CCK-8-induced contractions, with devazepide being more than 1000 times

Table 8.2.   Apparent affinity estimates for devazepide and L-365,260 determined in isolated preparations of rat gastric mucosa, guinea pig fundus, and guinea pig ileum

		pA_2/pK_B	
Tissue	Agonist	Devazepide	L-365,260
RGM	Pentagastrin	5.77 ± 0.30 (4)[†]	7.62 ± 0.10 (12)
GPF	CCK-8	9.59 ± 0.25 (12)	5.8 (3)[†]
GPI	CCK-8	10.40 ± 0.30 (15)	7.72 ± 0.33 (16)
GPI	CCK-4	<7.3 (12)[†]	8.35 ± 0.26 (15)[a]

Apparent affinities of antagonists are pA_2 values or pK_B estimates (indicated by daggers) expressed as mean $\pm$ SEM with the number of determinations shown in parentheses. All pA_2 values had Schild slope estimates that were not significantly different from unity.
[a] The pA_2 value for L-365,260 against CCK-4 in GPI was determined in the presence of devazepide to exclude the possibility of interference by CCK_A-receptor-mediated effects. (The concentrations of devazepide used in these studies (100 nM) had no effect on the response to CCK-4, producing concentration ratios of <2.)

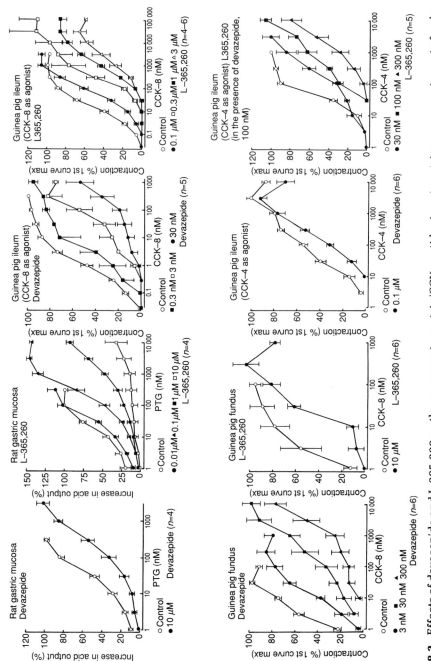

Fig. 8.3. Effects of devazepide and L-365,260 on the response to gastrin/CCK peptides in rat gastric mucosa, guinea pig fundus, and guinea pig ileum.

more potent than L-365,260 (Table 8.2). The affinity estimates for devazepide and L-365,260 in GPF agreed well with reported binding and functional studies that demonstrate interactions with CCK_A receptors (Bock *et al.* 1989; Huang *et al.* 1989).

In GPI, when CCK-8 was used as agonist, devazepide was a potent antagonist with a pA_2 value (10.40) similar to that reported previously for guinea pig ileum (Chang and Lotti 1986) and for inhibition of CCK-8-induced amylase release from rat pancreatic acini (Louie *et al* 1988). These studies were consistent with the view that CCK-8 induced contraction of GPI by interaction with CCK_A receptors. However, L-365,260 was a more potent antagonist of CCK-8 than would have been predicted from studies in GPF and from previous reports of CCK_A binding affinity (Bock *et al.* 1989) and other CCK_A-receptor-mediated responses (Marshall *et al.* 1990). Since two types of contractile response were displayed by gastrin/CCK peptides in GPI, the effects of these antagonists were studied further using CCK-4 as agonist. As shown in Table 8.2, different affinity estimates for devazepide and L-365,260 were obtained in GPI when CCK-4 was used as agonist. Furthermore, antagonist affinity estimates versus CCK-4 agreed with reported affinities of L-365,260 and devazepide from binding studies using guinea pig cortex (Bock *et al.* 1989) and were consistent with an involvement of CCK_B receptors in these contractile responses in GPI.

The affinity estimate for L-365,260 versus CCK-4 in GPI was significantly different from that obtained for L-365,260 against pentagastrin in RGM (no overlap of 95 per cent confidence limits), providing evidence for the contention that gastrin and CCK_B receptors are distinct entities (Jensen 1989). Whilst agonists have supported this hypothesis (Bohme *et al.* 1988; Menozzi *et al.* 1989), binding studies comparing antagonists in guinea pig cortex and gastric glands have failed to disclose receptor differences in these two preparations (Freidinger 1989). In the present functional studies in RGM, L-365,260 had a lower apparent affinity than predicted from binding studies. It is possible that binding studies in gastric glands do not reflect interaction with gastrin receptors and may actually estimate affinity for CCK_B receptors present in gastric mucosa (Praissman and Brand 1990).

Conclusions

The present studies are consistent with the view that stimulation of acid secretion by pentagastrin in RGM is mediated by gastrin receptors, whilst the contractile effects of CCK-8 in GPF and GPI are mediated by CCK_A receptors. In addition, antagonist affinity estimates suggest that the contractile effects of CCK-4 in GPI are mediated by CCK_B receptors and these receptors appear distinct from gastrin receptors present in RGM.

These studies show that isolated preparations from the gastrointestinal tract appear useful for the functional characterization of gastrin/CCK receptor subtypes and have extended the information gained from binding studies with the demonstration that gastrin and CCK_B receptors are different subtypes.

References

Bitar, K. N., Saffouri, B., and Makhlouf, G. M. (1982). Cholinergic and peptidergic receptors on isolated human antral smooth muscle cells. *Gastroenterology* **82**, 832–7.

Bock, M. G., Dipardo, R. M., Evans, B. E., Rittle, K. E., Whitter, W. L., Veber, D. F., Anderson, P. S., and Freidinger, R. M. (1989). Benzodiazepine gastrin and brain cholecystokinin receptor ligands: L-365,260. *J. Med. Chem.* **89**, 13–16.

Boden, P. and Hill, R. G. (1988). Effects of cholecystokinin and related peptides on neuronal activity in the ventromedial nucleus of the rat hypothalamus. *Br. J. Pharmacol.* **94**, 246–52.

Bohme, G. A., Stutzmann, J. M., and Blanchard, J. C. (1988). Excitatory effects of cholecystokinin in rat hippocampus: pharmacological response compatible with 'central' or B-type CCK receptors. *Brain Res.* **451**, 309–18.

Chang, R. S. L. and Lotti, V. J. (1986). Biochemical and pharmacological characterisation of an extremely potent and selective non-peptide cholecystokinin antagonist. *Proc. Natl Acad. Sci. USA* **83**, 4923–6.

Dockray, G. J. and Gregory, R. A. (1990) Gastrin. In *Handbook of physiology*, Section 6, Vol. II (ed. S. G. Schultz, G. M. Makhlouf, and B. B. Rauner), pp. 311–36. American Physiological Society, Baltimore.

Freidinger, R. M. (1989). Development of selective non-peptide CCK_A and CCK_B/gastrin receptor antagonists. In *The neuropeptide cholecystokinin*, (ed. J. Hughes, G. J. Dockray, and G. N. Woodruff), pp. 23–32. Ellis Horwood, Chichester.

Hill, D. R. and Woodruff, G. N. (1989). Differentiation of central CCK receptor binding sites using the non-peptide antagonists MK-329 and L-365,260. *Br. J. Pharmacol.* **98**, 629P.

Huang, S. C., Zhang, L., Chiang, H. C. V., Wank, S. A., Maton, P. N., Gardner, J. D., and Jensen, R. T. (1989). Benzodiazepine analogues L-365,260 and L-364,718 as gastrin and pancreatic CCK receptor antagonists. *Am. J. Physiol.* **257**, G169–74.

Jensen, R. T. (1989). Pancreatic cholecystokinin receptors: comparison with other classes of CCK receptors. In *The neuropeptide cholecystokinin* (ed. J. Hughes, G. J. Dockray, and G. N. Woodruff), pp. 150–62. Ellis Horwood, Clichester.

Jensen, R. T., Lemp, G. F., and Gardner, J. D. (1982). Interactions of COOH-terminal fragments of cholecystokinin with receptors on dispersed acini from guinea-pig pancreas. *J. Biol. Chem.* **257**, 5554–9.

Kemp, J. A., Marshall, G. R., and Woodruff, G. N. (1989). Antagonism of CCK-induced excitation of rat ventromedial hypothalamic neurones by a new selective CCK_B receptor antagonist. *Br. J. Pharmacol.* **98**, 630P.

Louie, D. S., Liang, J. P., and Owyang, C. (1988). Characterization of a new CCK antagonist, L-364,718: *in vitro* and *in vivo* studies. *Am. J. Physiol.* **255**, G261–6.

Marshall, F. H., Barnes, S., Pinnock, R. D., and Hughes, J. (1990). Characterisation of cholecystokinin octapeptide-stimulated endogenous dopamine release from rat nucleus accumbens *in vitro*. *Br. J. Pharmacol.* **99**, 845–8.

Menozzi, D., Gardner, J. D., Jensen, R. T., and Maton, P. N. (1989). Properties of receptors for gastrin and CCK on gastric smooth muscle cells. *Am. J. Physiol.* **257**, G73–9.

Moran, T. H., Robinson, P. H., Goldrich, M. S., and McHugh, P. R. (1986). Two brain cholecystokinin receptors: implications for behavioural actions. *Brain Res.* **362**, 175-9.

Praissmann, M. and Brand, D. L. (1990). Evidence for two different CCK binding sites in human gastric oxyntic mucosa. *Gastroenterology* **98**, A518.

Reeves, J. J. and Stables, R. (1985). Effects of indomethacin, piroxicam and selected prostanoids on gastric acid secretion by the rat isolated gastric mucosa. *Br. J. Pharmacol.* **86**, 677-84.

Soll, A. H., Amirian, D. A., Thomas, L. P., Reedy, T. J., and Elashoff, J. D. (1984). Gastrin receptors on isolated canine parietal cells. *J. Clin. Invest.* **73**, 1434-47.

Walsh, J. H. (1987). Gastrointestinal hormones. In *Physiology of the gastrointestinal tract* (2nd edn) (ed. L. R. Johnson), pp. 82-90. Raven Press, New York.

Yau, W. M., Makhlouf, G. M., Edwards, L. E., and Farrar, J. T. (1974). The action of cholecystokinin and related peptides on guinea-pig small intestine. *Can. J. Physiol. Pharmacol.* **52**, 298-303.

9. Blood–brain transfer of the CCK antagonists L-365,260 and devazepide

Richard J. Hargreaves and Juinn H. Lin

The concept of a blood–brain barrier was first formulated at the turn of the century as a result of experiments which showed that some dyes and pharmacologically active compounds did not stain or affect the brain when injected into the bloodstream, but did so when injected into the cerebrospinal fluid. It was not until the 1960s that it became clear that the anatomical basis of the blood–brain barrier comprised the cerebral capillary endothelium in which adjacent cells were joined together at all points of contact by tight intercellular junctions making a virtually continuous monocellular layer. In an early paper on ionic exchange across living membranes, Krogh (1946) suggested that the permeability properties of the cerebral capillaries were those of the endothelial cell membrane and that the design of drugs intended to act in the CNS should concentrate on lipid solubility as one of the primary factors determining rate of entry into the brain. Most solutes that pass through the blood–brain barrier must do so by dissolution in the membranes and cytoplasm of the capillary endothelial cells. As a result, transcellular movement is influenced by the lipid (membrane) solubility of a compound and its diffusion rate. In turn, these factors are dependent upon the fraction of drug in capillary plasma that is not ionized at physiological pH and upon molecular size. Brain uptake of solutes may also be affected by plasma protein binding. It was hypothesized that protein bound drug was not available for transfer from capillary plasma to brain since the high molecular weight, and sometimes electrical charge, of plasma proteins prevented their movement with bound drug from the vascular compartment. However, studies have now indicated that this may be an oversimplification since the brain penetration of many drugs and hormones that are plasma protein bound is greater than would be predicted from free drug concentrations (Pardridge and Landaw 1984).

In the present studies we have examined, in conscious and anaesthetized rats, the blood–brain barrier transfer of the benzodiazepine derivatives L-365,260, a CCK$_B$ selective antagonist (Lotti and Chang 1989), and devazepide, a potent highly selective antagonist at the CCK$_A$ receptor (Chang *et al.* 1986). The plasma and brain concentration profiles for these compounds have also been determined. Such information is valuable in the selection of systemic doses that optimize the selectivity of the antagonists and the separation between them when attempting to define the sites of action of CCK and the subtypes of receptors involved in its actions in the CNS.

The method used in conscious rats was essentially that described by Ohno *et al.* (1978). Male Sprague–Dawley rats (approximately 280 g) were anaesthetized with isoflurane and a tail artery and vein were cannulated. The animals were allowed to recover for at least 2 h before experiments in a cage which prevented turning to gain access to the cannulae. Devazepide or L-365,260 was administered intravenously (i.v.) at a dose of 1 mg/kg in a vehicle comprising 10 per cent ethanol, 60 per cent propylene glycol, and 30 per cent saline using a dose volume of 1 ml/kg. At various times after injection, dependent upon the duration of the experiment, small arterial blood samples were taken to determine the i.v. decay of the test compounds. Plasma samples were centrifuged to obtain a plasma fraction. At 1, 3, 10, or 30 min after injection the animals were killed by an overdose of sodium pentobarbitone and the brain was removed. In each experiment the residual plasma in brain tissue at the end of the experiment was measured using hydroxy[^{14}C]methyl inulin. This enabled the estimates of drug in brain tissue to be corrected for the contribution of drug remaining in residual plasma. Drug in plasma and brain was measured using high pressure liquid chromatography (HPLC) techniques.

The brain uptake index (BUI) technique of Oldendorf (1971) was used for studies in anaesthetized rats. Male Sprague–Dawley rats were anaesthetized with sodium pentobarbitone and the left carotid artery was exposed. [^{14}C] L-365,260 or [^{3}H]devazepide was given by rapid intracarotid injection in either buffered Ringers solution (pH 7.4, 5 mM Hepes) or rat serum in a dose volume of 200 μl. The freely diffusable internal reference compounds included in the injectate were [^{3}H]water for [^{14}C] studies and [^{14}C]butanol for [^{3}H] studies. The animals were decapitated 15 s after the injection and the cerebral hemisphere ipsilateral to the injection was removed, homogenized, and solubilized for dual-isotope liquid scintillation counting. Injectate solutions were also counted.

Plasma protein binding was measured *in vitro* using the filtration method. Rat serum was incubated with radio-labelled test drug at 37 °C for 5 min, transferred to an Amicon micropartition system, and centrifuged at 1500 g for 15 min, and 150 μl of ultrafiltrate was collected. The unbound fraction of test drug was calculated from the counts in the ultrafiltrate and those in the original serum sample after liquid scintillation counting of both fractions.

Figures 9.1(a) and 9.1(b) show the time-dependent concentration profiles of L-365,260 and devazepide in the plasma and brain of conscious rats after dosing at 1 mg/kg. Both compounds entered the brain readily, and equilibrium between plasma and brain was rapidly attained. The brain-to-plasma ratio for devazepide at equilibrium was 2.42 and that for L-365,260 was 0.38. At 10 and 30 min after dosing the brain concentrations were 170 and 65 nM L-365,260 and

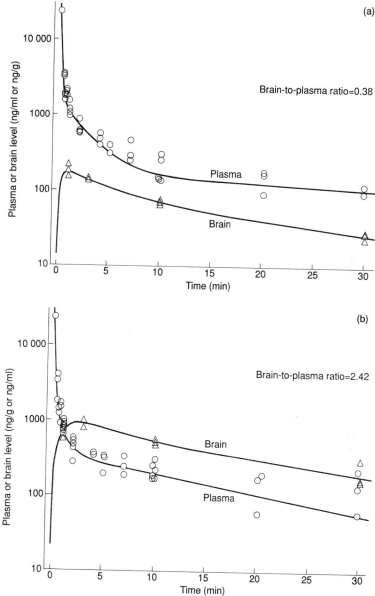

Fig. 9.1. The time-dependent concentration profiles of (a) L-365,260 and (b) devazepide in plasma and brain of conscious rats after i.v. dosing at 1 mg/kg. Plasma decays shown are from 10 and 30 min experiments only and are data from sequential samples from each of eight animals. Curves are the best fit to all experimental data (1, 3, 10, and 30 min experiments) using non-linear least mean squares regression analysis. Brain concentrations have been corrected for drug remaining in residual plasma in the brain. The brain-to-plasma ratio is the value at equilibrium (10 min).

1.25 and 0.5 μM devazepide. The clearance K of the compounds from arterial plasma into brain and the calculated initial extraction fraction $E_{T(0)}$ and permeability–surface area (PS) products are given in Table 9.1. The data show that the permeability of the blood–brain barrier to devazepide is greater than

Table 9.1.   Blood–brain barrier transfer of L-365,260 and devazepide in conscious rats

	K^a (ml/min/g)	$E_{T(0)}^{bc}$ (%)	$PS^{c,d}$ (ml/min/g)
L-365,260			
Mean	0.05	8.0	0.05
SEM	0.01	2.0	0.01
Devazepide			
Mean	0.19	31.1	0.22
SEM	0.02	2.8	0.03

Values are mean ± SEM (n = 3)
[a] If negligible backflux of test compound is assumed

$$K = \frac{C_{br}(T)}{\int C_{pl}\, dt}$$

where $C_{br}(T)$ is the amount of drug in brain at the end of the experiment after correction for drug remaining in plasma and $\int C\, dt$ is the area under the plasma concentration decay curve (Ohno et al. 1978).
[b] The clearance constant K is related to the initial extraction ratio $E_{T(0)}$ by the relationship

$$K = Ff E_{T(0)}.$$

Hence

$$E_{T(0)} = \frac{K}{Ff}$$

where F is the rate of blood flow and the f is the fractional volume of blood or intracapillary fluid that is the source of drug for uptake into brain during the measurement period.
[c] A plasma flow of 0.6ml/min/g brain is assumed for calculations.

$$E_{T(0)} = 1 - \exp\left(-\frac{PS}{Ff}\right) \qquad \text{(Crone 1963)}.$$

[d] Thus from footnote b,

$$K = Ff\left[1 - \exp\left(-\frac{PS}{Ff}\right)\right]$$

$$PS = Ff \ln\left(1 - \frac{K}{Ff}\right)$$

where PS is the permeability–surface area product for a solute at the cerebral capillary wall.

that to L-365,260 by a factor of about 4. Thus during the first passage of the compounds through the cerebral circulation about 31 per cent of devazepide and 8 per cent of L-365,260 are extracted into brain.

The results of the BUI studies in anaesthetized rats are given in Table 9.2. The initial extractions of devazepide and L-365,260 from buffer injectate were 42 per cent and 6 per cent respectively. These values are similar to those found in the conscious animals (see above). When the test compounds were administered in rat serum the extraction fraction fell markedly for devazepide to 26.7 per cent and to 2.8 per cent for L-365,260. However, this reduction in extraction after injection in serum compared with buffer was not as great as might have been predicted from the fraction of drug observed to remain unbound in serum *in vitro*. The plasma protein binding studies showed that only 2.8 per cent of devazepide and 2.6 per cent of L-365,260 remained unbound in rat serum (Table 9.2).

Table 9.2. Blood–brain barrier transfer of [^{14}C]L-365,260 and [^{3}H]devazepide in anaesthetized rats

Parameters	Composition of injectate solution	[^{14}C]L-365,260	[^{3}H]devazepide
$E_{T(0)}$[a] (%)	Buffer	6.0 ± 2.0	42.0 ± 5.0
	serum	2.8 ± 0.5	26.7 ± 5.6
PS[b,c] (ml/min/g)	Buffer	0.04 ± 0.01	0.33 ± 0.05
	serum	0.02 ± 0.003	0.19 ± 0.05
Plasma protein binding[d] (% unbound)	Buffer	100	100
	serum	2.6	2.8
Octanol–buffer partition coefficient		3738 ± 650	112 ± 35

Values are mean ± SEM (n = 5–7)
[a] Brain uptake $E_{T(0)}$ % was calculated using the BUI equation

$$\frac{\text{dpm test drug in brain/dpm reference in brain}}{\text{dpm test drug in injectate/dpm reference in injectate}} \times 100\%$$

since for benzodiazepines BUI at 15 s approximates $E_{T(0)}$ %.
[b] A blood flow of 0.6 ml/min/g brain was assumed for calculation.
[c] Given a whole brain blood flow rate

$$PS = -Ff \ln\left(1 - E_{T(0)}\right)$$

where PS is the permeability–surface area product at the cerebral capillary wall (Crone 1963). For studies using buffer, the fractional volume f of intracapillary fluid, which is the source of drug for uptake into brain during the period of measurement, is equal to 1.
[d] Mean of duplicate measurements.
Reproduced from Lin and Lin (1990).

At present there is considerable use of L-365,260 to determine the involvement of central CCK_B receptors in various physiological responses. Data from the present studies can be used to predict doses of L-365,260 that will be required to antagonize central CCK_B sites whilst remaining selective for CCK_B over CCK_A receptors. By using the equilibrium brain-to-plasma ratio (0.38) for L-365,260 found in the present studies, brain concentrations can be predicted from plasma data regardless of the route of dosing, with the assumptions that entry of the drug into the brain is linear with concentration and that an equilibrium between plasma and brain exists. For i.v. dosing protocols the present data can be scaled to estimate plasma and brain concentrations so that appropriate doses and time intervals can be chosen for i.v. CCK_B antagonism studies. In other studies with L-365,260 (results not given) we have found that plasma levels of up to 100 nM are attained 20–30 min after dosing subcutaneously (s.c.) at 400 μg/kg in an ethanol-methylcellulose vehicle. From this, using the brain-to-plasma ratio of 0.38, a brain concentration of 38 nM can be predicted. This level is 10 times the IC_{50} for L-365,260 displacement of [^{125}I]CCK-8 from rat brain (Lotti and Chang 1989) and similar to the apparent dissociation constant K_D (33 nM) for the functional antagonism of excitatory responses evoked by CCK-8S in the rat ventromedial hypothalamic slice (Chapter 7 of this volume). Table 9.3 shows predicted plasma and brain concentrations of L-365,260 at 20 min after s.c. administration in ethanol-methylcellulose at various dose levels. The table indicates a useful working range for functional selective antagonism at central CCK_B receptors of between 40 μg/kg and 10 mg/kg L-365,260 administered s.c.

The results of the present studies in conscious and anaesthetized rats using different techniques to assess the blood–brain barrier transfer of devazepide and L-365,260 from circulating plasma and serum were in good agreement. The small differences that were observed are attributable to differing limitations on the accuracy of measurements using the BUI and integral techniques. BUI

Table 9.3. Predicted relationship between subcutaneous doses of L-365,260[a] and plasma and brain drug levels 20–30 min after dosing

Dose (mg/kg)	Plasma (nM)	Brain (nM)	Comment
10	2500	950	= IC_{50} CCK_A^b
0.4	100	38	= K_D VMH slice[c]
0.04	10	3.8	= IC_{50} CCK_B^d

[a] Vehicle 10 per cent ethanol, 90 per cent (0.5 per cent) methylcellulose.
[b] [^{125}I] CCK8 displacement rat pancreas: IC_{50} = 1100 nM (Lotti and Chang 1989).
[c] K_D versus CCK-8S in rat ventromedial hypothalamic (VMH) slice: K_D = 33 nM (Chapter 7 of this volume).
[d] [^{125}I] CCK-8 displacement rat brain: IC_{50} = 3.9 nM (Lotti and Chang 1989).

studies lack accuracy with compounds such as L-365,260 which have an extraction of less than 10 per cent but are excellent with compounds such as devazepide which have higher extraction (20–90 per cent). This is mainly because the amount of a poorly extracted test compound entering the brain during the brief (15 s) measurement period is too low to assay accurately without making corrections for drug remaining in residual plasma in brain tissue at the end of experiment. The converse is true of the integral technique used in the conscious rat studies since experimental times are longer. However, with highly extracted or diffusable compounds errors occur because of backflux of the test compound from the brain to blood during the measurement period. This disrupts the assumption implicit in the method that movement of the compound has been unidirectional during the measurement period and leads to an underestimation of influx. Thus measurements are likely to be best for L-365,260 in the conscious preparation but best for devazepide in the shorter BUI experiments.

Measurements of brain uptake of devazepide from rat serum gave values below that found for extraction from buffer. The reduction in extraction from serum compared with buffer was not as great as might have been expected from the small unbound fraction (2–3 per cent) measured *in vitro* if only the unbound fraction is available for transfer. Therefore it is clear that the source of drug *in vivo* is much greater than the unbound fraction estimated *in vitro*. It is noteworthy that the free drug fraction determined by *in vitro* protein binding studies is thought to be the minimum that is likely to be available for transfer from plasma to brain. Pardridge and Landaw (1984) mathematically analysed the transport and dissociation of protein-bound compounds during a single passage through the cerebral capillaries. Their results showed that the dissociation constant of the protein complex could be up to 50-fold greater than the dissociation constant measured *in vitro*. With highly extracted protein-bound compounds the unbound fraction in capillary plasma is reduced by movement into brain but may then be restored by dissociation of the drug–protein complex during a single passage through the cerebral capillary bed. Brain uptake of protein bound drug thus depends upon the ratio between the association and dissociation constants for drug at each binding site on plasma protein relative to cerebral capillary transit times and the intrinsic permeability of the blood-brain barrier to unbound drug. Taken together, these observations suggest that *in vitro* protein binding alone is of limited value for predicting the brain penetration of protein-bound drugs *in vivo*.

Studies of the octanol–buffer partition coefficient of L-365,260 and devazepide (Lin and Lin 1990) indicate that both compounds are extremely lipid soluble (Table 9.2). Cornford *et al.* (1982) studied the relationship between the lipid solubility of a compound and its brain entry, taking into account molecular size. The correlation between these parameters was shown to be significant ($r = 0.86$, $p < 0.001$) and to be described by the linear relationship

$$(\log \text{BUI}) \times (\sqrt{\text{mol. wt}}) = 6.02 \times (\log \text{partition coefficient}) + 14.5.$$

Substituting into this relationship $E_{T(0)}$ for BUI and using the partition coefficients (Tables 9.1 and 9.2), it can be calculated that devazepide entry into brain from serum and buffer in conscious and anaesthetized rats fits this correlation. The relationship predicts a much higher brain entry for L-365,260 than reported in the BUI studies. The reason for this discrepancy is unknown. In conscious rats, L-365,260 after i.v. injection may exist in circulating plasma in a form that is less bioavailable than measurements of total drug in the injectate suggest. The apparent brain extraction observed for L-365,260 may therefore be an underestimate of its true permeability at the blood–brain barrier.

Acknowledgments

Thanks are due to Bindi Sohal of DEVLAB for HPLC analyses during the conscious rat studies and to Tsu-Han Lin for work on the anaesthetized rat studies.

References

Chang, R. S. L., Lotti, V. J., Chen, T. B., and Kunkel, K. A. (1986). Characterisation of the binding of [^{3}H]($\pm$)L-364,718: a new potent non-peptide cholecystokinin antagonist radioligand selective for peripheral receptors. *Mol. Pharmacol.* **30**, 212–17.

Cornford, E. M., Braun, L. D., Oldendorf, W. H., and Hill, M. A. (1982). Comparison of lipid mediated blood–brain barrier penetrability in neonates and adults. *Am. J. Physiol.* **243**, C161–8.

Crone, C. (1963). The permeability of capillaries in various organs as determined by the use of the indicator diffusion method. *Acta Physiol. Scand.* **58**, 292–305.

Krogh, A. (1946). The active and passive exchanges of inorganic ions through the surface of living cells and through membranes generally. *Proc. R. Soc. Lond. Ser. B* **133**, 140–200.

Lin, T.-H. and Lin, J. H. (1990). Effects of protein binding and experimental disease states on brain uptake of benzodiazepines in rats. *J. Pharmacol. Exp. Ther.* **253**, 45–50.

Lotti, V. J. and Chang, R. S. L. (1989). A new potent and selective non-peptide gastrin antagonist and brain CCK-B receptor ligand. *Eur. J. Pharmacol.* **162**, 273–80.

Ohno, K., Pettigrew, K. D., and Rapoport, S. I. (1978) Lower limits of cerebrovascular permeability to nonelectrolytes in the conscious rat. *Am. J. Physiol.* **235**, H299–307.

Oldendorf, W. H. (1971). Brain uptake of radiolabelled amino acids amines and hexoses after arterial injection. *Am. J. Physiol.* **221**, 1629–39.

Pardridge, W. M. and Landaw, E. M. (1984). Tracer kinetic model of blood–brain barrier transport of protein bound ligands. Empiric testing of the free hormone hypothesis. *J. Clin. Invest.* **74**, 745–52.

Part II

CCK and anxiety

10. CCK and anxiety: introduction

Jens F. Rehfeld

This section is devoted to the anxiogenic and panicogenic effect of CCK and the potential anxiolytic and antipanic properties of CCK antagonists. As an introduction to the following detailed chapters by Hendrie and Neill, and Bradwejn and Koszycki, I shall first describe the fortuitous and unexpected discovery of this novel activity of CCK. On second thoughts, it is perhaps less surprising that CCK peptides are also involved in central anxiogenic mechanisms. Thus the facts that CCK is the most widespread and abundant peptide system in the CNS (Rehfeld 1978; Larsson and Rehfeld 1979), that CCK peptides are synthesized at a very fast rate in the brain (Goltermann *et al.* 1980; Stengaard-Pedersen *et al.* 1984), and that CCK peptides are highly potent transmitters (Emson *et al.* 1980; Dodd and Kelly 1981) raises the question of whether there are CNS activities in which CCK peptides are *not* involved.

In addition to the early history of the panicogenic and anxiogenic effects of CCK, I shall also reflect on the curious fact that so far only one molecular form of CCK, i.e. the small CCK-4, has displayed the capacity to induce anxiety attacks by exogenous administration. This is an important observation, which also needs further examination. It may have direct implications for our concept of the multiplicity of CCK receptors in the brain.

The discovery of the anxiogenic/panicogenic effect of CCK

In 1969, I was studying the effect of gastrin on insulin secretion in man. At the same time, I had to do my military service. In an attempt to make use of the abundant spare time in the army, I spent some time injecting gastrin intravenously into the private soldiers during measurement of their insulin release. Since at that time it was difficult to obtain and purchase sufficient amounts of pure gastrin-17, I also used the synthetic C-terminal bioactive tetrapeptide amide of gastrin as an analogue. Gastrin-4 constitutes the active site which is also common to CCK. In other words, gastrin-4 is identical with CCK-4, i.e. Trp–Met–Asp–Phe–NH_2. Immediately after injection of the tetrapeptide, it was obvious that the soldiers became strikingly quiet and looked fearful when the high dose (70 μg) of CCK-4 was used. I noted this effect, but did not follow it up. Later, I published only the results on insulin secretion (Rehfeld 1971).

Then, after detection of CCK in the hypothalamus and the median eminence

(Rehfeld 1978; Larsson and Rehfeld 1979), I became interested in the possibility that CCK peptides might influence the secretion of pituitary hormones, especially the growth hormone. Having in mind the unique effect of CCK-4 on insulin secretion (see above) in pilot experiments conducted in 1978, I injected a bolus of CCK-4 into myself and my research fellow, Thue Schwartz. The stimulation of growth hormone secretion was not impressive, but nevertheless CCK-4 showed a remarkable effect. Half a minute after the injection both of us felt an unexplained and intense anxiety. It became very unpleasant, and was accompanied by palpitations, sweating, and faintness. It peaked after 5–8 min and then declined gradually. After approximately 20 min the attack was over. During the attack, I had a distinct feeling of being scared with a hallucination that the room around me was sliding away. Afterwards, I occasionally thought that the observations made during this attack of anxiety, which we had never observed during CCK-8 infusions (unpublished), ought to be followed up. However, working in biochemistry departments far from psychiatry and changing university shortly afterwards slowed down the initiative.

However, at the conference on neuronal cholocystokinin held in Brussels in 1984 (see Vanderhaeghen and Crawley 1985), during the discussion session on the clinical significance of neuronal CCK, I described the experience with the CCK-4 injections into the soldiers, and into my research fellow and myself. Fortunately, Bradwejn and De Montigny from Montreal were present in the audience. Because of their background and interest in the mechanisms of anxiety and their recent observation of the interaction of CCK and benzodiazepines (Bradwejn and De Montigny 1984), they became interested in the perspectives and putative clinical significance of my description of the CCK-4-induced anxiety. Their subsequent careful studies have convincingly illustrated and emphasized the close link between CCK and anxiety (De Montigny 1989; Bradwejn *et al.* 1990). In the following chapter, Bradwejn and Koszycki review their latest results on patients and normal subjects. Also, the role of CCK in anxiety has recently been reviewed with emphasis on experimental and animal data (Ravard and Dourish 1990).

Why is the anxiogenic effect induced only by CCK-4 and not by longer CCK peptides?

CCK exists in a number of different molecular forms with transmitter activity, i.e. the bioactive CCK peptides all have the tetrapeptide amide (Trp–Met–Asp–Phe–NH$_2$) at their C-terminus. This C-terminal sequence is crucial for binding to both the CCK$_A$ and CCK$_B$ receptors. The forms and concentrations of transmitter-active CCK peptides present in the cerebral cortex is outlined in Table 10.1. Notably, the brain contains small amounts of CCK-8US as well as substantial amounts of CCK-5 and a little CCK-4 in addition to the predominant sulphated CCK-8 and further N-terminally extended peptides (CCK-22, CCK-33, CCK-39, CCK-58, and one even further extended) (Table 10.1)

Table 10.1. Molecular forms of transmitteractive ($\sim\alpha$-carboxyamidated) CCK peptides in the porcine cerebral cortex

Peptide	Mean concentration (n = 4) (pmol/g tissue (wet weight))
CCK component I	24.2
CCK-58	142.1
CCK-39 and CCK-33	37.6
CCK-22	4.7
CCK-8 (S and a little US)	429.6
CCK-7 (S and a little US)	3.1
CCK-5	102.6
CCK-4	13.3

Data from Rehfeld and Hansen (1986).

(Rehfeld and Hansen 1986). However, we do not know whether the small CCK-5 and CCK-4 peptides are synthesized as major transmitter products by cell-specific processing in a limited number of neurons, or whether they are merely degradation products from CCK neurons in which the main product is CCK-8S (Rehfeld 1985). Indeed, there are reasons for believing that the brain contains at least three subpopulations of CCK neurons with different post-translational processing pathways (Rehfeld and Hansen 1986). Following this line of reasoning, we also assume that the post-synaptic cell membranes contain a class of CCK receptors corresponding to the particular CCK peptide released from the appropriate presynaptic CCK neuron.

As mentioned, it is noteworthy that so far only CCK-4 (or CCK-5) has induced anxiety and panic on exogenous administration. There are three possible explanations for this phenomenon. First, it is possible that CCK-4 (and CCK-5) peptides can pass the blood–brain barrier easily owing to their relatively high hydrophobicity. In contrast, the longer hydrophilic CCK peptides such as CCK-8S and CCK-33S pass the blood–brain barrier less readily. Second, it is possible that the profuse gastrointestinal symptoms induced by high doses of CCK-8S and longer sulphated peptides may overshadow mild anxiety and panic symptoms. Finally, the existence of a third class of CCK receptors (a putative CCK_C receptor) that binds CCK-4 with higher affinity than CCK-8S and gastrin might explain the specific panicogenic effect of CCK-4. Presumably, the occurrence of such CCK_C receptors would be sparse in comparison with the abundant CCK_A receptors in the periphery and the CCK_B receptors in the brain.

Conclusion

It now remains to be shown which of the three theories is correct for the specific anxiogenic effect of CCK-4. It is obvious that comparative studies of the passage of different molecular forms of CCK through the blood–brain barrier have to be performed. Also, it is necessary to determine whether the panicogenic effect of CCK-4 is mediated by peripheral receptors. Finally, the intriguing possibility of the existence of a third class of CCK receptors deserves closer inspection. In fact, there exists evidence based on structure function studies of different CCK peptides which suggests that islet cells in the porcine and human pancreas and in canine thyroid C-cells may be equipped with receptors that recognize CCK-4 and CCK-5 with an affinity higher than that of CCK-8S (Rehfeld *et al.* 1980; Laurberg and Rehfeld 1987). Further studies of CCK binding to these cells may be rewarding.

References

Bradwejn, J. and de Montigny, C. (1984). Benzodiazepines antagonize cholecystokinin-induced activation of rat hippocampal neurons. *Nature* **312**, 363–4.
Bradwejn, J., Koszycki, D., and Meterissian, G. (1990). Cholecystokinin-tetrapeptide induces panic attack in patients with panic disorder. *Can. J. Psychiat.* **35**, 83–5.
De Montigny, C. (1989). Cholecystokinin-tetrapeptide induces panic attacks in healthy volunteers. *Arch. Gen. Psychiat.* **46**, 511–17.
Dodd, J. and Kelly, J. S. (1981). The actions of cholecystokinin and related peptides on pyramidal neurones of the mammalian hippocampus. *Brain Res.* **205**, 337–50.
Emson, P. C., Lee, C. M., and Rehfeld, J. F. (1980). Cholecystokinin octapeptide: vesicular localisation and calcium dependent release from rat brain *in vitro*. *Life Sci.* **26**, 2157–63.
Goltermann, N., Rehfeld, J. F., and Petersen, H. R. (1980). *In vivo* biosynthesis of cholecystokinin in rat cerebral cortex. *J. Biol. Chem.* **255**, 6181–5.
Larsson, L. I. and Rehfeld, J. F. (1979). Localisation and molecular heterogeneity of cholecystokinin in the central and peripheral nervous system. *Brain Res.* **165**, 201–18.
Laurberg, P. and Rehfeld, J.F. (1987). Cholecystokinin peptides as local modulators of thyroidal calcitonin secretion. *J. Endocrinol.* **115**, 77–82.
Ravard, S. and Dourish, C. T. (1990). Cholecystokinin and anxiety. *Trends Pharmacol. Sci.* **11**, 271–3.
Rehfeld, J. F. (1971). Effect of gastrin and its C-terminal tetrapeptide on insulin secretion in man. *Acta Endocrinol.* **66**, 169–76.
Rehfeld, J. F. (1978). Immunochemical studies on cholecystokinin II. Distribution and molecular heterogeneity in the central nervous system and small intestine of man and hog. *J. Biol. Chem.* **253**, 4022–30.
Rehfeld, J. F. (1985). Neuronal cholecystokinin: one or multiple transmitters? *J. Neurochem.* **44**, 1–10.
Rehfeld, J. F. and Hansen, H. F. (1986). Characterization of preprocholecystokinin products in the porcine cerebral cortex: evidence of different processing pathways. *J. Biol. Chem.* **261**, 5832–40.
Rehfeld, J. F., Larsson, L. I., Goltermann, N., Schwartz, T. W., Holst, J. J., Jensen, S. L., and Morley, J. S. (1980). Neural regulation of pancreatic hormone secretion by the C-terminal tetrapeptide of CCK. *Nature* **284**, 33–9.
Stengaard-Petersen, K. Larsson, L. I., Fredens, K., and Rehfeld, J. F. (1984). Modulation of the cholecystokinin concentration in rat hippocampus by chelation of heavy metals. *Proc. Natl Acad. Sci. USA* **81** 5876–80.
Vanderhaeghen, J. J. and Crawley, J. N. (ed.) (1985). Neuronal Cholecystokinin. *Ann. NY Acad. Sci.* **448**, 1–697.

11. CCK receptors and panic attacks in man

Jacques Bradwejn and Diana Koszycki

Panic attacks and their relation to psychiatric disorders

Panic attacks and panic disorders

The *Diagnostic and statistical manual of mental disorders* (*DSM–III–R*) (American Psychiatric Association 1987) describes a panic attack as the sudden onset of intense anxiety or fear, accompanied by at least four of the following 12 symptoms: dyspnea, choking, palpitations, or tachycardia: chest pain or discomfort; sweating; faintness; dizziness; light-headedness; unsteady feeling; nausea or abdominal distress; depersonalization or derealization; flushes or chills; trembling or shaking and fear of dying; fear of going insane; fear of doing something uncontrolled. The particular cluster of panic symptoms varies across individuals and may differ slightly from one attack to another within individuals.

The *DSM–III–R* describes two classes of panic attacks: specifically triggered and 'unexpected'. In the first, the attack is triggered by specific phobic cues and occurs predictably and invariably upon exposure to certain situations or objects. Situation-specific panic represents an intense form of phobic anxiety and is associated primarily with social phobia (e.g. fear of eating in public) and simple phobias (e.g. fear of heights). By definition, situation-specific panics can be circumvented provided that cues that regularly trigger them are avoided.

The second class of attacks is an abrupt crescendo of fear and autonomic distress for which there is no identifiable cue or trigger. These attacks may occur in situations associated with previous unexpected attacks but, unlike situation-specific panic, the individual is uncertain when the attack will occur or if it will occur at all. Unexpected panic attacks can last for anything from a few minutes to several hours. Typically, they build in intensity over a period of 10–15 min after which symptoms reach a peak. It is not uncommon for individuals to report a sensation of relief or intense fatigue following an attack. Outbursts of dramatic behaviour, other than an appearance of unease, seldom occur in the midst of an attack.

Unexpected panic attacks are known to occur sporadically in the general population (Craske *et al.* 1986). Usually, these attacks encompass fewer and less severe symptoms than those seen in the psychiatic population. When unexpected panic attacks occur with high frequency and are associated with persistent fears of having another attack a diagnosis of panic disorder is made. Panic attacks may develop in the context of other psychiatric syndromes, notably major depressive disorder, somatization disorder, and generalized anxiety disorder. However, the attacks do not constitute the central feature of these disorders.

In some individuals, panic attacks and the fear of their occurrence do not greatly impair overall functioning (Marks 1987). However, when anticipatory anxiety becomes sufficiently marked, extensive phobic avoidance and impairment of functioning may occur. At this stage a diagnosis of panic disorder with agoraphobia is made. Agoraphobia, as currently described in the *DSM-III-R*, is a fear of being in situations from which escape with dignity may be impeded, or help not available, in the event of an unexpected panic attack. Avoidance may be limited, with some restriction in occupational and social activities, or extensive, with significant impairment of overall functioning. In its extreme form the individual becomes housebound and is unable to leave home unless accompanied.

Without treatment, the course of panic disorder with or without agoraphobia is variable. The course may be phasic, with periods of complete remission of panic and agoraphobic symptoms after their first onset, or unremitting, with either a definite deterioration of symptoms or no change in symptoms. The onset of panic and agoraphobic symptoms can also lead to several complications, including hypochondriasis, depression, social phobia, generalized anxiety disorder, and drug and alcohol-related problems (Tyrer 1986).

Panic disorders represent a significant psychiatric illness in the general population. A comparative report of two large epidemiological studies, the National Institute of Mental Health Epidemiological Catchment Area Program (ECA) and the Munich Follow Up Study (MFS), shed light on the prevalence of panic disorder and panic disorder with agoraphobia in the community (Wittchen 1988). It revealed a lifetime prevalence of panic disorder and agoraphobia of approximately 2 per cent and 5 per cent respectively. The 6-month prevalence of this disorder was approximately 1 per cent and 3.5 per cent.

A growing body of evidence suggests a familial pattern in panic disorder and in panic disorder with agoraphobia. Crowe *et al.* (1983) reported a lifetime morbidity risk of panic disorder for first-degree relatives of probands of 25 per cent compared with 2 per cent for normal controls. Torgensen (1983) examined the prevalence of anxiety disorders in twins suffering from panic disorder or panic disorder with agoraphobia. Anxiety disorders were five times as frequent in monozygotic twins as in dizygotic twins.

Neurobiology of panic attacks

General research strategies

Several clinical strategies have been used to investigate the neurobiology of panic attacks, including the measurement of neurotransmitters and their metabolites, evaluation of the effect of antipanic medication on the CNS, and the provocation of panic attacks with pharmacological agents. Although the cause of panic attacks still remains unknown, it has been postulated that they could be attributed to neurotransmitter abnormalities in brain neurochemistry, metabolism, and receptor physiology. There is some evidence that panic attacks are associated with a CNS dysfunction in mid-brain structures that involve noradrenergic and serotoninergic neurotransmitter systems. Several comprehensive reviews of biological models of panic disorder exist in the literature (Redmond 1987; Gorman *et al*. 1989). Panic provocation has been extensively used as a tool to study the neurobiology of panic disorders. As this chapter is concerned with the role of CCK as a potential provocation agent in the study of panic disorder, we shall elaborate on the pharmacological provocation of panic attacks as a neurobiological research tool.

Pharmacological provocation studies

The use of provocation agents (i.e. lactate, CO_2, yohimbine, isoproterenol) represents a viable way of investigating the biochemical and physiological events that occur during a panic attack (Gorman *et al*. 1987). It is hoped that these agents will uncover the neurobiological systems involved in the aetiology of panic attacks. As with any psychiatric disorder or symptom, it is unlikely that only one neurochemical system is implicated directly in the occurrence of a panic attack. Rather, it is more likely that several neurochemical systems act concurrently and/or sequentially to produce an attack.

The development of a provocation agent which can affect neurochemical systems selectively, and at a specific locus in a chain of neurochemical events, might help to clarify the neurobiological basis of panic attacks and related disorders. At this point, some provocation agents such as yohimbine have been associated with neurochemical systems while others have not demonstrated a clear association with any system. Nevertheless, the increased use of a variety of panic-inducing agents in research has prompted researchers to establish specific criteria for an ideal provocation agent (Guttmacher *et al*. 1983; Gorman *et al*. 1987). These criteria can now be used to evaluate and develop potential provocation agents.

Criteria for an ideal provocation agent

Guttmacher *et al*. (1983) and Gorman *et al*. (1987) have proposed the following seven criteria for a pharmacological agent to be accepted as an ideal provocation agent. They are listed in an order pertinent to this chapter.

1. The induced panic attacks should encompass both physical symptoms of panic and subjective symptoms of anxiety, terror, fear, etc.

2. Patients should judge the induced attack to be symptomatically identical or very similar to natural panic attacks.

3. The agent, in the panicogenic dose, should be safe for routine administration to human subjects.

4. The induction of panic attacks should be specific to patients with a history of panic attacks. This can be expressed in one of two ways: the induced attack occurs exclusively in patients with a history of spontaneous panic attacks (absolute specificity), or the induced attack occurs in patients with a history of spontaneous panic attacks at a lower dose relative to other subjects (threshold specificity).

5. The effects of the provocation agent should be consistent in a given patient. If a desensitization effect occurs this should be predictable.

6. Drugs that block spontaneous panic attacks, such as tricyclic antidepressants, monoamine oxidase inhibitors, or benzodiazepines, should also block the pharmacologically induced attack.

7. Agents that do not block clinical panic attacks should not block the pharmacologically induced attack.

A panic-inducing agent which satisfies the above criteria, is endogenous to the CNS, and fulfils criteria for a neurotransmitter could be of theoretical and clinical importance in the study of panic disorders. We propose that CCK might be such an agent. The neurobiological properties of CCK and the evidence suggesting its possible role in the neurobiology of panic disorders will now be reviewed.

CCK and the neurobiology of panic attacks

The initial link between CCK, anxiety, and panic: micro-iontophoretic studies in the rat

Our interest in the role of CCK in anxiety and panic disorder stems from micro-iontophoretic studies of the effect of benzodiazepines on CCK-induced excitation of rat hippocampal pyramidal neurons (Bradwejn and de Montigny 1984, 1985).

Benzodiazepines are specific and efficacious psychotropics in the treatment of generalized anxiety disorder and panic disorders (Charney *et al.* 1986; Liebowitz *et al.* 1986). Moreover, several researchers have reported the presence of receptor sites for benzodiazepines in the CNS (Braestrup and Squires 1978; Mohler *et al.* 1978).

Benzodiazepines (flurazepam, chlordiazepoxide, lorazepam, or diazepam), administered intravenously or injected by micro-iontophoresis, antagonize CCK-8S-induced excitation of rat hippocampal pyramidal neurons (Bradwejn and de Montigny 1984). This antagonism is specific for CCK-8S-induced excitations and is not found with excitations induced by acetylcholine, metenkephalin, aspartate, or glutamate. Also, this antagonism seems to be selective for benzodiazepines as it is not observed following intravenous (i.v.) administration of phenobarbital, haloperidol, or meprobamate. This action of benzodiazepines is mediated by their specific receptors; the administration of flumazenil, the specific benzodiazepine antagonist, blocks or prevents this action completely. From dose–response curves of the (i.v.) effect of lorazepam or diazepam during excitations produced by CCK-8S, ED_{50} values of 32 μg/kg and 106 μg/kg were obtained for lorazepam and diazepam respectively. These values fall within the range of clinically used doses of these medications.

The above experiments provided the first demonstration of an antagonism of the central action of a peptide by benzodiazepines. Several authors have since reported an antagonism by benzodiazepines of the action of CCK in the periphery (Kubota *et al.* 1985). Taken together, these findings provided support for the premise that CCK might be involved in the neurobiology of anxiety disorders. This impression was further enhanced by Rehfeld (personal communication, 1985; see Chapter 10 of this volume), who noted that i.v. injection of CCK-4, conducted in the context of endocrinological studies, induced symptoms of anxiety, choking, and unreality. Rehfeld's descriptions reminded us of symptoms reported by patients with panic disorder. Thus, we decided to study the effect of CCK-4 in panic disorder.

Clinical evidence of a panicogenic action of cholecystokinin

A pilot study was carried out in 11 untreated patients with panic disorder (Bradwejn *et al.* 1990). Five women and six men ranging in age from 20 to 51 years were recruited from the Anxiety Disorders Clinic of St Mary's Hospital. In a double-blind trial, each patient received one injection of 2.5 ml of CCK-4 (50 μg in 0.9 M NaCl) and one injection of 2.5 ml placebo (0.9 M NaCl) on two separate days. All 11 patients panicked with CCK-4 while none of the patients panicked with placebo. Using the same methodology, similar results have been obtained with a larger sample of patients. Of 35 patients with either panic disorder or panic disorder with agoraphobia, 93 per cent panicked with CCK-4 while only 6 per cent panicked with placebo. These preliminary findings raised the possibility that CCK-4 might be implicated in the neurobiology of panic attacks and led us to evaluate its panicogenic properties.

Evaluation of CCK as a provocation agent

We have begun a systematic evaluation of CCK-4 as a panicogenic agent using the set of criteria discussed earlier. The data for each criterion studied so far are described below.

1. Can CCK-4 induce physical and emotional symptoms of panic attacks?

Essentially, CCK-4 can induce anxiety, apprehension, tension, and fear in addition to somatic and cognitive symptoms characteristic of panic attacks. We have studied the specific effects of CCK-4 by systematically evaluating the responses of subjects. Immediately after each injection subjects are asked to describe fully the various symptoms they experience subsequent to the injection. Once the overall effects of the injection diminishes, subjects are evaluated using a panic inventory derived from *DSM–III–R*. This procedure permits the itemization of symptoms induced by CCK-4.

Table 11.1 illustrates the symptom profiles of 35 patients with panic disorder and panic disorder with agoraphobia following i.v. injections of either 25 μg or 50 μg of CCK-4. The emotional symptoms appear under anxiety/fear/

Table 11.1. Symptoms induced with CCK-4 in patients with panic disorder and panic disorder with agoraphobia

Symptom	Severity Total percentage reporting	Mild	Moderate	Severe	Extremely severe
Dyspnea/ smothering sensation	88.5	0	25.7	37.1	25.7
Palpitation/ rapid heart	77.1	11.4	17.1	22.9	25.7
Sweating	80.0	5.7	25.7	25.7	22.9
Faintness	68.6	14.3	20.0	22.9	11.4
Unsteadiness	65.6	14.3	17.1	17.1	17.1
Dizziness	80.0	22.9	17.1	31.4	8.6
Shaking/trembling	71.5	8.6	20.0	22.9	20.0
Nausea	65.7	2.9	17.1	11.4	34.3
Andominal distress	57.1	20.0	14.3	11.4	11.4
Choking	45.7	5.7	14.3	8.6	17.1
Chest pain/ discomfort	57.2	5.7	28.6	8.6	14.3
Paraesthesia	74.3	14.3	20.0	17.1	22.9
Hot flushes/ cold chills	74.3	5.7	17.1	28.6	22.9
Depersonalization/ derealization	51.4	5.7	17.1	22.9	5.7
Anxiety/fear/ apprehension	97.1	2.9	17.1	37.1	40.0
Fear of dying	28.6	11.4	2.9	11.4	2.9
Fear of losing control	68.6	14.3	22.9	20.0	11.4
Fear of going insane	37.2	5.7	8.6	20.0	2.9

Values are based on percentage reporting in 35 patients.

apprehension. All patients who panicked with CCK-4 have reported both physical and emotional symptoms. The same pattern of response has also been observed in healthy volunteers who have panicked with CCK-4. In fact, we have noted that the symptom profile of healthy volunteers who panicked with CCK-4 is indistinguishable from that of patients who panicked with CCK-4 (Bradwejn *et al.* 1991*a*,*b*).

These studies provide evidence that CCK-4 fulfils the first criterion for a panicogenic agent.

2. Can CCK induce panic attacks symptomatically identical or very similar to unprovoked panic attacks?

The phenomenological nature of panic attacks is relatively stable over time except for changes in the duration and intensity of symptoms. Therefore it is possible to ask patients to compare induced panic attacks with unprovoked ones. The CCK-4 induced attacks have been judged to be very similar or identical with unprovoked attacks with regard to quality and type of symptoms. Response to CCK-4 is not stereotypic; rather, it varies from one patient to another, mimicking the attacks normally experienced by each patient. Most patients describe the CCK-4-induced attack as being more abrupt in onset. The intensity and duration of CCK-4-induced attacks ranges from being milder or shorter than unprovoked attacks to being more severe and longer. This finding appears to be correlated with the dose administered. Patients who experience mild or short attacks with CCK-4 have compared these attacks with the 'start of a typical attack'. These results suggest that CCK-4 fulfils the second criterion.

3. Is CCK-4 safe for research in humans?

We have attempted to determine whether CCK-4 can result in short-term or long-term adverse events. After each CCK-4 injection, patients were asked to report symptoms that have never appeared during unprovoked panic attacks. So far, symptoms such as a bitter taste or a tickling sensation in the throat have been reported. These new symptoms have been relatively mild, of short duration, and generally overshadowed by the experience of panic. De Montigny (1989) reported a vasovagal reaction of transient nature in one subject. We have observed two incidences of prolonged panic attacks (over 1 h) which required administration of a benzodiazepine. All patients who participate in the CCK-4 studies are followed up at our anxiety disorders clinic. Some patients have been followed for up to 3 years. We have not observed adverse events in any of these patients, and the peptide does not appear to alter the nature of the illness.

4. Are the effects of CCK-4 specific to patients suffering from panic attacks?

Rehfeld suggested that CCK-4 might be panicogenic in healthy volunteers in addition to a clinical population. It remained to be determined whether the effects of CCK-4 would be more severe or intense in patients, demonstrating a relative specificity in those experienced with panic attacks. De Montigny (1989), in an open uncontrolled study, administered CCK-4 to 10 healthy volunteers at doses ranging from 20 to 100 µg. 'Panic-like attacks' were reported to occur in seven subjects. Methodological differences between De Montigny's study and our initial report on panic disorder patients precluded any comparison of the differential effects of CCK-4 in panic disorder and healthy subjects.

Thus we conducted a study to determine the specificity of action of CCK-4 (Bradwejn et al. 1991a). The effects of different doses of CCK-4 in panic disorder patients and healthy volunteers were compared using a double-blind placebo (saline) controlled design with a randomized sequence of injection. DSM–III–R criteria, including moderate to severe anxiety, were used to define the occurrence of a panic attack in healthy volunteers. The panic rate with CCK-4 was significantly higher in patients than in healthy volunteers. For example, 25 µg CCK-4 induced a panic attack in 10 out of 11 (91 per cent) patients compared with two out of 12 (17 per cent) healthy volunteers. These results support a specificity of action of CCK-4 in panic disorder and indicate that the peptide fulfils the third criterion.

5. Are the effects of CCK-4 consistent?

This criterion essentially questions the reliability of action of CCK-4 in a test–retest situation. Preliminary results in nine panic disorder patients indicate that the panicogenic effects of CCK-4 can be replicated within a given patient (Bradwejn et al., unpublished data). We have also conducted a dose-ranging study in healthy volunteers (Bradwejn et al. 1991b). The effects of three i.v. doses of CCK-4 (9, 25, and 50 µg) were studied using the same double-blind procedure and criteria for a panic attack described earlier. The percentages of subjects who panicked with 9, 25, and 50 µg of CCK-4 were 11 (1/9), 17 (2/12), and 47 (7/15) respectively. The 50 µg dose produced a significantly greater number of symptoms, more intense symptoms, and a more rapid onset of symptoms than the 9 µg dose, and also produced a more rapid onset of symptoms than the 25 µg dose. These findings support a dose-related panicogenic effect in healthy volunteers.

In conclusion, the results described above suggest that CCK-4 might be a useful challenge paradigm with which to study panic attacks in a controlled situation. CCK-4 satisfies at least five of the seven proposed criteria for an ideal panicogenic agent: the sixth and seventh criteria are currently under investigation.

CCK-4 might be more potent than other provocation agents. Bradwejn and Koszycki (1991) found that the incidence of panic attacks tended to be higher with CCK-4 than with a single-breath inhalation of 35 per cent CO_2. Similarly, in a review of nine lactate infusion studies, Cowley and Arana (1990) concluded that the average response rate in panic patients was 67 per cent, which is lower than the rate observed with CCK-4. CCK-4 might also be more suitable for research in panic disorder. The peptide can be administered in a small volume of saline by rapid i.v. injection. This mode of administration might have an advantage over the slow infusion procedure necessary for sodium lactate administration. Lactate infusion has been associated with non-specific physiological factors such as volume overload or metabolic changes which may introduce non-specific psychological effects (Margraf *et al.* 1986). The technical advantage of CCK-4 administration, in addition to its presence in the mammalian CNS, renders it more suitable for research in panic disorder.

Additional evidence for the role of CCK in panic disorder

Recently, Abelson and Nesse (1990) have described the effects of pentagastrin, a peptide chain which incorporates the indentical four amino acid sequence of CCK-4, in five panic disorder patients and four matched controls. The incidence of panic attacks ranges from 60 per cent to 100 per cent in patients compared with 0 to 25 per cent in controls, depending on the criteria used to judge the occurrence of an attack.

In animal studies, injection of CCK in the amygdala has been shown to produce an anxiogenic effect in rats using a defensive burying paradigm; this effect was antagonized by benzodiazepines (Conska *et al.* 1988). Harro and associates have reported that CCK receptor agonists (caerulein, pentagastrin, and CCK-4) are anxiogenic in rodents (Harro *et al.* 1990*b*; Chapter 13 of this volume). Rats with high (non-anxious) and low (anxious) exploratory activity in an elevated plus maze paradigm have been reported to differ in their brain CCK and benzodiazepine receptor characteristics (Harro *et al.* 1990*a*). Non-anxious rats had a significantly lower density of CCK receptors and a higher density of benzodiazepine receptors in the frontal cortex (Harro *et al.* 1990*a*; Chapter 13 of this volume). Administration of CCK-4 to green vervet monkeys was found to elicit behavioural responses characteristic of anxiety and fear (Palmour and Bradwejn, unpublished results). Studies of the anxiolytic activity of CCK antagonists in several animal models are summarized in Chapters 5, 12, 13, and 14 of this volume.

Future research

A question of central importance concerns the site(s) of panicogenic action of CCK-4. The lack of anxiogenic effects of CCK-8S, a larger molecule which penetrates the CNS only minimally at tolerable doses, suggests that CCK-4 exerts its panicogenic action via a central mechanism (De Montigny 1989). It will be important to elucidate in future research whether CCK-4 can penetrate the central nervous system and/or whether treatment with either a selective peripheral or a central CCK antagonist differentially effects response to CCK-4. Cerebral imaging studies in humans might help in answering this question. The role of neurotransmitters in anxiety in general and in CCK-4-induced panic attacks in particular will need to be established. Finally, it will be important to determine the therapeutic efficacy of CCK antagonists in panic disorder.

Acknowledgements

The authors thank Gabriella Boiardi for providing secretarial services. The studies were funded by the Medical Research Council of Canada (MA-10502), St. Mary's Hospital Foundation, and St. Mary's Psychopharmacology Fund.

References

Abelson, J. L. and Nesse, R. M. (1990). Cholecystokinin-4 and panic. *Arch. Gen. Psychiat.* **47**, 395.
American Psychiatric Association (1987) Diagnostic and Statistical Manual of Mental Disorders (3rd edn, revised). American Psychiatric Association, Washington, DC.
Bradwejn J. and De Montigny, C. (1984). Benzodiazepines antagonize cholecystokinin-induced activation of rat hippocampal neurons. *Nature* **312**, 363–4.
Bradwejn J., and De Montigny, C. (1985). Effects of PK 8165, a partial benzodiazepine receptor agonist, on cholecystokinin-induced activation of hippocampal pyramidal neurons: a microiontophoretic study in the rat. *Eur. J. Pharmacol.* **112**, 415–18.
Bradwejn, J., Koszycki, D., and Meterissian, G. (1990). Cholecystokinin-tetrapeptide induces panic attacks in patients with panic disorder. *Can J. Psychiat.* **35**, 83–5.
Bradwejn, J. and Koszycki, D. (1991). Comparison of the panicogenic effect of cholecystokinin and carbon dioxide in panic disorder. *Prog. Neuropsychopharmacol. Biol. Psychiat.*, **15**, 237–9.
Bradwejn, J., Koszycki, D., and Shriqui, C. (1991*a*). Enhanced sensitivity to cholecystokinin tetrapeptide in panic disorder: clinical and behavioral findings. *Arch. Gen. Psychiat.* **48**, 603–10.
Bradwejn, J., Koszycki, D., and Bourin, M. (1991*b*). Dose ranging study of CCK-4 in healthy volunteers. *J. Psychiat. Neurosci.* **16**, 91–5.
Braestrup, C. and Squires, R. F. (1978). Pharmacological characterization of benzodiazepine receptors in the brain. *Eur. J. Pharmacol.* **48**, 263–70.
Charney, D. S., Woods, S. W., Goodman, W. K., Rifkin, B., Kinch, M., Aiken, B., Quadrino, L., and Heninger, G. (1986). Drug treatment of panic disorder: the comparative efficacy of imipramine, alprazolam, and trazodone. *J. Clin. Psychiat.* **47**, 580–6.
Cowley, D. S. and Arana, I. W. (1990). The diagnostic utility of lactate sensitivity in panic disorder. *Arch. Gen. Psychiat.* **47**, 227–84.
Craske, M. G., Rachman, S. J., and Tallman, K. (1986). Mobility, cognitions and panic. *J Psychopathol. Behav. Assess.* **8**, 199–210.
Crowe, R. R., Noyes, R., Pauls, D. I., and Slyman, D. J. (1983). A family study of panic disorder. *Arch. Gen. Psychiat.* **40**, 1065–9.
Conska, E., Fekete, M., Nagy, G., Szanto-Fekete, M., Teledgy, G., Penke, B., and Kovaks, K. (1988). Anxiogenic effects of cholecystokinin in rats. In Peptides (ed. B. Penke and A. Torok), pp. 249–52. Walter DeGruyter, New York.
De Montigny, C. (1989) Cholecystokinin-tetrapeptide induces panic-like attacks in healthy volunteers. *Arch. Gen. Psychiat.* **46**, 511–17.
Gorman, J. M., Fyer, M. R., Liebowitz, M. R., and Klein, D. F. (1987). Pharmacologic

provocation of panic attacks. In *Psychopharmacology: The third generation of progress* (ed. H. Y. Meltzer), pp. 985–93. Raven Press, New York.

Gorman, J. M., Liebowitz, M. R., Fyer, A. J., and Stein, J. (1989). Neuro-anatomical hypothesis for panic disorder. *Am J Psychiat.* **146**, 148–61.

Guttmacher, L. B., Murphy, D. L., and Insel, T. R. (1983). Pharmacologic models of anxiety. *Compr. Psychiat.* **24**, 312–26.

Harro, J., Kiivit, R. A., Lang, A., and Vasar, E. (1990a). Rats with anxious or non-anxious type of exploratory behaviour differ in the brain CCK-8 and benzodiazepine receptor characteristics. *Behav. Brain. Res.* **39**, 63–71.

Harro, J., Pold, M., and Vassar, E. (1990b). Anxiogenic-like action of caerulein, a CCK-8 receptor agonist, in the mouse: influence of acute and subchronic diazepam treatment. *Naunyn-Schmiedebergs Arch. Pharmacol.* **341**, 62–7.

Kubota, K., Sugaya, K., Kunagane, N., and Matsuda, I. (1985). Cholecystokinin antagonism by benzodiazepines in the contractile response of the isolated guinea pig gallbladder. *Eur. J. Pharmacol.* **110**, 225–31.

Liebowitz, M. R., Fyer, A. J., Gorman, J. M., Campeas, R., Levin, A., Davies, S. R., Goetz, D., and Klein, D. F. (1986). Alprazolam in the treatment of panic disorder. *J. Clin. Psychopharmacol.* **6**, 13–20.

Margraf, J., Ehlers, A., and Roth, W. T. (1986). Sodium lactate infusions and panic attacks: a review and critique. *Psychosom. Med.* **48** 23–51.

Marks, I. M. (1987). Behavioural aspects of panic disorder. *Am. J. Psychiat.* **144**, 1160–5.

Mohler, H., Okada, T., Heiz., P., and Ulrich, J. (1978). Biochemical identification of the site of action of benzodiazepines in human brain by ^{3}H-diazepam binding. *Life Sci.* **22**, 985–96.

Redmond, D. E., Jr (1987). Studies of the nucleus locus coeruleus in monkeys and hypothesis for neuropsychopharmacology. In *Psychopharmacology: The third generation of progress* (ed. H. Y. Meltzer), pp. 967–75. Raven Press, New York.

Torgensen, S. (1983). Genetic factors in anxiety disorders. *Arch. Gen. Psychiat.* **40**, 396–400.

Tyrer, P. (1986). Classification of anxiety disorders: a critique of DSM-III. *J. Affect. Disorder* **11**, 99–104.

Wittchen, H. U. (1988). Natural course and spontaneous remissions of untreated anxiety disorders: Results of Munich follow-up study (MFS). In *Panic and phobias 2* (ed. I. Hand and H. U. Wittchen), pp. 3–17. Springer-Verlag, Berlin.

12. Ethological analysis of the role of CCK in anxiety

C. A. Hendrie and J. C. Neill

The involvement of neurotransmitters in the expression of defence behaviour

Behavioural systems develop in response to evolutionary forces, and for many species one of the most important selection pressures has been predation. Species that are preyed upon have therefore developed various strategies to defend against predation which include, as the distance between a predator and a prey species closes, (1) avoiding situations where predation is likely, (2) freezing upon detection of a predator, (3) attempted escape, (4) active defence, (5) highly intense defensive threat and attack (e.g. the 'jump attack' in rats), and (6) tonic immobility ('death feigning') (Gallup 1974; Ratner 1977).

Whilst these behaviours represent essential prerequisites for successful defence, they clearly do not operate in isolation and influence the expression of other behaviours. For example, both attack (Miczek *et al.* 1982; Rodgers and Hendrie 1983) and threat of attack (Rodgers and Randall 1985) have been found to activate endogenous analgesia mechanisms, but in different ways. In the case of attack, the analgesia is opioid mediated, as indicated by its duration of 45–60 min, the ability of naloxone to block and reverse the response, and its bidirectional cross-tolerance with morphine (Miczek and Thompson 1984; Rodgers and Randall 1986). In contrast, the threat of attack induces short duration (< 10 min) non-naloxone-reversible analgesia which is thought not to be mediated by opioid mechanisms (Rodgers and Randall 1987).

Interestingly, both forms of analgesia have been shown to be under psychological, rather than physiological, control. That is, opioid analgesia mechanisms are activated in circumstances where the animal has perceived that it has lost control over the source of the aversive stimulation (Maier *et al.* 1982), whilst stimuli signalling danger produce non-opioid analgesia. Evidence also seems to suggest that non-opioid analgesia is activated following low intensity threat (Rodgers and Randall 1987). In contrast, opioid mechanisms appear to be activated during or in anticipation of actual attack, functioning to inhibit wound-orientated behaviour, to promote defensive responses concerned with minimizing physical injury, and thereby to increase the possibility of escape (Liebeskind and Paul 1977; Ratner 1977).

Within such a highly developed system it is also evident that certain behaviours will be inhibited as a consequence of near contact with a predator. Thus it follows that consummatory behaviour (and associated behaviours such as territorial defence, foraging, courtship display, etc.) should be suppressed during predator–prey interactions to allow the animal to engage in active defence. Rebound increases in these responses may follow an encounter, which, in the case of feeding, may represent an aspect of recuperative behaviour and may also be a relevant analysis when applied to 'stress-induced' feeding (Teskey *et al.* 1984). Within this context, it is interesting to note that glucoprivation and food deprivation *per se* both activate endogenous analgesia mechanisms (Bodnar *et al.* 1978, 1979), perhaps suggesting a closed response loop since hungry animals will forage more and hence expose themselves to a greater extent to the possibility of predation.

Therefore, in terms of antipredation strategies, a highly adaptive cluster of responses would be increased vigilance/avoidance, promotion of defensive behaviour, and suppression of behaviours competing with defence. This cluster may be (and probably is) achieved in various ways. However, one possibility is that increased vigilance/avoidance translates into anxiety/fear. Promotion of defensive behaviours may be expressed as analgesia, since this inhibits recuperative behaviour (Bolles and Fanselow 1980), and further suppression of competing behaviour induced by active inhibition of consummatory and associated responses.

Complex behavioural interactions of this kind are likely to involve interplay between the chemical messenger systems of the brain. It is already known that a number of anatomical and chemical systems are involved in the interplay and expression of the anxiety–analgesia–suppression of consummatory behaviour cluster. First, there is clear evidence to suggest an involvement of opioid systems since agonists and antagonists can influence feeding, drinking (e.g. Sanger and McCarthy 1980; Tseng and Cheng 1980), and sexual behaviour (Meyerson and Terenius 1977). Opiates are, of course, potent analgesics and in the clinic are known to modify the affective response to pain.

Serotonergic mechanisms have also been investigated in the same context. Serotonin is involved in the control of ingestive behaviour (Samanin 1983; Blundell 1986), and drugs which increase serotonergic neurotransmission, such as fenfluramine, reduce feeding in animals and man. Further, an increase in serotonergic activity is related to a decrease in sexual activity (Meyerson *et al.* 1974; Mendelson and Gorzalka 1989). Decreases in 5-hydroxytryptamine (5-HT) neurotransmission also produce anxiolytic-like effects, suggesting that 5-HT *per se* may be anxiogenic (e.g. Iversen 1984). To complete the evidence that 5-HT may be involved in the expression of the behavioural interplay between pain, anxiety, and consummatory behaviour, it has been reported that serotonergic mechanisms are involved in the expression of analgesia (e.g. Fasmer *et al.* 1986). However, as with opioids, evidence is equivocal as to the exact nature of this effect.

Finally, CCK also has many properties and effects that suggest a strategic position in the expression of the anxiety–analgesia–suppression of consummatory behaviour cluster. Of course, it has been known for many years that CCK is implicated in the control of feeding behaviour (Gibbs *et al.* 1973; Chapter 17 of this volume). Further, CCK has been reported to be involved in the expression of opiate analgesia. For example, in rats, CCK-8S has been shown to induce analgesia *per se* at high doses (Zetler 1980) and to inhibit opiate analgesia at doses more within the physiological range (Faris *et al.* 1983). In this context, it is important to note that CCK appears not to influence the onset of opiate analgesia but to reduce its duration (Faris 1985), and, whilst evidence suggests that CCK antagonists enhance morphine analgesia (Baber *et al.* 1989; Chapter 44 of this volume), CCK_A antagonists block social conflict analgesia (Hendrie *et al.* 1989). This apparent contradiction may be explained by the hypothesis that social conflict analgesia is primarily under psychological rather than physiological control (see above). Thus it is possible that the blockade of social conflict analgesia by CCK antagonists may be due to an anxiolytic action. Indeed, a role for CCK in anxiety has recently been proposed. Early suggestions of a role for the peptide came from studies demonstrating that acute stress influences CCK concentrations in discrete areas within the limbic system and hypothalamus (Seigel *et al.* 1985, 1987). Further, the ability of benzodiazepines (BZPs) to attenuate CCK-8-induced activation of rat hippocampal pyramidal neurons led to the suggestion that the anxiolytic actions of BZPs may be due to their infuence on the effects of this peptide (Bradwejn and De Montigny 1984, 1985). These findings indicated that CCK may act as an endogenous anxiogen, a view which is supported by observations that CCK-4 and pentagastrin produce panic-like behaviour in patients who normally suffer such attacks and in healthy volunteers (De Montigny 1989; Abelson and Nesse 1990; Bradwejn *et al.* 1990; Chapter 11 of this volume).

Behavioural studies of the anxiolytic properties of CCK antagonists

Since CCK is anxiogenic in humans, CCK antagonists would be predicted to attenuate these responses and consequently to have anxiolytic properties. The following series of studies were conducted, initially using the black–white exploration model (Costall *et al.* 1987) to evaluate this hypothesis. In this apparatus, developed from the original Crawley design (Crawley 1981), an animal's normal preference for the dark, dimly illuminated portion of the test apparatus is altered by anxiolytics such that behaviour is distributed approximately 50:50 between the dark section and the brighty illuminated white section. Measures taken in this model are locomotion (line crossing), rearing, and time spent in the two sections. Animals used in the present studies were male DBA/2 mice.

Antagonists were chosen on the basis of their affinity for the two currently identified CCK receptor subtypes. Devazepide has nanomolar affinity for CCK_A receptors and micromolar affinity for CCK_B receptors. L-365,031 has proportionally less affinity for both CCK_A and CCK_B receptors, whilst L-365,260 is a specific CCK_B receptor antagonist (Chang and Lotti 1986; Evans *et al*. 1986; Bock *et al*. 1989; Lotti and Chang 1989). The effects of devazepide, L-365,031, and L-365,260 were evaluated in the black–white box.

In all the studies presented, results were initially analysed using ANOVA with Dunnett's *t* as follow-up tests. Devazepide was tested across a wide dose range (0.05–5000 µg/kg) and the results are presented in Fig. 12.1. Significant anxiolytic activity was found at 0.05, 0.5, and 5.0 µg/kg on measures of line crossing and rearing. The greatest effect was obtained with 5.0 µg/kg where significant effects were seen on all measures. A significant anxiolytic effect re-emerged at 500.0 µg/kg, although no effect was seen at 50.0 µg/kg. There is no explanation for the shape of the dose-response curve at present, although the results were fully replicated in a second study (Fig. 12.1).

L-365,031 over a wide dose range (0.005–5000 µg/kg) gave no consistent anxiolytic activity in this test.

L-365,260 also failed to produce a consistent anxiolytic effect under the

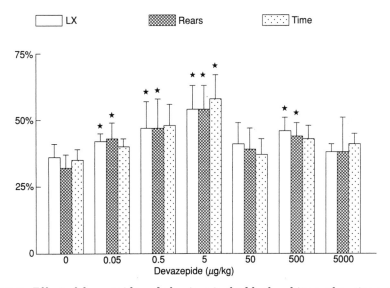

Fig. 12.1. Effect of devazepide on behaviour in the black–white exploration model. Data are presented as mean (± SEM) percentage of total behaviour spent in the white section of the apparatus (i.e. white/total). An anxiolytic-like effect is indicated when activity in the white approaches 50 per cent of the total. The asterisks indicate where there were no differences in activity in each section of the apparatus. Results suggest that devazepide has a inverted U-shaped dose-response curve centred around 5 µg/kg with a re-emergence of anxiolytic-like activity at 500 µg/kg.

testing conditions used. In further experiments combinations of L-365,031 and L-365,260 were studied but no significant anxiolytic effects were found, suggesting that the action of devazepide in the black–white box cannot be reproduced by simultaneous blockade of CCK_A and CCK_B receptors.

Interestingly, these findings conflict with data produced in other anxiolytic models. Significant anxiolytic effects of L-365,260 have been reported in a conditioned suppression of drinking model (Dourish *et al.* 1990), a conditioned emotional response task and the elevated plus maze (Ravard *et al.* 1990; Chapter 5 of this volume). Further, in view of demonstrations showing that selective CCK_B agonists induce panic in volunteers and patients (De Montigny 1989; Bradwejn *et al.* 1990) and that L-365,260 appears to be equipotent with or more potent than devazepide in these animal models, it will be important to continue to explore the effect of CCK antagonists on a range of anxiolytic tests in a number of different species. There are marked species differences in the density and anatomical distribution of the CCK receptor subtypes. In addition, we do not fully understand the behavioural bases of the anxiolytic tests. For example, studies showing anxiolytic properties of L-365,260 (Dourish *et al.* 1990; Ravard *et al.* 1990) have used procedures where the aversive properties of the test apparatus are clearly evident (i.e. open elevated arm of maze or tone signalling shock), whereas this would not appear to be the case with the black–white exploration model.

Behavioural effects of CCK agonists in the black–white box

In order to evaluate further the roles of CCK_A and CCK_B receptor mechanisms in anxiety, the behavioural actions of CCK-8S (which has good affinity for both receptor subtypes) and of CCK-8US and pentagastrin (which have a high affinity for CCK_B receptors only) were examined in the black–white box. Since CCK_B antagonists are anxiolytic in some behavioural tests, CCK agonists would be predicted to have anxiogenic activity. This hypothesis was investigated in the black–white box. Both pentagastrin and CCK-8US produced effects on line crossing and time spent in each section consistent with anxiolysis, with a minimum effective dose of 50 μg/kg (Fig. 12.2). It is important to note that these effects were seen in the absence of any effect on overall activity. In contrast, CCK-8S produced a profile that appeared anxiogenic-like. However, this conclusion was compromised by an overall reduction in activity suggesting sedation rather than inhibition of specific behaviours as an explanation. Thus CCK agonists produce a profile in the black–white exploration model which, whilst not unequivocally supporting CCK_A receptor mediation of anxiolysis, severely question the exclusive involvement of CCK_B mechanisms. Furthermore, it is difficult to reconcile how both an agonist and an antagonist at the CCK_A receptor can produce the same effect without suggesting the existence of a CCK-inverse agonist. Clearly, further work on the behavioural effects of

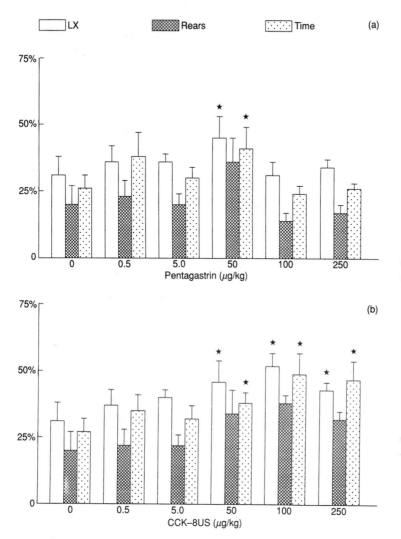

Fig. 12.2. Effect of CCK$_B$ agonists on behaviour in the black–white exploration model. (a) Pentagastrin produces a modest but significant anxiolytic profile at 50 μg/kg. (b) Summary of the effects of CCK-8US, where significant anxiolysis was found at doses of 50 μg/kg and above (it is important to note that this profile was produced in the absence of measurable sedation within the same test).

a range of CCK agonists with varying efficacy at CCK$_A$ and CCK$_B$ receptors would be worthwhile.

Finally, the possibility remains that models showing anxiolytic effects of L-365,260 (conditioned suppression of drinking, elevated plus maze) and the black–white exploration model of anxiety are measuring related, but qualitatively different, aspects of 'anxiety'. Thus, as CCK fragments have been shown to produce panic in humans, it is conceivable that a model specifically developed to detect anxiolytics may simply be inappropriate for the identification of antipanic agents. This hypothesis is difficult to evaluate, however, as there is no specific animal model of panic as yet (but see below).

Effects of CCK antagonists and benzodiazepines in a novel ethological model for measuring behavioural responses to predation

Recent evidence suggests that the calls of night-hunting aerial predators activate endogenous (defensive) analgesia mechanisms in mice (Hendrie 1991). Behaviourally, the response to the call is intense. Mice show high levels of defensive and escape-orientated behaviours, as well as a degree of stretched attend postures directed towards the source of the call, a small loudspeaker placed high on the wall of an observation arena. These effects have been shown to be specific to the tawny and barn owls, and probably have a genetic basis since defensive responses are observed in laboratory mice.

The hunting pattern of the tawny owl (*Strix aluco*) is such that, after making territorial calls, it flies to a 'hunting perch' where it remains silent until such time as it spots a potential prey animal. Thus, for mice, the time of highest probability and hence risk of predation is shortly after it has heard an owl territorial call.

On the basis of these observations, a test apparatus was constructed in an attempt to differentiate and quantify the behavioural strategies seen in response to the predator call. The apparatus provided a dimly lit burrow into which the mouse could retreat and an open arena above. It was hypothesized that different intensities of anxiety and defensive behaviour induced by the predator call would generate different patterns and distribution of behaviour in the open arena and the lower burrow. Mice were individually exposed in the arena to 2 min of tawny owl territorial call and given the opportunity to enter the burrow during the next 5 min. Pilot studies revealed that, under these circumstances, mice show very little behaviour in the burrow and, instead, concentrate their behaviour on the surface area, showing high levels of defensive and escape-oriented behaviours. The interpretation of these responses is that mice are showing 'panic'-like behaviour to facilitate escape from an area of potential danger.

Work regarding this model is very much in its early stages; however,

---

Fig. 12.3. Effect of alprazolam and CCK antagonists on behaviour in the 'call box'. (a) Alprazolam significantly increases time spent in a strategically defensible burrow location following exposure to the call of the tawny owl, indicating that 'panic'-like escape-oriented behaviour has been replaced by a more organized, less protean defensive strategy. (b) The effects of devazepide, which had no influence

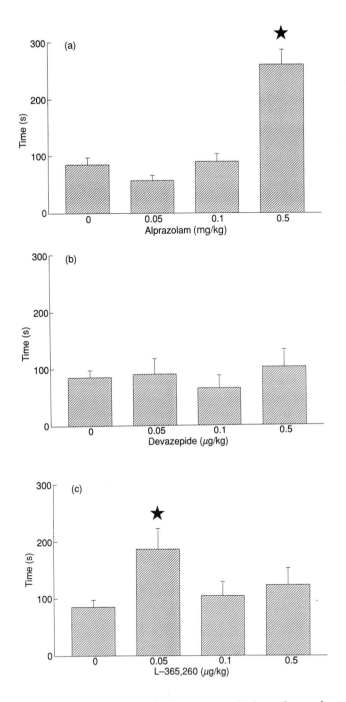

in this test. (c) The contrasting effects of L-365,260 which produces changes in the same direction as alprazolam. These data suggest that L-365,260 may have anti-panic properties. See text for further details. The star indicates $p < 0.05$ from control.

preliminary results have now been obtained with BZPs and L-365,260. The BZP alprazolam, a clinically prescribed antipanic agent, diverts behaviour from surface escape-oriented responses to increased time spent in the strategically defensible burrow location (Fig. 12.3). In terms of the distance-dependent defence hierarchy discussed in the introduction to this chapter, this would represent an expression of a less intense form of defence (i.e. that 'panic' has been reduced and replaced by a fully organized, less protean defensive reaction). Similarly, L-365,260 at 0.05 μg/kg significantly increases time spent in the burrow (Fig. 12.3). In contrast, devazepide has no effect on the time spent in the burrow (Fig. 12.3).

More studies are required to establish the 'Callbox' as an animal model of panic. However, studies to date indicate that L-365,260 and devazepide have different effects in this model. These findings reinforce the view that the anxiolytic profile of novel compounds needs to be investigated in a wide range of animal models which measure different aspects of anxiolytic behaviour.

In conclusion, a number of neurotransmitter systems may be involved in the mediation of the anxiety–analgesia-suppression of consummatory behaviour cluster. Further studies are required to establish the nature and function of these neurotransmitter interactions. However, at least one possibility is that they are all involved, to some extent, with their relative contributions to behaviour changing as a consequence of the perception of the intensity of threat. A further possibility is that different receptor subtypes for a given transmitter may be recruited as the perception of the intensity of threat increases or decreases. The results with CCK antagonists in the tests used in these studies suggest that CCK_A mechanisms are involved in mediating anxiety (low intensity) whilst CCK_B receptors appear to mediate the response to high intensity threat-inducing panic. The interactions between chemical messenger systems in the mediation of the cluster of behaviours seen in response to stress and anxiety will be complex, but are likely to be more meaningful than the one receptor–one behaviour philosophical standpoint that has been favoured in the past.

References

Abelson, J. L. and Nesse, R. M. (1990). Cholecystokinin-4 and panic. *Arch. Gen. Psychiat.* **47**, 395.

Baber, N. S., Dourish, C. T., and Hill, D. R. (1989). The role of CCK, caerulein and CCK antagonists in nociception. *Pain* **39**, 307–28.

Blundell, J. E. (1986). Serotonin manipulations and the structure of feeding behaviour. *Appetite* **7**, 39–56.

Bock, M. G., DiPardo, R. M., Evans, B. E., Rittle, K. E., Whitter, W. L., Veber, D. F., Anderson, P. S., and Freidinger, R. M. (1989). Benzodiazepine, gastrin and brain cholecystokinin receptor ligands: L-365,260. *J. Med. Chem.* **32**, 13–16.

Bodnar, R. J., Kelly, D. D., Spiaggia, A., and Mansour, A. (1978). Biphasic alterations of nociceptive thresholds induced by food deprivation. *Physiol. Psychol.* **6**, 391–5.

Bodnar, R. J., Kelly, D. D., Mansour, A., and Glusman, M. (1979). Differential effects of hypophysectomy upon analgesia induced by two glucoprivic stressors and morphine. *Pharmacol. Biochem. Behav.* 11, 303–8.

Bolles, R. C. and Fanselow, M. S. (1980). A perceptual–defensive–recuperative model of fear and pain. *Psychol. Rev.* 77, 32–48.

Bradwejn, J. and De Montigny, C. (1984). Benzodiazepines antagonise cholecystokinin-induced activation of rat hippocampal neurones. *Nature* 312, 363–4.

Bradwejn, J. and De Montigny, C. (1985). Effects of PK 8165, a partial benzodiazepine receptor agonist, on cholecystokinin-induced activation of hippocampal pyramidal neurons: a microionophoretic study in the rat. *Eur. J. Pharmacol* 112, 415–18.

Bradwejn, J., Koszycki, D., and Meterissian, G. (1990). Cholecystokinin-tetrapeptide induces panic attacks in patients with panic disorder. *Can. J. Psychiat.* 35, 83–5.

Chang, R. S. L. and Lotti, V. J. (1986). Biochemical and pharmacological characterisation of an extremely potent and selective non-peptide cholecystokinin antagonist. *Proc. Natl Acad. Sci. USA* 83, 4923–6.

Costall, B., Hendrie, C. A., Kelly, M. E., and Naylor, R. J. (1987). Actions of sulpiride and tiapride in a simple model of anxiety in mice. *Neuropharmacology* 26, 511–17.

Crawley, J. N. (1981). Neuropharmacologic specificity of a simple model for the behavioural actions of benzodiazepines. *Pharmacol. Biochem. Behav.* 15, 695–9.

De Montigny, C. (1989). Cholecystokinin tetrapeptide induces panic like attacks in healthy volunteers. *Arch. Gen. Psychiat.* 46, 511–17.

Dourish, C. T., Rycroft, W., Dawson, G. R., Tattersall, F. D., and Iversen, S. D. (1990). Anxiolytic effects of the CCK antagonists devazepide and L-365,260 in a conditioned suppression of drinking model. *Eur. J. Neurosci.* 2, (Suppl. 3), 38.

Evans, B. E., Bock, M. G., Rittle, K. E., DiPardo, R. M., Whitter, W. L., Veber, D. F., Anderson, P. S., and Freidinger, R. M. (1986). Design of potent, orally effective, non-peptidal antagonists of the peptide hormone cholecystokinin. *Proc. Natl Acad. Sci. USA* 83, 4918–22.

Faris, P. L. (1985). Opiate antagonistic function of cholecystokinin in analgesia and energy balance systems. *Ann. NY Acad. Sci.* 448, 437–47.

Faris, P. L., Komisaruk, B. R., Watkins, L. R., and Mayer, D. (1983). Evidence for the neuropeptide cholecystokinin as an antagonist of opiate analgesia. *Science* 219, 310–12.

Fasmer, O. B., Berge, O. G., Post, C., and Hole, K. (1986). Effects of the putative 5-HT$_{1A}$ receptor agonist 8-OH-2-(di-n-propylamino) tetralin on nociceptive sensitivity in mice. *Pharmacol. Biochem. Behav.* 25, 883–8.

Gallup, G. G. (1974). Animal hypnosis: factual status of a fictional concept. *Psychol. Bull.* 81, 836–53.

Gibbs, J., Young, R. C., and Smith, G. P. (1973). Cholecystokinin decreases food intake in rats. *J. Comp. Physiol. Psychol.* 84, 488–95.

Hendrie, C. A. (1991). The calls of murine predators activate endogenous analgesia mechanisms in laboratory mice. *Physiol. Behav.*, 49, 569–73.

Hendrie, C. A., Shepherd, J. K., and Rodgers, R. J. (1989). Differential effects of the CCK antagonist, MK-329, on analgesia induced by morphine, social conflict (opioid) and defeat experience (non-opioid) in male mice. *Neuropharmacology* 28, 1025–32.

Iversen, S. D. (1984). 5-HT and anxiety. *Neuropharmacology* 23, 1553–60.

Leibeskind, J. C. and Paul, L. A. (1977). Psychological and physiological mechanisms of pain. *Ann. Rev. Psychol.* 28, 41–60.

Lotti, V. J. and Chang, R. S. L. (1989). A new potent and selective non-peptide gastrin

antagonist and brain cholecystokinin (CCK-B) ligand: L-365,260. *Eur. J. Pharmacol.* **162**, 273.

Maier, S. F., Drugan, R. C. and Grau, J. W. (1982). Controllability, coping behaviour and stress-induced analgesia. *Pain* **12**, 47–56.

Mendelson, S. D. and Gorzalka, B. B. (1989). Differential roles of 5-HT receptor subtypes in female sexual behaviour. In *Behavioural pharmacology of 5-HT* (ed. P. Bevan, A. R. Cools, and T. Archer), pp. 55–72. Lawrence Erlbaum, Hillsdale, NJ.

Meyerson, B. J. and Terenius, L. (1977). β-endorphin and male sexual behaviour. *Eur. J. Pharmacol.* **42**, 191–2.

Meyerson, B. J., Carrer, H., and Eliasson, M. (1974). 5-Hydroxytryptamine and sexual behaviour in the female rat. In *Advances in biochemical psychopharmacology* (ed. E. Costa, G. L. Gessa, and M. Sandler), Vol. 11, pp. 229–42. Raven Press, New York.

Miczek, K. A., Thompson, M. L., and Shuster, L. (1982). Opioid-like analgesia in defeated mice. *Science* **215**, 1520–2.

Miczek, K. A. and Thompson, M. L. (1984). Analgesia resulting from defeat in a social confrontation: the role of endogenous opioids in brain. In *Modulation of sensorimotor activity during altered behavioural states* (ed. R. J. Bandler), pp. 431–56. A. R. Liss, New York.

Ratner, S. C. (1977). Immobility in invertebrates: what can we learn? *Psychol. Rec.* **1**, 1–14.

Ravard, S., Dourish, C. T., and Iversen, S. D. (1990). Evidence that the anxiolytic-like effects of the CCK antagonists devazepide and L-365,260 in the elevated-plus maze paradigm in rats are mediated by CCK receptors. *Br. J. Pharmacol.* **101**, 576P.

Rodgers, R. J. and Hendrie, C. A. (1983). Social conflict activates status dependent endogenous analgesic or hyperalgesic mechanisms in male mice. *Physiol. Behav.* **30**, 775–80.

Rodgers, R. J. and Randall, J. I. (1985). Social conflict analgesia: studies on naloxone antagonism and morphine cross-tolerance in male DBA/2 mice. *Pharmacol. Biochem. Behav.* **23**, 883–7.

Rodgers, R. J. and Randall, J. I. (1986). Acute non-opioid analgesia in defeated mice. *Physiol. Behav.* **36**, 947–50.

Rodgers, R. J. and Randall, J. I. (1987). On the mechanisms and adaptive significance of intrinsic analgesia systems. *Rev. Neurosci.* **1**, 185–200.

Samanin, R. (1983). Drugs affecting serotonin and feeding. In *Biochemical pharmacology of obesity* (ed. P. B. Curtis-Prior), pp. 339–56. Elsevier, Amsterdam.

Sanger, D. J. and McCarthy, P. S. (1980). Increased food and water intake produced in rats by opiate receptor agonists. *Psychopharmacology* **74**, 217–20.

Seigel, R. A., Duker, E. M., Fucher, E., Panke, U., and Wuttke, W. (1985). Responsiveness of mesolimbic, mesocortical, septal and hippocampal cholecystokinin and Substance P neuronal systems to stress in the male rat. *Neurochem. Int.* **6**, 783–9.

Seigel, R. A., Duker, E. M., Pahnke, U., and Wuttke, W. (1987). Stress induced changes in cholecystokinin and substance P concentrations in discrete regions of the rat hypothalamus. *Neuroendocrinology* **46**, 75–81.

Teskey, G. C., Kavaliers, M., and Hirst, M. (1984). Social conflict activates opioid analgesia and ingestive behaviours in male mice. *Life Sci.* **35**, 303–15.

Tseng, L. F. and Cheng, D. S. (1980). Acute and chronic administration of β-endorphin and naloxone on food and water intake in rats. *Fed. Proc.* **39**, 606.

Zetler, G. (1980). Analgesia and ptosis caused by caerulein and cholecystokinin octapeptide (CCK-8). *Neuropharmacology* **19**, 415–22.

13. CCK receptors and anxiety in rats

Eero Vasar, Jaanus Harro, Anu Pôld, and Aavo Lang

Agonists at CCK receptors have been shown to induce or potentiate fear-related behaviour in rats. Thus very low doses of caerulein and pentagastrin have, an anxiogenic-like effect on rats in an elevated plus maze (Harro *et al.* 1989). The CCK antagonist proglumide completely reverses the anxiogenic-like action of caerulein and pentagastrin in rodents (Harro *et al.* 1989, 1990). Intra-cerebroventricular administration of CCK-4 increases the intensity of foot-shock-elicited aggressiveness in male rats (Vasar *et al.* 1984). The intravenous administration of CCK-4 has recently been shown to cause very severe anxiety and panic-like attacks in healthy volunteers (De Montigny 1989). However, it is not clear whether the primary target of the anxiogenic-like action of CCK-related peptides lies in the brain or the periphery. Therefore the aim of the present study is to reveal the primary target of the anxiogenic-like effect of CCK agonists in the rat.

Male and female rats (weighing 180–220 g) were used throughout the study. The animals were used only once. The anxiogenic-like effect of CCK agonists (caerulein, pentagastrin, and CCK-4) was studied using an elevated plus maze according to the method of Pellow *et al.* (1985). The lowest dose of caerulein to cause an anxiogenic-like effect in the rat was 100 ng/kg. Pentagastrin had a similar effect after administration of a dose of 500 ng/kg. Subcutaneous treatment with 10 μg/kg of CCK-4 in some experiments also significantly decreased the exploratory activity of rats. The maximal reduction of the animals' behaviour was seen after injection of 25 and 50 μg/kg of CCK-4 (Table 13.1). At higher doses (100 μg/kg and 1 mg/kg) CCK-4 failed to affect the rats' behaviour. The anxiogenic-like effect of CCK agonists was in good agreement with their potency to inhibit [^{3}H]propionylated-CCK-8 ([^{3}H]pCCK-8, 0.3 nM) binding in the cerebral cortex, but not in the pancreas (Table 13.2). According to these results it is probable that CCK_B (central subtype) receptors play a significant role in mediating anxiogenic-like action of CCK agonists in the rat.

The ability of different CCK antagonists (lorglumide, proglumide, devazepide, and L-365,260) to block the anxiogenic-like effect of CCK-4 (50 μg/kg) was also studied. Pretreatment with 0.01 mg/kg L-365,260, a selective antagonist at CCK_B/gastrin receptors, caused a statistically significant antagonism of the anxiogenic effect of CCK-4 (Fig. 13.1). L-365,260 antagonized the action of CCK-4 at a dose 10 times smaller than that of lorglumide

Table 13.1.   Effect of CCK-4 on the exploratory activity of rats in an elevated plus maze

Test dose		Latency of first open arm entry (s)	No. of lines crossed in open arms	Total time spent in open arms (s)
Vehicle		42 ± 8	10.3 ± 1.1	57 ± 6
CCK-4	1 μg/kg	36 ± 24	11.3 ± 2.6	69 ± 12
	10 μg/kg	57 ± 10	7.4 ± 2.6	45 ± 6
	25 μg/kg	81 ± 8[a]	6.2 ± 1.2[a]	39 ± 5
	50 μg/kg	86 ± 12[a]	5.5 ± 1.0[a]	33 ± 5
	100 μg/kg	69 ± 14	9.0 ± 2.2	63 ± 10
	1000 μg/kg	39 ± 24	10.9 ± 2.1	58 ± 10

All values are means ± SEM. The test time was 4 min. CCK-4 was administered subcutaneously 15 min prior to the experiment.
[a] $p < 0.05$ (Significantly different from the vehicle-treated group. Newman–Keuls test after ANOVA, one-way ANOVA: for entries $F_{(1.153)} = 2.18$, $p < 0.05$; for crossings $F_{(1.153)} = 2.36$, $p < 0.05$; for total time $F_{(1.153)} = 1.92$, $p = 0.08$).

Table 13.2.   Correlation between the anxiogenic-like effect of CCK-agonists and their affinity at CCK-receptors in the rat cerebral cortex and pancreas

CCK-8 agonists	Suppression of exploratory activity in elevated plus maze (pmol/kg)	IC_{50} values versus [^{3}H] pCCK-8 (nM)	
		Cerebral cortex	Pancreas
Caerulein	0.074	1.1	(0.6)
Pentagastrin	0.670	10	6200
CCK-4	43.3	411	>10 000
Pearson's γ		0.99998	0.797
p		0.004	0.41

The doses of CCK agonists are those inducing a statistically significant anxiogenic-like effect in the elevated plus maze. The radio-ligand binding studies were performed according to the method of Praissman et al. (1983).

(effective dose, 0.1 mg/kg). The CCK_B/gastrin antagonist was 100 times more effective than the selective CCK_A (peripheral subtype) antagonist devazepide (1 mg/kg) and proglumide (1 mg/kg). Nevertheless, the compounds, which blocked both CCK receptor subtypes (lorglumide, proglumide, and devazepide) had a stronger anxiolytic-like effect than L-365,260. It is also worth noting that the antagonism of the CCK response by the glutaramic acid derivatives (proglumide and lorglumide) was more pronounced than the effect of the 1,4-benzodiazepines (devazepide and L-365,260).

CCK-8 has been shown to be localized in some brain regions (cerebral cortex, hippocampus) with high densities of GABAergic neurons (Kosaka et al. 1985).

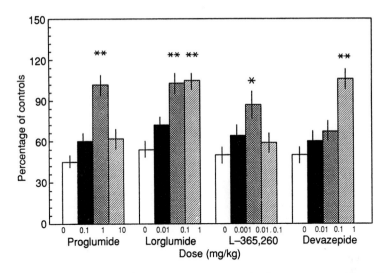

Fig. 13.1. The interaction of CCK antagonists with the anxiogenic-like effect of CCK-4 on rats in an elevated plus maze. The data presented are the percentage of the control values for each separate experiment. The scores are the mean (± SEM) percentage of crossed lines in the open arms. In the control group the mean value of crossed lines during a period of 4 min was between 10 and 14 in the different experiments. Proglumide (0.1–10 mg/kg i.p.) and lorglumide (0.01–1 mg/kg i.p.) were administered 10 min before the CCK-4 (50 μg/kg s.c.) treatment, whereas devazepide (0.01–1 mg/kg i.p.) and L-365,260 (0.001–0.1 mg/kg i.p.) were given 15 min prior to the CCK-4 dose. Lorglumide, devazepide, and L-365,260 were suspended in distilled water containing a few drops of Tween-85. The same vehicle (a few drops of Tween-85 in distilled water) was given to control rats. $\star$, $p < 0.05$; $\star\star$, $p < 0.01$ compared with respective CCK-4 group values.

A clear antagonistic interaction between benzodiazepine traquillizers and CCK-8 has been observed in electrophysiological experiments (Bradwejn and De Montigny 1984). An anxiogenic dose (0.5 mg/kg) of picrotoxin, a potent antagonist at the GABA receptor chloride channel, increased the density of CCK-receptors in the frontal cortex and the hippocampus (Fig. 13.2). Higher doses (1 and 2.5 mg/kg) of picrotoxin induced seizures and apparently decreased the density of CCK receptors (Fig. 13.2). In the acute experiments picrotoxin failed to affect binding to benzodiazepine receptors. However anxiogenic doses of the CCK-agonists caerulein (100 ng/kg) and pentagastrin (500 ng/kg) did not change the affinity and density of CCK receptors, but decreased the apparent number of benzodiazepine receptors in the frontal cortex by 34 per cent and 38 per cent respectively. It is quite probable that, at least in the frontal cortex, there is a negative interaction between CCKergic and GABAergic systems in the regulation of anxiety. This hypothesis is supported by experiments using rats selected according to their behaviour in the elevated

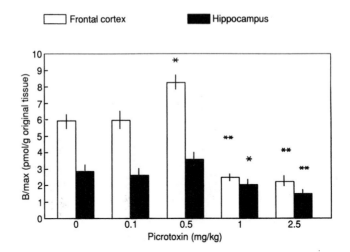

Fig. 13.2. The effect of picrotoxin on [³H]pCCK-8 binding in the rat brain. Data are expressed as the apparent number of binding sites (B/max) in pmol/g original tissue wet weight of the rat frontal cortex and hippocampus. The mean values of three independent studies (± SEM) are presented. Picrotoxin (0.1–2.5 mg/kg s.c.) was injected 30 min before decapitation. ⋆, $p < 0.05$; ⋆⋆, $p < 0.01$ compared with the control group (Student's t test).

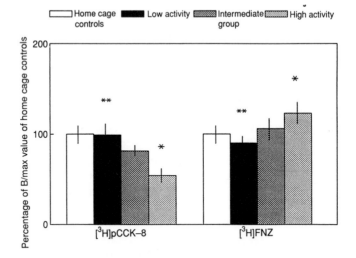

Fig. 13.3. [³H]pCCK-8 and [³H]-flunitrazepam (FNZ) binding in the frontal cortex of rats selected according to their response in the elevated plus maze. In the low activity group the number of crossed lines in the open part was 0.9 ± 0.6. In the intermediate group it was 8.5 ± 0.5 and in the high activity group it was 18.2 ± 1.5. The differences between the three selected groups are statistically significant ($p < 0.01$, Mann–Whitney U test). The mean values of three independent selection experiments are presented. Data are the percentage of home cage control values

plus maze. It was possible to find subgroups of animals in the rat population in which exploratory activity differed very significantly. Radio-ligand binding experiments with [^{3}H]pCCK-8 and [^{3}H]flunitrazepam revealed differences between the high activity ('non-anxious') and low activity ('anxious') rats in the density of CCK and benzodiazepine receptors in the frontal cortex (Fig. 13.3). 'Non-anxious' rats had a lower density of CCK receptors and a higher density of benzodiazepine receptors in the frontal cortex compared with 'anxious' animals.

In conclusion, the anxiogenic-like effect of peripherally administered CCK agonists in the rat is probably related to CCK$_B$ receptors. In the frontal cortex at least CCK$_B$ receptors appear to undergo a strong negative interaction with the GABA–benzodiazepine receptor complex. Thus the balance between CCKergic and GABAergic systems appears to be significant in the genesis of anxiety.

Acknowledgements

Devazepide and L-365,260 were generous gifts from Merck Sharp & Dohme. Proglumide and lorglumide were donated by Rotta Pharmaceuticals.

References

Bradwejn, J. and De Montigny C. (1984). Benzodiazepines antagonize cholecystokinin-induced activation of rat hippocampal neurones. *Nature* **312**, 363–4.

De Montigny, C. (1989). Cholecystokinin tetrapeptide induces panic-like attacks in healthy volunteers: preliminary findings. *Arch. Gen. Psychiat.* **46**, 511–17.

Harro, J., Põld, M., Vasar, E., and Allikmets, L. (1989). The role of CCK-8-ergic mechanisms in the regulation of emotional behaviour in rodents, *Zh. Vyss. Nerv. Deyat.* **39**, 877–83.

Harro, J., Põld, M., and Vasar, E. (1990). Anxiogenic-like action of caerulein, a CCK-8 receptor agonist, in the mouse: influence of acute and subchronic diazepam treatment. *Naunyn-Schmiedebergs Arch. Pharmacol.* **341**, 62–7.

Kosaka, T., Kosaka, K., Tateishi, K., Hamaoka, Y., Yanaihara, N., Wu, J-Y., and Hama, K. (1985). GABAergic neurons containing CCK-8-like and/or VIP-like immunoreactivities in the rat hippocampus and dentate gyrus. *J. Comp. Neurol.* **239**, 420–30.

Pellow, S., Chopin, P., File, S., and Briley, M. (1985). Validation of open:closed arm entries in an elevated plus-maze as a measure of anxiety in the rat. *J. Neurosci. Methods* **14**, 149–67.

---

in the frontal cortex. The apparent number of [^{3}H]pCCK-8 binding sites in the frontal cortex of home cage controls was 3.66 ± 0.34 pmol/g original tissue, and for [^{3}H]flunitrazepam it was 137 ± 12 pmol/g original tissue. $*$, $p < 0.05$, significant difference from home-cage controls; $**$, $p < 0.05$, significant difference from high activity group (Student's t test).

Praissman, M., Martinez, P. A., Saldino, C. F., Berkowitz, J. M., Steggles, A. M., and Finkelstein, J. A. (1983). Characterization of cholecystokinin binding sites in rat cerebral cortex using a [^{125}I]-CCK-8 probe resistant to degradation. *J. Neurochem.* **40**, 1406–13.

Vasar, E., Maimets, M., and Allikmets, L. (1984). Role of the serotinin$_2$-receptors in regulation of aggressive behaviour. *Zh. Vyss. Nerv. Deyat.* **34**, 283–9.

14. Anxiogenic-like effect of CCK-8 micro-injected into the dorsal periaqueductal grey of rats in the elevated plus maze

F. S. Guimarães, A. S. Russo, J. C. De Aguiar, G. Ballejo, and F. G. Graeff

Experiments involving electrical stimulation of subcortical regions of the brain have established that interconnected structures comprising the dorsal peria-queductal grey (DPAG) of the mid-brain, the medial hypothalamus, and parts of the amygdala constitute an important neural substrate for the expression of defensive/aversive behaviour and the associated neurovegetative and neuroen-docrine changes (Graeff 1989). These brain structures are also responsible for the development of aversive motivational states, and thus constitute the so-called behavioural aversive system (BAS) (Graeff 1989).

The neuropharmacology of the BAS, particularly the DPAG, has been systematically investigated using intracerebral micro-injection of drugs often coupled with brain electrical stimulation. The results obtained so far indicate that many neurotransmitters, including γ-aminobutyrate (GABA), endogen-ous opioids, 5-hydroxytryptamine (5-HT), excitatory amino acids, and ace-tylcholine, are involved in the modulation of aversive states in the DPAG (Graeff 1984, 1987, 1989, 1991; Schmitt *et al.* 1984, 1986; Graeff *et al.* 1986; Carobrez 1987).

Role of DPAG in anxiety

Anxiolytic drugs such as benzodiazepines, barbiturates, and ethanol, as well as the 5-HT re-uptake inhibitors fluvoxamine and zimelidine, share the property of attenuating aversion due to DPAG stimulation, which suggests that this structure is involved in pathophysiological processes, particularly pain, depres-sion, and anxiety (Graeff 1984, 1989, 1991; Jenck *et al.* 1989). Interestingly, it has been shown in patients undergoing neurosurgery that electrical stimulation of the DPAG elicits symptoms and autonomic changes that closely resemble a panic attack (Nashold *et al.* 1974).

5-HT re-uptake inhibitors are the best drug treatment for panic disorder, and systemic as well as direct injection of these drugs reduces the effects of DPAG stimulation (Kiser and Lebowitz 1975; Schüetz *et al.* 1985; Jenck *et al.* 1989). In contrast, 5-HT antagonists facilitate aversion elicited by DPAG stimulation (Schenberg and Graeff 1978; Jenck *et al.* 1989), which may explain why the clinical anxiolytic ritanserin, a $5\text{-}HT_2/5HT_{1c}$ antagonist, is at best, ineffective in panic (Den Boer and Westenberg 1990).

These findings suggest that the BAS may be involved in the expression of panic, and it has been proposed that panic symptoms are a consequence of pathological activation of this system (Deakin *et al.* 1991; Graeff 1991).

The use of the elevated plus maze to investigate the role of DPAG in anxiety

Many experimental investigations of the role of DPAG in anxiety have been based on the effects of direct electrical or chemical stimulation. To examine this role in a more naturalistic way we decided to carry out experiments employing

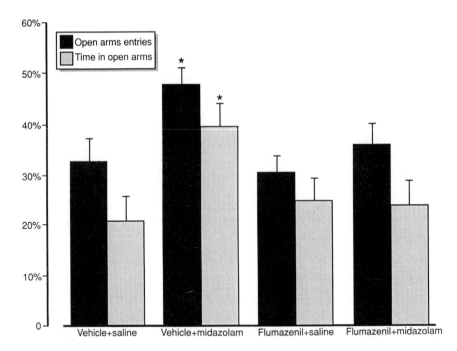

Fig. 14.1. Anxiolytic effect of midazolam (80 nmol/0.5 μl) and its antagonism by flumazenil (80 nmol/0.5 μl 10 min before) determined in the elevated plus maze. Both drugs were micro-injected into the DPAG of rats. Asterisks indicate signifi-cant differences from other groups (Newman–Keuls test) in both the percentage of entries into the open arms ($F(3, 36) = 4.54$, $p < 0.01$) and the percentage of time spent there ($F(3, 36) = 3.74$, $p < 0.05$). No effect was detected in the total number of entries ($F(3, 36) = 0.52$, not significant).

the elevated plus maze test. The elevated plus maze is an animal model of anxiety that has been validated behaviourally, physiologically, and pharmacologically (Pellow et al. 1985). It consists of a roofless plus-shaped maze elevated 50 cm from the floor and comprising two opposite arms, of dimensions 50 cm × 10 cm, crossed at right angles by two arms, of the same dimensions and enclosed by walls 40 cm high. The test we have used is slightly modified by the addition of a Plexiglass edge 1 cm high surrounding the open arms to avoid falls. In this test indices of exploration of the open arms are usually employed as measures of anxiety.

Results from our laboratory show that the test is able to detect a specific anxiolytic effect of midazolam, a benzodiazepine agonist, micro-injected into the DPAG (Fig. 14.1). In addition, employing the same test, we have recently demonstrated dose-dependent anxiolytic effects of the N-methyl-D-aspartate (NMDA) receptor antagonist AP7 micro-injected into the same region (Guimarães et al. 1991). These results support the involvement of this brain structure in anxiety and show that the test is sensitive to local micro-injection of drugs.

CCK and anxiety

Recent experimental evidence appears to implicate CCK-mediated mechanisms in anxiety disorders. Benzodiazepine anxiolytics antagonize the behavioural effects induced by CCK-8 or caerulein administration (Kubota et al. 1985, 1986; Sugaya et al. 1985) and inhibit the activation of hippocampal neurons caused by CCK (Bradwejn and De Montigny 1984). CCK_A antagonists suppress black-side preference in mice exploring a black–white box (Hendrie and Dourish 1990) and block the anxiogenic effect in the elevated plus maze induced by systemic injection of caerulein in mice (Harro et al. 1990) or by intracerebral micro-injection of CCK-8S, but not CCK-8US, into the posteromedial part of the nucleus accumbens of rats (Dauge et al. 1989). In a conditioned suppression of drinking paradigm, however, CCK_B antagonists have a much more potent anxiolytic effect than CCK_A antagonists making it unclear (Dourish et al. 1990) which CCK receptor (A and/or B) is more important in the expression of anxiety (Ravard and Dourish 1990).

Human experiments also support a role for CCK in anxiety. Systemic administration of its C-terminal tetrapeptide CCK-4 can induce panic attacks in both healthy volunteers and patients with panic disorders (Abelson and Nesse 1990; Bradwejn et al. 1990; Chapter 11 of this volume). Since there is a high concentration of CCK in the periaqueductal grey (De Belleroche et al. 1990), we decided to investigate the effects of CCK-8 micro-injected into the DPAG on the behaviour of rats in the elevated plus maze test.

The effects of CCK-8 micro-injected into the DPAG

Male Wistar rats weighing 250–300 g were anaesthetized with sodium pentobarbitone and implanted with a stainless steel guide cannula (outside diameter, 0.7 mm) aimed at the DPAG. The guide cannula was introduced 1.9 mm lateral to the right side of lambda at an angle of 22° with the sagittal plane until the cannula tip was 4.0 mm below the surface of the skull. Stainless steel screws and acrylic cement were used to attach the cannula to the bone.

The rats were randomly assigned to treatment groups 5–7 days after surgery and received a micro-injection of either saline solution or CCK-8S (500 ng/0.5 μl). In order to increase maze exploration, each rat was left inside a wooden arena (dimensions 60 cm × 60 cm × 35 cm) for a 5 min period starting 5 min after the injection before being placed in the centre of the plus maze facing the same enclosed arm for 5 min. The number of entries and the time spent on each arm was scored.

The results obtained for the rats in which histological analysis confirmed the site of injection as being located within the limits of the DPAG showed that CCK-8S significantly decreased the percentage of entries into the open arms (d.f. = 43, $t = 2.67$, $p = 0.011$) and the percentage of time spent there

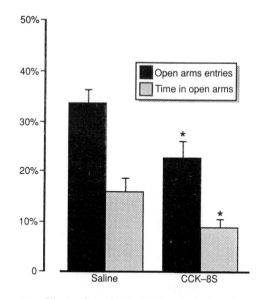

Fig. 14.2. Anxiogenic effect of CCK-8S (500 ng/0.5 μl) micro-injected into the DPAG of rats determined in the elevated plus maze. Asterisks indicate significant differences from saline group in the percentage of entries into the open arms ($t = 2.67$, d.f. = 43, $p = 0.011$) and the percentage of time spent there ($t = 2.08$, d.f. = 43, $p = 0.043$). No effect was detected in the total number of entries ($t = -0.14$, d.f. = 43, not significant).

(d.f. $= 43$, $t = 2.08$, $p = 0.043$) (Fig. 14.2) without affecting the total number of entries (d.f. $=43$, $t = -0.14$, $p = 0.891$).

Although these results are preliminary, they suggest that CCK in the DPAG has an anxiogenic effect which is expressed as a selective decrease in exploration of the open arms in an elevated plus maze. Therefore this region may be a site for the effects of CCK agonists and antagonists in anxiety (see Chapters 11, 12, and 13 of this volume). Since CCK-8S is a non-specific CCK agonist, further studies are needed to determine the role of CCK_A and CCK_B receptors in the induction of anxiety.

Acknowledgements

Financial support was received from FINEP and CAPES.

References

Abelson, J. L. and Nesse, R. (1990). Cholecystokinin-4 and panic. *Arch. Gen. Psychiat.* **47**, 395.

Bradwejn, J, and De Montigny, C. (1984) Benzodiazepines antagonize cholecystokinin-induced activation of rat hippocampal neurones. *Nature* **312**, 363–4.

Bradwejn, J., Koszycki, D., and Meterissian, G. (1990). Cholecystokinin-tetrapeptide induces panic attacks in patients with panic disorders. *Can. J. Psychiatry* **35**, 83–5.

Carobrez, A. P. (1987). Excitatory aminoacid mediation of the defence reaction. In *Neuroscience and behaviour* (ed. M. L. Brandao), pp. 21–29. UFES-Grafica, Victoria, Brazil.

Daugé, V., Steimes, P., Derrien, M., Beau, N., Roques, B. P., and Feger, J. (1989). CCK8 effects on motivational and emotional states of rats involve CCKA receptors of the postero-median part of the nucleus accumbens. *Pharmacol. Biochem. Behav.* **34**, 157–63.

Deakin, J. F. W., Guimarães, F. S., and Wang, M. (1991). 5 HT receptor mechanisms in human anxiety. In *New concepts in anxiety* (ed. M. Briley and S. E. File), pp. 74–93. Macmillan, New York.

De Belleroche, J., Bandopadhyay, R., King, A., Malcolm, A. D. B., O'Brien, K., Premi, B. P., and Rashid, A. (1990). Regional distribution of cholecystokinin messenger RNA in rat brain during development: quantification and correlation with cholecystokinin immunoreactivity. *Neuropeptides* **15**, 201–12.

Den Boer, J. A. and Westenberg, H. G. M. (1990). Serotonin function in panic disorder: a double blind placebo controlled study with fluvoxamine and ritanserin. *Psychopharmacology* **102**, 85–94.

Dourish, C. T., Rycroft, W., Dawson, G. R., Tattersall, F. D., and Iversen, S. D. (1990). Anxiolytic effects of the CCK antagonist devazepide and L-365,260, in a conditioned suppression of drinking model. *Eur. J. Neurosci.*, **2** (Suppl. 3), 38.

Graeff, F. G. (1984). The anti-aversive action of minor tranquillizers. *Trends Pharmacol. Sci.* **5**, 230–3.

Graeff, F. G. (1987). The anti-aversive action of drugs. In *Advances in behavioral pharmacology* (ed. T. Thompson, P. B. Dews, and J. E. Barrett), Vol. 6, pp. 129–56. Lawrence Erlbaum, Hillsdale, NJ.

Graeff, F. G. (1989). Serotonin and aversion. In *Behavioral pharmacology of 5 HT* (ed.

P. Bevan, A. R. Cools, and T. Archer), pp. 425–43, Lawrence Erlbaum, Hillsdale, NJ.

Graeff, F. G. (1991). Neurotransmitters in the periaqueductal gray and animal models of panic anxiety. In *New concepts in anxiety* (ed. M. Briley and S. E. File), pp. 288–312. Macmillan, New York.

Graeff, F. G., Brandao, M. L., Audi, E. A., and Schuetz, M. T. (1986). Modulation of the brain aversive system by GABAergic and serotonergic mechanisms. *Behav. Brain Res.* 21, 65–72.

Guimarães, F. S., Carobrez, A. P., de Aguiar, J. C., and Graeff, F. G, (1991). Anxiolytic effect on the elevated plus-maze of the NMDA receptor antagonist AP7 microinjected into the dorsal periaqueductal gray. *Psychopharmacology*, 103, 91–4.

Harro, J., Pold, M., and Vasar, E. (1990). Anxiogenic-like action of caerulein, a CCK-8 receptor agonist, in the mouse: influence of acute and subchronic diazepam treatment. *Naunyn-Schmiedebergs Arch. Pharmacol.* 341, 62–7.

Hendrie, C. A. and Dourish, C. T. (1990). Anxiolytic profile of the cholecystokinin antagonist devazepide in mice. *Br. J. Pharmacol.* 99, 137P.

Jenck, F., Broekkamp, C. L. E., and Van Delft, A. M. L. (1989). Effects of serotonin receptor antagonists on PAG stimulation induced aversion: different contributions of $5HT_1$, $5HT_2$ and $5HT_3$ receptors. *Psychopharmacology* 97, 489–95.

Kiser, K. R. and Lebowitz, R. M. (1975). Monoaminergic mechanism in aversive brain stimulation. *Physiol. Behav.* 15, 47–53.

Kubota, K., Matsuda, Y., Sugaya, K., and Uruno, T. (1985). Cholecystokinin antagonism by benzodiazepines in food intake in mice. *Physiol. Behav.* 36, 175–8.

Kubota, K., Sun, F. Y., Sugaya, K., and Sunagane, N. (1986). Reversal of antinociceptive effect of cholecystokinin by benzodiazepines and a benzodiazepine antagonist Ro 15-1788 *Jpn. J. Pharmacol.* 37, 101–5.

Nashold, B. S., Jr, Wilson, N. P., and Slaughter, G. S. (1974). The midbrain and pain. In *Advances in neurology* (ed. J. J. Bonica), Vol. 4, *International Symposium on Pain*, pp. 191–6. Raven Press, New York.

Pellow, S., Chopin, P., File, S. E., and Briley, M. (1985). Validation of open:closed arm entries in an elevated plus-maze as a measure of anxiety in the rat. *J. Neurosci. Methods* 14, 149–67.

Ravard, S. and Dourish, C. T. (1990). Cholecystokinin and anxiety. *Trends Pharmacol. Sci.* 11, 271–3.

Schenberg, L. C. and Graeff, F. G. (1978). Role of the periaqueductal gray substance in the antianxiety action of benzodiazepines. *Pharmacol. Biochem. Behav.* 9, 287–95.

Schmitt, P., Di Scala, G., Jenck, F., and Sandner, G. (1984). Periventricular structures, elaboration of aversive effects and processing of sensory information. In *Modulation of sensorimotor activity during alterations in behavioural states* (ed. R. Bandler), pp. 393–414. Alan R. Liss, New York.

Schmitt, P., Carrive, P., Di Scala, G., Jenck, F., Brandao, M. L., Bagri, A., Moreau, J. L., and Sadner, G. (1986). A neuropharmacological study of the periventricular neural substrate involved in flight. *Behav. Brain Res.* 22, 181–90.

Schüetz, M. T. B., de Aguiar, J. C., and Graeff, F. G. (1985). Anti-aversive role of serotonin in the periaqueductal grey matter. *Psychopharmacology* 85, 340–5.

Sugaya, K., Matsuda, I., and Kubota, K. (1985). Inhibition of hypothermic effect of cholecystokinin by benzodiazepines and a benzodiazepine antagonist RO 15-1788, in mice. *Jpn J. Pharmacol.* 39, 277–80.

15. Central actions of CCK: modulation of GABA release by CCK

J. de Belleroche and R. Bandopadhyay

Introduction

CCK is extremely abundant in the mammalian cerebral cortex, hippocampus, caudate nucleus, and hypothalamus. CCK messenger RNA (mRNA) is concentrated in the cerebral cortex in cells lying in layers II–III and V–VI (Fig. 15.1). The presence of CCK-like immunoreactivity and CCK mRNA in the cerebral cortex is clearly evident from day 1 after birth in the rat and shows a rapid increase in concentration up to adult levels by 4 weeks (Fig. 15.2) (de Belleroche *et al.* 1990), in keeping with its putative role as a transmitter in the adult cerebral cortex. High levels of CCK mRNA are also found in the periaqueductal grey and the thalamus. The physiological actions of central CCK neurons are poorly understood, although specific brain regions such as the nucleus tractus solitarius and the periaqueductal grey are likely to be important in the processing and regulation of satiety and pain responses in co-ordination with peripheral mechanisms (Chapters 17 and 44 of this volume). The behavioural effects of CCK in intracranial self-stimulation experiments (Hamilton *et al.* 1990) and in animal models of anxiety point to important CCK actions involving central mechanisms, notwithstanding the potential significance of peripheral systems,

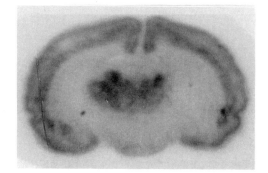

Fig. 15.1. Distribution of CCK mRNA in adult rat brain. Frozen sections of rat brain were hybridized with a CCK cDNA probe as described in detail by de Belleroche *et al.* (1990). Note the high levels of labelling in the cerebral cortex, the thalamus, and the claustrum.

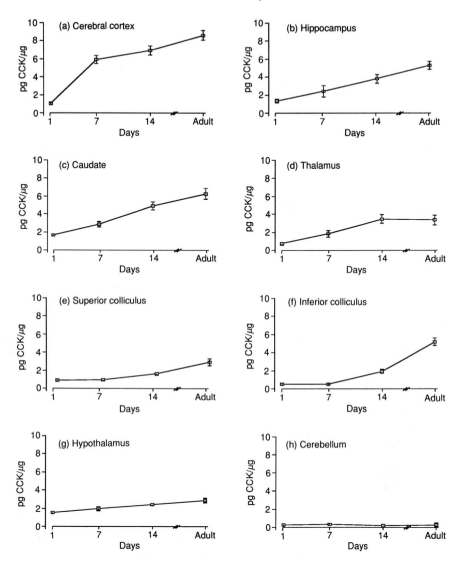

Fig. 15.2. Levels of CCK-like immunoreactivity in eight regions of rat brain during development. CCK was extracted in 90 per cent methanol and assayed by radio-immunoassay. Values (pg/μg protein) are means ± SEMs for n experiments in (a) cerebral cortex (n = 7–9), (b) hippocampus (n = 6–8), (c) caudate (n = 6–9), (d) thalamus (n = 7–10), (e) superior colliculus (n = 6), (f) inferior colliculus (n = 6–8), (g) hypothalamus (n = 5–6), and (h) cerebellum (n = 4). (Reproduced with permission from de Belleroche *et al.* 1990.)

e.g. vagal afferents, in modulating these effects. This is demonstrated by the observation that the effect of intraperitoneally injected caerulein on intra-cranial self-stimulation is partially reduced by the CCK_A receptor antagonist devazepide (Hamilton *et al.* 1990), indicating that both central and peripheral mechanisms may modulate self-stimulation.

Little is known about the physiological function of CCK in the cerebral cortex. The release of CCK is sensitive to modulation by glutamate agonists (Yaksh *et al.* 1987), which involves an *N*-methyl-D-aspartate receptor-mediated mechanism (Bandopadhyay and de Belleroche 1991). The coexistence of CCK with glutamic acid decarboxylase (GAD) indicates a possible interaction bet-ween these two neurotransmitters (Hendry *et al.* 1984). In previous studies (Sheehan and de Belleroche 1983) we have demonstrated a facilitatory effect of CCK and the CCK analogue caerulein on the release of γ-aminobutyrate (GABA) which supports this proposal. The development of selective CCK receptor antagonists has now made it possible to characterize the nature of the receptors involved in this action of CCK. We have therefore made use of the selective CCK receptor antagonists devazepide and L-365,260 to investigate further the facilitatory effect of CCK on GABA release in the cerebral cortex.

Modulation of GABA release by CCK receptors

In previous studies, we have shown that the K^+-evoked release of [^{14}C]GABA from tissue slices of rat cerebral cortex is facilitated by low concentrations of CCK (0.1 μM) and the CCK analogue caerulein (1 nM–0.1 μM). This effect is sensitive to proglumide, a CCK antagonist of low potency and selectivity. In this study, we tested the effect of selective CCK_A and CCK_B receptor antago-nists on this facilitation of the K^+-evoked release of endogenous GABA from tissue slices of cerebral cortex. The effect of caerulein was tested at concentra-tions between 1 nM and 2 μM and was shown to produce a dose-dependent facilitation of GABA release at low concentrations although this effect was absent at higher concentrations (2 μM). The concentrations of other amino acids present in the superfusate (glutamate, glycine, and alanine) were unaf-fected by the presence of caerulein at a concentration of 100 nM.

The presence of the selective CCK_B antagonist L-365,260 alone in the incubation medium significantly inhibited the K^+-evoked release of GABA at concentrations of 100 nM ($p < 0.05$), 300 nM ($p < 0.02$), and 1000 nM ($p < 0.05$). This effect of L-365,260 was specific for GABA release, as the release of glutamate, glycine, and alanine was unaffected at all three concentra-tions of the drug. Further, facilitation of K^+-evoked GABA release by caerulein was significantly ($p < 0.05$) reduced by the presence of L-365,260 (100 nM), but was unaffected by devazepide at a concentration of 10 nM which is selective for the CCK_A receptor.

Conclusion

In this study, we have been able to demonstrate the facilitation of the K^+-evoked release of endogenous GABA from cerebral cortex by caerulein. The magnitude of the maximal effect (31 per cent) was similar to that obtained when measuring the release of preloaded [^{14}C]GABA (36 per cent). This property was used as an index of central CCK receptors to investigate their sensitivity to selective antagonists. The results indicate that the CCK_B antagonist L-365,260 is able to block the facilitation of GABA release by caerulein and to reduce the release of GABA induced by K^+ in the absence of caerulein which indicates that endogenously released CCK is also able to modulate GABA release.

References

Bandopadhyay, R. and de Belleroche, J. (1991). Regulation of CCK release in cerebral cortex by *N*-methyl-D-aspartate receptors: sensitivity to APV, MK-801, kynurenate, magnesium and zinc ions. *Neuropeptides* **18**, 159–63.

de Belleroche, J., Bandopadhyay, R. King, A., Malcolm, A. D. B., O'Brien, K., Premi, B. P., and Rashid, A. (1990). Regional distribution of cholecystokinin messenger RNA in rat brain during development: quantitation and correlation with cholecystokinin immunoreactivity. *Neuropeptides* **15**, 201–12.

Hamilton, M. H., Rose, I. C., Herberg, L. J., and de Belleroche, J. (1990). Effect of intracerebroventricular and systemic injections of caerulein, a CCK analogue, on electrical self-stimulation and its interaction with the CCKA receptor antagonist, L-364,718 (MK-329). *Psychopharmacology* **101**, 384–9.

Hendry, S. H. C., Jones, E. G., De Felipe, J. Schmechel, D., Brandon, C., and Emson, P. C. (1984). Neuropeptide-containing neurones of the cerebral cortex are also GABAergic. *Proc. Natl Acad. Sci. USA* **81**, 6526–30.

Sheehan, M. J. and de Belleroche, J. (1983). Facilitation of GABA release by cholecystokinin and caerulein in cerebral cortex. *Neuropeptides* **3**, 429–34.

Yaksh, T. L., Furu, T., Kanawati, L. S., and Go, V. L. W. (1987). Release of cholecystokinin from rat cerebral cortex *in vivo*: role of GABA and glutamate receptor systems. *Brain Res.* **406**, 207–14.

Part III

CCK and satiety

16. CCK and satiety: introduction

G. P. Smith

It is almost 20 years since CCK was first reported to inhibit food intake (Gibbs *et al.* 1973). The chapters in this part of the volume provide a reasonably complete review of the current understanding of this phenomenon. Given that the editors chose the topics and the authors did not consult one another, it is interesting that the chapters read like a family album — each chapter shares common features with the others, but each is distinctly different.

The family resemblance of these chapters lies in their concern with the physiological role of CCK in the control of food intake, the problem of aversive side-effects, the type and site of receptor mechanism, and the role of gastric emptying and vagal afferent fibres in the satiating effect. There is now considerable agreement on these topics. This represents real progress from the controversy surrounding them just a short time ago. It is clear that much of this progress was made possible by the availability of potent and selective antagonists of CCK receptors.

The differences between the chapters are the growth points in this experimental area. These include interactions between peripheral CCK and central CCK, between peripheral CCK and central serotonin and dopamine, and between peripheral CCK and the hormonal state of females.

Despite the consensus about core issues and the interesting ways in which new relationships are being investigated, these chapters leave a residue of problems. I shall take up the consensus first, review the growth points next, and finally discuss the problems.

Consensus

Devazepide, a specific antagonist of CCK_A receptors, increases food intake in many situations. This appears to be decisive evidence for the hypothesis that endogenous CCK has a satiating effect. Although many of these results do not specify whether the source of the endogenous CCK that was antagonized by devazepide was peripheral or central, some implicate CCK released from the small intestine and these are the best evidence for the original hypothesis (see Chapter 17).

The positive results with devazepide also bear directly on the possibility that the satiating effect of exogenously administered CCK is entirely due to aversive side-effects. When devazepide increases food intake by antagonizing the inhibitory effect of endogenous CCK released from the small intestine, explaining satiation by aversion is no longer plausible (see Chapters 17, 18, 19, 21, and 23). Furthermore, in the case of exogenously administered CCK, it is now clear that aversive side-effects can be reliably produced only when CCK is administered by intravenous (i.v.) bolus injection and/or in relatively large doses (see Chapters 17, 19, and 23). This is probably due to the abnormally large transient increase in circulating CCK achieved by such administrations. When lower doses are administered slowly either i.v. or intraperitoneally (i.p.), aversive side-effects are minimized or eliminated (see Chapters 17, 19, 21, and 23).

Positive results with devazepide also implicate type A receptors in the mediation of the satiating effect of endogenous and exogenous CCK. This is consistent with the relative potency and efficacy of peptides structurally related to sulphated CCK.

Afferent fibres of the vagus nerve are necessary for the satiating effect of exogenous CCK. There is some evidence that this is also true for endogenous CCK, but more experiments are necessary to be certain of this (see Chapter 17).

The explicit withdrawal by Moran and McHugh of the strong form of their original hypothesis that the inhibition of food intake by CCK was an indirect effect of the inhibition of gastric emptying by CCK removes one of the important controversies in this area. The modified hypothesis that the increase in gastric distension produced by the decrease of gastric emptying contributes to the satiating effect of relatively large non-aversive doses of exogenous CCK is consistent with all the available data for the rat and is accessible to experiment.

Growth points

CCK_B receptors also appear to be involved in the satiating effect of endogenous CCK, but not of exogenous CCK (see Chapter 21). The primary situation in which this has been demonstrated is after a large oral preload. Since the relative satiating potency of sulphated and unsulphated CCK administered centrally or peripherally did not predict the results obtained with L-365,260, the specific antagonist of type B receptors, further experiments to determine the site of these receptors and their functional relationship to endogenous CCK will be very instructive. The fact that under these experimental conditions the type A antagonist also reduced the satiating potency of the preload suggests that the satiating effect involves both type A *and* type B receptors. This is an example of one of the growth points of the work on the satiating effect of CCK.

Södersten and Lindén (Chapter 20) suggest another type of interaction — that between CCK acting at peripheral receptors and CCK acting at central receptors. They demonstrated that proglumide administered into the lateral cerebral ventricle (LCV) blocked the satiating effect of CCK (15 μg/kg i.p.) in rats deprived of food for 48 h. Using less severe deprivation in pigs, however, Baldwin did not show such an interaction (Chapter 19). Administration of devazepide into the LCV antagonized the inhibition of food intake produced by CCK-8 (LCV), but not the inhibition produced by CCK-8 (i.p.). Despite the failure to confirm an interaction between central and peripheral CCK receptors in the pig, Baldwin's results implicate the central type A receptor rather than the type B receptor. Thus the results of Södersten and Lindén and of Baldwin do not help us to understand Dourish's results (Chapter 21) that implicate a central type B receptor in the satiating effect of CCK released by a large preload.

The third growth point is the interaction that Moran and McHugh (Chapter 18) proposed between type A receptors on the circular smooth muscle of the pyloric sphincter and type A receptors on the vagal afferent fibres in the satiating effect of peripherally administered CCK-8. On the basis of a variety of experimental results, Moran and McHugh suggest that these two receptor populations account for different regions of the dose-related inhibition of food intake by CCK-8 (i.p.): the vagal afferent receptors mediate the effect of doses of 2 μg/kg or less; the pyloric receptors are necessary for the full effect of larger doses, presumably because of increased gastric distension caused by decreased gastric emptying.

A fourth growth point is the interaction of CCK with serotonergic (5-HT$_1$) and dopaminergic (D$_1$) mechanisms (see Chapter 23). Bringing CCK into contact with the more developed tools and ideas relating these monoamines to satiation will be catalytic for the rate of investigation of the role of central CCK in the control of meal size.

The final growth point is the relationship between oestrogen or the hormones of lactation and the satiating effect of peripherally administered CCK-8 (see Chapter 20). Not only does lactation decrease the inhibitory potency of CCK-8 (i.p.), it also dissociates the relationship between the concentration of CCK in the plasma from the concentration of CCK in the cerebrospinal fluid (CSF). In contrast oestrogen increases the satiating potency of CCK-8 (i.p.) in ovariectomized rats. The observation that low oestrogen decreases the satiating potency of CCK-8 may be relevant to the clinical observation that binge eating is much more common in women than men and can be associated with decreased plasma oestrogen, particularly in those women with a history of anorexia nervosa.

These interactions are the major growth points of the field. They are likely to attract considerable experimental attention in the next few years.

Problems

Both Moran and McHugh (Chapter 18) and Robinson (Chapter 27) appeal to a vagally mediated gastric distension mechanism as contributing to the satiating effect of CCK. This assumes that the distension-sensitive activity in gastric vagal afferent fibres is used by the central neural network that processes information for satiation in addition to being used for other autonomic non-behavioural functions. This assumption predicts that removal of gastric vagal afferent fibres will result in abnormally large meals. This result has only been reported twice — once in adult dogs (Towbin 1955) and once in preweanling rats (Lorenz 1983). No other reports of results in dogs, rabbits, and human confirm the prediction (Smith and Gibbs 1979). In fact, when the contribution of gastric distension to satiety was maximized by preventing food from emptying from the stomach by closing the pylorus throughout an intake test, vagotomized rats ate less, not more (Kraly and Gibbs 1980; Gonzalez and Deutsch 1981). Thus more work is required to demonstrate the relative satiating potency of vagal afferent activity produced by gastric distension before we can use this activity as a mechanism for mediating part of the satiating effect of CCK.

A second problem emphasized in two of the chapters (Chapters 17 and 18) is the current failure to identify the site of the type A receptors that are necessary and/or sufficient to produce the satiating effect of low doses of CCK administered peripherally or released from the small intestine. Evaluating the receptors on the vagal afferent fibres is particularly difficult because it requires an evaluation of the receptor function of those fibres separate from their sensory function.

A third problem is the difference between rats and humans. For example, devazepide did not increase gastric emptying in humans even though exogenous CCK decreased gastric emptying (see Chapters 26 and 27).

Finally, how should we interpret the measurement of increased immunoreactive CCK in the CSF in the experiments of Lindén and Södersten (Chapter 20)? Does this represent increased release or decreased degradation? Is the immunoreactive CCK biologically active? Where in the brain does this CCK come from?

All these problems are a sign of the rapid pace of recent research. They are the types of problem that investigators thrive on, and their successful solution will represent a significant advance in scientific understanding.

Conclusion

The chapters in this part demonstrate how the systematic investigation of a behavioural effect of CCK over the past two decades has carried researchers into the deep issues of the relationships between cellular receptor effects and the control of eating. They also demonstrate the power of potent and specific antagonists to extract physiological facts. Given the current investigative momentum, the next decade should produce increasingly detailed answers to the question of how CCK decreases food intake.

References

Gibbs, J., Young, R. C., and Smith, G. P. (1973). Cholecystokinin decreases food intake in rats. *J. Comp. Physiol. Psychol.* **84**, 377–82.

Gonzalez, M. F. and Deutsch, J. A. (1981). Vagotomy abolishes cues of satiety produced by gastric distention. *Science* **212**, 1283-4.

Kraly, F. S. and Gibbs, J. (1980). Vagotomy fails to block the satiating effect of food in the stomach. *Physiol. Behav.* **24**, 1007-10.

Lorenz, D. N. (1983). Effects of gastric filling and vagotomy on ingestion, nipple attachment, and weight gain by suckling rats. *Dev. Psychobiol.* **16**, 469-83.

Smith, G. P. and Gibbs, J. (1979). Postprandial satiety. In *Progress in psychobiology and physiological psychology*, Vol. 8 (ed. J. M. Sprague and A. N. Epstein), pp. 179-242. Academic Press, New York.

Towbin, E. J. (1955). Thirst and hunger behaviour in normal dogs and the effects of vagotomy and sympathectomy. *Am. J. Physiol.* **182**, 377-82.

17. The development and proof of the CCK hypothesis of satiety

G. P. Smith and J. Gibbs

In 1973, we reported that the intraperitoneal (i.p.) administration of CCK to rats shortly before a test meal decreased the size of that meal (Gibbs *et al.* 1973*a*). The effect was obtained with an impure extract of CCK as well as with the synthetic C-terminal octapeptide CCK-8 (Gibbs *et al.* 1973*b*). The effect was dose-related and did not appear to be non-specific or toxic. On the basis of these results, we proposed that endogenous CCK released from the mucosal cells of the upper small intestine by food ingested during a meal acted to terminate that meal by a hormonal mode of action. We now believe that sufficient evidence has been obtained to consider the essence of the hypothesis *proven* although the mode of action of CCK – endocrine or paracrine – has not been settled. In this chapter, we review the evidence that we consider crucial. It comes from two kinds of experiments – experiments with agonists and experiments with antagonists.

Experiments with agonists

Rank order of potency

The synthetic C-terminal octapeptide is approximately equipotent with an impure extract of CCK that contains CCK-8 and larger forms of CCK. The potency of pure preparations of larger forms, such as CCK-22 and CCK-33, has not been assessed. Unsulphated CCK-8 and CCK-4 are at least 100 times less potent (Gibbs *et al.* 1973*b*), and the sulphated and unsulphated forms of gastrin, alone or together, have no detectable activity in doses as large as 1 mg/kg i.p. (Lorenz *et al.* 1979). On the other hand, ceruletide is significantly more potent than CCK-8 (Gibbs *et al.* 1973*a*). Thus the rank order of the satiating potency of agonists administered i.p. is ceruletide $>$ CCK-8 $\geqslant$ unsulphated CCK-8, CCK-4, gastrin. This is the same as the rank order of potency for the agonist action of CCK-8 at receptors in gut organs, such as those on the pancreatic acinar cell and gall bladder smooth muscle, now called type A receptors (Moran *et al.* 1986). Note that the efficacy and satiating potency of CCK is more variable in female rats (Wager-Srdar *et al.* 1986; Strohmayer and Smith 1987).

Specificity of the satiating effect

To call the inhibitory effect of CCK on food intake a satiating effect is an interpretation. The interpretation claims that within a certain range of doses and in the context of an impending meal, the inhibitory effect of CCK is specific in the sense that it only acts when the animal ingests commodities with the sensory characteristics associated with food (innately or by prior association). There is compelling evidence for this. Doses of CCK-8 ($< 8 \mu g/kg$ i.p.) produce dose-related inhibitions of liquid food intake when rats are food-deprived, but do not inhibit water intake when rats are water-deprived (Gibbs *et al.* 1973*a*; Smith, 1984). However, when small amounts of sugar are added to water, the inhibitory effect of CCK is restored (Waldbillig and O'Callaghan 1980), presumably through the orosensory effects of the sugar that are associated with ingesting food, but not water.

These results show that the inhibitory effect is specific in the sense that it takes place only when the animal is ingesting food. But critics like Deutsch and Hardy (1977) and Stricker and Verbalis (1990) have argued that these results are not conclusive because it is still possible that an aversive mildly toxic effect of CCK ('nausea' or 'malaise') only occurs when animals are ingesting food, and this accounts for the apparently specific inhibitory effect of CCK. This is an important point and three types of experiments have been used to evaluate it.

The most frequent test of this idea has been to determine whether CCK could produce an unconditional stimulus that was sufficient to produce a conditioned taste aversion (CTA). This work was based on the assumption that a CTA is always produced by a stimulus that can be characterized as sickness, but this assumption remains unproven (Moore and Deutsch 1985). But even if the interpretation of a CTA were straightforward, the cumulative experimental evidence is not consistent with such a role for CCK. Although Deutsch and Hardy (1977) reported that CCK produced a CTA, there are now numerous published failures to replicate this effect with doses of CCK sufficient to inhibit food intake by 30–50 per cent. Kulkosky (1985) summarized the evidence as follows: 'In summary, CCK-like peptides are typically ineffective unconditioned stimuli in conditioned food aversion designs. Aversions can be demonstrated with high doses of CCK, if administered after completion of a consummatory session, but when CCK acts to decrease a meal, no aversion is observed'.

The second test has been pharmacological. If CCK produced a subclinical toxic effect in the rat that was similar to nausea in the human, then pretreatment with an anti-emetic, e.g. trimethobenzamide, should abolish the inhibitory effect of CCK on food intake. Moore and Deutsch (1985) reported that 5 mg/kg of trimethobenzamide attenuated, but did not abolish, the inhibitory effect of CCK. But when we attempted to reproduce this effect, we failed to replicate their result despite using a much larger range of doses of trimethobenzamide (2.5–20.0 mg/kg) (Avilion, Smith, and Gibbs, unpublished).

The third test has been neuroendocrinological and has been pursued by

Stricker, Verbalis, and their colleagues. Their strategy was based on the observation that vomiting is correlated with marked increases of arginine vasopressin (AVP) release in humans (Rowe *et al.* 1979) and in subhuman primates (Verbalis *et al.* 1986*a*). However, the AVP response is dependent on the stimulus used to produce emesis: apomorphine and other drugs release AVP, but self-induced emesis (Kaye *et al.* 1989) and the emesis produced by ipeca-cuanha syrup (Nussey *et al.* 1988) do not.

Rowe *et al.* (1979) made the important observation that AVP increased when human subjects were nauseated but did not vomit. They considered the relationship between nausea and increased AVP to be tightly correlated because administration of dopaminergic antagonists prior to apomorphine prevented nausea and abolished the increase of AVP.

This correlation between increased AVP release and nausea without emesis suggested that increases of plasma AVP could be an indication of visceral distress, 'malaise', or 'nausea' in the rat, an animal that does not vomit. However, when AVP responses to apomorphine, lithium chloride, and copper sulphate were measured in rats, AVP changed very little but the plasma concentration of oxytocin (OT) increased markedly in a dose-dependent manner (Verbalis *et al.* 1986*a*). Since these three drugs produced CTAs, Verbalis *et al.* concluded that high peripheral concentrations of OT may represent a quantifiable marker of visceral illness.

Verbalis *et al.* (1986*b*) then measured the plasma OT response to CCK (0.1–100 µg/kg i.v. or i.p.) and observed a dose-related increase in OT. However, they also observed a significant increase in OT when rats ate large meals of liquid food. In this and subsequent studies, the maximal increase in OT produced by a meal was about 10 µU/ml (Flanagan *et al.* 1988; McCann *et al.* 1988, 1989). CCK doses of 10 µg/kg or more produced increases of OT that were larger than 10 µU/ml, but smaller doses did not. For example, McCann *et al.* (1989) produced a mean inhibition of food intake of 64 per cent with a CCK dose (5 µg/kg i.p.) that produced a mean concentration of plasma OT equal to 8.9 µU/ml. Thus, even if it is assumed that plasma OT is a valid correlate of visceral distress or 'malaise' in the rat when other stimuli for OT release (Lang *et al.* 1983) are not present (Stricker and Verbalis (1990) now warn against such an intepretation), CCK doses that produce large inhibitions of food intake do not produce increases in plasma OT that could be considered evidence of sickness or distress because they do not exceed the plasma OT concentration produced by the ingestion of a meal.

Verbalis *et al.* (1987) found a similar result in monkeys: a low dose of CCK (1 µg/kg i.v.) produced a small increase in plasma AVP and no emesis, while a larger dose (10 µg/kg i.v.) produced a much larger increase in AVP and emesis occurred in five of eight tests. It is relevant that doses of CCK in the range of 1 µg/kg produced significant inhibitions of food intake in monkeys without retching or emesis or other behavioural signs of discomfort (Falasco *et al.* 1979; Moran and McHugh 1982; Metzger and Hansen 1983).

Finally, in 14 humans the threshold CCK dose for increasing plasma AVP and for the occurrence of self-reports of nausea and, in one subject, vomiting was $0.05\,\mu g/kg$ i.v., given as a bolus (Miaskiewicz et al. 1989). It should be noted that the subjects knew that they were receiving CCK injections and they were only asked to assess feelings of nausea on a scale that has not been validated (Morrow 1984; Melzack et al. 1985). Nevertheless, this CCK dose was still larger than doses that have been previously reported to inhibit food intake in humans when CCK was administered by slow continous i.v. infusion (Kissileff et al. 1981; Pi-Sunyer et al. 1982; Stacher et al. 1982; Shaw et al. 1985).

Thus this series of neuroendocrinological experiments in rats, monkeys, and humans is strong evidence that there are doses of CCK ($\leq 5\,\mu g/kg$ in the rat) that inhibit food intake without producing abnormal increases of AVP or OT. The inhibition of intake after larger doses ($\geq 10\,\mu g/kg$ in the rat) probably includes a toxic component.

It is relevant to add that i.p. injection of CCK-8 (2–$4\,\mu g/kg$) produced circulating concentrations of biaoctive CCK in the plasma that were not significantly larger than those produced by a test meal (Smith et al. 1989).

On the basis of the results from CTA experiments, experiments with an anti-emetic drug, and measurement of neurohypophyseal hormones, we conclude that there is no substantial evidence that doses of CCK $\leq 5\,\mu g/kg$ i.p. produce an aversive or subclinical toxic effect that accounts for their inhibition of food intake. When this conclusion is coupled with the facts that behavioural analysis demonstrates that these doses of CCK elicit the complete sequence of behaviours that characterize satiety in rats which would not otherwise show it during sham feeding (Gibbs et al. 1973b; Antin et al. 1975), do not inhibit water intake (Gibbs et al. 1973a), do not elicit abnormal behaviours during real feeding (Gibbs et al. 1973a), do not produce an alteration in body temperature (Gibbs et al. 1973a), and, when administered every other day for months, do not produce any observable sign of clinical toxicity such as weight loss or aversion to the test diet, we feel that we are justified in referring to the inhibition of food intake by CCK as a specific satiating effect of the peptide.

In contrast with the large number of studies in rodents using these indirect measures of toxicity or abnormal visceral experiences, there are a smaller number of studies in humans in which the subjects have been probed by questionnaire and interview for self-reports of pain, nausea, or other abnormal effects of CCK. In 1987 we reviewed the literature and reported that slow continuous intravenous (i.v.) administration of CCK in doses that produced up to 50 per cent inhibition of a test meal was accompanied by some abnormal report in 16 of 121 subjects (13 per cent) (Smith and Gibbs 1987). The abnormal reports were often minor disturbances and no subject could reliably distinguish test days on which CCK was infused from test days on which vehicle infusions were given. It should be noted that when bolus i.v. administration was used in 11 subjects, all of them experienced mild nausea, presumably because of the high transient concentration of CCK obtained with this mode of administration (Shaw *et al.* 1985). Recall that Miaskiewicz *et al.* (1989) used bolus i.v. administration in the study of plasma AVP release, nausea, and vomiting. It is clear that slow continuous i.v. infusion of CCK is the mode of administration that produces inhibition of food intake with the smallest incidence of side-effects.

Thus the more limited data from human studies also supports the interpretation that the inhibition of food intake by low doses of CCK is a specific satiating effect of the peptide.

Site and mechanism of action

In 1981 we hypothesized that the site of action for the satiating effect of exogenous CCK was peripheral, located on alimentary receptors in abdominal organs such as the pancreas, stomach, or small intestine (Smith *et al.* 1981). This hypothesis was based on the rank order of potency of the agonists (see above) and the fact that CCK does not cross the blood–brain barrier (Oldendorf 1981).

If the site of action for the satiating effect is on receptors in the abdomen, then the information about the visceral event (smooth muscle contraction or acinar cell secretion) must be carried to the CNS by afferent fibres of the abdominal vagus nerves or by afferent fibres in the visceral branches of the dorsal root ganglion neurons that project to the spinal cord. Lesion of one or both of these visceral afferent pathways should abolish the satiating effect of CCK. This prediction was confirmed when we and others reported that bilateral abdominal vagotomy abolished the satiating effect of low doses of CCK ($\leq 4\,\mu g/kg$ i.p.) and markedly reduced the effect of higher doses ($8\,\mu g/kg$ i.p.) (Smith *et al.* 1981; Lorenz and Goldman 1982; Morley *et al.* 1982). In contrast, transection of the spinal cord above T_3, a lesion that eliminates all the information from spinal afferents from the gut, did not change the satiating potency of CCK (Lorenz and Goldman 1982).

Transection of the abdominal vagus nerves disconnected efferent as well as afferent fibres. Subsequent experiments demonstrated that the critical lesion involved the afferent fibres (Smith *et al.* 1985), particularly unmyelinated C-fibres sensitive to the toxic action of capsaicin (South and Ritter 1988).

Using selective transections of the various branches of the abdominal vagus, we initially reported that bilateral gastric vagotomies were as effective as total abdominal vagotomy (Smith *et al.* 1981), but this has not been replicated (Kraly 1984; LeSauter *et al.* 1988).

Niijima (1983) then demonstrated that CCK (i.p. or i.v.) increased activity in gastric afferent fibres and decreased activity in hepatic afferent fibres. Subsequent work revealed that some gastric afferent fibres sensitive to gastric distension were also activated by CCK (Davison and Clarke 1988), and that these fibres are destroyed by capsaicin (Raybould and Tache 1988). There is a similar relationship in neurons of the nucleus tractus solitarius (NTS) in the central projection sites of gastric afferent fibres—some of the units responsive to gastric distention are also responsive to CCK (Raybould *et al*. 1985, 1988).

Synergism between gastric distension and CCK for inhibiting food intake has been observed in the monkey (Moran and McHugh 1982). This was the basis for the hypothesis of Moran and McHugh that the satiating effect of CCK is mediated indirectly through the inhibition of gastric emptying which, in turn, inhibits eating through activation of a gastric distension mechanism, presumably mediated through vagal afferent fibres. This hypothesis has been much investigated in rats, and most of the evidence does not support it (Conover *et al*. 1989), although Moran and McHugh (1988) continue to find correlations between the inhibitory effects of CCK on food intake and gastric emptying that are consistent with the hypothesis. It is certainly clear from experiments with sham feeding rats that CCK can inhibit food intake without inhibiting gastric emptying or increasing gastric distension so that a *necessary* relation between gastric distension and satiation has been rejected. However, a modified hypothesis that emphasizes synergistic interaction between CCK and gastric mechanical stimuli is heuristic and can point to the electrophysiological evidence of convergence on units in the NTS described above as an afferent mechanism for such synergism. Such a modified hypothesis deserves more investigation.

What effect of CCK do vagal afferent fibres sense? The two abdominal sites in the rat with the densest population of type A receptors are the pyloric sphincter and the pancreatic acinar cells. Of the two sites, the receptors in the sphincter are more likely to be involved because of the known vagal afferent receptors sensitive to mechanical stimuli. Removal of the pyloric receptors, however, did not abolish the satiating effect of CCK. No change occurred if rats were tested 1 month or longer after pylorectomy (Moran *et al*. 1988; Smith *et al*. 1988), but if rats were tested 1 week after pylorectomy, the satiating potency of larger doses of CCK ($>2\mu$g/kg i.p.) was reduced (Moran *et al*. 1988). Since this pattern is the mirror image of the pattern observed after capsaicin-induced damage of vagal afferent fibres, Moran and McHugh (1988) suggested that these fibres are activated by small and large doses of CCK, presumably through direct activation of type A receptors in the vagal afferent terminals (Moran *et al*. 1987); larger doses of CCK also activate the afferent fibres indirectly by the mechanical stimuli produced by the direct action of CCK on type A receptors in the pyloric sphincter. This is a reasonable working hypothesis, but it neglects two aspects of the relevant data. The first is that pylorectomy removes vagal afferent terminals as well as the receptors in the pyloric sphincter, but the results of pylorectomy have been interpreted as the loss of only the receptors. The second aspect is the fact that transection of all vagal afferent fibres by total abdominal vagotomy reduces, but does not abolish, the satiating effect of CCK doses $\geq 8\,\mu$g/kg i.p. The residual effect could be due to an action of CCK carried over spinal visceral afferents or to an effect of CCK on type A receptors in the area postrema. We have preliminary evidence for the role of spinal visceral afferents (Mindell, Smith, and Gibbs, unpublished data), but lesion of the area postrema does not reliably reduce the satiating potency of CCK (Edwards *et al.* 1986).

Thus we conclude that the site of action of CCK is in the abdomen. The mechanism of action involves type A receptors, but their relative role on muscle cells in the pyloric sphincter, pancreatic acinar cells, and vagal afferent terminals remains to be determined. Vagal afferent fibres are the most important pathway for carrying the information about the abdominal effect of CCK to the brain. Spinal visceral afferents may also carry such information, particularly after large pharmacological doses of CCK. This peripheral mechanism is enhanced by pancreatic glucagon (LeSauter and Geary 1987) and requires an intact serotonin system (Stallone *et al.* 1989; Chapter 23 of this volume), but is not altered by cholinergic muscarinic activity because it is not atropine-sensitive (Smith *et al.* 1981).

Mode of action

Our original hypothesis specified a hormonal mode of action for the satiating effect of CCK. The effect of i.v. (Lorenz *et al.* 1979) and i.p. (Gibbs *et al.* 1973*a*) administration of CCK was consistent with this prediction. Note, however, that CCK released from the small intestine must first enter the portal vein and pass through the liver before gaining access to the systemic circulation. When we infused exogenous CCK-8 into the portal vein, its satiating potency was reduced at least four- to eightfold (Greenberg *et al.* 1987). This result, which has been confirmed (Strubbe *et al.* 1989), means that endogenous CCK-8 could not act to inhibit food intake through a hormonal mode, but would have to operate through a local paracrine mode. This may not be true of larger forms of CCK, such as CCK-22 or CCK-33, because the hepatic mechanism that produces the loss of potency may not affect them (Doyle *et al.* 1984). This possibility is based on the study of the loss of bioactivity of a series of structurally similar gastrin peptides — the potency for gastric acid secretion of peptides larger than eight amino acids was not affected by their entering the liver through the portal vein (Strunz *et al.* 1978).

Gores *et al.* (1986) identified an uptake mechanism for CCK-8 in the isolated perfused liver that could be blocked by taurocholate. We are currently testing the possibility that this mechanism is responsible for the loss of the satiating potency of CCK-8 when it is infused intraportally.

Thus, at least in the case of CCK-8, our original hypothesis must be modified to specify a paracrine mode of action. This is consistent with the report (Reidelberger and Solomon 1986) that the doses of CCK-8 given by slow i.v. infusion to inhibit food intake must be larger than those required to stimulate pancreatic enzyme secretion. This is probably because the concentration of circulating CCK-8 required to reproduce the local concentrations responsible for a paracrine effect is larger than is the case for a hormonal effect.

Experiments with antagonists

The development of specific and potent antagonists for type A and type B CCK receptors has enriched the analysis of the satiating effect of CCK in two important ways. First, the antagonists could be used to test the conclusion derived from experiments with agonists that the satiating effect of exogenous CCK was mediated by type A receptors. Second, the antagonists could be used to test the hypothesis that endogenous CCK released from the small intestine by food stimuli had a significant satiating effect.

Experiments with exogenous CCK

The satiating effect of peripherally administered CCK is abolished by pretreatment with type A antagonists (Schneider *et al.* 1988; Dourish *et al.* 1989a), but not by pretreatment with type B antagonists (Dourish *et al.* 1989a) in the preweanling (Smith *et al.* 1991) and in the adult rat. These results confirm the conclusion from experiments with agonists that the satiating effect of peripherally administered CCK is mediated through type A, but not type B, receptors.

Experiments with endogenous CCK

Experiments with antagonists are crucial for the identification of physiological controls because antagonist experiments test the fundamental prediction of the hypothesis that blockade of the hypothesized control results in degraded function. In the case of CCK released from the small intestine, administration of an antagonist should decrease the hypothesized satiating effect of food ingested during a meal, producing an increase in meal size. In the simplest case, a larger meal size should occur under a wide range of conditions that differ in deprivation, diet, ecology, and/or experience. However, it may be that meal size increases in certain situations, but not in others. This would indicate that the satiating potency of CCK is dependent on other physiological contingencies, centrally or peripherally. In somatosensory systems, we talk about this as the difference between the presence of the stimulus and behavioural effects of the central perceptual processing of that stimulus. Viscerosensory systems follow the same rules; for example, the identical baroreceptor stimulus has a different effect on blood pressure in the resting animal than it does in the exercising animal. This variable relationship also appears to be the case for the satiating effect of endogenous CCK because the efficacy of a CCK antagonist for increasing food intake depends on experimental conditions.

The available evidence can be summarized according to whether the stimuli for CCK release are restricted to the stimuli from the food that is being ingested *during* the test meal or whether the stimuli for CCK release also include stimuli from preloads of nutrients or other treatments that release CCK *prior to* the test meal. We refer to these two kinds of experiments as meal experiments and preload experiments respectively. Although both kinds of experiments test the satiating effect of endogenous CCK released from the small intestine, the critical test of our hypothesis comes from experiments in which the only stimuli for CCK release are derived from the food ingested during the test meal.

There are also relevant data from a third type of experiment in which CCK is released from the small intestine by non-nutrient stimuli. We now review the evidence from these three kinds of experiments.

Meal experiments

Hewson *et al.* (1988) reported a significant increase in test meal size after pretreatment with devazepide, the most specific and potent type A antagonist currently available. This effect has now been replicated in seven other experiments in lean male rats, mice, and pigs (Silverman *et al.* 1987; Strohmayer *et al.* 1988; Watson *et al.* 1988; Reidelberger and O'Rourke 1989; Silver *et al.* 1989; Ebenezer *et al.* 1990; Garlicki *et al.* 1990).Under conditions in which devazepide increased meal size in lean Zucker rats, however, it had no effect on meal size in obese Zucker rats (Strohmayer *et al.* 1989). This result may have important implications for the analysis of this genetic form of obesity.

An important defect in this evidence is the lack of any published report of this kind of experiment in female rats. Such a test is of considerable interest because the results of experiments with agonists demonstrated more variable potency and efficacy of CCK in female rodents than in male rodents (see above).

There are two other related experiments with positive results. The infusion of oleic acid and of Intralipid, a mixture of fats, into the duodenum during sham feeding decreased the volume of liquid diet that was sham fed (Greenberg *et al.* 1989; Yox *et al.* 1989). Since oleate and, to a lesser degree, a mixture of fats release CCK from the small intestine (Lewis and Williams 1990), Greenberg *et al.* (1989) and Yox *et al.* (1989) investigated the role of CCK in the inhibition of sham feeding by administering lorglumide, a type A antagonist, prior to the intraduodenal infusion of oleic acid or Intralipid. Lorglumide reduced, but did not abolish, the satiating effect of these infusions in both experiments. Thus CCK released from the small intestine mediates part of the satiating effect of fatty acids in the small intestine.

We consider 10 positive published reports from eight different laboratories under different experimental conditions to be compelling evidence for the hypothesis that we proposed in 1973. It should be noted, however, that the effect under a specific condition may not be robust. For example, we failed to replicate the result obtained by Hewson *et al.* (1988) despite a long series of experiments in our laboratory and an on-site experiment in Hewson's laboratory (Schneider, Smith, and Gibbs, unpublished data).

We have also failed to demonstrate this effect despite continuous i.v. infusion of devazepide in non-deprived rats in the light phase (Miesner *et al.* 1989), and Cooper *et al.* (1990) reported a similar failure in rats eating a highly palatable sweetened mash in the light phase or laboratory chow in the dark phase.

The effect is also much harder to obtain after more than a few hours of deprivation (Schneider *et al.* 1988), although Garlicki *et al.* (1990) reported a positive result after 20 h of food deprivation.

Since the release of CCK from the small intestine has not been measured in these failures to replicate, it is possible to explain them as failures to release sufficient amounts of CCK. Alternatively, the failures to replicate could be due to the failure to reproduce the other unknown physiological contingencies that are necessary for the central perceptual processing of the CCK stimulus into a measurable satiating effect. We consider the second possibility to be much more likely, but only further work can evaluate the relative power of these two explanations.

Preload experiments

Shillabeer and Davison (1984, 1985) were the first to use oral preload experiments to investigate the satiating effect of small intestinal CCK. They reported that proglumide, a weak antagonist of CCK and gastrin, decreased the satiating effect of a small oral preload but not of a large oral preload. Furthermore, abdominal vagotomy blocked this effect of proglumide (Shillabeer and Davison 1985). Shillabeer and Davison interpreted the effect of vagotomy as evidence that the effect of proglumide was due to its blockade of endogenous CCK because there would be no satiating effect of CCK to be antagonized by proglumide in the vagotomized rat.

Despite a scrupulous attempt to mimic the conditions of these experiments, we could not replicate the results even though we showed that the proglumide treatment *abolished* the satiating effect of exogenous CCK under these conditions (Schneider *et al.* 1986).

Dourish *et al.* (1989*b*), however, demonstrated that administration of the more potent and selective antagonist, devazepide, abolished or reduced the satiating effect of an oral preload. The effect was an inverted U-function of dose and the threshold was low − 100 ng/kg. They also reported that even lower doses (threshold of 1 ng/kg) of a selective and potent type B antagonist, L-365,260, produced the same effect. Since L-365,260 was more potent, they interpreted their results as evidence that the satiating effect of CCK was mediated through type B receptors and not through type A receptors. These results and their interpretation were startling because all the previous evidence from agonist and antagonist experiments pointed to mediation of the CCK effect by a type A receptor, not by a type B receptor.

Garlicki *et al.* (1990) also blocked the satiating effect of a preload of food on a subsequent test meal with devazepide. They did not attempt experiments with L-365,260.

Three subsequent reports do not support the suggested importance of type B receptors. First, Dourish *et al.* (1989*a*) demonstrated that devazepide blocked the satiating effect of exogenous CCK in adult rats, but L-365,260 did not. Second, we observed the same differential effect in preweanling rats (Smith *et al.* 1991). Third, Corwin *et al.* (1991) demonstrated that an even lower dose of devazepide (10 ng/kg i.p.) attenuated the satiating effect of an oral preload, but L-365,260 (10 ng/kg–100 μg/kg) had no effect in experiments that were very similar in design to those of Dourish *et al.* (1989*b*).

We conclude that these preload experiments provide further evidence for the satiating effect of endogenous CCK, presumably released from the small intestine. The results are consistent with the mediation of this effect through a type A receptor. The role of type B receptors requires further investigation.

Since these experiments test the satiating effect of endogenous CCK released *before* a meal on the size of that meal, they suggest an additional role for endogenous CCK beyond the one we originally hypothesized. This role would be that CCK is one of the mechanisms by which the satiating effect of food ingested at one meal affects the onset or size of the next meal. Miesner *et al.* (1989) recently observed an example of this. When devazepide was infused continuously for 2.5 h during the light phase, the size of the first meal increased and the duration of the interval until the second meal began was shorter than normal. Although neither of these effects reached statistical significance, the satiating effect of the food ingested in the first meal expressed as the ratio of the intermeal interval (minutes) over the volume of liquid food ingested (millilitres) was significantly decreased. This effect is consistent with the fact that exogenous CCK has produced the opposite result (Hsiao *et al.* 1979; Wiener *et al.* 1984; West *et al.* 1987).

Wolkowitz *et al.* (1990) have recently reported a related effect in humans. When devazepide was administered after an overnight fast, subjects reported significantly more hunger and a trend toward less fullness 90 min later.

Release of CCK from the small intestine by non-nutrient stimuli

Experiments that involve the release of CCK by non-food stimuli have also produced results that support the hypothesized satiating effect of CCK. For example, intragastric intubation of trypsin inhibitor, the most potent stimulus for CCK release currently available (Liddle *et al.* 1984), decreased the size of a test meal in 9–12-day-old rats (Weller *et al.* 1990). This inhibition of food intake was abolished by pretreatment with devazepide (Weller *et al.* 1990).

Garlicki *et al.* (1990) obtained similar results in adult rats that were deprived of food for 20 h. However, we failed to inhibit food intake by infusing trypsin inhibitor into the stomach or duodenum in adult rats deprived of food for 17 h despite releasing significant amounts of CCK to increase the concentration of plasma CCK to about four times the fasting concentrations as measured by bioassay (Smith *et al.* 1989).

Garlicki *et al.* (1990) also released CCK by diverting pancreatic and biliary juice from the small intestine. This produced an inhibition of food intake that was reversed by pretreatment with devazepide.

Conclusion

Recent antagonist experiments with single meals, oral preloads, and non-nutrient stimuli that release CCK provide convergent evidence with the earlier results from experiments with agonists for the satiating effect of endogenous CCK released from the small intestine. This convergent evidence not only proves the essence of our 1973 hypothesis, but it also suggests that endogenous CCK released by one meal can alter the onset and size of the next meal.

The mode of action of endogenous CCK-8 is paracrine. The mode of action of larger forms of CCK, such as CCK-22 or CCK-33, is likely to be endocrine, as we hypothesized, but further work is required to demonstrate this.

The evidence from both agonist and antagonist experiments is also compelling for mediation of the satiating effect by type A receptors, but the critical abdominal site of these receptors has not been identified. The results obtained by Dourish *et al.* (1989*b*) also suggest a role for type B receptors, but more work is required to determine the conditions under which type B receptors are involved and the site of their action (see Chapter 21 of this volume for further discussion).

Vagal afferent fibres are necessary to carry the information about type A receptor stimulation from the abdominal site to the brain where it is processed into a satiating effect on eating behaviour. The mechanisms of this central perceptual processing of the abdominal signal are completely unknown, but it seems likely that CCK released from neurons in sites like the paraventricular hypothalamus (Schwartz *et al.* 1988) and lateral hypothalamus (Schick *et al.* 1987) will be involved. The investigation of these mechanisms will be the work of the next decade.

Acknowledgements

The authors thank Jane Magnetti for her expert processing of this manuscript. The writing of the manuscript was supported by RSA MH00149 and DK33248.

References

Antin, J., Gibbs, J., Holt, J., Young, R. C., and Smith, G. P. (1975). Cholecystokinin elicits the complete sequence of satiety in rats. *J. Comp. Physiol. Psychol.* **89**, 784–90.

Conover, K. L., Collins, S. M., and Weingarten, H. P. (1989). Gastric emptying changes are neither necessary nor sufficient for CCK-induced satiety. *Am. J. Physiol.* **256**, R56–62.

Cooper, S. J., Dourish, C. T., and Barber, D. J. (1990). Reversal of the anorectic effect of (+)-fenfluramine in the rat by the selective cholecystokinin receptor antagonist MK-329. *Br. J. Pharmacol.* **99**, 65–70.

Corwin, R. L., Gibbs, J., and Smith, G. P. (1991). Increased food intake after type A but not type B cholecystokinin receptor blockade. *Physiol. Behav.* **50**, 255-8.

Davison, J. S. and Clarke, G. D. (1988). Mechanical properties and sensitivity to CCK of vagal gastric slowly adapting mechanoreceptors. *Am. J. Physiol.* **255**, G55-61.

Deutsch, J. A. and Hardy, W. T. (1977). Cholecystokinin produces bait shyness in rats. *Nature* **266**, 196.

Dourish, C. T., Ruckert, A. C., Tattersall, F. D., and Iversen, S. D. (1989*a*). Evidence that decreased feeding induced by systemic injection of cholecystokinin is mediated by CCK-A receptors. *Eur. J. Pharmacol.* **173**, 233-4.

Dourish, C. T., Rycroft, W., and Iversen, S. D. (1989*b*). Postponement of satiety by blockade of brain cholecystokinin (CCK-B) receptors. *Science* **245**, 1509-11.

Doyle, J. W., Wolfe, M. M., and McGuigan, J. E. (1984). Hepatic clearance of gastrin and cholecystokinin peptides. *Gastroenterology* **87**, 60-8.

Ebenezer, I. S., de la Riva, C., and Baldwin, B. A. (1990). Effects of the CCK receptor antagonist MK-329 on food intake in pigs. *Physiol. Behav.* **47**, 145-8.

Edwards, G. L., Ladenheim, E. E., and Ritter, R. C. (1986). Dorsal hindbrain participation in cholecystokinin-induced satiety. *Am. J. Physiol.* **251**, R971-7.

Falasco, J. D., Smith, G. P., and Gibbs, J. (1979). Cholecystokinin suppresses sham feeding in the rhesus monkey. *Physiol. Behav.* **23**, 887-90.

Flanagan, L. M., Verbalis, J. G., and Stricker, E. M. (1988). Nalaxone potentiation of effects of cholecystokinin and lithium chloride on oxytocin secretion, gastric motility and feeding. *Neuroendocrinology* **48**, 668-73.

Garlicki, J., Konturek, P. K., Majka, J., Kwiecien, N., and Konturek, S. J. (1990). Cholecystokinin receptors and vagal nerves in control of food intake in rats. *Am. J. Physiol.* **258**, E40-5.

Gibbs, J., Young, R. C., and Smith, G. P. (1973*a*). Cholecystokinin decreases food intake in rats. *J. Comp. Physiol. Psychol.* **84**, 488-95.

Gibbs, J., Young, R. C., and Smith, G. P. (1973*b*). Cholecystokinin elicits satiety in rats with open gastric fistulas. *Nature* **245**, 323-5.

Gores, G. J., La Russo, N. F., and Miller, L. J. (1986). Hepatic processing of cholecystokinin peptides. I. Structural specificity and mechanism of hepatic extraction. *Am. J. Physiol.* **250**, G344-9.

Greenberg, D., Smith, G. P., and Gibbs, J. (1987). Infusion of CCK-8 into the hepatic-portal vein fails to reduce food intake in rats. *Am. J. Physiol.* **252**, 1015-18.

Greenberg, D., Torres, N. I., Smith, G. P., and Gibbs, J. (1989). The satiating effects of fats is attenuated by the cholecystokinin antagonist lorglumide. *Ann. NY Acad. Sci.* **575**, 517-20.

Hewson, G., Leighton, G. E., Hill, R. G., and Hughes, J. (1988). The cholecystokinin receptor antagonist L364,718 increases food intake in the rat by attenuation of the action of endogenous cholecystokinin. *Br. J. Pharmacol.* **93**, 79-84.

Hsiao, S., Wang, C. H., and Schallert, T. (1979). Cholecystokinin, meal pattern, and the intermeal interval: Can eating be stopped before it starts? *Physiol. Behav.* **23**, 909-14.

Kaye, W. H., Gwirtsman, H. E., and George, D. T. (1989). The effect of bingeing and vomiting on hormonal secretion. *Biol. Psychiat.* **25**, 768-80.

Kissileff, H. R., Pi-Sunyer, F. X., Thornton, J., and Smith, G. P. (1981). Cholecystokinin-octapeptide (CCK-8) decreases food intake in man. *Am. J. Clin. Nutr.* **34**, 154-60.

Kraly, F. S. (1984). Vagotomy does not alter cholecystokinin's inhibition of sham feeding. *Am. J. Physiol.* **246**, R829-31.

Kulkosky, P. J. (1985). Conditioned food aversions and satiety signals. In Experimental assessments and clinical applications of conditioned food aversions (ed. N. S. Braveman and P. Bronstein). *Ann. NY Acad. Sci.* **443**, 330-47.

Lang, R. E., Heil, J. W. E., Ganten, D., Herman, K., Unger, T., and Rascher, W. (1983). Oxytocin unlike vasopressin is a stress hormone in the rat. *Neuroendocrinology* **37**, 314-16.

LeSauter, J. and Geary, N. (1987). Pancreatic glucagon and cholecystokinin synergistically inhibit sham feeding in rats. *Am. J. Physiol.* **253**, R719-25.

LeSauter, J., Goldberg, J. B., and Geary, N. (1988). CCK inhibits real and sham feeding in gastric vagotomized rats. *Physiol. Behav.* **44**, 527-34.

Lewis, L. D. and Williams, J. A. (1990). Regulation of cholecystokinin secretion by food, hormones, and neural pathways in the rat. *Am. J. Physiol.* **258**, G512-18.

Liddle, R. A., Goldfine, I. D., and Williams, J. A. (1984). Bioassay of plasma cholecystokinin in rats: effects of food, trypsin inhibitor, and alcohol. *Gastroenterology* **87**, 542-9.

Lorenz, D. N. and Goldman, S. A. (1982). Vagal mediation of the cholecystokinin satiety effect in rats. *Physiol. Behav.* **29**, 599-604.

Lorenz, D. N., Kreielsheimer, G., and Smith, G. P. (1979). Effect of cholecystokinin, gastrin, secretin and GIP on sham feeding in the rat. *Physiol. Behav.* **23**, 1065-72.

McCann, M., Verbalis, J. G., and Stricker, E. M. (1988). Capsaicin pretreatment attenuates multiple responses to cholecystokinin in rats. *J. Auton. Nerv. Syst.* **23**, 265-72.

McCann, M., Verbalis, J. G., and Stricker, E. M. (1989). LiCl and CCK inhibit gastric emptying and feeding and stimulate OT secretion in rats. *Am. J. Physiol.* **256**, R463-8.

Melzack, R., Rosberger, Z., Hollingworth, M. L., and Thirlwell, M. (1985). New approaches to measuring nausea. *Can. Med. Assoc. J.* **133**, 755-61.

Metzger, B. L. and Hansen, B. C. (1983). Cholecystokinin effects on feeding, glucose and pancreatic hormones in rhesus monkeys. *Physiol. Behav.* **30**, 509-18.

Miaskiewicz, S. L., Stricker, E. M., and Verbalis, J. G. (1989). Neurohypophyseal secretion in response to cholecystokinin but not meal-induced gastric distention in humans. *J. Clin. Endocrinol. Metab.* **68**, 837-43.

Miesner, J. A., Smith, G. P. Gibbs, J., and Tyrka, A. (1991). Intravenous infusion of CCK$_A$ receptor antagonist increases food intake in rats. *Am. J. Physiol.*, in press.

Moore, B. O. and Deutsch, J. A. (1985). An antiemetic is antidotal to the satiety effects of cholecystokinin. *Nature* **315**, 321-2.

Moran, T. H. and McHugh, P. R. (1982). Cholecystokinin suppresses food intake by inhibiting gastric emptying. *Am. J. Physiol.* **242**, R491-7.

Moran, T. H. and McHugh, P. R. (1988). Gastric and nongastric mechanisms for the satiety action of cholecystokinin. *Am. J. Physiol.* **254**, R628-32.

Moran, T. H., Robinson, P. H., Goldrich, M. S. and McHugh, P. R. (1986). Two brain cholecystokinin receptors: implications for behavioral actions. *Brain Res.* **362**, 175-9.

Moran, T. H., Smith, G. P., Hostetler, A. M., and McHugh, P. R. (1987). Transport of cholecystokinin (CCK) binding sites in subdiaphragmatic vagal branches. *Brain Res.* **415**, 149-52.

Moran, T. H., Shnayder, L., Hostetler, A. M. and McHugh, P. R. (1988). Pylorectomy reduces the satiety action of cholecystokinin. *Am. J. Physiol.* **255**, R1059-63.

Morley, J. E., Levine, A. S., Kniep, J., and Grace, M. (1982). The effect of vagotomy on the satiety effects of neuropeptides and naloxone. *Life Sci.* **30**, 1943-7.

Morrow, G. R. (1984). The assessment of nausea and vomiting; past problems, current issues, and suggestions for future research. *Cancer* 53, 2267–80.

Niijima, A. (1983). Glucose sensitive afferent nerve fibers in the liver and their role in food intake and blood glucose regulation. In *Vagal nerve function: behavioral and methodological considerations* (ed. J. G. Kral, T. L. Powley, and C.McC. Brooks), pp. 207–20. Elsevier, New York.

Nussey, S. S., Hawthorn, J., Page, S. R., Ang, V. T. Y., and Jenkins, J. S. (1988). Responses of plasma oxytocin and arginine vasopressin to nausea induced by apormorphine and ipecacuanha. *Clin. Endocrinol. (Oxf.)* 28, 297–304.

Oldendorf, W. H. (1981). Blood-brain barrier permeability to peptides: pitfalls in measurement. *Peptides* 2 (Suppl. 2), 109–11.

Pi-Sunyer, X., Kissileff, H. R., Thornton, J. and Smith, G. P. (1982). C-terminal octapeptide of cholecystokinin decreases food intake in obese men. *Physiol. Behav.* 29, 627–30.

Raybould, H. E. and Taché, Y. (1988). Cholecystokinin inhibits gastric motility and emptying via a capsaicin-sensitive vagal pathway in rats. *Am. J. Physiol.* 255, G242–6.

Raybould, H. E., Gayton, R. J., and Dockray, G. J. (1985). CCK-8: involvement of brainstem neurones responding to gastric distension. *Brain Res.* 342, 187–90.

Raybould, H. E., Gayton, R. J., and Dockray, G. J. (1988). Mechanisms of action of peripherally administered cholecystokinin octapeptide on brainstem neurons in the rat. *J. Neurosci.* 8, 3018–24.

Reidelberger, R. D. and O'Rourke, M. F. (1989). Potent cholecystokinin antagonist L-364,718 stimulates food intake in rats. *Am. J. Physiol.* 257, R1512–18.

Reidelberger, R. D. and Solomon, T. E. (1986). Comparative effects of CCK-8 on feeding, sham feeding, and exocrine pancreatic secretion in rats. *Am. J. Physiol.* 251, R97–105.

Rowe, J. W., Shelton, R. L., Helderman, J. H., Vistal, R. E., and Robertson, G. L. (1979). Influence of the emetic reflex on vasopression release in man. *Kidney Int.* 16, 729–35.

Schick, R. R., Reilly, W. M., Roddy, D. R., Yaksh, T. L., and Go, V. L. W. (1987). Neuronal cholecystokinin-like immunoreactivity is postprandially released from primate hypothalamus. *Brain Res.* 418, 20–6.

Schneider, L. H., Gibbs, J., and Smith, G. P. (1986). Proglumide fails to increase food intake after an ingested preload. *Peptides* 7, 135–40.

Schneider, L. H., Murphy, R. B., Gibbs, J., and Smith, G. P. (1988). Comparative potencies of CCK antagonists for the reversal of the satiating effect of cholecystokinin. In *Cholecystokinin antagonists* (ed. R. Y. Wang and R. Schoenfeld), pp. 263–84. Alan R. Liss, New York.

Schwartz, D. H., Dorfman, D. B., Hernandez, L. and Hoebel, B. G. (1988). Cholecystokinin: 1. CCK antagonists in the PVN induce feeding. 2. Effects of CCK in the nucleus accumbens on extracellular dopamine turnover. In *Cholecystokinin antagonists* (ed. R. Y. Wang and R. Schoenfeld), pp. 285–305. Alan R. Liss, New York.

Shaw, M. J., Hughes, J. J., Morley, J. E., Levine, A. S., Silvis, S. E., and Shafer, R. B. (1985). Cholecystokinin action on gastric emptying and food intake in normal and vagotomized man. *Ann. NY Acad. Sci.* 48, 640–1.

Shillabeer, G. and Davison, J. S. (1984). The cholecystokinin antagonist, proglumide, increases food intake in the rat. *Regul. Pept.* 8, 171–6.

Shillabeer, G. and Davison, J. S. (1985). The effect of vagotomy on the increase in food intake induced by the cholecystokinin antagonist, proglumide. *Regul. Pept.* **12**, 91–9.

Silver, A. J., Flood, J. F., Song, A. M., and Morley, J. E. (1989). Evidence for a physiological role for CCK in the regulation of food intake in mice. *Am. J. Physiol.* **256**, R646–52.

Silverman, M., Bank, S., and Lendvai, S. (1987). The cholecystokinin receptor antagonist, L-364,718 increases food consumption. *Dig. Dis. Sci.* **32**, 1188.

Smith, G. P. (1984). Gut hormone hypothesis of postprandial satiety. In *Eating and its disorders* (ed. A. J. Stunkard and E. Stellar), pp. 67–75. Raven Press, New York.

Smith, G. P. and Gibbs, J. (1987). The effect of gut peptides on hunger, satiety, and food intake in humans. *Ann. NY Acad. Sci.* **499**, 132–6.

Smith, G. P., Jerome, C., Cushin, B. J., Eterno, R., and Simansky, K. J. (1981). Abdominal vagotomy blocks the satiety effect of cholecystokinin in the rat. *Science* **213**, 1036–7.

Smith, G. P., Jerome, C., and Norgren, R. (1985). Afferent axons in abdominal vagus mediate satiety effect of cholecystokinin in rats. *Am. J. Physiol.* **249**, R638–41.

Smith, G. P., Falasco, J., Moran, T. H., Joyner, K. M. S., and Gibbs, J. (1988). CCK-8 decreases food intake and gastric emptying after pylorectomy and pyloroplasty. *Am. J. Physiol.* **255**, R113–16.

Smith, G. P., Greenberg, D., Falasco, J. D., Avilion, A. A., Gibbs, J., Liddle, R. A., and Williams, J.A. (1989). Endogenous cholecystokinin does not decrease food intake or gastric emptying in fasted rats. *Am. J. Physiol.* **257**, R1462–6.

Smith, G. P., Tyrka, A., and Gibbs, J. (1991). Type-A CCK receptors mediate the inhibition of food intake and activity by CCK-8 in 9-to 12-day old rat pups. *Pharmacol. Biochem. Behav.* **38**, 207–10

South, E. H. and Ritter, R. C. (1988). Capsaicin application to central or peripheral vagal fibers attenuates CCK satiety. *Peptides* **9**, 601–12.

Stacher, G., Steinringer, H., Schmierer, G., and Winklehner, C. (1982). Cholecystokinin octapeptide decreases intake of solid food in man. *Peptides* **3**, 133–6.

Stallone, D., Nicolaidis, S., and Gibbs, J. (1989). Cholecystokinin-induced anorexia depends on serotoninergic function. *Am. J. Physiol.* **256**, R1138–41.

Stricker, E. M. and Verbalis, J. G. (1990). Control of appetite and satiety: insights from biologic and behavioral studies. *Nutr. Rev.* **48**, 49–56.

Strohmayer, A. J. and Smith, G. P. (1987). A sex difference in the effect of CCK-8 on food and water intake in obese (ob/ob) and lean (+/+) mice. *Peptides* **8**, 845–8.

Strohmayer, A. J., Greenberg, D., von Heyn, R., Dornstein, L., and Balkman, C. (1988). Blockade of cholecystokinin (CCK) satiety in genetically obese Zucker rats. *Soc. Neurosci. Abstr.* **14**, 1196.

Strohmayer, A., von Heyn, R., Dornstein, L., and Greenberg, D. (1989). CCK receptor blockade by L-364,718 increases food intake and meal taking behavior in lean but not obese Zucker rats. *Appetite* **12**, 240.

Strubbe, J. H., Wolsink, J. G., Schutte, A. M., and Prins, A. J. A. (1989). Hepatic-portal and cardiac infusion of CCK-8 and glucagon induce different effects on feeding. *Physiol. Behav.* **46**, 643–6.

Strunz, U. T., Thompson, M. R., Elashoff, J., and Grossman, M. I. (1978). Hepatic inactivation of gastrins of various chain lengths in dogs. *Gastroenterology* **74**, 550–3.

Verbalis, J. G., McCann, M. J., McHale, C. M., and Stricker, E. M. (1986a). Oxytocin secretion in response to cholecystokinin and food: differentiation of nausea from satiety. *Science* **232**, 1417–19.

Verbalis, J. G., McHale, C. M., Gardiner, T. W., and Stricker, E. M. (1986b). Oxytocin and vasopressin secretion in response to stimuli producing learned taste aversions in rats. *Behav. Neurosci.* **100**, 466–75.

Verbalis, J. G., Richardson, D. W., and Stricker, E. M. (1987). Vasopressin release in response to nausea-producing agent and cholecystokinin in monkeys. *Am. J. Physiol.* **252**, R749–53.

Wager-Srdar, S. A., Morley, J. E., and Levine, A. S. (1986). The effect of cholecystokinin, bombesin and calcitonin on food intake in virgin, lactating, and post-weaning female rats. *Peptides* **7**, 729–34.

Waldbillig, R. J. and O'Callaghan, M. (1980). Hormones and hedonics, cholecystokinin and taste: A possible behavioral mechanism of action. *Physiol. Behav.* **25**, 25–30.

Watson, C. A., Schneider, L. H., Corp, E. S., Weatherford, S. C., Shindledecker, R., Murphy, R. B., Smith, G. P., and Gibbs, J. (1988). The effects of chronic and acute treatment with the potent peripheral cholecystokinin antagonist L-364,718 on food and water intake in the rat. *Soc. Neurosci. Abstr.* **14**, 1196.

Weller, A., Smith, G. P., and Gibbs, J. (1990). Endogenous cholecystokinin reduces feeding in young rats. *Science* **247**, 1589–91.

West, D. B., Greewood, M. R. C., Sullivan, A. C., Prescod, L., Marzullo, L. R., and Triscari, J. (1987). Infusion of cholecystokinin between meals into free-feeding rats fails to prolong the intermeal interval. *Physiol Behav.* **39**, 111–15.

Wiener, S. M., Gibbs, J., and Smith, G. P. (1984). Prolonged satiating effects of cholecystokinin and bombesin in rats. *Soc. Neurosci. Abstr.* **10**, 553.

Wolkowitz, O. M., Gertz, B., Weingartner, H., Beccaria, L., Thompson, K., and Liddle, R. A. (1990). Hunger in humans induced by MK-329, a specific peripheral-type cholecystokinin receptor antagonist. *Biol. Psychiat.* **28**, 169–73.

Yox, D. P., Stokesberry, H., and Ritter, R. C. (1989). Suppression of sham feeding by intraintestinal oleate: blockade by a CCK antagonist and reversal of blockade by exogenous CCK-8. In *The neuropeptide cholecystokinin (CCK), anatomy and biochemistry, receptors, pharmacology and physiology* (ed. J. Hughes, G. Dockray, and G. N. Woodruff), pp. 218–22. Ellis Horwood, Chichester.

18. Gastric mechanisms in CCK satiety

Timothy H. Moran and Paul R. McHugh

Introduction

A role for the brain–gut peptide CCK in the control of food intake was originally proposed in 1973 by Gibbs *et al.* (1973*a*). Since that time, a range of experiments has been carried out detailing the actions of CCK on food intake, examining the mechanisms by which it may exert these effects, and addressing the potential relationship of the actions of exogenously administered CCK to a physiological role for the endogenously released peptide. In this chapter, we shall attempt to place CCK into an integrative chain of physiological events produced by the entrance of ingested food into the gastrointestinal tract and resulting in the termination of food intake. We shall review our own experimental results and those from other laboratories and evaluate the hypothesis that at least part of the inhibition of food intake produced by exogenous CCK results from mimicking the actions of the endogenous peptide in the inhibition of gastric emptying.

To anticipate, we are proposing that endogenous CCK, released from the gastrointestinal tract by the intraluminal presence of the digestion products of fats and proteins, serves as a negative feedback signal for the inhibition of gastric emptying through its interaction with specific receptors localized both to afferent vagal fibres and on circular muscle cells localized to the pyloric sphincter. This action of CCK on the stomach results in a state of gastric distension as feeding continues. Signals arising from this gastric distension are transmitted to the brain along vagal afferent fibres. These signals combine with and amplify vagal signals generated by CCK directly through the activation of vagal CCK receptors to produce the inhibitory actions of CCK on food intake. Thus CCK affects food intake through both gastric and non-gastric actions.

In the following sections we shall outline experimental results which support or challenge this hypothesis and which have resulted in our ideas becoming more focused and more quantitative.

The inhibition of gastric emptying by CCK

Peripheral administration of CCK results in a slowing of the emptying of liquids from the stomach. This action of CCK has been demonstrated in a

Table 18.1. CCK inhibition of gastric emptying of 150 ml saline test meals in rhesus monkeys

Dose (pmol/kg/min)	Volume emptied in 10 min (ml)	Inhibition (%)
0	98.5 ± 3.4	
0.75	74.7 ± 2.7	24.2
1.5	47.8 ± 2.4	51.5
3.0	6.3 ± 3.1	93.6

variety of species including primates (Moran and McHugh 1982) and rodents (Conover *et al.* 1988; Moran and McHugh 1988*a*).

In the primate, CCK exerts a dose-related increase in percentage inhibition of gastric emptying of saline test meals (Table 18.1). The temporal characteristics and magnitude of this inhibition are such that emptying is rapidly inhibited following the initiation of an intravenous (i.v.) infusion of 3 pmol/kg/min, remains inhibited for the duration of infusion, and rapidly returns to baseline levels when the infusion is terminated (Fig. 18.1). The magnitude of the inhibitory action of CCK on gastric emptying at this dose in the primate is such that saline emptying is essentially prevented.

CCK also inhibits gastric emptying in the rat. However, in contrast with results in the primate, inhibition of gastric emptying by CCK in the rodent is

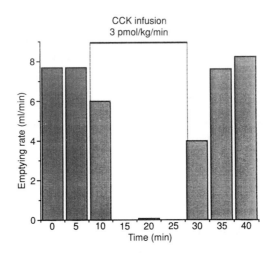

Fig. 18.1. Temporal characteristics of the inhibition of gastric emptying of 100 ml saline test meals by CCK. Intravenous CCK infusion began at 10 min and continued until 30 min. Emptying was rapidly inhibited at onset of infusion (inhibition was complete within 5 min) and remained inhibited for the duration of the infusion. Emptying rapidly returned to baseline levels at the termination of the CCK infusion. (Reproduced with permission from Moran and McHugh 1982.)

only partial. That is, although CCK slows the emptying of saline test meals from the rodent's stomach, emptying is never completely blocked by CCK. Furthermore, other than in a few experiments, there has been little evidence of a dose–response relationship in the gastric inhibitory action of CCK in the rat. The magnitude of suppression produced by 2 and 8 μg/kg of CCK in a 10 min emptying period is essentially the same (Smith *et al.* 1988). This may be due to the rapid degradation of CCK and the administration route which has been used for these experiments.

Mechanisms of gastric inhibition

The particular mechanisms by which CCK inhibits gastric emptying appear to depend upon actions of both the proximal and the distal stomach. CCK is known to relax the proximal stomach and contract the pyloric sphincter. Either one or both of these actions could play a role in the mediation of the gastric inhibitory actions of CCK. A pyloric contraction would reduce the size of the gastric outflow and in this way inhibit the passage of gastric contents to the duodenum. Relaxing the proximal stomach would reduce the pressure gradient between the stomach and the small intestine and thus result in a decrease in the rate at which gastric contents were emptied. CCK appears to act through both of these methods to decrease gastric emptying. Early work by Yamagishi and Debas (1978) in the dog, in which they examined the mechanism of action of CCK in gastric emptying, demonstrated roles for both the distal stomach and the vagus. In their studies, either pyloraplasty or antrectomy resulted in an elimination of the effectiveness of low but not high dosages of CCK to affect gastric emptying. Furthermore, vagotomy completely eliminated the ability of CCK to inhibit the emptying of test meals from the dog's stomach (Yamagishi and Debas 1978).

Results suggesting multiple mechanisms for CCK's gastric inhibitory actions have also been found in the rat. Vagal capsaicin lesions completely eliminate CCK's ability to inhibit the emptying of non-nutrient cellulose test meals (Raybould and Tache 1988) and glucose test meals (McCann *et al.* 1988). Together these results imply that an intact sensory vagus is necessary for CCK's inhibition of gastric emptying to occur. Raybould *et al.* (1987) have also demonstrated that CCK produces a dose-related decrease in intragastric pressure and this decrease is also eliminated by capsaicin treatment (Raybould and Tache 1988). Thus it appears that the vagal mechanism through which CCK affects gastric emptying is by interacting with a vagal afferent pathway mediating reflex decreases in intragastric pressure.

We have demonstrated that the role of the distal stomach in the inhibition of gastric emptying by CCK in the rat is complex (Moran *et al.* 1989). Removal of the pyloric sphincter and its population of CCK receptors does not alter the ability of CCK to affect the emptying of non-nutrient liquids. In contrast, pylorectomy eliminates the ability of CCK to inhibit the emptying of glucose

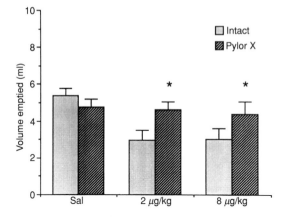

Fig. 18.2. Effect of pylorectomy on CCK's ability to inhibit gastric emptying of 10 ml glucose test meals (0.125 g/ml) over a 10 min emptying period. Injections of 2 and 8 µg/kg of CCK inhibit glucose emptying relative to saline injection in intact but not in pylorectomized rats. Asterisks indicate significant differences from intact values.

test meals beyond the slow rate of emptying produced by glucose itself. As demonstrated in Fig. 18.2, while intraperitoneally (i.p.) administered CCK inhibits the emptying of glucose test meals in intact animals, CCK administration has no effect on the emptying of glucose in animals who have undergone pylorectomy. This result, in combination with those examining the effects of vagotomy on gastric emptying, imply that both an intact vagus and a functional pylorus are necessary for the inhibition of gastric emptying of glucose test meals by exogenous CCK. Either vagal capsaicin lesions or removal of the pylorus blocks CCK's ability to inhibit emptying beyond that produced by the nutrients themselves, i.e. CCK appears to affect the gastric emptying of nutrients at multiple sites and through multiple mechanisms.

Physiological significance of CCK in the regulation of gastric emptying

Through the utilization of specific CCK antagonists, it has been demonstrated that endogenous CCK plays a role in the regulation of the emptying of various nutrients from the stomach. Results obtained by Green *et al.* (1988) have demonstrated that the emptying of peptone test meals in the rat is at least partially under the control of an endogenous release of CCK. Peripheral administration of the type A receptor antagonist devazepide results in an acceleration of the emptying of peptone test meals from the rat's stomach. We have recently employed a similar experimental approach to identify the role of endogenous CCK in the control of gastric emptying in the primate (Ameglio *et al.* 1990).

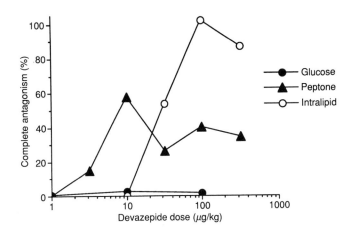

Fig. 18.3. Ability of various doses of the CCK_A receptor antagonist devazepide to alter the gastric emptying of 100 ml glucose, peptone and Intralipid test meals. Different results are found for the three nutrients. Devazepide does not affect the feedback inhibition on gastric emptying produced by glucose test meals, partially antagonizes the inhibition of peptone test meals, and completely antagonizes the inhibition of Intralipid test meals. Therefore endogenous CCK does not mediate the inhibition of glucose emptying, partially mediates the inhibition of peptone emptying, and appears to be the primary mediator of the inhibition of Intralipid emptying from the primate stomach.

Intragastric administration of the CCK_A antagonist devazepide 30 min before the administration of glucose, peptone, or Intralipid test meals had a differential effect on gastric emptying depending upon the type of nutrient in the stomach (Fig. 18.3). The CCK_A antagonist had no effect on the emptying of glucose test meals, but varying doses of devazepide produced dose-related increases in the rate at which protein or fat emptied from the stomach. However, the effect of the antagonist on protein emptying was incomplete, i.e. while the rate at which protein solutions emptied from the stomach was increased by administration of the CCK antagonist, the magnitude of the increase was such that the inhibitory action of the nutrient on gastric emptying was never completely eliminated. In contrast, devazepide completely eliminated the inhibitory action of Intralipid on gastric emptying, i.e. fat emptied from the stomach as if it were saline. These results have a variety of implications. First, the inhibition of gastric emptying produced by exogenously administered CCK appears to be mimicking a physiological action of the peptide in the regulation of gastric emptying. Second, the role of CCK in the regulation of emptying is nutrient-specific. CCK plays no role in the regulation of gastric emptying of glucose test meals, plays a partial role in the feedback inhibition of gastric emptying produced by protein, and seems to be the primary mediator of the inhibitory effect on gastric emptying of fats from the primate's stomach. CCK released by the intraluminal presence of the digestion products of fats and proteins plays a role as a negative feedback signal regulating their further emptying from the stomach and sustaining a gastric volume. Within a meal, this action of the peptide would contribute to the accumulation of the ingested substances within the stomach leading to the production of a state of gastric distention.

Identification of candidate CCK receptor populations for the mediation of gastric inhibitory and satiety actions

Our attempt to identify the particular site of action through which the gastric and feeding inhibitory actions of CCK could be mediated led to work localizing and characterizing CCK receptor populations. In an initial study (Smith *et al.* 1984), we examined the rat gastrointestinal tract for the presence of specific CCK binding sites. Utilizing receptor autoradiography and computerized microdensitometry in an initial screen of the upper gastrointestinal tract, we demonstrated specific binding in the areas of the proximal pyloric sphincter, distal pyloric sphincter, and proximal duodenum. As demonstrated

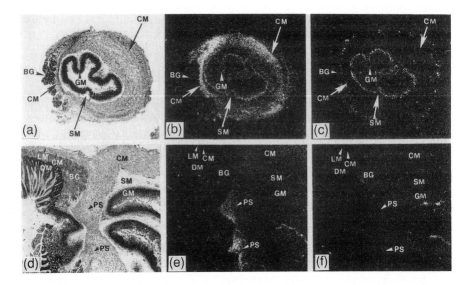

Fig. 18.4. Photomicrographs of toluidine-blue-stained tissue section of rat pyloric sphincter (PS) and corresponding autoradiographs showing total and non-specific CCK binding sites: (a) cross-section of distal pyloric sphincter; (b) autoradiograph of section in (a) showing total CCK binding; (c) autoradiograph of adjacent tissue section showing non-specific CCK binding; (d) longitudinal section of pyloroduodenal junction; (e) autoradiograph showing total CCK binding; (f) autoradiograph showing non-specific CCK binding. LM, longitudinal smooth muscle; CM, circular smooth muscle; SM, submucosa; BG, Brunner's glands; GM, gastric mucosa; DM, duodenal mucosa. (Reproduced with permission from Smith *et al* 1984.)

in Fig. 18.4, the particular anatomical localization of the CCK binding sites corresponds to an area of thickening of the gastric circular smooth muscle occurring in the most distal portion of the pyloric sphincter. Binding to this region was specific, reversible, and of high affinity. Some non-specific binding was evident in the gastric mucosa and the submucosal layer.

These pyloric CCK receptors fit the pharmacological profile for CCK_A receptors (Moran and McHugh 1988b), i.e. while they demonstrate a high affinity for sulphated CCK, unsulphated CCK and gastrin have only a low affinity for these receptors. Furthermore, the specific CCK_A antagonist devazepide has a very high affinity for these receptors.

Work by Murphy *et al.* (1987) and in our laboratory (Margolis *et al.* 1989; Schwartz *et al.* 1989) has demonstrated dose-related contractions of the isolated neonatal and adult rat pylorus in response to CCK. CCK's actions on the isolated pylorus are not mimicked by unsulphated CCK or gastrin, and pyloric contractions produced by CCK are blocked by pretreatment with type A antagonists. Together these results indicate that CCK-induced pyloric contractions are mediated by type A receptors and it is likely that the receptors identified in autoradiographic studies are those that are producing these contractile events. Furthermore, work by Murphy *et al.* (1987) has demonstrated that these *in vitro* contractions have a physiological relevance in that intravenous (i.v.) administration of CCK results in dose-related alterations in pyloric perfusion pressure. Thus these CCK receptors appear to play a role in the regulation of pyloric sphincter function.

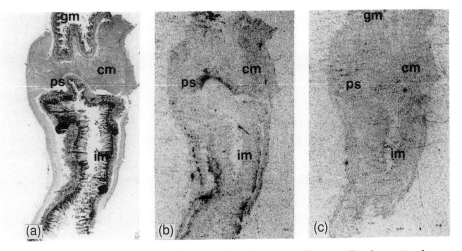

Fig. 18.5. Photomicrographs of (a) cresyl-violet-stained longitudinal section from rhesus monkey pylorus and corresponding autoradiographs showing (b) total and (c) non-specific binding. Specific binding is localized to the most distal portion of the pyloric circular muscle and to the antral submucosal layer.

CCK receptors have also been identified in the region of the pyloric sphincter of the rhesus monkey. As shown in Fig. 18.5, specific CCK binding sites are localized to the most distal portion of the pyloric circular muscle. In contrast with the rat, specific binding is also found in the duodenal submucosal region. Thus pyloric CCK receptors are found in both rat and primate species. Other work has demonstrated the presence of CCK receptors in mouse pylorus (Strohmayer *et al.* 1987) and human pylorus (Robinson *et al.* 1988) as well.

CCK binding sites have also been identified in the vagus nerve. Initial studies by Zarbin *et al.* (1981) demonstrated that CCK binding sites were being transported distally in the cervical vagus. We extended these findings to demonstrate that this transport occurs in all the subdiaphragmatic vagal branches and that these receptors meet the pharmacological profile of CCK_A receptors (Moran *et al.* 1987). We have also demonstrated that this transport of CCK binding sites occurs in afferent rather than efferent fibres (Moran *et al.* 1990), i.e. infraganglionic vagotomy but not supraganglionic vagotomy eliminated the peripheral transport of vagal sites (Fig. 18.6). Both vagotomy procedures disconnect cervical efferent fibres from the cell bodies of the dorsal motor nucleus, and if CCK binding sites were present in efferent fibres both procedures should eliminate CCK binding in the distal axons. Since only the infraganglionic vagotomy disconnects afferent axons from their cell bodies in the nodose ganglion and only this procedure eliminated the peripheral transport

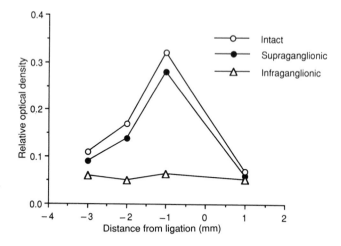

Fig. 18.6. Effect of vagotomies above (supraganglionic) and below (infraganglionic) the nodose ganglion on the vagal transport of CCK binding sites. Transport is indicated by a significant increase in optical density proximal to a ligation site. Infraganglionic vagotomy, which disconnects both afferent and efferent fibres from their cell bodies, eliminates CCK binding site transport. Supraganglionic vagotomy, which disconnects only efferent fibres from their cell bodies, does not affect transport. CCK binding sites are contained and transported in vagal afferent fibres.

of CCK binding sites, it is apparent that transport occurs in afferent fibres.

CCK$_A$ receptors are also present in the brainstem regions of the nucleus tractus solitarius (NTS) and the area postrema (AP) (Moran *et al.* 1986). The localization of CCK$_A$ receptors in both the NTS and vagus suggested the possibility that NTS CCK receptors are of vagal origin and that the NTS represents the termination for central transport of CCK binding sites within vagal afferent axons arising from cell bodies in the nodose ganglion. Data from work by Ladenheim *et al.* (1988), demonstrating a reduction in NTS CCK binding following unilateral nodosectomy, provided initial support for this view. We have further demonstrated that unilateral afferent rootlet transaction results in an ipsilateral depletion in CCK binding in the NTS (Moran *et al.* 1990). In addition, horse-radish peroxidase (HRP) injections after efferent rootlet transections demonstrated that the input of the afferent vagus into the NTS has a significant contralateral component. Furthermore, after a unilateral afferent transection, the distribution of remaining NTS CCK receptors demonstrated a close correspondence to the pattern of HRP reaction product found with unilateral labelling of vagal different input. Thus it appears that the CCK binding sites found in the NTS are presynaptic and of vagal origin. In contrast with this result in the NTS, unilateral nodosectomy, unilateral supraganglionic vagotomy of specific afferent, or afferent rootlet transection had no effect on CCK receptors in the AP. These AP receptors are not of vagal origin and are postsynaptic to any vagal input into this area. Since the AP is a circumventricular organ with a permeable blood–brain barrier, these receptors may be accessible to circulating CCK.

CCK and food intake

As outlined in the Introduction, CCK has been demonstrated to inhibit food intake in a variety of species over a range of experimental paradigms (see Chapter 17 of this volume). Rather than attempting to detail the results from all these experiments, we shall identify a number of general conclusions which can be drawn. First, the behavioural correlates of the inhibition of food intake by CCK mimics the cessation of feeding produced by the ingestion of food, i.e. CCK elicits the complete behavioural sequence of satiety (Antin *et al.* 1975; Chapter 21 of this volume). Animals stop eating, engage in grooming and exploratory behaviour for a period of time, and rest or sleep. Rats receiving exogenous CCK entered into this behavioural sequence after having consumed smaller volumes of food than rats who did not receive CCK. These data provided the initial suggestion that the state produced by exogenous CCK which led to meal termination had some relationship to normal satiety. Controversy continues over the affective state produced by CCK and CCK's overall physiological relevance to food intake. Experiments aimed at identifying the relationship of CCK to normal satiety and nausea have produced results ranging from conditioned aversion to conditioned preference (Holt *et al.* 1974;

Deutsch and Hardy 1977; Swerdlow 1983; Mehiel 1989; Perez and Sclafani 1989). In addition, recent work examining the neurohypophyseal secretion of vasopressin/oxytocin has demonstrated that doses of exogenous CCK produce levels of secretion similar to those found with nausea-producing agents such as lithium chloride (Verbalis *et al.* 1986, 1987). These results have been interpreted by some as demonstrating little physiological relevance for CCK in the control of food intake, and as such stand in contrast with results detailed in the present volume (Chapters 17, 19, and 21) which were obtained using potent CCK antagonists and demonstrate a significant role for endogenous CCK in normal satiety.

Second, CCK affects sham intake as well as normal food intake (Gibbs *et al.* 1973*b*). That is, in animals with an open gastric fistula in which ingested food drains immediately from the stomach, and therefore does not accumulate in the stomach or pass into the intestine, intake is inhibited by exogenous CCK. Thus CCK can affect food intake in the absence of other post-ingestive signals which would normally occur within a meal. Other signals produced by food in the intestine or the stomach are not necessary for CCK's inhibitory actions on food intake to occur.

Finally, CCK's effects on food intake are eliminated by total sub-diaphragmatic vagotomy (Smith *et al.* 1981). Furthermore, combining unilateral afferent rootlet transection with unilateral subdiaphragmatic vagotomy on the contralateral side eliminates the satiety effect of CCK (Smith *et al.* 1985). In contrast, unilateral efferent rootlet transection combined with a unilateral subdiaphragmatic vagotomy does not have this effect. Together these results demonstrate that an intact afferent vagus is necessary for CCK to inhibit feeding.

While it has been demonstrated that the accumulation of food in the stomach or its entrance into the small bowel is not necessary for CCK to influence food intake and that an intact afferent vagus is necessary for these satiety actions, the exact mechanisms by which CCK affects food intake, either when exogenously administered or when endogenously released, are not clear. We originally proposed that CCK's effects on food intake are secondary to its effects on gastric emptying (Moran and McHugh 1982). This was a proposal made on the basis of experiments in rhesus monkeys. In this section we shall outline the evidence which led to this proposal, detailing experiments which have challenged it, and provide a revision of that original proposal.

Work in our laboratory with rhesus monkeys has demonstrated that the ability of an intragastric saline load to affect subsequent food intake depended upon whether the distension produced by that load was maintained or whether the saline emptied rapidly from the stomach. Experiments were conducted in which the intragastric saline load was given either alone or in combination with an intra-intestinal glucose load. We demonstrated that the presence of the intra-intestinal glucose, in volumes that alone had no effect on food intake, provided a negative feedback signal preventing the rapid emptying of the saline test load from the stomach and produced an inhibition of food intake (McHugh and Moran 1985). These experiments demonstrated that the production and maintenance of gastric distension could provide a physiological influence on food intake. Since we had demonstrated that CCK exerted a powerful, and apparently physiological, inhibitory influence on gastric emptying, we sought to determine whether the mechanisms by which CCK influences food intake may have been secondary to this gastric inhibitory action. We designed experiments with rhesus monkeys where the feeding responses were tested in one of four conditions. The first was the control condition, in which neither a gastric saline preload nor an i.v. infusion of CCK was used. In the second condition, monkeys received a 150 ml load of saline into the stomach prior to the 4 h feeding period. In the third condition, the monkeys received an i.v. infusion of CCK beginning 1 h before the onset of feeding and continuing for 10 min into the feeding period. In the fourth condition, the CCK infusion and the saline load to the stomach were combined. As demonstrated in Fig. 18.7, the saline gastric load or the CCK infusion by themselves had no effect on subsequent food intake. However, there was an obvious and significant interaction between these two stimuli, resulting in a significant inhibition of food intake, i.e., a subthreshold dosage of CCK exerted a powerful inhibitory action on food intake when combined with a gastric saline load which was being prevented from emptying by the CCK. We reasoned that the combination of

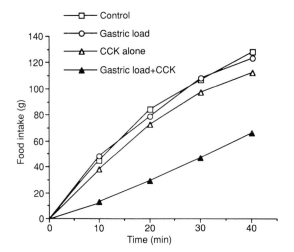

Fig. 18.7. Cumulative food intake in rhesus monkeys under four conditions: control; following 150 ml gastric saline load; feeding while receiving an i.v. infusion of 3 pmol/kg/min of CCK which began 60 min prior to and continued for 10 min into the feeding period; following combined gastric load and CCK infusion. Only the combination inhibited food intake. (Adapted with permission from Moran and McHugh 1982.)

the gastric load and CCK led to the production of a state of gastric distension which in turn was providing the inhibitory signal for food intake.

This original demonstration led to the strongly stated hypothesis that CCK's effects on food intake were completely secondary to its effects on gastric emptying. Our idea was that low physiological dosages of CCK only affected food intake when gastric emptying was inhibited.

This hypothesis has received a variety of challenges. Conover *et al.* (1988) have examined the kinetics of the suppression of feeding and gastric emptying by CCK and have demonstrated that exogenous CCK reduces feeding even when administered 15 min before testing, a time by which it no longer caused an inhibition of gastric emptying. That is, CCK could inhibit food intake at times when it was not able to inhibit gastric emptying. The conclusion was that the inhibition of gastric emptying was not necessary for CCK-induced satiety. Clearly, as we have stated earlier, since CCK inhibits sham feeding, a situation in which an inhibition of gastric emptying is meaningless, the production of a gastric distension signal by CCK is not necessary for CCK to inhibit food intake. Thus, as indicated by these and other experiments, the strong statement that CCK's actions in feeding are completely secondary to its actions in gastric emptying cannot be supported.

Second, since CCK's effects on gastric emptying in the monkey are of a greater magnitude than they appear in the rat, the demonstration of significant synergy between CCK and gastric distension found in the monkey may not even be applicable to the rat. In recent experiments in our laboratory, we have taken advantage of the availability of various CCK analogues which have alterations

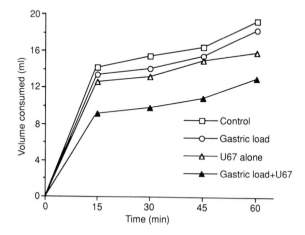

Fig. 18.8. Cumulative glucose intake in rats under four conditions: control; following 5 ml intragastric saline load; following i.p. injection of 0.5 μg/kg of U-67; following gastric load and U-67 injection. Only the combination affected glucose consumption.

in their peptide structure to address this issue. We used the compound U-67, a norleucine-substituted CCK analogue, which has a significantly enhanced *in vivo* bioavailability and a longer duration of action than CCK (Sawyer *et al.* 1988). Rats received one of four experimental manipulations prior to access to a liquid nutrient following food deprivation for 17 h. The four conditions were administration of a 5 ml saline load into the stomach 5 min prior to access to the liquid nutrient, administration of U-67 10 min before access to the liquid nutrient, administration of both the saline load and the peptide, and a control condition in which neither the gastric load nor U-67 was administered. The results were similar to those that we have obtained in the primate. As shown in Fig. 18.8, the saline load by itself had no inhibitory affect on subsequent food intake. A dose of 0.5 μg/kg of U-67 alone also had no effect on food intake. However, U-67 in combination with a saline intragastric load significantly suppressed subsequent intake. Therefore there is a significant synergy between U-67 and intragastric load.

Quantitative contribution of gastric mechanisms to CCK satiety

The experiments demonstrating a synergy between gastric load and exogenous CCK in the inhibition of food intake point to a causal relationship between CCK's action on gastric emptying and the production of a gastric distension signal and CCK's inhibitory influences on food intake. Although this hypothesized mechanism for the satiety effects of CCK has qualitative behavioural and physiological support, the experiments do not provide a quantitative estimate of the contribution of the inhibition of gastric emptying to the inhibition of feeding produced by CCK. We have attempted to quantify this contribution by comparing the magnitude of the inhibition of gastric emptying with the magnitude of the inhibition of food intake produced by a dose range of exogenous CCK.

The results of this experiment are shown in Fig. 18.9. The dark shaded bars demonstrate the mean inhibition of gastric emptying across the four CCK doses. The combination of the dark and light bars shows the mean inhibition of food intake produced by the same dosages. In both cases, increasing dosages of CCK result in increasing levels of inhibition. However, the magnitude of inhibition of feeding is always greater than the magnitude of inhibition of gastric emptying, and is greater by a roughly constant amount. Thus there is a significant inhibition of food intake which is not accounted for by gastric inhibitory mechanisms. Thus both gastric emptying and feeding are inhibited by CCK and these effects are both maximal at a dose of 4 μg/kg. However, there appears to be an initial inhibition of feeding present from the lowest dosages on to which the gastric inhibitory effect adds in a dose-related fashion. This initial and abiding inhibition is documented from the value of the Y intercept of a regression between the two inhibitions produced by CCK, i.e. CCK produces a 28 per cent inhibition of feeding distinct from and not

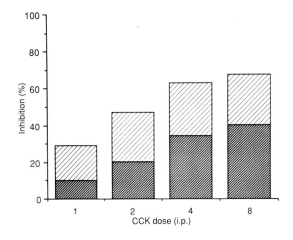

Fig. 18.9. Effects of various doses of CCK on gastric emptying (dark shading) and food intake (total of both shadings). At all doses, the magnitude of inhibition of food intake produced by CCK is greater than the magnitude of the inhibition of gastric emptying. CCK exerts a 28 per cent inhibition of food intake which is not attributable to its inhibition of gastric emptying.

attributable to its inhibition of gastric emptying.

Again, these results demonstrate a few points. CCK inhibits the emptying of the ingested glucose and the magnitude of this inhibition is directly related to and accounts for a significant portion of the magnitude of the suppression of intake produced by the various dosages of CCK. The results also demonstrate that an inhibition of gastric emptying is not necessary for an inhibition of food intake by CCK. Even at the lower dosages of CCK used in this experiment, there is an inhibition of food intake on to which any inhibitory effect of gastric emptying mediated by CCK adds. Thus there appear to be both gastric and non-gastric components of the satiety effect of CCK.

These two components may be mediated by CCK acting at the same or separate receptor sites. For example, the non-gastric component may be mediated by direct action of CCK on vagal CCK receptors and the gastric action may be mediated by a direct effect of CCK on pyloric CCK receptors. Alternatively, since the gastric inhibitory actions of CCK seem to depend, in part, on both vagal and pyloric CCK receptors, these two components of CCK satiety may represent a synergistic interplay of signals and mechanisms along the pathway mediating CCK satiety. These ideas are investigated in subsequent sections.

Effect of pylorectomy on CCK satiety

The demonstration of a gastric component of CCK satiety allows a specific search for the receptor population involved in the mediation of this component.

As we have discussed, the inhibition of gastric emptying by CCK involves both the vagal and the pyloric component. Vagotomy essentially removes the satiety effect of CCK, presumably eliminating the transmission of the information relevant to both the gastric and the non-gastric components of CCK satiety. In an effort to determine whether pyloric CCK receptors play a role in the mediation of CCK satiety, as they do in the mediation of CCK gastric inhibitory actions, and to quantify their contribution to this action of CCK, we compared the ability of CCK to inhibit food intake in the intact and pylorectomized rat (Moran *et al.* 1988). Figure 18.10 demonstrates the results. Prior to pylorectomy, following a 6 h daytime deprivation, CCK exerts a dose-related inhibition of the volume of glucose consumed. Two to three weeks after pylorectomy, the satiety effect of CCK is significantly attenuated. The specific nature of the attenuation of CCK's ability to inhibit food intake again supports the existence of multiple mechanisms in the mediation of CCK satiety. In the intact animal, the effect of CCK on glucose consumption appears to be a smoothly graded progressive reduction of intake with increasing dosages of CCK. Following pylorectomy, the suppression produced by the lower dosages of CCK (i.e. 1 or 2 µg/kg) was unchanged. Higher dosages of CCK, while still exerting a significant suppression on food intake, do not suppress intake further. The further suppression normally produced by these dosages is eliminated by pylorectomy. The elimination of this high dose segment of the dose–response curve supports the view that CCK receptors localized in the gastroduodenal junction provoke some aspect of CCK satiety, an aspect that in the intact animal adds to the low dose mechanisms in a smoothly graded incremental fashion. Critical to this

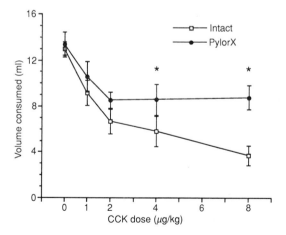

Fig. 18.10. Suppression of glucose intake in intact and pylorectomized rats by 1, 2, 4, and 8 µg/kg CCK. Pylorectomy results in a truncation of the CCK dose–response curve. Asterisks indicate significant differences from the intact level of intake at those doses. (Adapted with permission from Moran *et al.* 1988.)

attenuation is that following pylorectomy there is not a general loss of sensitivity to CCK, but the CCK dose–response curve is truncated as though a major segment of the satiety action of CCK has simply been eliminated.

We cannot say with certainty that what is eliminated by the pylorectomy procedure is the gastric component of CCK satiety. As we have discussed, the gastric inhibitory actions of CCK are complex. CCK's inhibition of gastric emptying depends upon both an intact vagus and an intact pylorus when the gastric contents are nutrients. However, as we have discussed, elimination of pyloric CCK receptors does eliminate the ability of CCK to inhibit the emptying of glucose. Thus, in the context of a loss of the gastric inhibitory action of CCK, the feeding inhibitory actions are significantly attenuated.

Vagal lesions and CCK satiety

Systemic administration of the neurotoxin capsaicin results in the lesion of small diameter afferent fibres. This lesion is reported to be relatively specific in that it affects small diameter unmyelinated afferent fibres while leaving other afferents intact (Jansco 1981). Work by Ritter and Ladenheim (1985) has demonstrated that such a lesion results in an alteration in the ability of CCK to affect food intake. In contrast with the results with pylorectomy, however, the nature of this attenuation is such that the dose–response inhibition of food intake produced by CCK is shifted to the right, i.e. the effects of low dosages of CCK on food intake were eliminated while higher dosages continued to inhibit food intake. We have replicated their finding, utilizing the same paradigm used in the pylorectomy experiments of rats ingesting a glucose solu-

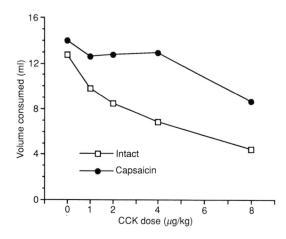

Fig. 18.11. Suppression of glusoce intake in intact rats and rats receiving systemic capsaicin administration by 1, 2, 4, and 8 μg/kg CCK. Capsaicin administration results in a loss of sensitivity to CCK.

tion following a 6 h daytime deprivation. These results are shown in Fig. 18.11. Following capsaicin administration, the 1, 2, and 4 μg/kg dosages of CCK are no longer effective in suppressing intake. The highest dose (8 μg/kg) begins to exert a dose-related inhibition on food intake. Capsaicin administration produces a qualitatively different attenuation of the feeding inhibitory actions of CCK than that produced by pylorectomy. Rather than truncating the dose-response curve, capsaicin produces a loss of sensitivity.

Although capsaicin administration significantly attenuates the satiety affect of CCK, the feeding inhibitory actions of gastric distension are not affected by these lesions. Results obtained by Ritter and Ladenheim (1985) demonstrated that a nutrient preload administered prior to access to food intake was equally effective in suppressing food intake in intact and capsaicinized rats. These results with capsaicin lesions can be contrasted with those found following complete subdiaphragmatic vagotomy. Subdiaphragmatic vagotomy elimi-nates the satiety effect of both CCK (Smith *et al.* 1981) and gastric distension (Gonzalez and Deutsch 1981). Capsaicin lesions attenuate the satiety effect of CCK but do not affect the feeding inhibition produced by gastric distension. An implication of this pattern of results is that the residual satiety effect of CCK evident at the higher dosages following the capsaicin lesion may be secondary to the production of the gastric distension signal. It is this component which is eliminated by pylorectomy.

Integration of gastric and non-gastric mechanisms in CCK satiety

Although a role for vagal afferents in the mediation of CCK satiety has been documented, the issue of how signals resulting from the gastric and non-gastric

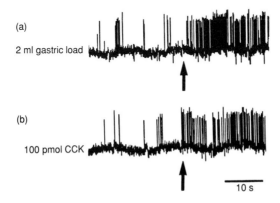

Fig. 18.12. Electrophysiological response of a gastric vagal afferent fibre to (a) an intragastric 2 ml load of saline and (b) coeliac artery infusion of 100 pmol of CCK. The fibre increases its rate of firing in response to both stimuli.

actions of CCK are transmitted over the vagus remains to be determined. There are some electrophysiological data that are relevant to this point.

It has been known for over 30 years that a population of gastric vagal afferent fibres increase their rate of firing in response to gastric distension (Paintal 1953). Furthermore, Davison and Clarke (1988) and Raybould and Davison (1989) have demonstrated that at least some of these vagal afferents also increase their rate of firing in response to i.v. or close arterial injection of CCK. Recent work in our laboratory has documented that at least two populations of gastric vagal afferents responding to gastric distension can be identified (Schwartz *et al.* 1991). As shown in Fig. 18.12, one subset is responsive to both phasic and static gastric distension, is slowly adapting, and shows an off response when a distension signal is removed. This population also demonstrates a significant increase in firing in response to discrete local intra-aortic infusions of CCK lasting for as long as 15 min. These gastric vagal afferents which respond to both distension and CCK also show an enhanced response to gastric distension following their activation by CCK. That is, as shown in

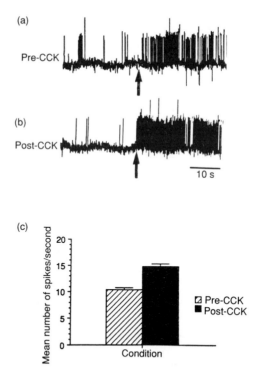

Fig. 18.13. Effect of arterial infusion of CCK on the response of a gastric vagal afferent fibre to a 2 ml intragastric load: (a) response to load prior to CCK; (b) response to load following CCK when the baseline rate of activity was reacquired; (c) graphical representation of pre- and post-CCK firing rates.

Fig. 18.13, while a 2 ml gastric distension load increases vagal firing in these fibres prior to an administration of CCK, once the response to CCK has ended and the baseline firing rate is re-acquired, their response to the same 2 ml load is significantly enhanced. The other population of vagal afferents which respond to distension responds only to phasic changes and adapts rapidly. These fibres fail to respond to CCK. Since the response to distension was used to identify fibres for further study, we do not know if there is a third set that responds to CCK but not to gastric distension. This possibility is suggested by the work of Nijiima (1983) demonstrating fibres in the hepatic vagus which respond to CCK.

The above pattern of results greatly advances our understanding of the relationship of CCK to gastric distension. Thus, not only does CCK produce gastric distension through the inhibition of gastric emptying, but its actions on vagal afferents mimic those produced by gastric distension even in the absence of food in the stomach. Therefore CCK's ability to inhibit sham feeding may not indicate that its mechanism of action is completely independent of gastric events — CCK directly mimics gastric events. Finally, CCK has a long duration of action on vagal fibres and enhances the vagal afferent response to subsequent gastric distension. This duration and enhancement may explain why CCK can affect food intake at times at which it no longer inhibits gastric emptying.

There are a number of points that can be made from these results in relation to the behavioural and physiological experiments we have already discussed. The results of the behavioural experiments suggest that information relevant to the gastric and non-gastric actions of CCK may have been carried to the brain by separate vagal fibres. In this view, one set of vagal afferents would respond to and transmit information relevant to gastric distension, i.e. activity in these fibres is provoked indirectly by CCK as a result of CCK's inhibition of gastric emptying. Another set of vagal fibres containing CCK receptors may have been directly stimulated by CCK. The co-ordination of these signals as reflected in the inhibition of food intake would depend on interactions that occurred in the brain beyond the first synapses of vagal afferents. This kind of organization would explain such results as our findings of the truncation of the CCK dose response by pylorectomy, the attenuation of CCK satiety but the retention of the inhibitory action of gastric distension on feeding with capsaicin administration, and the loss of CCK satiety with total subdiaphragmatic vagotomy. However, the electrophysiological data demonstrating that at least some fibres respond to both gastric distension and CCK make this an unlikely possibility.

An alternative possibility is that CCK's inhibition of food intake is the outcome of a set of distributed functions within a common vagal circuit. Intestinally released CCK could activate the circuit through several portals simultaneously and exert its effects through the activity of one set of vagal fibres. Thus by inhibiting gastric emptying through both pyloric and vagal mechanisms, CCK produces gastric distension resulting in the activation of vagal afferent fibres sensitive to distension. CCK activating CCK receptors on these same fibres could enhance their basal firing and amplify their response to stomach-generated distension discharges. There might even be further amplification of these vagal signals at their place of reception in the brainstem since CCK receptors are also present in the NTS. Some aspects of this model seem likely.

However, it may be that CCK inhibits food intake through both these mechanisms. For example, recent data evaluating the effects of various types of vagotomy on CCK satiety have demonstrated that, while a total sub-diaphragmatic vagotomy eliminates CCK satiety, if any subdiaphragmatic vagal branch is left intact, major portions of the feeding inhibitory actions of CCK remain (Le Sauter *et al.* 1988). That is, some of the aspects of CCK satiety may be independent of the gastric vagal branches. CCK receptors localized to these other branches may interact directly with intestinally derived CCK and information relevant to this interaction may play a role in CCK satiety. Further-more, since the possibility remains that there is a population of gastric vagal afferents which respond to CCK but not to gastric distension, interactions of this type may also occur in the gastric vagus. Thus the feeding inhibitory actions of CCK may depend upon the actions of CCK in the production and modula-tion of gastric distension signals as well as upon the direct stimulation of other vagal afferent fibres within the abdominal cavity with little or no linkage to the stomach. The summation of these actions may result in the full magnitude of CCK's inhibition of food intake.

Thus the model that we are proposing goes far beyond our original idea that there was a unitary function to CCK satiety. Our original hypothesis that all the satiety actions of CCK were secondary to the gastric inhibitory actions of CCK cannot be supported. However, we have presented evidence that the gastric inhibitory actions of CCK substantially contribute to the feeding inhibitory actions of the peptide. Furthermore, CCK appears to act in the inhibition of food intake at a multiplicity of sites, engaging a variety of mechanisms which share in common their direct or indirect transmission through the abdominal vagus.

Summary

Our aim in this work has been to place the physiological functions of CCK into a sequence or pattern that would indicate how such an intestinally derived peptide would facilitate the meal-taking characteristics of normal feeding. The gastric hypothesis has the important feature of suggesting how CCK, promptly released from the intestine at essentially the first moment of entry of food into the gastrointestinal tract, might have a delayed inhibitory action on satiety permitting the ingestion of a meal. Although our focus has been on the peripheral physiology of satiety and we may well have overlooked important central events, our several observations of critical anatomical and physiological linkages between the stomach and the vagus with respect to CCK continue to encourage us in the view that the dynamic action of CCK on the stomach and on the stomach's afferent linkages to the brain play an important role in the mediation of CCK satiety.

Acknowledgements

This work has been supported by NIH grant DK 19302 and by a generous gift from the Lorraine and Leonard Levin Fund for Research in Biological Psychiatry.

References

Ameglio, P. J., Seeb, D. H., McHugh, P. R., and Moran, T. H. (1990). Endogenous cholecystokinin (CCK) inhibits gastric emptying in rhesus monkeys. *Eastern Psychol. Assoc.* **61**, 43 (abstr.).

Antin, J., Gibbs, J., Holt, J., Young, R. C., and Smith, G. P. (1975). Cholecystokinin elicits the complete behavioral sequence of satiety in rats. *J. Comp. Physiol. Psychol.* **89**, 784–90.

Conover, K. L., Collins, S. M., and Weingarten, H. P. (1988). A comparison of CCK-induced changes in gastric emptying and feeding in the rat. *Am. J. Physiol.* **225**, R21–6.

Davison, J. S. and Clark, G. D. (1988). Mechanical properties and sensitivity to CCK of vagal gastric slowing adapting mechanoreceptors. *Am. J. Physiol.* **225**, G55–61.

Deutsch, J. A. and Hardy, W. T. (1977). Cholecystokinin produces bait shyness in rats. *Nature* **226**, 196.

Gibbs, J., Young, R. C., and Smith, G. P. (1973a) Cholecystokinin decreases food intake in rats. *J. Comp. Physiol. Psychol.* **84**, 488–95.

Gibbs, J., Young, R. C., and Smith, G. P. (1973b). Cholecystokinin elicits satiety in rats with open gastric fistulas. *Nature* **245**, 323–5.

Gonzalez, M. F. and Deutsch, J. A. (1981). Vagotomy abolishes cues of satiety produced by gastric distension. *Science* **212**, 1283–4.

Green, T., Dimaline, R., Peikin, S. and Dockray, G. J. (1988). Action of the cholecystokinin antagonist L364,718 on gastric emptying in the rat. *Am. J. Physiol.* **255**, G685–9.

Holt, J., Antin, J., Gibbs, J., Young, R. C., and Smith, G. P. (1974). Cholecystokinin does not produce bait shyness in rats. *Physiol. Behav.* **12**, 497–8.

Jancso, G. (1981). Intracisternal capsaicin: Selective degeneration of chemosensitivity primary sensory afferents in the adult rat. *Neurosci. Lett.* **27**, 41–5.

Ladenheim, E. E., Speth, R. C., and Ritter, R. C. (1988). Reduction of CCK-8 binding in the nucleus of the solitary tract in unilaterally nodosectomized rats. *Brain Res.* **474**, 125–9.

Le Sauter, J., Goldberg, B., and Geary, N. (1988). CCK inhibits real and sham feeding in gastric vagotomised rats. *Physiol. Behav.* **44**, 527–34.

Margolis, R. L., Moran, T. H., and McHugh, P. R. (1989). *In vitro* response of rat gastrointestinal segments to cholecystokinin and bombesin. *Peptides* **10** 157–61.

McCann, M. J., Verbalis, J. G., and Stricker, E. M. (1988). Capsaicin pretreatment attenuates multiple responses to cholecystokinin in rats. *J. Autonom. Nerv. System* **23**, 265–72.

McHugh, P. R. and Moran, T. H. (1985). A conception of its dynamic role in satiety. In *Progress in psychobiology and physiological psychology*, Vol. 11 (ed. A. Epstein and J. Sprague), pp. 197–232. Academic Press, New York.

Mehiel, R. (1989). Rats learn to like flavors with CCK. *Eastern Psychol. Assoc.* **60**, 39 (abstr.).

Moran, T. H. and McHugh, P. R. (1982). Cholecystokinin suppresses food intake by inhibiting gastric emptying. *Am. J. Physiol.* **242**, R491–7.

Moran, T. H. and McHugh, P. R. (1988*a*). Gastric and non-gastric mechanisms for satiety action of cholecystokinin. *Am. J. Physiol.* **254**, R628–32.

Moran, T. H. and McHugh, P. R. (1988*b*). Anatomical and pharmacological differentiation of pyloric, vagal and brainstem cholecystokinin (CCK) receptors. In *CCK Antagonists* (ed. R. Wang and R. Schoenfeld), pp. 117–32. Alan Liss, New York.

Moran, T. H., Robinson, P. H., Goldrich, M. S., and McHugh, P. R. (1986). Two brain cholecystokinin receptors: implications for behavioral action. *Brain Res.* **362**, 175–9.

Moran, T. H., Smith, G. P., Hostetler, A. M., and McHugh, P. R. (1987). Transport of cholecystokinin (CCK) binding sites in subdiaphragmatic vagal branches. *Brain Res.* **415**, 149–52.

Moran, T. H., Shnayder, L., Hostetler, A. M., and McHugh, P. R. (1988). Pylorectomy reduces the satiety action of cholecystokinin. *Am. J. Physiol.* **225**, R1059–63.

Moran, T. H., Crosby R. J., and McHugh, P. R. (1989). Effects of pylorectomy on the inhibition of gastric emptying by cholecystokin in (CCK). *Neurosci Abstr.* **14**, 1280.

Moran, T. H., Norgren, R., Crosby, R. J., and McHugh, P. R. (1990). Central and peripheral transport of vagal cholecystokinin binding sites occures in afferent fibres. *Brain Res.*, **526**, 95–102.

Murphy, R. B., Smith, G. P., and Gibbs, J. (1987). Pharmacological examination of cholecystokinin (CCK-8)-induced contractile activity in the rat isolated pylorus. *Peptides* **8**, 127–34.

Niijima, A. (1983). Glucose-sensitive afferent nerve fibres in the liver and their role in blood glucose regulation. *J. Auton, Nerv. Syst.* **9**, 207–16.

Paintal, A. S. (1953). Impulses in vagal afferent fibres from stretch receptors in the stomach and their role in the peripheral mechanism of hunger. *Nature* **172** 1194–5.

Perez, C. and Sclafani, A. (1989). Is satiety rewarding? The case of CCK. *Eastern Psychol. Assoc.* **60**, 39 (abstr).

Raybould, H. E. and Davison, J. S. (1989). Perivagal application of capsaicin abolishes the response of vagal gastric mechanoreceptors to cholecystokinin. *Soc. Neurosci. Abstr.* **15**, 973.

Raybould, H. E. and Tache, Y. (1988). Cholecystokinin inhibits gastric motility and emptying via a capsaicin-sensitive vagal pathway in rats. *Am. J. Physiol.* **255**, G242–6.

Raybould, H. E., Roberts, M. E., and Dockray, G. J. (1987). Reflex decreases in intragastric pressure in response to cholecystokinin in rats. *Am. J. Physiol.* **253**, G165–70.

Ritter, R. C. and Ladenheim, E. E. (1985). Capsaicin pretreatment attenuates suppression of food intake by cholecystokinin. *Am. J. Physiol.* **248**, R501–4.

Robinson, P. H., McHugh, P. R., Moran, T. H., and Stephenson, J. D. (1988). Gastric control of food intake. *J. Psychosom. Res.* **32**, 593–606.

Sawyer, T. K., Jensen, R. T., Moran, T. H., Schreur, P. J. K. D., Staples, D. J., de

Veux, A. E., and Hsi, A. (1988). Structure-activity of cholecystokinin-8 analogs: comparison of pancreatic, pyloric sphincter and brainstem CCK receptor activities with *in vivo* anorexigenic effects. In *Peptides: chemistry and biology. Proceedings of X American Peptide Symposium* (ed. G. R. Marshall), pp. 503–5. Escom, Leiden.

Schwartz, G. J., McHugh, P. R., and Moran, T. H. (1989). Effects of sulfated and non-sulfated cholecystokinin on pyloric contraction in the developing rat. *Soc. Neurosci. Abstr.* **14**, 1279.

Schwartz, G. J., McHugh, P. R., and Moran, T. H. (1991). Integration of vagal afferent responses to gastric loads and cholecystokinin in rats. *Am. J. Physiol.* **261**, 264–9.

Smith, G. P., Jerome, C., Cushin, B. J., Eterno, R., and Simansky, K. J. (1981). Abdominal vagotomy blocks the satiety effect of cholecystokinin in the rat. *Science*, **213**, 1036–7.

Smith, G. P., Jerome, C., and Norgren, R. (1985). Afferent axons in abdominal vagus mediate satiety effect of cholecystokinin in rats. *Am. J. Physiol.* **249** R638–41.

Smith, G. P., Falasco, J., Moran, T. H., Joyner, K. M. S., and Gibbs, J. (1988). CCK-8 decreases food intake and gastric emptying after pylorectomy or pyloroplasty. *Am. J. Physiol.* **225**, R113–6.

Smith, G. T., Moran, T. H., Coyle, J. T., Kuhar, M. J., O'Donahue, T. L., and McHugh, P. R. (1984). Anatomical localization of cholecystokinin receptors in the pyloric sphincter. *Am. J. Physiol.* **246**, R127–30.

Strohmayer, A. J., Greenberg, D., Moran, T. H., and McHugh, P. R. (1987). CCK receptor binding in the pylorus of genetically obese mice (ob/ob). *Soc. Neurosci. Abstr.* **13**, 587.

Swerdlow, N. R., van der Kooy, D., Koob, G. F., and Wenger, J. R. (1983). Cholecystokinin produces conditioned place-aversions not place-preferences, in food-deprived rats: Evidence against involvement in satiety. *Life Sci.* **32**, 2087–93.

Verbalis, J. G., McCann, M. J., McHale, C. M., and Stricker, E. M. (1986). Oxytocin secretion in response to cholecystokinin and food: Differentiation of nausea from satiety. *Science* **232**, 1417–19.

Verbalis, J. G., Richardson, D. W., and Stricker, E. M. (1987). Vasopressin release in response to nausea-producing agents and cholecystokinin in monkeys. *Am. J. Physiol.* **252**, R749–53.

Yamagishi, T. and Debas, H. T. (1978). Cholecystokinin inhibits gastric emptying by acting on both proximal stomach and pylorus. *Am. J. Physiol.* **234**, E375–8.

Zarbin, M. A., Wamsley, J. K., Innis, R. B., and Kuhar, M. J. (1981). CCK receptors: presence and axonal flow in the rat vagus nerve. *Life Sci.* **29**, 697–705.

19. CCK as a putative satiety factor in farm animals

B. A. Baldwin

In addition to the very extensive work using rats as experimental animals (see Chapter 17 of this volume), the function of CCK as a putative satiety factor has been studied in three farm animals — sheep, pigs, and chickens. From a comparative mammalian viewpoint the choice of sheep and pigs is fortunate, since the sheep is a typical ruminant with a complex digestive system and, under natural conditions, a grazing mode of nutrition, while the pig is a representative omnivore with a monogastric digestive system. Pigs, particularly the miniature breeds, are increasingly being used as experimental animals in biomedical research (Pond and Houpt 1978; Tumbleson 1986). The role of neuropeptides in the regulation of food intake is well established (Baldwin 1988), and in this chapter the evidence for CCK as a putative satiety factor in pigs, sheep, and chickens will be examined.

CCK and food intake in sheep

Central (intracerebroventricular) administration of CCK in sheep

Most of the work on the role of CCK as a putative satiety factor in sheep has been done by Baile and his collaborators (for reviews see Della-Fera and Baile 1984; Baile and Della-Fera 1985; Baile *et al.* 1986).

CCK is found in several regions of the brain, and it has been demonstrated in rats that the brain contains two types of CCK receptors, a type B (brain) receptor which is widely distributed and a type A (alimentary) receptor with a more restricted distribution (Moran *et al.* 1986). Species differences exist in the distribution of CCK in the brain (Hill *et al.* 1987), and it is very likely that the relative numbers and distributions of the two receptor types will vary between species. It would be of interest to determine their relative distribution in the brains of pigs and sheep. In sheep and pigs the main type of CCK found in the brain is the octapeptide CCK-8 (Dockray *et al.* 1978; Eng *et al.* 1983).

It has been shown that continuous intracerebroventricular (i.c.v.) injections of sulphated CCK-8 in picomolar concentrations can reduce food intake in sheep (Della-Fera and Baile 1979). The sheep were deprived of food for 2, 4, 8, or 24 h and after these periods synthetic cerebrospinal fluid (CSF) containing various concentrations of CCK-8 was injected continuously into the lateral cerebral ventricle for 3 h at 0.1 ml/min. It was demonstrated that, for all these periods of food deprivation, CCK-8 reduced food intake in a dose-related manner, and in sheep deprived for 2 h, which is a normal intermeal interval in this species, CCK delivered at the low rate of 0.01 pmol/min suppressed feeding by 35 per cent during the first hour of treatment. This rate of delivery has been estimated to produce CCK concentrations in the CSF which are within the physiological range (Della-Fera and Baile 1980; Baile and Della-Fera 1985). It was shown that the longer the period of fasting, the smaller was the effect of CCK-8 on food intake. (After 3 h at 0.64 pmol/min, feeding by sheep deprived of food for 2, 4, 8, and 24 h was reduced by 95 per cent, 48 per cent, 34 per cent, and 19 per cent respectively). Intravenous (i.v.) injections of CCK, even at 2.5 pmol/min, had no effect on food intake when injected over a 3 h period, thus confirming that it was not acting on peripheral mechanisms. Evidence for the specificity of i.c.v. injection of CCK-8 on food intake was provided by experiments in which sheep deprived of water for 2 h did not decrease water intake when injected for 3 h at 0.64 pmol/min i.c.v.

The unsulphated form of CCK-8 had no effect on food intake, even at relatively high doses, when injected into the lateral ventricle (0.64 pmol/min — a dose that causes a large decrease in feeding when the C-7 tyrosine is sulphated) (Della-Fera and Baile 1981). It was concluded that CCK receptors in the brain similar to those in the periphery play a part in satiety.

In contrast with the results obtained in sheep, Della-Fera and Baile (1979) found that, in rats, continuous intraventricular injections of sulphated CCK-8 at much higher doses than those used in sheep had no effect on food intake. They concluded that rats were considerably less sensitive than sheep to inhibition of feeding by i.c.v. injection of CCK-8.

The effects of i.c.v. injection of CCK-8 on food intake in sheep were not due to suppression of drinking behaviour or an increase in body temperature (Della-Fera and Baile 1980), both of which are known to reduce food intake. Della-Fera and Baile (1980) also demonstrated that CCK-8 injected into the CSF could cause alterations in rumen motility consisting of a decrease in the amplitude of rumen contractions in non-fed sheep which was similar to that seen during the first 30 min of a meal following a 2 h fast. Bueno et al. (1983) have obtained similar results.

The most convincing early evidence that endogenous CCK might play an important part in satiety in the sheep came from experiments in which, following the injection of CCK antibodies into the lateral ventricles of sheep, food intake increased during a 2 h period (Della-Fera et al. 1981). Antisera to synthetic unsulphated CCK-8 produced in two rabbits and shown to react to CCK-8, human gastrin 1, and porcine CCK-33 were injected over a 2 h period in doses calculated to bind 24, 6, and 1.5 pg/min of CCK injection. Food intake was significantly increased during injection of the antisera compared with injection of normal sera controls. The pattern of the increase in food intake suggested that satiety was inhibited by injection of the antibodies since a typical post-meal interval occurred during control injections, whereas during CCK antibody injections the sheep continued to eat at a rapid rate for at least an hour after the injections were started. The intake of food was significantly increased for 1 h after the end of injection and there was a trend for intake to remain increased for up to 2 h after the end of the injection.

The results of the above experiments using antisera suggest that brain CCK has a physiological role in satiety in sheep. It is unlikely that the antibodies would penetrate the ependyma lining the ventricles and therefore it is argued that CCK would have to enter the CSF before reacting with CCK receptors. Della-Fera *et al.* claim that CSF could have an active role in the movement of some peptides. They calculate, on the basis of measurement of CCK-8-like immunoreactivity in CSF from satiated sheep with concentrations of approximately 12 pg/ml and an estimated rate of CSF production of 0.2 ml/min (Pappenheimer *et al.* 1962), that CCK-8 would be secreted at 2.4 pg/min and that the highest dose of rat CCK antibody injected would have bound 10 times this amount of CCK-8, while the lowest effective dose would have been enough to bind 2.5 times the amount of CCK secreted per minute.

It was possible that, since the antibody used could bind gastrin, this could have contributed to the increase in food intake, but Della-Fera *et al.* (1981) do not consider this likely since previous work (Della-Fera and Baile 1979) demonstrated that only very high doses of pentagastrin injected i.c.v. decreased food intake in sheep and these doses caused abnormal behaviour. They believe that the increase in food intake was caused by the binding of CCK peptides in the CSF by the antibody molecules, preventing the peptides from binding to receptors involved in the satiety mechanism.

Baile and Della-Fera (1985) have summarized the evidence that, in sheep, the CSF plays an active rather than a passive role in the movement of neuropeptides in the brain and, in particular, that ventricular transport mechanisms play an essential part in satiety mechanisms involving CCK. Their hypothesis is based upon the results of a number of experiments.

1. Sequestration inactivation of CCK in the CSF by infusion of CCK antibodies into the lateral ventricle resulted in an increase in food intake (Della-Fera *et al.* 1981).

2. Injection of artificial CSF into the lateral ventricle at an injection rate of 0.1 ml/min increased food intake, while injections at 0.03 ml/min or no injection at all did not alter intake (Della-Fera *et al.* 1982). This suggests that dilution of endogenous CCK present in the CSF might increase food intake.

3. The low recovery (20–40 per cent) of CCK-8 measured in the CSF collected from the cisterna magna during a 3 h continuous perfusion of the lateral ventricle with CCK-8 and inulin (used as a marker for bulk CSF absorption) was indicative of a large loss which could not readily be accounted for by

degradation in the CSF or non-specific absorption of CCK-8 from the CSF. Della-Fera *et al*. (1982) consider that the findings are most easily explained by the specific uptake of CCK-8 from the CSF.

In an attempt to ascertain the approximate regional site of action of CCK-8, the effect of injections into the cisterna magna was examined and no significant effect on food intake was seen (Della-Fera *et al*. 1982). The authors concluded that the rostral but not the caudal ventricular system and the adjacent brain regions were involved in the putative satiety function of CCK-8.

CCK in the sheep hypothalamus

Studies have been carried out on the immunohistochemical distribution of cholecystokinin, dynorphin A, and met-enkephalin neurons in the sheep hypothalamus (Marson *et al*. 1987). CCK-containing neurons were found in the suprachiasmatic and supra-optic nuclei. Most of the cell bodies containing CCK immunoreactivity were small oval or fusiform neurons. A few CCK-containing neurons were found scattered in the lateral and ventral hypothalamic regions or intermingled with dense CCK-containing fibres in the ventromedial hypothalamus (VMH). In contrast with the rat (Innis *et al*. 1979; Loren *et al*. 1979), no CCK-containing neurons were found in the paraventricular nucleus (PVN). The concentrations of CCK in various hypothalamic nuclei of sheep have been examined by Scallet *et al*. (1985). CCK levels were highest in the VMH, and the peptide was also present in high levels in the anterior hypothalamic region which contains the suprachiasmatic nucleus. The authors point out that species differences in the distribution of CCK may explain behavioural differences in response to CCK.

Peripheral administration of CCK in sheep

Grovum (1981) infused sheep i.v. with porcine CCK-33 or with CCK-8 and examined the effects on food intake and motility of the reticulo-rumen. CCK-33, infused into sheep deprived of food for 6 h, produced a 39 per cent decrease in food intake during the first 10 min of feeding. The threshold for the effect was between 5 and 15 Ivy Dog Units (IDU). It is difficult to compare this result with the findings of Della-Fera and Baile (1979) who observed no decrease in food intake following i.v. infusion as they used CCK-8 and did not test CCK-33. Grovum (1981) found that the frequency of reticular contractions was depressed by doses of 5 and 15 IDU kg/h. CCK-8 depressed food intake at doses of 1.5 and 3 μg/kg/h but the effect was not significant.

In a subsequent study, Grovum (1982) examined the effects of injection of porcine CCK-33 into various veins and arteries in sheep in order to ascertain the site of action of CCK in reducing food intake. In sheep deprived of food for 6 h he infused CCK-33 at a dose of 15 IDU/kg/h for 30 min into the jugular vein, the common carotid artery, the portal vein, and the aortic arch before and during a meal. The intracarotid and portal vein infusions did not suppress intake significantly more than the jugular infusions although they would have produced much higher concentrations of CCK-33 in the cephalic circulation and liver. The injections into the aortic arch were the least effective in reducing food intake.

The problem of whether the effects of exogenous CCK on food intake may be due to aversive factors such as nausea has been examined in sheep by Ebenezer *et al.* (1989) who carried out experiments to determine whether i.v. injection of CCK-8 (bolus injection of 0.85 µg/kg) in sheep produced endocrine changes similar to those produced by stress or nausea. Measurements were made of plasma cortisol, prolactin, oxytocin, and vasopressin. Following CCK treatment, plasma cortisol levels were raised after 10 and 20 min while prolactin and vasopressin concentrations were increased after 10 and 20 min and oxytocin secretion was unaffected. Ebenezer *et al.* consider that the endocrine changes indicate that the sheep were stressed by the CCK injection, as similar changes in cortisol and prolactin occur when these animals are subjected to acute stress (Parrott *et al.* 1988), and probably also experienced nausea, as suggested by the increase in vasopressin (Rowe *et al.* 1979). It has been shown in rats that peripherally administered CCK stimulates the release of corticosterone (Sander and Porter 1988), prolactin (Vijayan *et al.* 1979), and oxytocin (Verbalis *et al.* 1986). In monkeys (Verbalis *et al.* 1987) and man (Miaskiewitz *et al.* 1989) plasma levels of vasopressin increased following peripheral injection of CCK.

It is interesting to note in terms of comparative endocrinology that, while oxytocin is a stress hormone in the rat (Gibbs 1988), it is not in the sheep, while vasopressin is released in response to nausea in man (Rowe *et al.* 1979) but not in the rat.

CCK and food intake in pigs

Studies on central administration of CCK in pigs

The effect of bolus injections of CCK-8 into the lateral cerebral ventricles of pigs trained to perform operant responses (pushing switch panels with their snouts) to obtain food and water has been investigated (Parrott and Baldwin 1981). The pigs were deprived of food for 17 h and then treated with 0, 0.17, 0.33, 0.66, or 1.33 µg of CCK-8. This experiment demonstrated that food intake was reduced in a dose-related manner but drinking was not affected. The lowest dose which reduced food intake in pigs was 0.17 µg. On separate occasions the pigs were deprived of water for 22 h and then treated with 0, 0.33, or 1.33 µg of CCK-8, but no effect on water intake was seen. As a control for possible peripheral effects of CCK due to leakage from the CSF into the blood, the pigs were injected i.v with 1.33 µg of CCK-8 following food deprivation for 17 h, but this dose had no effect on food intake.

Peripheral administration of CCK in pigs

There have been several studies of the effect of peripherally administered CCK on satiety mechanisms in the pig. Anika *et al.* (1981) carried out experiments in which young pigs were fasted for 4 h and then the effect of injections of CCK-33 and CCK-8, administered as bolus injections by various routes, was examined while the animals ate a meal. The experiments demonstrated that CCK injected intraperitoneally (i.p.), intraportally, or via the jugular vein can inhibit food intake in hungry pigs.

CCK-33 injected i.p. reduced food intake in a dose-related manner as did i.v. injections. I.v. injections (jugular) of synthetic CCK-8 also inhibited feeding in a dose-related manner. Intraportal injections of CCK-8 suppressed feeding, but it was not significantly more effective than a similar dose administered via the jugular vein. Anika *et al.* also attempted, without success, to demonstrate taste aversion to a novel flavour associated with intrajugular injection of CCK-33 and CCK-8. In their study an interesting experiment was carried out in which the latency to commence feeding was measured after the pigs had received a bolus injection of saline or CCK-33. It was shown that the latency to feed was proportional to dose but that the rate of eating, once it started, was not significantly different from that of the controls. It was concluded that the bolus injection acted as an 'off–on' switch for feeding behaviour. It is possible that the duration of any aversive effects of CCK would be dose-related and produce a similar result.

The site of action of CCK in decreasing meal size in pigs has also been investigated (Houpt 1983). In these experiments CCK-8 was infused at a rate of 67 ng/kg/min into the jugular vein, the carotid artery, the portal vein, the gastric branch of the splenic artery, and also the aorta both cranial and caudal to the origins of the cranial mesenteric and coeliac arteries. It was shown that jugular, carotid, and cranial aortic infusions resulted in a significant reduction in meal size to 65 per cent, 67 per cent, and 71 per cent respectively of control meal size. Perfusions into the portal vein and caudal aortic artery infusions had no significant effects on meal size. An interesting finding was that infusions into the gastric artery resulted in a significant increase of meal size to 122 per cent of control meal size. Houpt (1983) considers it possible that CCK caused relaxation of the gastric musculature and this may have reduced afferent flow to the CNS from gastric stretch receptors. Stretch stimuli have been considered to play a part in the inhibition of food intake. This result does not provide support for the idea that the stomach is the site of action of CCK in reducing meal size in pigs and the overall results of the experiment were interpreted as indicating that a major site of action of CCK in the reduction of meal size is located in the bed of the cranial mesenteric of coeliac arteries but that the stomach and liver could be excluded. Houpt (1983) also considered that the brain was probably not a major site for the action of peripherally administered CCK on food intake, but that it may have a minor role. This conclusion was based on the fact that infusion of CCK into one carotid artery reduced meal size by the same amount as did jugular infusion. It is argued that intracarotid infusion would have produced a higher concentration of CCK in the cephalic circulation than jugular infusion and a greater depression of meal size might have been expected. It is known that CCK does not cross the blood–brain barrier except possibly at sites which have increased vascular permeability, such as the area postrema (Moran *et al.* 1986).

Houpt (1983) found it surprising that intraportal infusion of CCK did not reduce meal size, a result which does not agree with his earlier work on intraportal injection of CCK (Anika *et al.* 1981). He attributes the discrepancy to a technical difference in that the bolus injections used in the earlier study would have caused a transient high concentration of CCK in the portal blood which was more effective than the lower and relatively constant level provided by the pump infusions. The pharmacokinetics and organ catabolism of CCK-8 have recently been investigated in pigs (Cuber *et al.* 1989) and it has been shown that the plasma half-life in conscious animals was 0.55 ± 0.03 min. They demonstrated that CCK-8 is only partially inactivated during hepatic transit and so may function as a hormone following release from intestinal stores.

Specificity of the effects of peripherally administered CCK in pigs

The specificity of the behavioural effects on food intake of i.v. injection of CCK-8 has been investigated in pigs trained to make operant responses for a variety of rewards (Baldwin *et al.* 1985). The effect of i.v. injection of saline and of 0.33, 0.66 and 1.3 μg/kg of CCK-8 was studied in pigs performing operant responses to obtain food, water sucrose solution, or radiant heat. In hungry pigs working for food, thirsty pigs responding for water, and nondeprived pigs working for sucrose solutions, 0.66 and 1.3 μg/kg of CCK-8 produced significant dose-related decreases in response rates lasting for about 5 min. In pigs responding to 10 bursts of infra-red radiant heat, at an ambient temperature of 5 °C, a CCK-8 dose of 1.3 μg/kg i.v. significantly reduced the response rate in the 5 min period following injection. It was of interest that, for all types of reinforcer, transient depression followed injection of CCK and then response resumed at the same rate as before.

The results of the above experiment indicate that peripherally administered CCK-8 does not produce effects which are specific to food intake and suggest that it might disrupt a variety of motivated behaviours by producing aversive effects. A recent study has shown that the disruption of operant feeding is accompanied by an increase in the release of cortisol but not of oxytocin (Ebenezer *et al.* 1990*b*). The possible source of some of the aversive effects could be the small intestine, since Houpt's (1983) experiments demonstrated that this appeared to be the peripheral site of action of i.v. injected CCK. It is well established that even low doses (0.1 μg/kg) of CCK injected i.v. can produce abnormal motility in the small intestine of dogs (Lang *et al.* 1988). In addition, there is evidence from human studies that exogenous CCK, even at low doses, can induce feelings of abdominal discomfort and nausea which are accompanied by increases in plasma vasopressin which appear to correlate with nausea in some subjects (Miaskiewicz *et al.* 1989). These authors consider that some component of the anorexic action of exogenous CCK in man could result from activation of the brainstem emetic centres by CCK. It would be of interest to determine the effect of i.c.v. injections of CCK-8 on motivated behaviours other than feeding and drinking, although the finding that i.c.v. CCK did not reduce water intake in thirsty pigs (Parrott and Baldwin 1981) supports the idea that i.c.v. administration of CCK may be more specific than peripheral administration in its action on feeding.

CCK and food intake in domestic fowls

Central administration of CCK in domestic fowls

CCK occurs in the brain and gut of birds (Dockray 1979; Gentle and Savory 1983). The effect of i.c.v. injection of CCK-8 on food intake in domestic chicks has been examined by Denbow and Myers (1982). They found that 100 or 150 ng of CCK-8 administered i.c.v. significantly reduced food intake in a dose-dependent manner in birds fasted for 24 h. Water intake was also significantly decreased, but the effect was not dose-dependent. They considered that CCK might be part of a central mechanisms involved in the neuronal control of short-term satiety in chickens.

Savory and Gentle (1983a) have also investigated whether brain CCK-8 acts as a satiety factor in fowls by examining the effects of i.c.v. injection of CCK-8. The birds had been deprived of food for 18 h, but in the first 30 min after CCK-8 injection (0.1 and 0.4 μg/kg), at the start of the meal, their food intake was reduced by 20 per cent and 30 per cent respectively. The birds were not unwell after injection. In a similar experiment Savory and Gentle demonstrated that birds injected i.c.v. with the CCK-8 antagonist dibutyryl cyclic guanosine monophosphate (CGMP) ate more in the first 15 min after injection. Birds injected i.c.v. with 0.5 μg/kg of CCK-8 did not show abnormal motility of the gizzard or duodenum, but two out of four birds became drowsy after CCK-8 injection. In addition, CCK-8 injections had no effect on blood glucose levels. Savory and Gentle concluded that the satiety effects of i.c.v. injections of CCK-8 cannot be explained by hyperglycaemia or gastric effects but may be associated with drowsiness. They considered that the results obtained with GMP supported the idea of a central satiety role for CCK.

Peripheral administration of CCK in fowls

Snapir and Glick (1978) failed to obtain a reduction in food intake in domestic fowls after i.p. injections of CCK. However, they only used one dose level and tested a small number of birds. In contrast, Savory and Gentle (1980) examined the effects of i.v. injections of CCK-8 on food intake in fowls and found that a dose-related reduction in food intake occurred in the first 15 min of access to food after deprivation for 18 h. In a subsequent study (Savory and Gentle 1983*b*) they demonstrated that with increasing food deprivation (0, 1, 2, or 3 h before injection), doses of 2 and 8 µg/kg of CCK-8 became progressively less effective in suppressing feeding. It had previously been shown (Savory *et al.* 1981) that i.v. injections of CCK-8 at doses of 0.5, 5, and 15 µg/kg body weight had strong short-term dose-related effects on gizzard and duodenal motility in fed and fasted turkeys and domestic fowls. The gizzard motility was inhibited, while the duodenum and proximal ileum were stimulated causing refluxes and segmenting contractions which were detected by means of strain gauge transducers attached to the gizzard, duodenum, and ileum. In some cases the effects of CCK on the gastrointestinal tract were observed by means of radiographic observations using an X-ray image intensifier to correlate the strain gauge data with the radiological data. Savory *et al.* (1981) and Savory and Gentle (1983*b*) consider that the effects on gastrointestinal motility may account for the reduced food intake following i.v. injections of CCK.

Savory and Hodgkiss (1984) investigated the effects of vagotomy on the action of exogenous CCK in domestic fowls and found, in contrast with results obtained for the rat (Smith *et al.* 1981), that the short-term suppression of food intake which followed the i.v. injection of CCK-8 was unaffected by vagotomy. They concluded that efferent information from the gastrointestinal tract may travel via the intestinal nerve, which is unique to birds.

Savory (1987) has carried out further experiments which support his hypothesis that peripherally administered CCK reduces food intake in fowls owing to abdominal discomfort as a result of abnormal motility of the gastrointestinal tract. Following i.v. injections of CCK-8, short periods of complete inhibition of feeding coincided with abnormal motility of the gizzard and longer periods of reduced feeding were associated with abnormal gastrointestinal motility. These changes were accompanied by increased heart rate. He demonstrated, using conditioned avoidance tests, that injections of CCK-8 (10 µg/kg) may be mildly aversive. Savory concluded that peripherally administered CCK-8 reduces feeding in chickens by producing abdominal discomfort. It would be of interest to examine the effects of specific CCK antagonists on food intake in chickens in order to reveal whether endogenous CCK plays a role in satiety in this species.

Experiments on the effects of CCK antagonists in sheep and pigs

As mentioned previously, the anorexic effects of exogenous CCK have been difficult to interpret because the observed reductions in food intake may have been partly due to aversive effects. Recently, the discovery of potent specific antagonists acting on the A or B receptors (Chang and Lotti 1986; Bock *et al.* 1989) has enabled experimenters to investigate the function of endogenous CCK in the brain or periphery without the added complication of possible aversive effects. If the hypothesis that endogenous CCK has a role as a satiety factor is correct, then it should follow that administration of a specific antagonist would block the effect of endogenous CCK released in the periphery or the brain during a meal and thus increase the amount of food consumed. Devazepide, which is a potent non-peptide antagonist to the CCK_A receptor, has been shown to increase food intake in rats (Hewson *et al.* 1988; Chapters 17 and 21 of this volume) and pigs (Ebenezer *et al.* 1990a), and also to increase hunger in man (Wolkowitz *et al.* 1990; Chapter 26 of this volume).

Sheep

Baile *et al.* (1989) injected the CCK_A receptor antagonist CR1409 (lorglumide) into the lateral cerebral ventricles of sheep at the end of a 15 min meal. Following the injection the amount of food eaten increased and was significantly greater than control injection by 2 h post-injection. In another experiment, sheep that had been fasted for 2 h were given injections of CR1409 into the lateral ventricle and a continuous injection of 0.64 pmol/min of CCK-8 which began 15 min before food was offered to animals. CCK-8 on its own significantly decreased food intake throughout the 3 h test period, but with the combination (CR1409 and CCK-8) the suppression of feeding did not occur. In the fasted sheep CR1409 on its own did not significantly increase food intake. The findings indicate that blocking brain CCK_A receptors can eliminate the satiety effects of centrally administered CCK and also that the antagonist could induce extra food intake in sheep that had recently fed but not in hungry fasted animals.

Pigs

The CCK receptor antagonist devazepide has been administered i.v. to pigs trained to make operant responses for food after 4 h deprivation. Devazepide is a potent specific antagonist of the CCK_A (alimentary) receptor which occurs mainly in the periphery although it has also been found in the brain (Moran *et al.* 1986; Dourish and Hill 1987). The antagonist produced a dose-related increase in food intake, when administered just before the meal started, with a maximal effect at 70 μg/kg and a diminishing effect at levels higher than this. The results lend support to the concept that the type A receptor and endogenous CCK are involved in the regulation of food intake. CCK-8 at a dose of 1 μg/kg i.v. produced a short-term reduction in feeding and this effect was completely abolished by pretreating the animals with devazepide (70 μg/kg). Devazepide readily crosses the blood–brain barrier (Pullen and Hodgson 1987) and could have acted on type A receptors in the brain and the periphery. Therefore it is impossible to conclude from this experiment whether the increased intake was due to faster gastric emptying via an effect on gastric type A receptors with consequent inhibition of the effect of CCK in delaying gastric emptying (Moran and McHugh 1982) or was due to a direct action on brain CCK_A receptors or resulted from a combination of both mechanisms.

Recent experiments in this laboratory have demonstrated that devazepide injected i.c.v. (50 μg) or i.v. (70 μg/kg) can abolish the effect on operant food intake of CCK-8 injected i.c.v. (1 μg), but that devazepide (50 μg) injected i.c.v. does not eliminate the reduction in food intake produced by i.v. injection of CCK-8 (1 μg/kg). This result suggests that peripheral CCK has an inhibitory action on food intake which is independent of the action of central CCK. It has also been demonstrated that devazepide injected i.v. or i.c.v. can induce feeding in satiated pigs (not feeding in the presence of food) and also increase meal size when injected 40 min after pigs started to feed after a 4 h fast. As the i.c.v. dose was effective centrally but ineffective when given i.v., this experiment indicates that blocking brain CCK_A receptors can increase food intake without affecting peripheral CCK_A receptors (Baldwin and De la Riva, unpublished).

It has been shown that the CCK_B receptor antagonist L-365,260 injected i.v. over a wide dose range did not eliminate the effect of intravenous CCK-8 on food intake, which is consistent with findings in rats (Dourish *et al.* 1989*a*). In addition in contrast with recent findings in rats (Dourish *et al.* 1989*b*), L-365,260 did not increase meal size when injected i.v. (1–1000 ng/kg) or i.c.v. (100 ng) in partially satiated pigs (allowed to feed for 40 min before injection) after 4 h deprivation of food. These results strongly suggest that the type B receptor is not involved in the action of exogenous CCK on food intake in pigs or in the possible function of endogenous CCK as a satiety factor in this species.

In contrast with the previous results in which devazepide increased food intake in satiated or partially satiated pigs (Baldwin and De La Riva, unpublished) and in pigs feeding after 4 h deprivation (Ebenezer *et al.* 1990), it has been shown by Gregory *et al.* (1989) that injection of devazepide before the pigs had a meal after 18 h deprivation did not increase the amount consumed. They speculate that this lack of effect may be due to the prolonged period of deprivation and consequent intense hunger. Devazepide did not increase food intake in rats deprived of food for 18 h (Hewson *et al.* 1988; see Chapter 21 of this volume). Animals which are very hungry are probably eating the maximum amount that they can and normal satiety signals may be overruled.

Gregory and his collaborators have carried out a series of experiments on the effects of infusion of various nutrients into the small intestine of pigs and have shown that food intake is strongly inhibited by duodenal but not ileal infusion of emulsified fat (Gregory and Raynor 1987). Since CCK is known to be released from the duodenum when fat is infused (Lilja *et al.* 1984), the inhibition of feeding could be due to release of CCK. Gregory *et al.* (1989) have investigated the effect of devazepide on the satiety induced by duodenal infusion of fat and fat digestion products and also the satiety induced by duodenal infusion of glucose. Injection of devazepide abolished the inhibition of feeding which follows the duodenal infusion of emulsified fat or infusion of monoglyceride. However, it did not alter the inhibition resulting from infusion of oleic acid, glycerol, or glucose. The authors suggest that the monoglyceride-induced CCK secretion is mainly responsible for the satiety to duodenal fat in the pig, but there also exists a satiety mechanism which is independent of CCK and is induced by fatty acids or glucose. They conclude that the nature of the diet may determine whether endogenously released CCK is important in the termination of normal meals.

Conclusions

There is evidence that i.c.v. administration of CCK-8 in low doses can reduce food intake in pigs and sheep without producing overt signs of illness or discomfort. I.c.v. administration also reduces food intake in chickens.

Food intake in pigs, sheep, and chickens is reduced following intravenous CCK injection but this may be partly due to aversive effects. In sheep and pigs endocrine responses resembling those seen in stressed animals are elicited by CCK, and, in pigs, operant tests revealed that the injection of CCK caused a transient reduction in several motivated behaviours and an increase in plasma cortisol. In chickens i.v. CCK caused abnormal motility of the gastro-intestinal tract coincidental with the reduction in feeding.

The difficulties inherent in the interpretation of the effects of exogenous CCK, which may reduce food intake by non-specific aversive effects, have recently been obviated by the use of specific potent antagonists acting on CCK receptors. CR1409, a type A receptor antagonist, increased food intake in sheep when injected i.c.v. at the end of a meal. In pigs, devazepide (a type A receptor antagonist) injected i.v. before a meal commenced after a 4 h fast resulted in increased food intake. It also increased intake when injected towards the end of a meal. Devazepide increased food intake when injected i.c.v. at doses too low to be effective peripherally. This indicates that it can increase food intake by acting on central type A receptors. Administration of the type B receptor antagonist L-365,260 did not affect food intake in pigs when administered i.v. or i.c.v. It is concluded that CCK and the type A receptors, which are located in the brain as well as the periphery, have a role in the regulation of food intake in pigs and sheep.

References

Anika, S. M., Houpt, T. R., and Houpt, K. A. (1981). Cholecystokinin and satiety in pigs. *Am. J. Physiol.* **240**, R310–18.
Baile, C. A. and Della-Fera, M. A. (1985). Central nervous system cholecystokinin and the control of feeding. In *Neuronal cholecystokinin* (ed. J. J. Vanderhaeghen and J. N. Crawley). *Ann. NY Acad. Sci.* **448**, 424–430.
Baile, C. A., McLaughlin, C. L., and Della-Fera, M. A. (1986). Role of cholecystokinin and opioid peptides in control of food intake. *Physiol. Rev.* **66**, 172–234.
Baile, C. A., Della-Fera, M. A., Brown, D. R., and Colman, B. D. (1989). CCK receptor blocker CR 1409 increases feeding and reverses CCK-8 mediated satiety after lateral cerebral ventricular (LV) injection in sheep. *Soc. Neurosci. Abstr.* **15**, 1279.
Baldwin, B. A. (1988). Neuropeptides and food intake. In *Progress in endocrinology,*

Vol. 1 (ed. H. Imura, K. Shizume, and S. Yoshida), pp. 705–10. Elsevier, Amsterdam.

Baldwin, B. A., Cooper, T. R., and Parrott, R. F. (1985). Intravenous cholecystokinin octapeptide in pigs reduces operant responding for food, water, sucrose solution or radiant heat. *Physiol. Behav.* **30**, 399–403.

Bock, M. G., Di Pardo, R. M., Evans, B. E., Rittle, K. E., Whitter, W. L., Veber, D. F., Anderson, P. S., and Freidinger, R. M. (1989). Benzodiazepine gastrin and cholecystokinin receptor ligands: L365,260. *J. Med. Chem.* **32**, 13–16.

Bueno, L., Duranton, A., and Ruckebush, Y. (1983). Antagonistic effects of naloxone on CCK-octapeptide induced satiety and rumino-reticular hypomotility in sheep. *Life Sci.* **32**, 855–63.

Chang, R. S. L. and Lotti, V. J. (1986). Biochemical and pharmacological characterization of an extremely potent and selective non-peptide cholecystokinin antagonist. *Proc. Natl Acad. Sci. USA* **83**, 4923–6.

Cuber, J. C., Barnard, C., Gibar, T., and Chayvialle, J. A. (1989). Pharmacokinetics and organ catabolism of cholecystokinin octapeptide in pigs. *Regul. Pept.* **26**, 203–13.

Della-Fera, M. A. and Baile, C. A. (1979). Cholecystokinin octapeptide: continuous picomole injections into the cerebral ventricles of sheep suppress feeding. *Science* **206**, 471–3.

Della-Fera, M. A. and Baile, C. A. (1980). CCK-octapeptide injected in CSF and changes in feed intake and rumen motility. *Physiol. Behav.* **24**, 943–50.

Della-Fera, M. A. and Baile, C. A. (1981). Peptides with CCK-life activity administered intracranially elicit satiety in sheep. *Physiol. Behav.* **26**, 979–83.

Della-Fera, M. A. and Baile, C. A. (1984). Control of feed intake in sheep. *J. Anim. Sci.* **59**, 1362–8.

Della-Fera, M. A., Baile, C. A., Schneider, B. S., and Grinker, J. A. (1981). Cholecystokinin antibody injected in cerebral ventricles stimulates feeding in sheep. *Science* **212**, 687–9.

Della-Fera, M. A., Baile, C. A., and Beinfeld, M. C. (1982). Cerebral ventricular transport and uptake: importance for CCK mediated satiety. *Peptides* **3**, 963–8.

Denbow, D. M. and Myers, R. D. (1982). Eating, drinking and temperature responses to intracerebroventricular cholecystokinin in the chicken. *Peptides* **3**, 739–43.

Dockray, G. I. (1979). Cholecystokinin-like peptides in avian brain and gut. *Experientia* **35**, 628–30.

Dockray, G. J., Gregory, R. A., Hutchison, J. B., Harris, J. I., and Runswick, M. J. (1978). Isolation structure and biological activity of two cholecystokinin octapeptides from sheep brain. *Nature* **274**, 711–13.

Dourish, C. T. and Hill, D. R. (1987). Classification and function of CCK receptors. *Trends Pharmacol. Sci.* **8**, 207–8.

Dourish, C. T., Ruckert, A. C., Tattersall, F. D., and Iversen, S. D. (1989a). Evidence that decreased feeding induced by systemic injection of cholecystokinin is mediated by CCK-A receptors. *Eur J. Pharmacol.* **173**, 233–4.

Dourish, C. T., Rycroft, W., and Iversen, S. D. (1989b). Postponement of satiety by blockade of brain cholecystokinin (CCK-B) receptors. *Science* **245**, 1509–11.

Ebenezer, I. S., Thornton, S. N., and Parrott, R. F. (1989). Anterior and posterior pituitary hormone release induced in sheep by cholecystokinin. *Am. J. Physiol.* **256**, R1355–7.

Ebenezer, I. S., De La Riva, C., and Baldwin, B. A. (1990a). Effects of the CCK receptor antagonist MK329 on food intake in pigs. *Physiol. Behav.* **47**, 145–8.

Ebenezer, I. S., Misson, B. H., De La Riva, C., Thornton, S. N., and Baldwin, B. A. (1990*b*). The effect of cholecystokinin (CCK) on oxytocin and cortisol release in pigs during operant feeding. *Br. J. Pharmacol.* **100**, 419P.

Eng, J., Shüna, Y., Pan, Y. C. E., Blacher, R., Chang, M., Stein, S., and Yalow, R. S. (1983). Pig brain contains cholecystokinin octapeptide and several cholecystokinin desoctapeptides. *Proc. Natl Acad. Sci. USA* **80**, 6381-5.

Gentle, M. J. and Savory, C. J. (1983). Regional distribution of cholecystokinin in the chick brain in relation to age and nutritional status. *IRCS Med. Sci.* **11**, 199-200.

Gibbs, D. M. (1988). Vasopressin and oxytocin, hypothalamic modulators of the stress response – a review. *Psychoneuroendocrinology* **11**, 131-40.

Gregory, P. C. and Rayner, D. V. (1987). The influence of the gastrointestinal infusion of fats on regulation of food intake in pigs. *J. Physiol.* **385**, 471-81.

Gregory, P. C., McFadyen, M., and Rayner, D. V. (1989). Duodenal infusion of fat, cholecystokinin secretion and satiety in the pig. *Physiol. Behav.* **45**, 1021-4.

Grovum, W. L. (1981). Factors affecting voluntary food intake by sheep. 3. The effect of intravenous infusions of gastrin, cholecystokinin and secretion on motility of the reticulo rumen and intake. *Br. J. Nutr.* **45**, 183-201.

Grovum, W. L. (1982). Cholecystokinin administered intravenously did not act directly on the central nervous system or on the liver to suppress food intake by sheep. *J. Physiol.* **326**, 55P.

Hewson, G., Leighton, G. E., Hill, R. G., and Hughes, J. (1988). The cholecystokinin receptor antagonist L364718 increases food intake in the rat by attenuation of the actions of endogenous cholecystokinin. *Br. J. Pharmacol.* **93**, 70-84.

Hill, D. R., Shaw, T. M., and Woodruff, G. (1987). Species differences in the localisation of peripheral type cholecystokinin receptors in rodent brain. *Neurosci. Lett.* **79**, 286-9.

Houpt, T. R. (1983). The sites of action of cholecystokinin in decreasing meal size in pigs. *Physiol. Behav.* **31**, 693-8.

Innis, R. B., Correa, F. M. A., Uhl, G. R., Schneider, B., and Snyder, S. H. (1979). Cholecystokinin octapeptide-like immunoreactivity: histochemical localization in rat brain. *Proc. Natl Acad. Sci. USA* **76**, 521-5.

Lang, I. M., Marvig, J., and Sarna, S. K. (1988). Comparison of gastrointestinal responses to CCK-8 and associated with vomiting. *Am. J. Physiol.* **254**, G254-63.

Lilja, P., Wiener, I., Inoue, K., Fried, G. M., Greely, G. H., and Thompson, J. C. (1984). Release of cholecystokinin in response to food and intraduodenal fat in pigs, dogs and man. *Surg. Gynecol. Obstet.* **159**, 557-61.

Loren I., Alumets, J., Hakanson, R., and Sandler, F. (1979). Distribution of gastrin and CCK-like peptides in rat brain: an immunocytochemical study. *Histochemistry* **59**, 249-57.

Marson, L., Lauterio, T. J., Della-Fera, M. A., and Baile, C. A. (1987). Immunohistochemical distribution of cholecystokinin, dynorphin A and met-enkephalin neurons in sheep hypothalamus. *Neurosci. Lett.* **81**, 35-40.

Miaskiewicz, S. L., Stricker, E. M., and Verbalis, J. G. (1989). Neurohypophysial secretion in response to cholecystokinin but not meal induced gastric distension in humans. *J. Clin. Endocrinol. Metab.* **68**, 837-43.

Moran, T. H. and McHugh, P. R. (1982). Cholecystokinin suppresses food intake by inhibitory gastric emptying. *Am. J. Physiol.* **242**, R491-7.

Moran, T. H., Robinson, P. H., Goldrich, M. S., and McHugh, P. R. (1986). Two brain cholecystokinin receptors: implications for behavioural actions. *Brains Res.* **362**, 175-9.

Pappenheimer, J. R., Hersey, S. R., Jordan, E. F., and Downer, C. (1962). Perfusion of the cerebral ventricular system in unanaesthetised goats. *Am. J. Physiol.* **203**, 763–74.

Parrott, R. F. and Baldwin, B. A. (1981). Operant feeding and drinking in pigs following intracerebroventricular injection of synthetic cholecystokinin octapeptide. *Physiol. Behav.* **26**, 419–22.

Parrott, R. F., Thornton, S. N., and Robinson, J. E. (1988). Endocrine responses to acute stress in castrated rams: no increase in oxytocin but evidence for an inverse relationship between cortisol and vasopressin. *Acta Endocrinol.* **117**, 381–6.

Pond, W. G. and Houpt, K. A. (1978). *The biology of the pig.* Comstock, Ithaca, NY.

Pullen, R. G. L. and Hodgson, O. J. (1987). Penetration of diazepam and the non-peptide CCK antagonist L364,718 into rat brain. *J. Pharm. Pharmacol.* **39**, 863–4.

Rowe, J. W., Shelton, R. L., Helderman, J. H., Vestal, R. F., and Robertson, J. L. (1979). Influence of the emetic reflex on vasopressin release in man. *Kidney Int.* **16**, 729–35.

Sander, L. D. and Porter, J. R. (1988). Influence of bombesin, CCK secretion and CRF on corticosterone concentration in the rat. *Peptides* **9**, 113–17.

Savory, C. J. (1987). An alternative explanation for apparent satiating properties of peripherally administered bombesin and cholecystokinin in domestic fowls. *Physiol. Behav.* **39**, 191–202.

Savory, C. J. and Gentle, M. J. (1980). Intravenous injections of cholecystokinin and caerulein suppresss food intake in domestic fowls. *Experientia* **36**, 1191–2.

Savory, C. J. and Gentle, M. J. (1983a). Brain cholecystokinin and satiety in fowls. *Appetite* **4**, 223.

Savory, C. J. and Gentle, M. J. (1983b). Effects of food deprivation, strain, diet and age on feeding responses of fowls to intravenous injections of cholecystokinin. *Appetite* **4**, 165–76.

Savory, C. J. and Hodgkiss, J. P. (1984). Influence of vagotomy in domestic fowls on feeding activity, food passage, digestibility and satiety effects of the peptides. *Physiol. Behav.* **33**, 937–44.

Savory, C. J., Duke, G. E. and Bertoy, R. W. (1981). Influence of intravenous injections of cholecystokinin on gastrointestinal motility in turkeys and domestic fowls. *Comp. Biochem. Physiol.* **70**, 179–89.

Scallett, A. C., Della-Fera, M. A. and Baile, C. A. (1985). Satiety, hunger and regional brain content of cholecystokinin/gastrin and met-enkephalin immunoreactivity in sheep. *Peptides* **6**, 937–43.

Smith, G. P., Jerome, C., Cuskin, B. J., Eterno, R., and Simansky, K. J. (1981). Abdominal vagotomy blocks the satiety effect of cholecystokinin in the rat. *Science* **213**, 1036–7.

Snapier, N. and Glick, Z. (1978). Cholecystokinin and meal size in the domestic fowl. *Physiol. Behav.* **21**, 1051–2.

Tumbleson, M. F. (ed.) (1986). *Swine in biomedical research.* Plenum, New York.

Verbalis, J. G., McCann, K. J., McHale, C. M., and Stricker, E. M. (1986). Oxytocin secretion in response to cholecystokinin and food: differentiation of nausea from satiety. *Science* **236**, 832–5.

Verbalis, J. G., Richardson, D. W., and Stricker, E. M. (1987). Vasopressin release in response to nausea-producing agents and cholecystokinin in monkeys. *Am. J. Physiol.* **252**, R749–53.

Vijayan, E., Samsom, W. K., and McCann, S. M. (1979). *In vivo* and *in vitro* effects of cholecystokinin on gonadotropin, prolactin, growth hormone and thyrotropin release in the rat. *Brain Res.* **172**, 295–302.

Wolkowitz, O. M., Gertz, B., Weingartner, H., Beccaria, L., Thompson, K., and Liddle, R. A. (1990). Hunger in humans induced by MK329, a specific peripheral-type cholecystokinin receptor antagonist. *Biol. Psychiat.* **28**, 169–73.

20. Plasma and cerebrospinal fluid concentrations of CCK and satiety

P. Södersten and A. Lindén

Introduction

The inhibitory effect on food intake of exogenously administered CCK octapeptide CCK-8, initially noted as a side-effect (Sjödin 1972), was first demonstrated experimentally by Gibbs *et al.* (1973) (see Chapter 17 of this volume). This observation drew attention to peripheral mechanisms in the control of feeding behaviour, a field which had previously been dominated by studies of the brain, notable hypothalamic regulatory mechanisms of hunger. A drawback in the classical experiments on hunger mechanisms had been that the stimuli which initiate a meal in a food-deprived animal are unknown. The study of the peripheral mechanisms which suppress feeding or induce satiety during a meal offered a new perspective because the stimulus which controls the behaviour (satiety) is known (food) and its site of action is also known: the gastro-intestinal tract, a major endocrine organ in the body (Smith 1982). The secretions of this organ may be involved in satiety because administration of food into stomach transplants devoid of nervous supply elicits satiety in animals (Koopmans 1981) and food intake in hungry rats is reduced after their blood is mixed with that of satiated rats (Davis *et al.* 1967). Thus it is necessary to study the secretions of the gastro-intestinal tract during ingestion of food, particularly as CCK-8, which is secreted from the duodenum during feeding, has been suggested to be a primary satiety-inducing hormone (see Chapter 17 of this volume). However, the study of CCK-8 in satiety has been hampered by the absence of sensitive assay methods. With the development of such methods (Lindén 1989), the role of CCK-8 in food intake and satiety has become accessible to direct study and our recent series of such investigations is discussed below.

Methodological considerations of CCK determination

Several problems have arisen in the development of a specific and sensitive radio-immunoassay (RIA) for the measurement of plasma CCK. The major one is that the C-terminal pentapeptide of CCK is identical with that of gastrin (Gregory *et al.* 1964). Antibodies recognizing the biologically active carboxyl terminus will therefore cross-react with gastrin. Since the concentration of gastrin in plasma is 10- to 100-fold higher than that of CCK, this problem assumes considerable importance. Thus separation of gastrin from CCK as well as analysis of the various molecular forms of CCK requires chromatographic methods (high pressure liquid chromatography (HPLC)) in combination with RIA. In view of the low levels of CCK in plasma RIA, sensitivity presents an additional problem.

The method used in this study is based on HPLC separation of CCK and gastrin peptides prior to detection of CCK with an unspecific antibody which identifies the carboxyl terminus of CCK and gastrin (Lindén 1989). Since in the rat complete separation of CCK and gastrin with HPLC is difficult, this problem was circumvented by enzyme digestion of gastrin before HPLC separation of CCK peptides from gastrin fragments. This method enabled us to measure different molecular forms of CCK separately. The CCK immuno-reactive peaks in rat plasma co-eluted with standards of CCK-8 and of CCK-33 and CCK-39 (which co-elute). However, when the HPLC system was optimized another immunoreactive peak corresponding to the CCK-22 standard was detected. It is possible that this peak was included in the immunoreactive peak corresponding to CCK-33 and CCK-39 before the method was improved. In the present study the different forms of CCK were calculated together and presented as total CCK in most experiments. The detection limit of the assay was 4 pM. In order to measure low levels of CCK, plasma samples (2 ml) were concentrated ten times before RIA.

CCK-like immunoreactivity (CCK-LI) was also analysed in the cerebrospinal fluid (CSF). After SEP-PAK extraction, CCK-LI concentrations were meas-ured with RIA using the carboxyl terminal antiserum (Lindén 1989). The HPLC step was omitted since no gastrin-specific immunoreactivity was found in the CSF. Thus different methods were used to measure CCK in plasma and in CSF, and therefore direct comparisons are inappropriate.

Characterization of the molecular forms of CCK-LI in CSF was performed using HPLC of a large pooled sample of CSF. Four peaks were found, cor-responding to CCK-4, CCK-5, sulphated CCK-8 and a late eluting peak. This peak contained no gastrin-specific immunoreactivity and eluted close to canine CCK-58, and therefore is probably rat CCK-58. Thus the molecular pattern of CCK-LI forms in CSF differs from that in plasma but is similar to that in brain extracts (Rehfeld *et al.* 1988). Hence it is probable that CCK-LI in the CSF is derived from the brain. The measurement of CCK-LI reported here is the total of the various molecular forms in the CSF.

Plasma levels of CCK-8 during ingestion of food and after CCK-8 injection

The physiological importance of the repeatedly demonstrated inhibitory effect of intraperitoneally (i.p.) administered CCK-8 on food intake has been

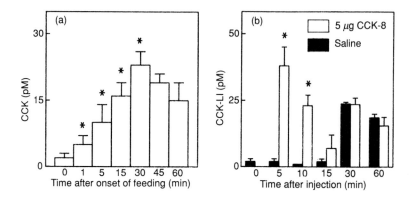

Fig. 20.1. Plasma concentrations of CCK (a) during feeding or (b) after i.p. injection of 5 µg CCK-8 (open bars) or saline (solid bars) in male rats deprived of food for 48 h. Animals injected with CCK-8 or saline and killed for CCK determination 30 and 60 min later were given food 10 min after the injection. Asterisks indicate $p \leqslant 0.05$ compared with the preceding value or with rats injected with saline.

questioned for many reasons, one of which is that i.p. injection of 5 µg of CCK-8, a commonly used dose, is likely to produce unphysiologically high levels of plasma CCK-8 (Lindén and Södersten 1990). Resolution of this issue requires direct comparison of the CCK concentrations in plasma after feeding with those after i.p. injection of CCK-8. Figure 20.1 shows such a comparison in which male rats, deprived of food for 48 h, were either allowed to eat food pellets or injected with 5 µg CCK-8 i.p. and killed at various time intervals after the onset of feeding or the injection of CCK-8. Whereas plasma levels of CCK increased gradually during ingestion of food, they increased abruptly and also declined rapidly after an injection of CCK-8 (Lindén *et al.* 1989*b*). Note that animals injected with CCK-8 had the same concentration of CCK in plasma as rats injected with saline after consumption of food pellets for 30 and 60 min.

Thus the changes in plasma levels of CCK after i.p. injection of CCK-8 differ from those observed after feeding. Additionally, injection of CCK-8, but not feeding, caused a marked increase in plasma levels of oxytocin. Furthermore, the concentrations of plasma gastrin and somatostatin increased during feeding, whereas after i.p. injection of CCK-8 the concentration of somatostatin, but not gastrin, increased (Lindén *et al.* 1989*b*). Obviously, and not surprisingly, the hormonal changes after i.p. injection of CCK-8 differ markedly from those after feeding. However, whereas injection of CCK-8 consistently reduced food intake in food-deprived rats, injection of somatostatin, gastrin, or oxytocin did not (Lindén *et al.* 1989*b*). It would appear, therefore, that of the cascade of hormonal changes which occur during feeding or after i.p. injection of CCK-8, the increase in plasma CCK levels is particularly important for suppression of food intake. However, the facts that the maximal plasma level of CCK produced by i.p. injection of CCK-8 is higher and the pattern of plasma levels is radically different from that observed during feeding might still undermine the suggestion that suppression of food intake after i.p. injection of CCK-8 is physiologically relevant. This criticism is rebutted by the demonstration that i.p. implantation of osmotic minipumps delivering 0.5 μg/h of CCK-8 and producing plasma CCK concentrations of 15.7 ± 4.8 pM, i.e. levels clearly within the range seen during feeding, inhibited food intake and produced short-term satiety in male rats (Lindén and Södersten 1990). Also, this manner of CCK-8 administration, in contrast with i.p. injection of CCK-8, only slightly increased plasma oxytocin levels (Lindén *et al.* 1989*a*). These data support a possible physiological role of peripheral CCK in the inhibition of food intake.

Role of brain CCK in food intake

Despite the fact that i.p. injection of CCK-8 increased plasma levels of CCK only during a time period of less than 15 min (Fig. 20.1), it inhibited food intake during each of three consecutive 15 min periods of a 60 min test of food intake (Lindén *et al.* 1989*b*). Thus a brief pulse of peripheral CCK exerts a triggering action on the mechanisms which control food intake. The central neural parts of these mechanisms (outlined by Lindén (1989)) are likely to have CCK as a transmitter. Thus sensory information from the gastro-intestinal tract is transmitted via the vagus nerve to the nucleus tractus solitarius, which is connected to the hypothalamic paraventricular nucleus (PVN) and the ventromedial nucleus (VMH) via the parabrachial nucleus (PBN) in the pons. These nuclei have high densities of CCK-binding sites and, with the exception of the pathway from the PBN to the PVN, these pathways are immunoreactive for CCK and lesions here interfere with the effect of i.p. CCK-8 on food intake (Lindén 1989). In addition, intracerebroventricular (i.c.v.) injection of a CCK antagonist before i.p. injection of CCK-8 was found to blunt the inhibitory effect of i.p. CCK-8 on food intake (Lindén *et al.* 1989*b*) and, moreover, injection of the CCK_B receptor antagonist L-365,260 stimulated food intake and postponed satiety in partially satiated rats (Dourish *et al.* 1989).

To investigate further the possibility that brain CCK may play a role in inhibition of food intake we measured the alterations in the concentration of CCK-LI in the CSF of rats under conditions of free access to food and water (freely fed) after deprivation of food (deprived) and after deprivation of food followed by a period of free access to food pellets (fed). The CSF is easily accessible and all transmitters which are present in the brain are also present in the CSF (Herkenham 1987). Furthermore, enzymes required for peptide transmitter activation and inactivation are present in the CSF (Terenius and Nyberg 1988), and it has been shown for some transmitters that their turnover in the CSF reflects their turnover in the brain (Hutson *et al.* 1984). In addition, overflow of CCK-LI into the extracellular fluid of the hypothalamus, which is a good index of release from nerve terminals (Starke *et al.* 1989), has been

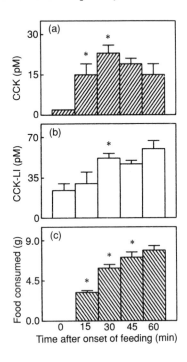

Fig. 20.2. (a) Plasma concentrations of CCK, (b) CSF concentrations of CCK-LI, and (c) amount of food consumed during feeding in male rats deprived of food for 48 h. Asterisks indicate $p \leqslant 0.05$ compared with the preceding value.

shown to occur during ingestion of food in cats (Schick *et al.* 1989).

Figure 20.2 shows that the concentration of CCK-LI increased gradually in the CSF during feeding in male rats deprived of food for 48 h (Lindén *et al.* 1990*a*). Note, however, that the increase in CCK-LI levels in the CSF did not occur until after CCK levels had increased in the plasma. Although peptides do not readily cross the blood–brain barrier (Pardridge 1986), the delay in the increase in CCK-LI in the CSF in response to feeding raises the possibility that the CCK-LI in the CSF was derived from a peripheral source rather than from the brain. We approached this problem by investigating the effect of blocking CCK_A receptors on the inhibitory effect of i.p. CCK-8 on food intake and on CCK-LI levels in the CSF.

Figure 20.3 shows that the inhibitory effect of i.p. CCK-8 on food intake was reversed in a close-dependent manner by injection of the CCK_A receptor antagonists devazepide or lorglumide. The figure also shows that the concentration of CCK-LI in the CSF increased after i.p. injection of CCK-8 and that this effect was eliminated by treatment with devazepide or lorglumide at doses which reversed the effect of CCK-8 on food intake (Lindén *et al* 1990*a*). There

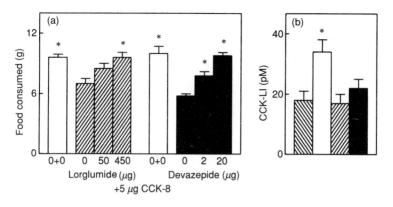

Fig. 20.3. (a) Amount of food consumed by male rats deprived of food for 48 h and injected i.p. with saline (open bars) or 5 µg CCK-8 in combination with lorglumide (hatched bars) or devazepide (solid bars); (b) concentrations of CCK-LI in CSF of male rats deprived of food for 48 h and injected with saline + vehicle (first hatched bar), 5 µg CCK-8 in combination with vehicle (open bar), 450 µg lorglumide (second hatched bar), or 20 µg devazepide (solid bar). Asterisks indicate $p \leqslant 0.05$ compared with (a) CCK-8-treated animals or (b) animals treated with saline + vehicle.

is no obvious reason to expect the CCK_A receptor antagonists to prevent i.p. injected CCK-8 from crossing the blood–brain barrier, if indeed CCK can cross this barrier. Therefore, these data provide evidence that i.p. injection of CCK-8 causes release of CCK-LI from the brain.

Dissociation of CCK levels in CSF and plasma

Lactational hyperphagia

If brain CCK plays a role in the development of satiety during a meal and if this is reflected in overflow of CCK-LI into the CSF as a result of feeding, there should be alterations in CSF CCK-LI levels during physiological states when food intake is altered. Lactating rats overeat because of the nutritional demands of their offspring (Fleming 1976). Figure 20.4 shows that freely fed lactating rats had higher levels of CCK in plasma than non-lactating controls, but the same concentration of CCK-LI in the CSF. After deprivation of food for 24 h there was a marked decrease in the concentration of CCK in both plasma and CSF of lactating rats, and subsequent feeding for 1 h increased the level in CSF but not in plasma (Fig. 20.4) (Lindén *et al.* 1990*b*).

Removal of the offspring reduced food intake to levels comparable with those of non-lactating rats within 24 h and, correlatively, increased the concentration of CCK in the CSF but not in plasma (Lindén *et al.* 1990*b*).

These results show that the high concentration of plasma CCK seen during lactation is not associated with a high level of CCK-LI in the CSF and that the

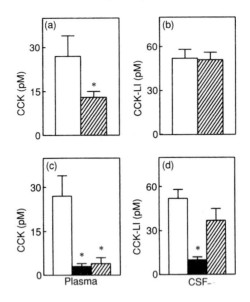

Fig. 20.4. Concentrations of (a) CCK in plasma and (b) CCK-LI in CSF in lactating (open bars) and regularly cycling (hatched bars) rats. Concentrations of (c) CCK in plasma and (d) CCK-LI in CSF in lactating rats: open bars, freely fed; solid bars, deprived of food for 24 h; hatched bars, deprived of food and allowed to eat food pellets for 60 min. Asterisks indicate $p \leqslant 0.05$ compared with lactating rats ((a), (b)) or freely fed rats ((c), (d)).

concentration of CCK-LI in the CSF can increase in the absence of an increase in CCK levels in plasma (Lindén *et al.* 1990*b*). This supports the suggestion that the increase in the level of CCK-LI in the CSF in response to feeding is caused by release of CCK from the brain. Since lactating rats deprived of food for 24 h were found to be behaviourally satiated after 1 h of feeding, the results also enforce the view that brain CCK is important for the development of satiety because, whereas CCK-LI increased in the CSF after this period of feeding, plasma CCK levels did not (Lindén *et al.* 1990*b*).

No information is available on how the neural CCK pathways (outlined by Lindén (1989)) are affected by the neuroendocrine state of lactation.

Because plasma levels of CCK were high in freely fed lactating rats and did not increase after feeding in deprived rats it appears that alterations of peripheral CCK levels play no role in inhibition of feeding during lactation. This view is supported by the observation that i.p. injection of CCK-8 fails to suppress feeding in lactating rats (McLaughlin *et al.* 1983; Wager-Srdar *et al.* 1986). Removal of the offspring reinstates sensitivity to the inhibitory effect of i.p. CCK-8 on food intake (Lindén *et al.* 1990*b*). The mechanisms whereby the offspring prevent peripheral CCK from suppressing feeding during lactation remain to be explored. However, activation of these mechanisms does not prevent peripheral CCK from affecting the behaviour of lactating rats. In fact evidence has been presented that CCK, acting from a peripheral site, may play a role in parental care in rats (Lindén *et al.* 1989*a*).

Oestradiol-induced hypophagia

The most conspicuous alteration in the level of CCK-LI in the CSF of lactating rats was its marked decrease after deprivation of food and subsequent restoration after feeding (Lindén *et al.* 1990*b*). On the basis of this observation we expected a less pronounced decrease in the concentration of CCK-LI in the CSF in a model of hypophagia. We tested this expectation in female rats made modestly obese by ovariectomy. Such rats will reduce their daily food intake and body weight in response to treatment with oestradiol as a consequence of activation of neural as well as peripheral receptors for this ovarian hormone (Wade 1976; Wade and Gray 1979).

Treatment of ovariectomized rats with oestradiol-filled constant-release implants caused temporary hypophagia and a permanent reduction of body weight (Lindén *et al.* 1990*c*). Irrespective of the time after implantation of

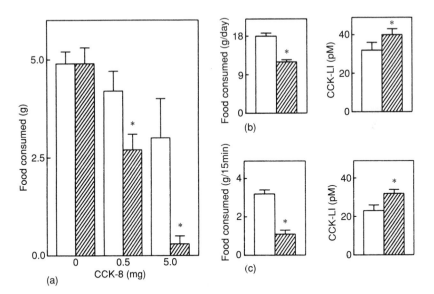

Fig. 20.5. Amount of food consumed by ovariectomized rats treated with empty implants (open bars) or with oestradiol-filled constant-release implants (hatched bars) and injected i.p. with CCK-8: (a) the rats were deprived of food for 24 h and allowed to eat for 60 min; (b) daily amount of food consumed and concentration of CCK-LI in the CSF 4 days after implantation of oestradiol; (c) amount of food consumed during a 15 min period and concentration of CCK-LI in the CSF after 6 h of food deprivation. Asterisks indicate $p \leqslant 0.05$ compared with controls.

oestradiol, i.p. injection of CCK-8 caused a marked reduction in food intake in the oestradiol-treated rats compared with controls (Fig. 20.5). Since ovariectomized rats treated with oestradiol and deprived of food for 24 h did not consume less food than controls in the absence of exogenous CCK-8, these data indicate that oestradiol-treated rats are more sensitive to exogenous, but not endogenous, CCK. If, however, the rats were deprived of food for 6 h during as well as after the temporary phase of hypophagia, the oestradiol-treated rats consumed considerably less food than the controls during a 15 min test (Lindén *et al.* 1990c). The concentration of CCK-LI in the CSF of the oestradiol-treated rats was higher than that in the CSF of controls both in the freely fed condition and after food deprivation (Fig. 20.5). No differences between oestradiol-treated and control rats were noted in plasma levels of CCK under these conditions (Lindén *et al.* 1990c).

Thus treatment of ovariectomized rats with oestradiol induces hypophagia, markedly potentiates the inhibitory effect of i.p. CCK-8 on food intake, increases the concentration of CCK-LI in the CSF, and counteracts the decrease in these levels in response to food deprivation. These results provide additional support for a role of brain CCK in the control of food intake. In this context it should be noted that oestradiol exerts pronounced effects on CCK-LI and CCK binding in the brain (Lindén 1989). For example, the binding of CCK to the VMH, a site on which oestradiol acts to decrease feeding (Wade 1976), is strongly affected by treatment with oestradiol (Akeson *et al.* 1987). Whether these alterations in brain CCK systems after treatment with oestradiol are functionally related to the alterations of ingestive behaviour remains to be determined. Although these correlations are conspicuous, possible peripheral sites of action of both oestradiol and CCK in the control of food intake should not be overlooked. Thus treatment with either osmotic minipumps filled with CCK-8 or oestradiol-filled constant-release implants reduces meal size (Lindén and Södersten 1990; Lindén *et al.* 1990c) and stimulates lipolysis (Wade and Gray 1979; Richester and Schwandt 1988). Since this peripheral action of oestradiol is important in food intake regulation (Wade and Gray 1979), it might also contribute to the inhibitory effect of CCK-8 on food intake.

Intracerebral administration of CCK fails to affect food intake

Although the evidence reviewed above supports a role of brain CCK in food intake regulation, an extensive series of attempts to inhibit food intake by intracerebral administration of CCK failed (Lindén *et al.* 1990a). Thus i.c.v. injection of even very high doses of CCK-8 (10 μg), i.c.v. injection of BC 264, a CCK$_B$ receptor agonist which resists metabolism (Charpentier *et al.* 1988a, b), injection of CCK-8 or BC 264 into the VMH, or continuous i.c.v. infusion of CCK-8 via osmotic minipumps, which increased the concentration of CCK-LI in the CSF to high levels, all failed to inhibit food intake (Lindén *et al.* 1990a). A review of the literature (Lindén *et al.* 1990a) also indicated that most attempts to inhibit feeding by intracerebral administration of CCK-8 have failed, and those investigators who have reported success have used very high doses (10 μg) and, even so, found relatively small decrements in the amount of food consumed. We are aware of only one study (Ritter and Ladenheim 1984) in which a modest dose of CCK-8 (10–50 ng), infused into the fourth ventricle of the brain, markedly inhibited food intake (by 28–94 per cent). Surprisingly, there is only one study in which the relative potency of i.p. CCK-8 has been compared with that of i.c.v. CCK-8 (Makovec *et al.* 1986), and from that study it is clear that i.c.v. CCK-8 is much less effective than i.p. CCK-8 in inhibiting food intake. The possibility that CCK interacts with another transmitter, e.g. dopamine (Crawley *et al.* 1985), to affect food intake has been suggested to account for the failure of intracerebral administration of CCK to affect feeding (Lindén 1989). For further discussion see Bednar *et al.* (1991) and Chapter 23 of this volume.

Conclusions

Since alterations in food intake which occur physiologically correlate with changes in the concentration of CCK-LI in the CSF, and since these changes can be dissociated from changes in plasma concentrations of CCK, it seems likely that brain CCK plays a role in satiety. Yet the failure of intracerebral administration of CCK to affect feeding argues against this possibility. However, it is possible that CCK interacts with some other transmitter to affect feeding.

Acknowledgements

This work was supported by the Swedish Medical Research Council (grant 7615), The Bank of Sweden Tercentenary Foundation (grant 84/108), and the Fund of the Karolinska Institute. We are grateful to Dr C. T. Dourish of Merck Sharp & Dohme and Dr L. Rovati of Rotta Research for generously supplying samples of devazepide and lorglumide. We thank the *Journal of Endocrinology* and Oxford University Press for permission to use previously published material.

References

Akeson, T. R., Mantyh, R. W., Mantyh, C. R., Matt, D. W., and Micevych, P. E. (1987). Estrous cyclicity of ^{125}I-cholecystokinin octapeptide binding in the ventromedial hypothalamic nucleus. *Neuroendocrinology* **45**, 257–62.

Bednar, I., Forsberg, G., Lindén, A., Qureshi, A., and Södersten, P. (1991). Involvement of dopamine in inhibition of food intake by cholecystokinin octapeptide in male rats. *J. Neuroendocrinol.*, in press.

Charpentier, B., Durieux, C., Pelaprat, D., Dor, A., Reibaud, M., Blanchard, J.-C., and Roques, B. P. (1988*a*). Enzyme-resistant CCK analogues with high affinity for central receptors. *Peptides* **9**, 835–42.

Charpentier, B., Pelaprat, D., Durieux, C., Dor, A., Reibaud, M., Blanchard, J.-C., and Roques, B. P. (1988*b*). Cyclic cholecystokinin analogues with high selectivity for central receptors. *Proc. Natl Acad. Sci. USA* **85**, 1968–72.

Crawley, J. N., Stivers, J. A., Blumstein, L. K., and Paul, S. M. (1985). Cholecystokinin potentiates dopamine-mediated behaviors: evidence for modulation specific to a site of coexistence. *J. Neurosci.* **5**, 1972–83.

Davis, J. D., Gallagher, R. L., and Ladove, R. (1967). Food intake controlled by a blood factor. *Science* **156**, 1247–8.

Dourish, C. T., Rycroft, W., and Iversen, S. D. (1989). Postponement of satiety by blockade of brain cholecystokinin (CCK-B) receptors. *Science* **245**, 1509–11.

Fleming, A. S. (1976). Control of food intake in the lactating rat: role of suckling and hormones. *Physiol. Behav* **17**, 841–8.

Gibbs, J., Young, R. C., and Smith, G. P. (1973). Cholecystokinin elicits satiety in rats with open gastric fistulas. *Nature* **245**, 243–5.

Gregory, H., Hardy, P. M., Jones, D. S., Kenner, G. W., and Sheppard, R. C. (1964). The antral hormone gastrin. *Nature* **204**, 931.

Herkenham, M. (1987). Mismatches between neurotransmitter and receptor localizations in the brain: observations and implications. *Neuroscience* **23**, 1–38.

Hutson, P. H., Sarna, G. S., Kantamanemi, B. D., and Curzon, G. (1984). Concurrent determination of brain dopamine and 5-hydroxytryptamine turnovers in individually freely moving rats using repeated sampling of cerebrospinal fluid. *J. Neurochem.* **43**, 151–9.

Koopmans, H. S. (1981). The role of the gastrointestinal tract in the satiation of hunger. In *The body weight regulatory system: normal and disturbed mechanisms* (ed. L. A. Cioffi, W. P. T. James, and T. B. Van Italie), pp. 45–55. Raven Press, New York.

Lindén, A. (1989). Role of cholecystokinin in feeding and lactation. *Acta Physiol. Scand.* **137** (Suppl. 585).

Lindén, A. and Södersten, P. (1990). Relationship between the concentration of cholecystokinin-like immunoreactivity in plasma and food intake in male rats. *Physiol. Behav.*, **48**, 859–63.

Lindén, A., Uvnäs-Moberg, K., Eneroth, P., and Södersten, P. (1989*a*). Stimulation of maternal behaviour in rats with cholecystokinin octapeptide. *J. Neuroendocrinol.* **1**, 389–92.

Lindén, A., Uvnäs-Moberg, K., Forsberg, G., Bednar, I., and Södersten, P. (1989*b*). Plasma concentrations of cholecystokinin octapeptide and food intake in male rats treated with cholecystokinin octapeptide. *J. Endocrinol.* **121**, 59–65.

Lindén, A., Uvnäs-Moberg, K., Forsberg, G., Bednar, I., and Södersten, P. (1990*a*). Role of cholecystokinin in food intake. I. Concentrations of cholecystokinin-like immunoreactivity in the cerebrospinal fluid of male rats. *J. Neuroendocrinol.* **2**, 783–9.

Lindén, A., Uvnäs-Moberg, K., Forsberg, G., Bednar, I. Eneroth, P., and Södersten, P. (1990*b*). Involvement of cholecystokinin in food intake. II. Lactational hyperphagia in the rat. *J. Neuroendocrinol.* **2**, 791–6.

Lindén, A., Uvnäs-Moberg, K., Forsberg, G., Bednar, I., and Södersten, P. (1990*c*). Role of cholecystokinin in food intake. III. Oestradiol potentiates the inhibitory effect of cholecystokinin on food intake. *J. Neuroendocrinol.* **2**, 797–801.

Makovec, F., Bani, M., Christie, R., Revel, L., Rovati, L. C., and Setnikar, I. (1986). Different peripheral and central antagonistic activity of new glutaramic acid derivatives on satiety induced by cholecystokinin in rats. *Regul. Pept.* **16**, 281–90.

McLaughlin, C. L., Baile, C. A., and Peikin, S. R. (1983). Hyperphagia during lactation: satiety response to CCK and growth of the pancreas. *Am. J. Physiol.* **244**, E61–5.

Pardridge, W. M. (1986). Receptor-mediated peptide transport through the blood brain barrier. *Endocr. Rev.* **7**, 314–20.

Rehfeld, J. F., Bardram, L., Cantor, P., Hilsted, L., and Schwartz, T. W. (1988). Cell-specific processing of pro-cholecystokinin and pro-gastrin. *Biochemie* **70**, 25–31.

Richester, W. O. and Schwandt, P. (1988). Cholecystokinin 1–21 stimulates lipolysis in human adipose tissue. *Horm. Metab. Res.* **21**, 216–7.

Ritter, R. C., and Ladenheim, E. E. (1984). Fourth ventricular infusion of cholecystokinin suppresses feeding in rats. *Soc. Neurosci. Abstr.* **10**, 191.4.

Schick, R. R., Yaksh, T. M., Roddy, D. R., and Go, V. L. W. (1989). Release of hypothalamic cholecystokinin in cats: effects of nutrients and volume loading. *Am. J. Physiol.* **256**, R248–54.

Sjödin, L. (1972). Influence of secretin and cholecystokinin on canine gastric secretion elicited by food and exogenous gastrin. *Acta Physiol. Scand.* **85**, 110–17.

Smith, G. P. (1982). Satiety and the problem of motivation. In *The physiological mechanisms of motivation* (ed. D. W. Pfaff), pp. 133–43. Springer-Verlag, New York.

Starke, K., Göthert, M., and Kilbinger, H. (1989). Modulation of neurotransmitter release by presynaptic autoreceptors. *Physiol. Rev.* **69**, 864–989.

Terenius, L. and Nyberg, F. (1988). Neuropeptide-processing, -converting, and -inactivating enzymes in the human cerebrospinal fluid. *Int. Rev. Neurobiol.* **30**, 101–21.

Wade, G. N. (1976). Sex hormones, regulatory behaviors, and body weight. In *Advances in the study of behavior* (ed. J. S. Rosenblatt, R. A. Hinde, E. Shaw, and C. G. Beer), Vol. 6, pp. 201–79. Academic Press, New York.

Wade, G. N. and Gray, J. M. (1979). Gonadal effects on food intake and adiposity: a metabolic hypothesis. *Physiol. Behav.* **22**, 583–93.

Wager-Srdar, S. A., Gannon, M., and Levine, A. S. (1986). The effect of cholecystokinin, bombesin and calcitonin on food intake in virgin, lactating and postweanling rats. *Peptides* **7**, 729–34.

21. Behavioural analysis of the role of CCK_A and CCK_B receptors in the control of feeding in rodents

Colin T. Dourish

Introduction

Since the discovery that CCK decreases food intake in rats (Gibbs *et al.* 1973*a*; see Chapter 17 of this volume) there has been considerable interest in the proposal that the peptide may act as an endogenous satiety signal. The 'CCK satiety' hypothesis has aroused much controversy, and there have been suggestions that the decrease in feeding induced by the peptide is not due to satiety but is caused by sedation, nausea, or mild toxicity. The major proponents of the nausea/toxicity hypothesis have been Deutsch and his colleagues and, more recently, Stricker and his colleagues. Deutsch has shown that CCK produces a conditioned taste aversion (CTA) to various novel-flavoured solutions (Deutsch and Hardy 1977) and that the decrease in feeding induced by CCK is attenuated by an anti-emetic drug (Moore and Deutsch 1985). However, the relationship between the ability of a drug to induce a CTA and to cause nausea is uncertain, and indeed many drugs that induce a CTA also support self-administration, indicating that they have rewarding properties (Goudie 1987). Furthermore, it appears that small doses of CCK which produce a reliable decrease in food intake do not induce a CTA, suggesting that the presence of a CTA is not a prerequisite for observing a satiety-like action of CCK (see Chapter 17 of this volume for further discussion).

Stricker and his colleagues have proposed, on the basis of endocrinological findings (Verbalis *et al.* 1986*a*), that CCK-induced satiety may be secondary to nausea. Two neurohypophyseal hormones, vasopressin and oxytocin, that are thought to be correlated with nausea (in rodents) and vomiting (in primates and humans) were measured in the plasma of rats and monkeys after administration of CCK. Dose-dependent increases in the plasma concentrations of oxytocin in rats (Verbalis *et al.* 1986*b*) and vasopressin in monkeys (Verbalis *et al.* 1987) were observed after injection of CCK. However, an increase in plasma oxytocin concentration was also observed in rats after ingestion of a large meal (Verbalis *et al.* 1986*b*). This increase was equivalent in magnitude to the increase in oxytocin caused by intraperitoneal (i.p.) injection of CCK doses of up to $10 \mu g/kg$. As CCK doses below $10 \mu g/kg$ were shown to decrease food intake in this study and in many others (McCann *et al.* 1989), this can be interpreted as evidence that an increase in plasma oxytocin is also not a prerequisite for the observation of a satiety-like action of CCK (see Chapter 17 of this volume for further discussion of the implications of these results).

The above discussion summarizes one of the major controversies that has surrounded the CCK satiety hypothesis. A second concern that is of major importance to the hypothesis is whether the inhibition of feeding, first observed by Gibbs *et al.* (1973*a*) after injecting *exogenous* CCK, reflects a physiological role of *endogenous* CCK in satiety. This issue can be addressed by using selective CCK antagonists which elicit their behavioural effects by preventing the action of endogenous CCK. If CCK acts as an endogenous satiety signal, then we would predict that a CCK antagonist given prior to, or during, a meal should prolong the meal and postpone the onset of satiety. The use of a CCK antagonist would also address the issue of nausea/toxicity as we would predict that if CCK decreases food intake by a non-specific action then a CCK antagonist would not increase food intake. In the first section of this chapter the results of recent studies of the effects of selective CCK antagonists on feeding behaviour in rodents are reviewed and the implications of these data for the CCK satiety hypothesis are considered.

The third major controversy associated with the CCK hypothesis involves the respective roles of peripheral and central mechanisms in mediating the satiety-like action of CCK. The original postulate of Smith and Gibbs was that the site of action for the satiating effect of exogenous CCK was peripheral, more specifically in the gastro-intestinal tract (Smith *et al.* 1981; Chapter 17 of this volume). Afferent fibres of the vagus nerve appear to play a crucial role in the satiety pathway as bilateral lesions of these neurons abolish the satiating effect of low doses of CCK (Smith *et al.* 1985; South and Ritter 1988). It is thought that CCK stimulates receptors in the abdomen (possibly located on the pyloric sphincter and/or pancreatic acinar cells) and that this causes activation of vagal afferents (see Chapters 17 and 18 of this volume for detailed discussion). This signal is then probably relayed via the medial region of the nucleus tractus solitarius (NTS), where the vagal afferents terminate, to the hypothalamus (Fig. 21.1).

Ligand binding studies have identified CCK receptors in the gastro-intestinal tract, on the vagus nerve, and throughout the brain (Zarbin *et al.* 1983; Moran and McHugh 1988), and it is clear that peripheral CCK receptors can be differentiated from brain CCK receptors. This distinction was first proposed by Innis and Snyder (1980) on the basis of the differential potencies of various CCK fragments in displacing the binding of radio-labelled CCK to peripheral and brain tissues. It was found that pentagastrin, unsulphated CCK octapeptide (CCK-8US), and the C-terminal tetrapeptide of CCK (CCK-4) had similar affinities to that of sulphated CCK octapeptide (CCK-8S) in displacing CCK binding to brain membranes, but that CCK-8S was three orders of magnitude more potent than the other fragments in displacing binding to peripheral

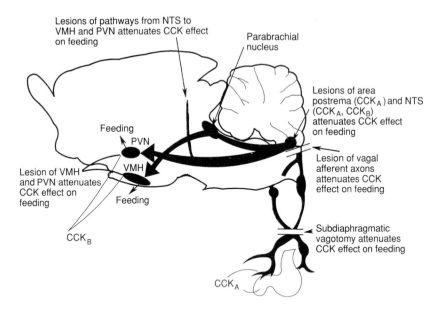

Lesions of pathways from NTS to VMH and PVN attenuates CCK effect on feeding

Parabrachial nucleus

Lesions of area postrema (CCK$_A$) and NTS (CCK$_A$, CCK$_B$) attenuates CCK effect on feeding

Feeding

PVN

VMH

Lesion of VMH and PVN attenuates CCK effect on feeding

Feeding

CCK$_B$

Lesion of vagal afferent axons attenuates CCK effect on feeding

Subdiaphragmatic vagotomy attenuates CCK effect on feeding

CCK$_A$

Fig. 21.1. Schematic illustration of the putative neuronal pathways involved in processing the CCK satiety signal: NTS, nucleus tractus solitarius; PVN, paraventricular nucleus of the hypothalamus; VMH, ventromedial nucleus of the hypothalamus.

tissues. Subsequently, autoradiographical studies by Moran *et al.* (1986) revealed a limited number of brain binding sites with low affinities for CCK-8US, CCK-4, and pentagastrin which resembled those found in the periphery. These peripheral sites were termed CCK$_A$ (alimentary) to distinguish them from the classical brain (CCK$_B$) receptor (Moran *et al.* 1986). The findings of Moran, and further studies using the potent and selective CCK antagonists devazepide and L-365,260 as displacing agents, indicated that CCK$_A$ receptors in rodents were confined to a few regions of the brainstem such as the medial NTS and the area postrema (Moran *et al.* 1986; Hill *et al.* 1987; Chapter 5 of this volume). More recently, however, functional evidence has indicated that CCK$_A$ receptors are also present in various regions of the rat forebrain (Vickroy and Bianchi 1989). Nevertheless, the majority of CCK receptors in the CNS of rodents are of the CCK$_B$ subtype and are distributed ubiquitously throughout the brain (Moran *et al.* 1986; Hill *et al.* 1987; Chapter 5 of this volume). The discovery of devazepide, a selective antagonist for the CCK$_A$ receptor (Chang and Lotti 1986; Evans *et al.* 1986), and L-365,260, a selective antagonist for the CCK$_B$ receptor (Bock *et al.* 1989; Lotti and Chang 1989), has greatly facilitated the examination of the respective roles of CCK$_A$ and CCK$_B$ receptors in mediating the satiating effect of exogenous and endogenous CCK. In the latter sections of this chapter I review studies which have examined the roles of the two CCK receptor subtypes in satiety and, in particular, attempt to delineate the circumstances under which one type appears to be predominant.

Does endogenous CCK act as a satiety signal in the rat?

Early examinations of the effects of CCK antagonists on food intake only appeared to add to the controversy aroused by results obtained using agonists. The first experiments used the weak non-selective CCK antagonist proglumide (Hahne *et al.* 1981). In some studies, it was reported that proglumide increased food intake in rats that had been given a food preload to induce partial satiety (Shillabeer and Davison 1984). However, others failed to replicate this effect using a very similar experimental procedure (Schneider *et al.* 1986). Initial studies using the potent selective CCK antagonist devazepide also yielded conflicting data. Thus Hewson *et al.* (1988) reported that devazepide produced large increases (50–100 per cent of baseline) in the intake of a palatable diet in rats. In contrast, Schneider *et al.* (1988) failed to find a robust increase in palatable diet intake after devazepide treatment under almost identical conditions, even when the studies were carried out in the laboratory of Hewson and his colleagues. Despite these initial setbacks, numerous subsequent studies using devazepide have established that CCK receptor blockade increases food intake in rodents and pigs (see Table 21.1 and Chapters 17 and 19 of this volume) and causes hunger in humans (Wolkowitz *et al.* 1990). Nevertheless, failures to replicate the hyperphagic effects of CCK antagonists in rodents persist and continue to accumulate, and the challenge remains to discover the physiological and environmental variables that are responsible for these apparent discrepancies (see below for further discussion).

Studies using the satiety sequence in food-deprived rats

The studies summarized above suggest that blockade of CCK receptors increases food intake in rodents. However, an increase in food intake does not, in itself, provide compelling evidence of an antisatiety action as other influences on feeding (e.g. an increase in appetite or general arousal) could account for such an effect. In an attempt to address this issue we carried out studies which examined the effects of CCK antagonists on the behavioural satiety sequence in rats. Rats that are deprived of food overnight and then allowed to feed to satiety on the following day exhibit a characteristic sequence of behaviour (Antin *et al.* 1975). Initially, feeding behaviour is predominant, but as the rat becomes satiated other behaviours appear in a characteristic and reproducable sequence. The sequence is feeding → activity/exploration → grooming → resting/sleep (Fig. 21.2). Use of the satiety sequence enables a more precise examination to be made of the effects of drugs on feeding. A specific effect of a drug on satiety should be reflected in an alteration in the behaviour of the rat on the temporal plane. Thus a putative satiety signal (such as CCK) should

Table 21.1. Effect of devazepide on food intake in animals

Species	Dose/route	Diet	FD (prefeed)	Effect	MED	Reference
Rat	0.001–10.0 p.o.	Pellet	20 h	Increase	1.0	Hanson and Strouse 1987
Rat	0.01–0.1 i.p.	Sweet mash	None	+50%–100%	0.01	Hewson et al. 1988
Rat	0.01–10.0 s.c.	Pellet	None/17 h	+10% in non-FD; none in FD	1.0	Dourish et al. 1988
Rat	0.03–1.0 i.p.	Sweet mash/milk	None	None	NA	Schneider et al. 1988
Mouse	0.0001–10.0 i.p.	Sweet mash	24 h	None	NA	Khosla and Crawley 1988
Rat	0.5 + 1.0 i.p. (a.m.) [p.m.]	Liquid	None	+24%–32%	0.5	Watson et al. 1988
Rat	0.75 i.p.	Pellet	None	+11%	0.75	Strohmayer et al. 1989
Rat	0.00001–1.0 s.c.	Pellet	17 h (40 min)	+100%–200%	0.0001	Dourish et al. 1989a
Mouse	0.05–0.5 i.p.	Pellet/sweet milk	None/18 h (120 min)	+75%–150%	0.05	Silver et al. 1989
Rat	0.03–1.0 i.p.	Liquid/chow	None/1 h	+9%–35%	0.1	Reidelberger and O'Rourke 1989
Rat	1.0 i.v.	Liquid	1 h	+17%–29%	1.0	Miesner et al. 1989
Pig	0.1 i.a.	Barley diet	18 h	None	NA	Gregory et al. 1989
Pig	0.017–0.14 i.v.	Pellet	4 h	+30%–50%	0.035	Ebenezer et al. 1990
Rat	0.0000001–1.0 s.c.	Liquid	Overnight (40 min)	+39%–54%	0.00001	Corwin et al. 1990
Rat	0.01–0.3 s.c.	Pellet/sweet mash	None	None	NA	Cooper et al. 1990
Rat	2.5 i.p.	Liquid	FD*	+21%	2.5	Garlicki et al. 1990

MED, minimum effective dose; FD, food deprivation; FD*, duration not stated; NA, not applicable; p.o., oral; i.p., intraperitoneal; s.c., subcutaneous; i.v., intravenous; i.a., intra-arterial.

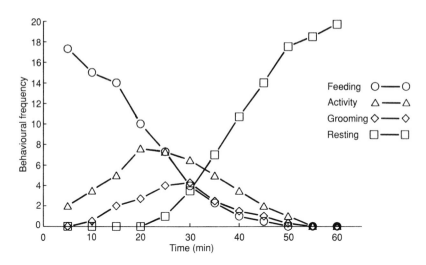

Fig. 21.2. An illustration of the satiety sequence in rats. The data were obtained from six rats that were deprived of food overnight prior to testing and then allowed to feed to satiety. An observer recorded their behaviour on 20 occasions during each 5 min time bin of a 60 min test. Behaviour was recorded in one of four mutually exclusive categories (feeding, active, grooming, resting) using the keypad of a microcomputer.

promote the appearance of the satiety sequence without disrupting its component parts. A non-specific drug action which disrupts feeding (e.g. locomotor stimulation or sedation) would also disrupt the components of the satiety sequence rather than producing a shift in the temporal pattern.

Effects of food preloads on the satiety sequence

In order to create a template with which to compare the effects of CCK and CCK antagonists we first examined the effects of food preloads on the satiety sequence. Rats were deprived of food overnight and then given access to food for periods of 0, 10, 20, 30, or 40 min prior to a 60 min satiety sequence test (see Fig. 21.2 for details). As expected, increasing the duration of the prefeeding period decreased food intake and accelerated the appearance of the satiety sequence during the test period. A clear temporal shift to the left in the appearance of activity, grooming, and resting was observed as the duration of the prefeed was increased (Fig. 21.3). A 40 min prefeeding period induced satiety almost immediately in the test period, and this condition was used to examine the potential antisatiety effects of the CCK antagonists (see below). In contrast, the putative satiety-like action of CCK was examined in rats given no prefeed.

Effects of CCK-8S on the satiety sequence

Doses of 0.1, 0.5, 1.0, 2.0, 5.0, and 10.0 µg/kg of CCK-8S were injected i.p. in rats that had been deprived of food overnight. The animals were immediately given access to food and their behaviour recorded during a 60 min test. There was no significant effect of CCK on total food intake during the test as the inhibitory effects of the peptide dissipated within 30 min and rebound eating occurred in the drug-treated animals. However, as anticipated from previous reports (Antin *et al.* 1975), significant drug effects were observed on the components of the satiety sequence. During the first 5 min time bin, all doses of CCK tested significantly decreased the frequency of feeding. There was a corresponding increase in the frequency of resting at all doses, with the exception of the lowest dose tested (0.1 µg/kg) (Fig. 21.4). The decrease in feeding caused by 0.1 µg/kg CCK during the first 5 min of the test appeared to be associated with an increase in active behaviour (Fig. 21.4). The pattern of behaviour induced by CCK contrasted with that induced by prefeeding (compare Fig. 21.3 and 21.4). Thus, although feeding decreased and resting increased, there was no clear indication of a shift to the left in the appearance of activity and groom-

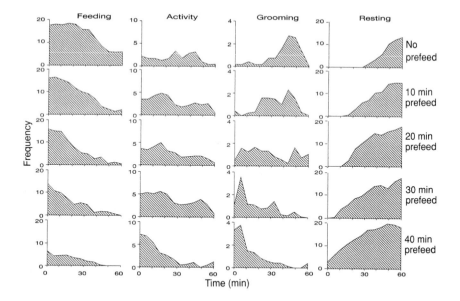

Fig. 21.3. Effects of prefeeding on the pattern of behaviour in the satiety sequence test. Note the shift to the left in the appearance of activity, grooming, and resting as the duration of the prefeed increases. There is a corresponding shift to the right in the presence of feeding as the duration of the prefeed decreases. Data are the mean of 12 rats per treatment group.

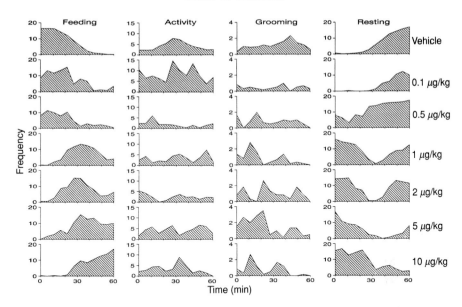

Fig. 21.4. Effects of CCK (0.1–10.0 μg/kg i.p.) on the pattern of behaviour in the satiety sequence test. Note that the decrease in feeding frequency at doses of 0.5–10.0 μg/kg appears to be almost entirely due to the accelerated appearance of resting at the apparent expense of the activity and grooming components of the sequence. The pattern of behaviour induced by CCK contrasts with that induced by prefeeding (see Fig. 21.3).

ing after injection of the peptide, a pattern which was apparent after prefeeding. These data do not provide convincing evidence of a satiating effect of the CCK. Nevertheless, there are two possible interpretations of the data. First, it is possible that the inhibitory effect of CCK is non-specific and animals are unable to feed as the peptide inhibits locomotion (possibly owing to nausea/mild toxicity (see above)). Thus doses of CCK in excess of 0.5 μg/kg induced resting and inhibited feeding during the first 20–30 min of the test. This interpretation is consistent with reports that CCK decreases locomotion and exploration in satiated rodents (Crawley *et al.* 1981; Chapter 22 of this volume). Alternatively, CCK may exert a potent satiating effect that causes a 'short-circuit' of the satiety sequence such that the animal advances to the final component of the sequence (resting) immediately following injection of the peptide. It is extremely difficult to determine which, if either, of these two interpretations is correct, and therefore the importance of examining the effects of CCK antagonists on the satiety sequence is reinforced.

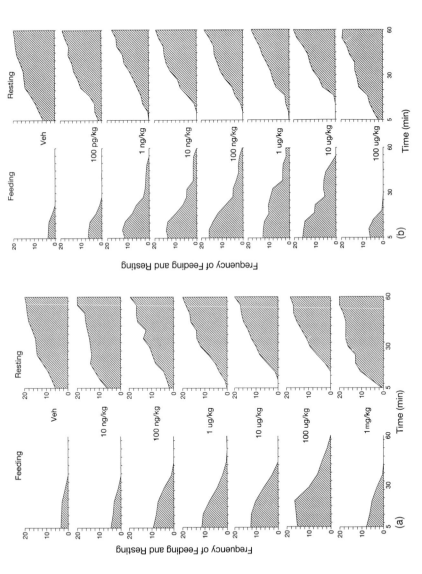

Fig. 21.5. Effects of (a) devazepide and (b) L-365,260 on the frequency of feeding and resting in rats deprived of food overnight prior to testing and then given a 40 min prefeed. Both drugs significantly increased the frequency of feeding and postponed the onset of resting, but L-365,260 was 100 times more potent than devazepide. The data are the mean of 12 or more rats per group. [From Dourish *et al.* 1989*a* (devazepide) and unpublished data (L-365,260).]

Effects of CCK antagonists on the satiety sequence

The effects of the CCK_A antagonist devazepide (at doses of 10 ng/kg–1 mg/kg subcutaneously (s.c.)) and the CCK_B antagonist L-365,260 (at doses of 100 pg–100 μg/kg s.c.) were examined in rats that had been deprived of food overnight and given a 40 min prefeed (to induce partial satiety (see Fig. 21.3)) prior to a 60 min satiety sequence test. Both devazepide and L-365,260 significantly increased the frequency of feeding and consequently delayed the onset of resting (Dourish *et al.* 1989*a*) (Fig. 21.5). The dose-response curves for both antagonists were bell-shaped and the explanation for this finding is unclear at present. However, there have been numerous previous reports of this phenomenon with CCK antagonists in studies on feeding (Schwartz *et al.* 1988) and analgesia (Watkins *et al.* 1984; Chapter 44 of this volume). In Fig. 21.6 the effects of devazepide (at a dose of 10 μg/kg) on the satiety sequence after a 40 min prefeeding period are compared with the effects of a 20 min or 40 min period of prefeeding in vehicle-treated rats. It is apparent that there is a striking similarity, both quantitatively and qualitatively, in the patterns of the two graphs. These data indicate that, in food-deprived rats, devazepide produces an effect that is equivalent to 20 min less access to food during the prefeeding

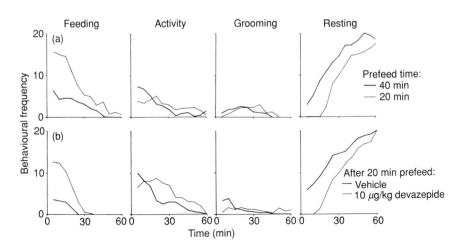

Fig. 21.6. Comparative responses in the satiety sequence of (a) vehicle-treated rats deprived of food overnight prior to testing and then given a 20 min or 40 min prefeed, and (b) devazepide-treated (10 μg/kg s.c.) rats and vehicle-treated rats deprived of food overnight prior to testing and then given a 40 min prefeed. The data are the mean of 12 or more rats per group. The effect of 10 μg/kg of devazepide on the satiety sequence closely resembles the effect of a 20 min decrease in the duration of the prefeed. (From Dourish *et al.* 1989*a*.)

period. A similar result was obtained when the effects of L-365,260 were compared with those of prefeeding. When considered together with the results of other recent studies with CCK antagonists (see above and Table 21.1), these data provide strong support for the hypothesis (see Chapter 17 of this volume) that endogenous CCK acts as a satiety signal in rats. Therefore the results of studies with CCK antagonists have apparently resolved two of the three controversies surrounding the CCK satiety hypothesis: the issue of behavioural specificity and the question of whether endogenous CCK (like exogenous CCK) induces satiety. However, the third issue—the role of peripheral and central mechanisms, and of CCK_A and CCK_B receptors—remains unresolved, and to date the results of studies with CCK antagonists have provided more questions than answers.

Contrasting effects of CCK_A and CCK_B receptor blockade on satiation induced by endogenous and exogenous CCK

The results of studies with CCK agonists have consistently suggested that the satiating effect of exogenous CCK is mediated by CCK_A receptors. Thus CCK-8S decreased food intake and exploratory behaviour, whereas CCK-8US, CCK-4, and gastrin were inactive (Gibbs *et al.* 1973*b*; Crawley *et al.* 1984; Schick *et al.* 1986). This pattern of activity correlates with the potency of agonists at CCK_A receptors but not with their potency at CCK_B receptors (Moran *et al.* 1986; Chapter 17 of this volume). The results of studies with CCK antagonists are consistent with mediation of the satiating effects of exogenous CCK by CCK_A receptors. Thus the hypophagic effect of CCK injected i.p. was blocked by the selective CCK_A antagonist devazepide (Hanson and Strouse 1987; Khosla and Crawley 1988; Dourish *et al.* 1988; 1989*b*; Schneider *et al.* 1988) but not by the selective CCK_B antagonist L-365,260 (Dourish *et al.* 1989*b*). Furthermore, it has been proposed that the hypophagic effect of intracerebroventricular (i.c.v.) injection of CCK in the rat is mediated by peripherally located CCK_A receptors (Crawley *et al.* 1990). Thus it was shown that the dose of CCK required to decrease feeding when injected by the i.c.v. route was greater than the dose required if the peptide was injected i.p. (Note that this finding is species dependent as CCK is more potent when injected i.c.v. than when injected i.v. in sheep and pigs (see Chapter 19 of this volume)). Measurement of the plasma levels of the peptide after i.c.v. injection of radiolabelled CCK indicated that the concentration which leaked to the periphery was in excess of that required to decrease feeding (Crawley *et al.* 1990). Furthermore, the decrease in food intake after injection of i.c.v. CCK was blocked by devazepide but not by L-365,260 (Crawley *et al.* 1990).

   The above findings, which strongly implicate CCK_A receptors in mediating the satiating effects of exogenous CCK, suggest that the same mechanism would be activated by post-prandial release of endogenous CCK. It was therefore startling to discover that the selective CCK_B antagonist L-365,260 was two orders of magnitude more potent than the selective CCK_A antagonist devazepide in postponing the onset of satiety (Dourish *et al.* 1989*a*) (see Fig. 21.5), suggesting mediation by CCK_B receptors. This finding was particularly surprising in view of the fact that L-365,260 had no effect on satiety induced by exogenous CCK-8S (Dourish *et al.* 1989*b*) (see above). Therefore we suggested that there may be two different CCK satiety mechanisms and that CCK-8S may be the endogenous ligand for only one of them (Dourish *et al.* 1989*b*). This interpretation poses the following question: if it is not CCK-8S,

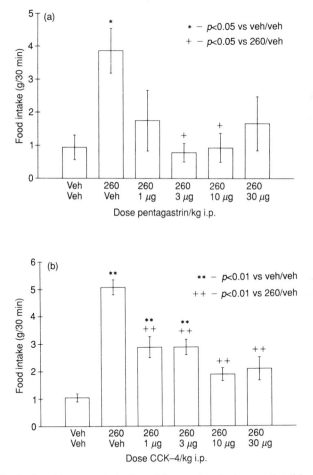

Fig. 21.7. Blockade of L-365,260-induced hyperphagia in a palatable diet test by the CCK_B agonists (a) pentagastrin and (b) CCK-4. Following a 40 min prefeed of the palatable diet, the agonists were injected i.p. 25 min after s.c. injection of 100 ng/kg of L-365,260. The rats were given access to the diet 5 min after injection of the agonist for a period of 30 min. The data are the mean (SEM shown as lines on each bar) of at least 12 rats per group.

then what is the endogenous ligand which is blocked by the action of L-365,260 in the satiety sequence studies? In an attempt to address this issue we have begun to examine the ability of various CCK agonist fragments to prevent the hyperphagic action of L-365,260 in partially satiated rats. The method used was a palatable diet test which was adapted from that described by Jackson and Cooper (1985). Non-deprived rats were adapted for a period of 6 days to access to a palatable sweet mash diet. The palatable diet was available to the animals for 1 h each day in addition to *ad libitum* access to their standard pellet diet. On test days the rats were given a 40 min prefeed of the palatable diet (based on results in the satiety sequence test (see above)) and were then injected with devazepide or L-365,260; 30 min later they were allowed access to the diet for a further period of 60 min. Both devazepide and L-365,260 significantly increased intake of the palatable diet and their relative potencies were similar to those observed in the satiety sequence test (i.e. L-365,260 was at least 100 times more potent than devazepide). Treatment with CCK-4 or pentagastrin (i.p. 25 min after the antagonist, test began 5 min after injection of the agonist) blocked the hyperphagia induced by L-365,260 (Fig. 21.7). As CCK-4 and pentagastrin are at least three orders of magnitude more potent at CCK_B receptors than at CCK_A receptors (Innis and Snyder 1980; Moran *et al.* 1986), these data provide further support for the idea that the hyperphagic effect of L-365,260 is mediated by CCK_B receptors. The data also suggest that CCK-4 and pentagastrin may act as endogenous satiety signals in the rat, and the physiological and environmental conditions under which these peptides may be released are discussed below.

A consideration of the physiological and environmental variables that determine that involvement of CCK_A and CCK_B receptors in the control of feeding

The results described above suggest the existence of two CCK satiety mechanisms, one mediated by CCK_A receptors and the other involving CCK_B receptors. This dual-system hypothesis poses a number of important questions that deserve consideration.

1. Why are two systems necessary?

2. Under what conditions does each system operate?

3. Which, if either, system is more important in the control of post-prandial satiety?

In order to answer these questions it is instructive to review the circumstances under which CCK_B agonists and antagonists have been observed to alter feeding behaviour. The majority of studies with endogenous (CCK-8US, CCK-4) or synthetic (BC 264) CCK_B agonists have been unsuccessful in demonstrating any effect of these peptides on food intake (Lindén 1989;

Chapters 17 and 20 of this volume), and CCK-8S has consistently been shown to be the minimum sequence to produce satiation. Following our discovery that CCK-4 and pentagastrin were able to block the hyperphagia induced by L-365,260 (see above), we examined the effects of CCK-4 on palatable diet intake. The peptide had no effect on food consumption in any of our standard tests. We took into account the fact that the effects of L-365,260 were only observed in animals given a large food preload and examined the effects of CCK-4 in rats given various durations of access to a palatable diet prior to testing. Again we were unsuccessful in demonstrating a hypophagic effect of CCK-4. To the best of my knowledge, there is only one paradigm in which an inhibition of feeding by CCK-4 or any other selective CCK_B agonist has been demonstrated. This paradigm involves the exposure of hungry animals to food in a novel environment. When rats are tested in a novel environment both CCK-8US and CCK-4 have been shown to decrease approaches to food and food intake (Halmy *et al.* 1982; Kadar *et al.* 1985, 1986). Kadar *et al.* (1986) have proposed that this is a central action of the peptides, and this conclusion is supported by the observation that CCK-4 is at least three orders of magnitude more potent when injected i.c.v. than when injected i.p. The authors conclude from their results that CCK-4 may decrease feeding by enhancing the emotionality and anxiety provoked by exposure to a novel environment (Kadar *et al.* 1985, 1986). This possibility is reinforced by the recent discovery that CCK-4 and pentagastrin elicit panic attacks in volunteers and patients suffering from panic disorder (DeMontigny 1989; Chapter 11 of this volume). Furthermore, i.p. or s.c. injection of CCK-4 (Dourish *et al.* 1991; Chapter 13 of this volume and i.c.v. injection of pentagastrin (Chapter 5 of this volume) have been shown to induce anxiogenic-like effects in an animal model of anxiety, the elevated plus maze. In complementary studies L-365,260 has been shown to have anxiolytic-like effects in a number of animal models (Ravard and Dourish 1990; and in two of these models, conditioned suppression of drinking and the elevated plus maze, the potency of the drug was almost identical with its potency in the satiety sequence test (Dourish *et al.* 1990, 1991). One possible interpretation of these findings is that the hyperphagia induced by L-365,260 is associated with the putative anxiolytic properties of the drug. Indeed, it is interesting to consider that most drugs from the two major classes of anxiolytics in current clinical use (i.e. benzodiazepines and $5-HT_{1A}$ agonists) have been shown to increase food intake in rodents (Cooper 1980; Dourish *et al.* 1986). A prediction of this anxiolytic hypothesis would be that the hyperphagia induced by the drug would not be observed under all experimental conditions and may be limited to situations in which there is some degree of novelty or stress. This prediction appears to be borne out by the failure of L-365,260 to block the satiating effect of CCK-8S in rats (Dourish *et al.* 1989*b*) or to increase 24 h food intake (Dourish, unpublished results) or nocturnal feeding (Cooper *et al.* 1991). In the studies described above, in which a robust increase in food intake was induced by L-365,260 (satiety sequence, palatable diet test), the rats used had been housed individually for a prolonged period prior to testing, a procedure that is known to be stressful in animals. Additional studies which examine the effects of L-365,260 under a wide range of stressful and non-stressful experimental conditions will be necessary to explore this hypothesis. One paradigm which could be of particular interest is tail-pinch-induced feeding (Antelman and Szechtman 1975). The increase in feeding induced by tail pinch is thought to involve an arousal mechanism (Robbins and Fray 1980) and appears to be mediated by brain dopaminergic mechanisms (Antelman and Szechtman 1975). Tail-pinch-induced feeding is blocked by CCK-8S and, more importantly, is also inhibited by i.c.v. injection of the CCK_B agonists CCK-8US and CCK-4 (Telegdy *et al.* 1984). Therefore an examination of the effects of L-365,260 on feeding induced by tail pinch may prove valuable.

The possible relationship between the hyperphagic and anxiolytic actions of L-365,260 may provide an answer to the first two questions posed above. Two CCK satiety systems appear to be necessary as they fulfil distinct roles and are likely to be activated under different circumstances. The CCK_A receptor mechanism probably mediates post-prandial satiety, and there is now convincing evidence for its activation under a wide variety of experimental conditions which produce satiation (see Chapter 17 of this volume). The CCK_B receptor mechanism appears to be activated in response to stress and anxiety, and there-

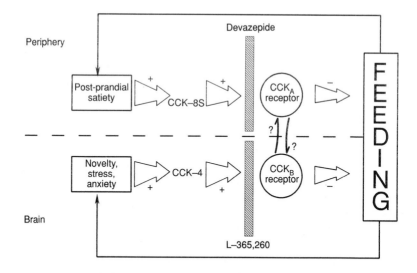

Fig. 21.8. Schematic representation of the dual system hypothesis of CCK-induced satiation. It is proposed that there are independent mechanisms mediated by CCK_A and CCK_B receptors which may be activated under different circumstances. The CCK_A receptor is located in the gastro-intestinal tract and is activated by CCK-8S released during feeding to induce post-prandial satiety. The CCK_B receptor is located in the brain and is activated by CCK-4 released during stress/anxiety. Feeding is decreased by CCK-4 under stressful conditions.

fore its influence on feeding is observed under more limited conditions than that of the CCK_A receptor. Figure 21.8 shows a schematic representation of this hypothesis. Inspection of this figure raises one further issue which merits attention: the respective locations of the CCK_A and CCK_B receptor subtypes.

There is compelling evidence that the CCK_A receptor population which mediates the satiating effect of CCK-8S in rats (this does not extend to all species; see Chapter 19 of this volume) is located in the gastrointestinal tract. Thus the hypophagic effect of CCK-8S is abolished by vagotomy and the peptide is relatively ineffective when injected i.c.v. compared with i.p. (see above and Chapter 17 of this volume). In contrast, the location of the CCK_B receptor population that mediates the effects of L-365,260 is much less certain. There are receptors with a high affinity for L-365,260 in both the periphery (gastrin receptors) and the brain (CCK_B receptors), and as yet there are no data which conclusively implicate either of these two populations. Crucial experiments will be to examine the effects of CCK-4 and L-365,260 in feeding and anxiety models in vagotomized rats and to compare the relative potency of both compounds in the same models when injected systemically and intracerebrally.

Conclusions

The use of potent and selective CCK antagonists during the past 4 years has permitted a definitive test of the CCK satiety hypothesis first proposed by Smith and Gibbs in 1973 (Gibbs *et al.* 1973*a*). The data reviewed in this chapter and elsewhere in this volume (see Chapters 17 and 19) provide convincing evidence for the proposal that CCK acts as an endogenous satiety signal. In the rat, the site of action of CCK in producing satiety appears to be peripheral, probably at one or more sites in the gastrointestinal tract (Chapter 17 of this volume). In contrast, in pig and sheep there is strong evidence for a central satiety-like action of the peptide (Chapter 19 of this volume). At present it is unclear which species is the better model of the situation in humans. Whether the location is central (pig and sheep) or peripheral (rat), there seems little doubt that the satiating action of CCK is primarily mediated by CCK_A receptors. However, CCK_B receptors can also control food intake and their influence appears to predominate under conditions of stress and anxiety. Indeed, it is possible that in certain feeding disorders which are associated with stress (e.g. anorexia nervosa, bulimia nervosa) CCK_B receptors may play an important role. The anatomical location of this CCK_B receptor population and its possible interaction with the CCK_A receptor population, which apparently signals satiety, remain to be discovered.

References

Antelman, S. M. and Szechtman, H. (1975). Tail pinch induces eating in sated rats which appears to depend on nigrostriatal dopamine. *Science* **189**, 731-3.

Antin, J., Gibbs, J., Holt, J., Young, R. C., and Smith, G. P. (1975). Cholecystokinin elicits the complete sequence of satiety in rats. *J. Comp. Physiol. Psychol.* **89**, 784–90.

Bock, M. G., DiPardo, R. M., Evans, B. E., Rittle, K. E., Whitter, W. L., Veber, D. F., Anderson, P. S., and Freidinger, R. M. (1989). Benzodiazepine gastrin and cholecystokinin receptor ligands: L-365,260. *J. Med. Chem.* **32**, 13–16.

Chang, R. S. L. and Lotti, V. J. (1986). Biochemical and pharmacological characterization of an extremely potent and selective non-peptide cholecystokinin antagonist. *Proc. Natl Acad. Sci. USA* **83**, 4923–6.

Cooper, S. J. (1980). Benzodiazepines as appetite enhancing compounds. *Appetite* **1**, 7–19.

Cooper, S. J., Dourish, C. T., and Barber, D. J. (1990). Reversal of the anorectic effect of (+)-fenfluramine in the rat by the selective cholecystokinin receptor antagonist MK-329. *Br. J. Pharmacol.* **99**, 75–90.

Cooper, S. J., Dourish, C. T., Barber, D. J., and Moseley, S. M. (1991). Reversal of the anorectic effect of systemically-administered 5-hydroxytryptamine by the selective cholecystokinin receptor antagonist devazepide. *Psychopharmacology*, in press.

Corwin, R. L., Gibbs, J., and Smith, G. P. (1990). Increased food intake after cholecystokinin-A-type receptor blockade. *FASEB J.* **4**, A941.

Crawley, J. N., Hays, S. E., Paul, S. M., and Goodwin, F. K. (1981). Cholecystokinin reduces exploratory behaviour in mice. *Physiol. Behav.* **27**, 407–11.

Crawley, J. N., St-Pierre, S., and Gaudreau, P. (1984). Analysis of the behavioral activity of C- and N-terminal fragments of cholecystokinin octapeptide. *J. Pharmacol. Exp. Ther.* **230**, 438–44.

Crawley, J. N., Fiske, S. M., Durieux, C., Derrien, M., and Roques, B. P. (1990). Peripheral receptor subtype specificity for cholecystokinin-induced inhibition of feeding. *Soc. Neurosci. Abstr.* **16**, P267.

De Montigny, C. (1989). Cholecystokinin tetrapeptide induces panic-like attacks in healthy volunteers. *Arch. Gen. Psychiat.* **46**, 511–17.

Deutsch, J. A. and Hardy, W. T. (1977). Cholecystokinin produces bait shyness in rats. *Nature* **266**, 196.

Dourish, C. T., Hutson, P. H., Kennett, G. A., and Curzon, G. (1986). 8-OH-DPAT-induced hyperphagia: its neural basis and possible therapeutic relevance. *Appetite* **7**, (Suppl.), 127–40.

Dourish, C. T., Coughlan, J., Hawley, D., Clark, M. L., and Iversen, S. D. (1988). Blockade of CCK-induced hypophagia and prevention of morphine tolerance by the CCK antagonist L-364,718. In *Neurology and neurobiology*, Vol. 47, *Cholecystokinin antagonists* (ed. R. Y. Wang and R. Schoenfeld), pp. 307–25. Alan R. Liss, New York.

Dourish, C. T., Rycroft, W., and Iversen, S. D. (1989a). Postponement of satiety by blockade of brain cholecystokinin (CCK-B) receptors. *Science* **245**, 1509–11.

Dourish, C. T., Ruckert, A. C., Tattersall, F. D., and Iversen, S. D. (1989b). Evidence that decreased feeding induced by systemic injection of cholecystokinin is mediated by CCK-A receptors. *Eur. J. Pharmacol.* **173**, 233–4.

Dourish, C. T., Rycroft, W., Dawson, G. R., and Iversen, S. D. (1990). Anxiolytic-effects of the CCK antagonists devazepide and L-365,260 in a conditioned suppression of drinking model. *Eur. J. Neurosci.* **2**, (Suppl. 3), 38.

Dourish, C. T., Kitchener, S. J. and Iversen, S. D. (1991). The anxiogenic-like effect of CCK-4 in the elevated plus-maze is blocked by the CCK-B antagonist L-365,260. *Biol Psychiat.*, **29**, 325S.

Ebenezer, I. S., De La Riva, C., and Baldwin, B. A. (1990). Effects of the CCK receptor antagonist MK-329 on food intake in pigs. *Physiol. Behav.* **47**, 145-8.

Evans, B. E., Bock, M. G., Rittle, K. E., DiPardo, R. M., Whitter, W. L., Veber, D. F., Anderson, P. S., and Freidinger, R. F. (1986). Design of potent, orally effective, non-peptidal antagonists of the peptide hormone cholecystokinin. *Proc. Natl Acad. Sci. USA* **83**, 4918-22.

Garlicki, J., Konturek, P. K., Majka, J., Kwiecien, N., and Konturek, S. J. (1990). Cholecystokinin receptors and vagal nerves in control of food intake in rats. *Am. J. Physiol.* **258**, E40-5.

Gibbs, J., Young, R. C., and Smith, G. P. (1973a). Cholecystokinin decreases food intake in rats. *J. Comp. Physiol. Psychol.* **84**, 488-95.

Gibbs, J., Young, R. C. and Smith, G. P. (1973b). Cholecystokinin elicits satiety in rats with open gastric fistulas. *Nature* **245**, 323-5.

Goudie, A. J. (1987). Aversive stimulus properties of drugs: The conditioned taste aversion paradigm. In *Experimental psychopharmacology* (ed. A. J. Greenshaw and C. T. Dourish), pp. 341-91. Humana Press, Clifton, NJ.

Gregory, P. C., McFadyen, M., and Rayner, D. V. (1989). Duodenal infusion of fat, cholecystokinin secretion and satiety in the pig. *Physiol. Behav.* **45**, 1021-4.

Hahne, W. H., Jensen, R. T., Lemp, G. F., and Gardner, J. D. (1981). Proglumide and benzotript; members of a different class of cholecystokinin antagonists. *Proc. Natl Acad. Sci. USA* **78**, 6304-8.

Halmy, L., Nyakas, C., and Walter, J. (1982). The C-terminal tetrapeptide of cholecystokinin decreases hunger in rats. *Experientia* **38**, 873-4.

Hanson, H. and Strouse, J. (1987). Effects of the CCK antagonist, L-364,718 on food intake and on the blockade of feeding produced by exogenous CCK in the rat. *Fed. Proc.* **46**, 1480.

Hewson, G., Leighton, G. E., Hill, R. G., and Hughes, J. (1988). The cholecystokinin receptor antagonist L-364,718 increases food intake in the rat by attenuation of the action of endogenous cholecystokinin. *Br. J. Pharmacol.* **93**, 79-84.

Hill, D. R., Campbell, N. J., Shaw, T. M., and Woodruff, G. N. (1987). Autoradiographic localisation and biochemical characterization of peripheral type CCK receptors in rat CNS using highly selective non-peptide CCK antagonists. *J. Neurosci.* **7**, 2967-76.

Innis, R. B. and Snyder, S. H. (1980). Distinct cholecystokinin receptors in brain and pancreas. *Proc. Natl Acad. Sci. USA* **77**, 6917-21.

Jackson, A. and Cooper, S. J. (1985). Effects of kappa opiate agonists on palatable food consumption in non-deprived rats, with and without food preloads. *Brain Res. Bull.* **15**, 391-6.

Kadar, T., Penke, B., Kovacs, K., and Telegdy, G. (1985). Depression of rat feeding in familiar and novel environment by sulfated and nonsulfated cholecystokinin octapeptide. *Physiol. Behav.* **34**, 395-400.

Kadar, T., Penke, B., Kovacs, K., and Telegdy, G. (1986). Inhibition of feeding by the C-terminal tetrapeptide fragment of cholecystokinin in a novel environment. *Neuropeptides* **7**, 97-108.

Khosla, S. and Crawley, J. N. (1988). Potency of L-364,718 as an antagonist of the behavioural effects of peripherally administered cholecystokinin. *Life Sci.* **42**, 153-60.

Lindén, A. (1989). Role of cholecystokinin in feeding and lactation. *Acta Physiol. Scand.* **137**, (Suppl. 585).

Lotti, V. J. and Chang, R. S. L. (1989). A new potent and selective non-peptide gastrin

antagonist and brain cholecystokinin receptor (CCK-B) ligand: L-365,260. *Eur. J. Pharmacol.* **163**, 273–80.

McCann, M., Verbalis, J. G., and Stricker, E. M. (1989). Lithium chloride and CCK inhibit gastric emptying and feeding and stimulate oxytocin secretion in rats. *Am J. Physiol.* **256**, R463–8.

Miesner, J. A., Yen, E. I., Gibbs, J., Weller, A., and Smith, G. P. (1989). Intravenous administration of cholecystokinin antagonist increases food intake in the rat. *Soc. Neurosci. Abstr.* **15**, P1280.

Moore, B. O. and Deutsch, J. A. (1985). An antiemetic is antidotal to the satiety effects of cholecystokinin. *Nature* **315**, 321–2.

Moran, T. H. and McHugh, P. R. (1988). Anatomical and pharmacological differentiation of pyloric, vagal and brain stem cholecystokinin receptors. In *Neurology and neurobiology*, Vol. 47, *Cholecystokinin antagonists* (ed. R. Y. Wang and R. Schoenfeld), pp. 117–32. Alan R. Liss, New York.

Moran, T. H., Robinson, P. H., Goldrich, M. S., and McHugh, P. R. (1986). Two brain cholecystokinin receptors: implications for behavioural actions. *Brain Res.* **362**, 175–9.

Ravard, S. and Dourish, C. T. (1990). Cholecystokinin and anxiety. *Trends Pharmacol. Sci.* **11**, 271–3.

Reidelberger, R. D. and O'Rourke, M. F. (1989). Potent cholecystokinin antagonist L364718 stimulates food intake in rats. *Am. J. Physiol.* **257**, R1512–18.

Robbins, T. W. and Fray, P. J. (1980). Stress-induced eating: fact, fiction or misunderstanding? *Appetite* **1**, 103–3.

Schick, R. R., Yaksh, T. L., and Go, V. L. W. (1986). Intracerebroventricular injections of cholecystokinin octapeptide suppress feeding in rats-pharmacological characterization of this action. *Regul. Pept.* **14**, 277–91.

Schneider, L. H., Gibbs, J., and Smith, G. P. (1986). Proglumide fails to increase food intake after an ingested preload. *Peptides* **7**, 135–40.

Schneider, L. H., Murphy, R. B., Gibbs, J., and Smith, G. P. (1988). Comparative potencies of CCK antagonists for the reversal of the satiating effect of cholecystokinin. In *Neurology and neurobiology*, Vol. 47, *Cholecystokinin antagonists* (ed. R. Y. Wang and R. Schoenfeld), pp. 263–84. Alan R. Liss, New York.

Schwartz, D. H., Dorfman, D. B., Hernandez, L., and Hoebel, B. G. (1988). Cholecystokinin: 1. CCK antagonists in the PVN induce feeding. 2. Effects of CCK in the nucleus accumbens on extracellular dopamine turnover. In *Neurology and neurobiology*, Vol. 47, *Cholecystokinin antagonists* (ed. R. Y. Wang and R. Schoenfeld), pp. 285–305. Alan R. Liss, New York.

Shillabeer, G. and Davison, J. S. (1984). The cholecystokinin antagonist, proglumide, increases food intake in the rat. *Regul. Pept.* **8**, 171–6.

Silver, A. J., Flood, J. F., Song, A. M., and Morley, J. E. (1989). Evidence for a physiological role for CCK in the regulation of food intake in mice. *Am. J. Physiol.* **256**, R646–52.

Smith, G. P., Jerome, C., Cushin, B. J., Eterno, R., and Simansky, K. J. (1981). Abdominal vagotomy blocks the satiety effect of cholecystokinin in the rat. *Science* **213**, 1036–7.

Smith, G. P., Jerome, C., and Norgren, R. (1985). Afferent axons in abdominal vagus mediate satiety effect of cholecystokinin in rats. *Am. J. Physiol.* **249**, R638–41.

South, E. H. and Ritter, R. C. (1988). Capsaicin application to central or peripheral vagal fibres attenuates CCK satiety. *Peptides,* **9**, 601–12.

Strohmayer, A., Von Heyn, R., Dornstein, L., and Greenberg, D. (1989). CCK receptor blockade by L364,718 increases food intake and meal taking behaviour in lean but not obese Zucker rats. *Appetite* **12**, 240.

Telegdy, G., Kadar, T., Kovacs, K., and Penke, B. (1984). The inhibition of tail-pinch-induced food intake by cholecystokinin octapeptides and their fragments. *Life Sci.* **35**, 163–70.

Verbalis, J. G., McCann, M. J., McHale, C. M., and Stricker, E. M. (1986a). Oxytocin secretion in response to cholecystokinin and food: differentiation of nausea from satiety. *Science* **232**, 1417–19.

Verbalis, J. G., McHale, C. M., Gardiner, T. W., and Stricker, E. M. (1986b). Oxytocin and vasopressin secretion in response to stimuli producing learned taste aversions in rats. *Behav. Neurosci.* **100**, 466–75.

Verbalis, J. G., Richardson, D. W., and Stricker, E. M. (1987). Vasopressin release in response to nausea-producing agents and cholecystokinin in monkeys. *Am. J. Physiol.* **252**, R749–53.

Vickroy, T. W. and Bianchi, B. R. (1989). Pharmacological and mechanistic studies of cholecystokinin-facilitated [^{3}H]dopamine efflux from rat nucleus accumbens. *Neuropeptides* **13**, 43–50.

Watkins, L. R., Kinscheck, I. B., and Mayer, D. J. (1984). Potentiation of opiate analgesia and apparent reversal of morphine tolerance by proglumide. *Science* **224**, 395–6.

Watson, C. A., Schnieder, L. H., Corp, E. S., Weatherford, S. C., Shindledecker, R., Murphy, R. B., Smith, G. P., and Gibbs, J. (1988). The effects of chronic and acute treatment with the potent peripheral cholecystokinin antagonist L-364,718 on food and water intake in the rat. *Soc. Neurosci. Abstr.* **14**, P1196.

Wolkowitz, O. M., Gertz, B., Weingartner, H., Beccaria, L., Thompson, K., and Liddle, R. A. (1990). Hunger in humans induced by MK-329, a specific peripheral-type cholecystokinin receptor antagonist. *Biol. Psychiat.* **28**, 169–73.

Zarbin, M. A., Innis, R. B., Walmsley, J. K., Snyder, S. H., and Kuhar, M. J. (1983). Autoradiographic localization of cholecystokinin receptors in rodent brain. *J. Neurosci.* **3**, 877–906.

22. Evidence that hypolocomotion induced by peripheral or central injection of CCK in the mouse is mediated by peripherally located CCK$_A$ receptors

Michael F. O'Neill and Colin T. Dourish

CCK reduces locomotor activity in rodents when given peripherally or centrally (Crawley *et al.* 1981; Van Ree *et al.* 1983). The receptor mechanism by which CCK exerts its influence on activity and the anatomical location of these receptors is as yet unclear. Initially, two subtypes of receptor were proposed on the basis of differential affinities of CCK fragments for binding to membranes from brain and peripheral tissues. Subsequent studies by Moran *et al.* (1986) showed that both peripheral type (CCK$_A$) and brain type (CCK$_B$) CCK receptors were present in brain tissue.

Both subdiaphragmatic vagotomy (Crawley *et al.* 1981) and lesions of vagal nerve fibres with capsaicin (Ritter *et al.* 1986) attenuate CCK-induced hypolocomotion, suggesting a peripheral site of action. Furthermore, intracerebroventricular (i.c.v.) infusions of the weak non-selective CCK antagonist proglumide failed to block the decrease in locomotion induced by peripherally administred CCK (Katsuura *et al.* 1984). In contrast, the selective CCK$_A$ antagonist A-65186 reversed the hypolocomotion induced by both peripherally and centrally administered CCK (Britton *et al.* 1989). Similarly, injection of low doses of CCK (3.0 fmol) directly into the nucleus accumbens decreased exploration in rats (Daugé *et al.* 1989). These results suggest that centrally administered CCK caused hypoactivity by acting on CCK$_A$ receptors in the brain (Britton *et al.* 1989; Daugé *et al.* 1989).

Until recently, research in this area was hampered by a lack of selective antagonists for the CCK$_A$ and CCK$_B$ receptor subtypes. Recently, however, a series of non-peptide CCK antagonists has been synthesized and characterized. Devazepide is a selective antagonist for the CCK$_A$ subtype and L-365,260 is a selective CCK$_B$ receptor antagonist (Chang and Lotti 1986; Evans *et al.* 1986; Bock *et al.* 1989; Lotti and Chang 1989). To determine whether CCK-induced hypoactivity is mediated by CCK$_A$ or CCK$_B$ receptors we examined the effects of devazepide and L-365,260 on the hypolocomotion induced by intraperitoneal (i.p.) injection of CCK in mice exposed to a novel environment. In addition, to investigate the site of action of the peptide we compared the ability of devazepide and L-365,260 to block hypolocomotion induced by peripherally and centrally injected CCK.

Sulphated CCK octapeptide (Bachem) was dissolved in distilled water in a volume of 1 mg/ml. All subsequent dilutions were made in 0.9 per cent saline. Devazepide and L-365,260 were suspended in 0.5 per cent methyl cellulose and sonicated for 1 min.

Male BKTO mice (20–25 g) were randomly assigned to groups and injected subcutaneously (s.c) with devazepide, L-365,260, or methyl cellulose vehicle. After 30 min they were injected (i.p or i.c.v) with either CCK or saline vehicle and immediately placed in test cages for a 30 min test. Peripherally injected drugs were administered in a volume of 10 ml/kg, while all i.c.v. doses were given in a volume of 5 μl. For i.c.v. injections mice were placed in an anaesthetic chamber and exposed to isoflurane (Abbott) for 30 s. When the mice were unconscious they were removed from the anaesthetic chamber. A 1 cm incision was made along the mid-line of the head to expose lambda on the skull. At lambda a needle was inserted 3.5 mm into the skull projecting forwards at an angle of 45° and the drug or vehicle was injected manually with a Hamilton microsyringe. The mice were then returned to their home cages and allowed to recover from the anaesthetic for 2 min prior to being placed in locomotor activity cages.

Testing of locomotor activity was carried out in individual Perspex boxes (27 cm × 21 cm × 27 cm) lined with a metal grid (0.5 cm × 0.5 cm) to facilitate climbing. The boxes were fitted with four pairs of infra-red photocells, positioned 2 cm from the top and bottom of the box. Activity was measured as the number of photocell interrupts. The locomotor cages were connected via a laboratory interface to a CUBE microcomputer system (Paul Fray Ltd, Cambridge) which recorded photocell counts at 5 min intervals. The results were downloaded to a VAX minicomputer for statistical analysis. Data were analysed by multifactorial analysis of variance and multiple comparison tests where appropriate.

CCK injected i.p. caused a dose-dependent decrease in locomotion in mice in a novel environment ($F(3, 28) = 14.1$, $p < 0.0001$). Significant effects were observed at doses of 10 and 30 μg/kg. The minimum effective dose (10 μg/kg) was used in subsequent antagonist studies. Devazepide had no intrinsic effect on mouse locomotion in a novel environment. However, when given 30 min prior to CCK (10 μg/kg i.p.) it reversed the hypolocomotion induced by CCK in a dose-dependent manner (devazepide main effect, $F(4, 118) = 3.55$, $p = 0.009$; CCK main effect, $F(1, 118) = 22.17$, $p < 0.00001$; interaction, $F(4, 118) = 4.32$, $p = 0.002$). The minimum effective dose of devazepide in this test was 0.1 mg/kg Fig. 22.1(a)).

L-365,260 had no intrinsic effect on spontaneous activity and did not block CCK-induced hypolocomotion at any of the doses tested (0.0001–0.1 mg/kg)

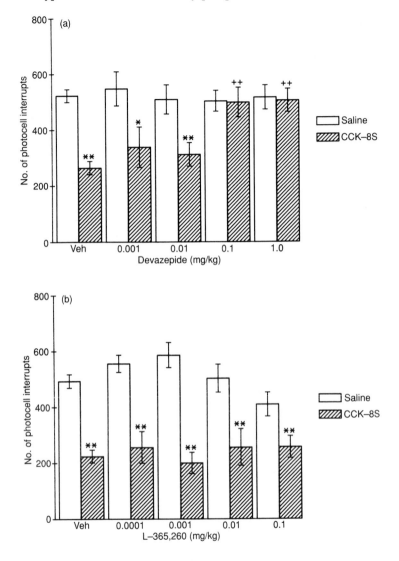

Fig. 22.1. (a) Effect of devazepide on CCK-induced (10 μg/kg i.p.) hypolocomotion in mice. The *x* axis shows the devazepide dose (mg/kg s.c.) and the *y* axis shows the number of photocell interrupts. The data are mean ± SEM of eight mice per group during a 30 min test. Significant differences were determined by Tukey tests following ANOVA: ∗∗, $p < 0.01$ versus saline/vehicle; ∗, $p < 0.05$ versus saline/vehicle; + +, $p < 0.01$ versus CCK/vehicle.

(b) Effect of L-365,260 on CCK-induced (10 μg/kg i.p.) hypolocomotion in mice. The *x* axis shows the L-365,260 dose (mg/kg s.c.) and the *y* axis shows the number of photocell interrupts. Details are as described for (a). 30 min test: ∗∗, $p < 0.01$ versus saline/vehicle i.c.v.

(L-365,260, $F(4, 113) = 0.86$, $p = 0.48$; CCK, $F(1, 113) = 98.2$, $p < 0.00001$; interaction, $F(1, 113) = 1.65$, $p = 0.16$) (Fig. 22.1(b)). The effect on CCK-induced hypolocomotion of doses of L-365,260 above 0.1 mg/kg were not tested as these doses of the antagonist decreased locomotion in a novel environment.

CCK injected i.c.v. also caused a dose-dependent decrease in locomotor activity of mice in a novel environment ($F(4, 44) = 4.56$, $p < 0.005$). The minimum effective dose was 3.5 μg per mouse and this dose was used in the subsequent antagonist challenge studies. The latency to onset of i.c.v. CCK-induced hypolocomotion in this test was 15 min. In contrast, peripherally administered CCK significantly reduced locomotor activity within 5 min of injection.

Devazepide (i.p.) reversed hypolocomotion induced by i.c.v. injection of CCK (3.5 μg) in a dose-dependent manner (devazepide main effect, $F(3, 84) = 4.38$, $p < 0.01$; CCK main effect, $F(1, 84) = 12.99$, $p < 0.0005$; interaction, $F(3, 84) = 4.95$, $p < 0.005$) (Fig. 22.2(a)). In contrast, L-365,260 had no significant effect on the CCK response (L-365,260 main effect, $F(2, 54) = 0.55$, $p < 0.57$; CCK main effect, $F(1, 54) = 14.33$, $p < 0.0005$; interaction, $F(2, 54) = 0.69$, $p < 0.5$) (Fig. 22.2(b)).

I.p. injection of CCK produced a dose-dependent decrease in the spontaneous activity of mice exposed to a novel environment. The selective CCK_A antagonist devazepide blocked hypoactivity induced by i.p. injection of CCK in mice. This finding is consistent with the results of Khosla and Crawley (1988) who demonstrated that devazepide reversed the reduction in activity caused by i.p. CCK. Similarly, Britton *et al.* (1989) reported that peripheral injection of the selective CCK_A antagonist A-65186 blocked the hypolocomotion induced by an i.p. injection of CCK in mice. In contrast, a wide range of doses of the selective CCK_B antagonist L-365,260 had no effect on hypoactivity induced by i.p. injection of CCK. These data suggest that hypolocomotion induced by an i.p. injection of CCK in mice is mediated by CCK_A receptors.

I.c.v. injection of CCK also decreased locomotor activity in mice. The hypolocomotion induced by centrally administered CCK was also dose-dependently blocked by devazepide but not by L-365,260, indicating CCK_A receptor mediation. Similarly, Britton *et al.* (1989) reported that the selective CCK_A antagonist A-65186 blocked the hypolocomotion induced by i.c.v. CCK and concluded that CCK acted via brain CCK_A receptors to reduce activity. However, the site of action of centrally administered peptides is controversial. Indeed, it has been demonstrated that CCK injected i.c.v. diffuses rapidly into the peripheral circulation (Passaro *et al.* 1982). In our studies, the minimum dose of CCK required to induce hypolocomotion when injected i.c.v. was 3.5 μg per mouse, whereas a dose of 10 μg/kg (i.e. approximately 0.2 μg per 20 g mouse) was sufficient to attenuate activity if given peripherally. Furthermore, the hypolocomotion induced by i.p. CCK was apparent within 5 min of injection, whereas in mice given CCK i.c.v. activity was not attenuated until 15 min after injection. The increased latency to onset of the behavioural

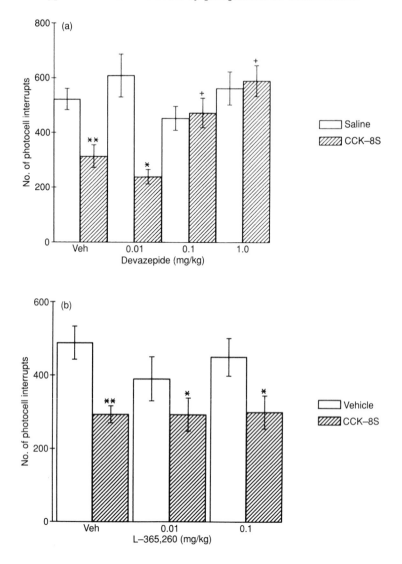

Fig. 22.2. (a) Effect of devazepide (s.c.) on CCK-induced (3.5 μg i.c.v.) hypolocomotion in mice. The x axis shows the devazepide dose (mg/kg) and the y axis shows the number of photocell interrupts. Data are mean ± SEM of N mice (N ⩾ 7) during a 30 min test: ∗, p 0.05 versus vehicle/saline; ∗∗, p < 0.01 versus vehicle/saline; +, p < 0.05 versus vehicle/CCK.

(b) Effect of L-365,260 (s.c.) on CCK-induced (3.5 μg i.c.v.) hypolocomotion in mice. The x axis shows the L-365,260 dose (mg/kg s.c.) and the y axis shows the number of photocell interrupts. Details as for (a).

response after i.c.v. injection is consistent with a delay in the penetration of
the drug to its site of action, and may reflect diffusion from the ventricles to
another site in the brain, such as the nucleus accumbens, or to the periphery.

In conclusion, the present results, when considered with previous findings,
suggest that both peripherally and centrally administered CCK induces hypolocomotion via CCK_A receptor mechanisms. The observation that peripherally
administered CCK had an almost immediate effect and was considerably more
potent than centrally administered CCK (which reduced activity only after a
considerable delay) suggests that both peripherally and centrally injected CCK
decreases locomotion by stimulating peripheral CCK_A receptors.

References

Bock, M. G., DiPardo, R. M., Evans, B. E., Rittle, K. E., Whitter, W. L., Veber,
D. F., Anderson, P. S., and Freidinger, R. M. (1989). Benzodiazepine gastrin and
cholecystokinin receptor ligands: L-365,260. *J. Med. Chem.* **32**, 13–16.

Britton, D. R., Yahiro, L., Cullen, M. J., Kerwin, J. F., Jr, Kopecka, H., and Nadzan,
A. M. (1989). Centrally administered CCK-8 suppresses activity in mice by a
'peripheral-type' CCK receptor. *Pharmacol. Biochem. Behav.* **34**, 779–83.

Chang, R. S. L. and Lotti, V. J. (1986). Biochemical and pharmacological characterisation if an extremely potent and selective nonpeptide cholecystokinin antagonist.
Proc. Natl Acad. Sci. USA **83**, 4923–6.

Crawley, J. N., Hays, S. E., Paul, S. M., and Goodwin, F. K. (1981). Cholecystokinin
reduces exploratory behaviour in mice. *Physiol. Behav.* **27**, 407–11.

Daugé, V., Steimes, P., Derrien, M., Beau, N., Roques, B. P., and Feger, J. (1989).
CCK8 effects on motivational states of rats involve CCKA receptors of the postero-
median part of the nucleus accumbens. *Pharmacol. Biochem. Behav.* **34**, 157–63.

Evans, B. E., Bock, M. G., Rittle, K. E., DiPardo, R. M., Whitter, W. L., Veber,
D. F., Anderson, P. S., and Freidinger, R. M. (1986). Design of potent, orally effec-
tive, nonpeptidal antagonists of the peptide hormone cholecystokinin. *Proc. Natl
Acad. Sci. USA* **83**, 4918–22.

Katsuura, G., Hsiao, S., and Itoh, S. (1984). Blocking of cholecystokinin octapeptide
behavioural effects by proglumide. *Peptides* **5**, 529–34.

Khosla, S. and Crawley, J. N. (1988). Potency of L-364,718 as an antagonist of the
behavioural effects of peripherally administered cholecystokinin. *Life Sci.* **42**, 153–9.

Lotti, V. J. and Chang, R. S. L. (1989). A new potent and selective non-peptide gastrin
antagonist and brain cholecystokinin receptor (CCK_B) ligand: L-365,260. *Eur. J.
Pharmacol.* **162**, 273–80.

Moran, T. H., Robinson, P. H., Goldrich, M. S., and McHugh, P. R. (1986). Two
brain cholecystokinin receptors: implications for behavioural actions. *Brain Res.*
362, 175–9.

Passaro, E., Debas, H., Oldendorf, W., and Yamada, T. (1982). Rapid appearance of
intraventricularly administered neuropeptides in the peripheral circulation. *Brain
Res.* **241**, 335–40.

Ritter, R. C., Kalivas, P., and Bernier, S. (1986). Cholecystokinin-induced suppression
of locomotion is attenuated in capsaicin pretreated rats. *Peptides* **7**, 587–90.

Van Ree, J. M., Gaffori, O., and De Wied, D. (1983). In rats, the behavioral profile
of CCK-8 related peptides resembles that of antipsychotic agents. *Eur. J. Pharmacol.*
93, 63–78.

23. CCK–monoamine interactions and satiety in rodents

Steven J. Cooper

Introduction

At first sight, the problem posed by *satiety* is quite straightforward. To illustrate this, I begin by drawing a comparison between the feeding habits of a laboratory rat, housed alone, and driving a car. To achieve the required degree of experimental control, the rat is normally given free access to diet of standard composition (or the food is given on a fixed schedule of restricted access) and environmental factors (such as temperature, lighting, capacity of the living space) are held constant. We can appreciate the analogy between the rat's living under such uniform conditions and its eating requirements, and the task of driving a vehicle under constant conditions and having to take account of fuelling requirements.

There are occasions when we drive a car on a straight flat road with a good surface; there may be little additional traffic, low wind resistance, a constant load, and a constant speed to maintain. If a fuel of known quality is used and the available quantity is monitored, then the question of anticipating fuel requirements on the journey is simple to answer. The decision to stop for fuel can occur at any time, provided that the fuel does not run out and there is a petrol station available, at that time.

In a similar way, approaching the question of satiety solely in relation to the normal laboratory-maintained rat encourages us to see a simple process operating by virtue of simple mechanisms. In the case of driving a car under constant conditions, 'satiety' signifies the end of refuelling and usually indicates that enough fuel has been taken for the next leg of a predictable journey. In the rat, satiety signifies the end of feeding and indicates that enough food has been consumed until the next meal.

The advantage of dealing with the rat's feeding under highly constrained conditions is that its behaviour is at its simplest and should be correspondingly straightforward to explain. There is then a temptation, to which we readily succumb, to assume that the underlying mechanisms involved in the control of feeding must also be simple. It seems reasonable to seek a correspondence between a predictable well-defined behavioural response on the one hand, and a mediating mechanism that is not necessarily complex or sophisticated on the other. Nevertheless, we have here the basis of a fallacy which should not go unnoticed. Thus the mediating mechanisms may be extremely sophisticated in their operation, but deliver a simple reliable response if the environmental features and contingencies are highly regulated. In other words, a complex machine could be asked to do nothing more than perform a simple operation; the simplicity of the operation would provide no clues to the complexity and full capabilities of the machine.

The example of car-driving given above deals with ideal conditions, yet cars are designed and drivers are trained to function reliably under sets of conditions that are far from optimal. When conditions can alter continually, predicting fuel requirements becomes a more difficult business open to error at any stage. By the same token, we should recognize that systems have evolved in the laboratory rat to allow it to function effectively under varied environmental conditions and to guarantee its fuel requirements. For convenience, we have chosen to approach the controls of feeding in its simplest terms, and we have been content to posit simple methods of control. When we accept fully the rat's endowment of sophisticated methods of control in its feeding habits, we shall look at the potential offered by more complex systems with much greater interest.

Simple satiety systems

Stellar's (1954) theoretical review, entitled 'The physiology of motivation', bears re-reading in detail. In it, he proposed that 'the amount of motivated behavior is a direct function of the amount of activity in certain excitatory centers of the hypothalamus'. Thus feeding behaviour is ultimately determined by activity in a hypothalamic hunger centre. A corresponding 'inhibitory center' functions by inhibiting the activity of the hunger centre, and, most importantly, Stellar proposed that the hypothalamic inhibitory centre 'may constitute a separate "satiation mechanism"'. As everyone working in the field must know, there followed years of intensive study of the role of the hypothalamus in the control of feeding, and interest in the hypothalamus remains at a high level today. Stellar's (1954) vision of a multifactor control of hypothalamic activity determining motivation is, at last, being realized in some interesting ways (Leibowitz 1992). Nevertheless, the idea which forms the kernel of his theory of satiation is nothing more than a hypothalamic 'off switch' which terminates feeding activity.

Stellar's anatomical model was succeeded by neurochemical models based largely on data derived from pharmacological experiments. Within the brain, 5-hydroxytryptamine (5-HT) has been postulated to be a neurotransmitter which is important in the satiation of feeding (Blundell 1977, 1984). In the periphery, CCK has been proposed as a physiological signal of satiety (Smith and Gibbs 1979; Chapter 17 of this volume). With so much information to accumulate on each of these two 'chemical messengers of satiety' (Crawley

1992; Dourish 1992), it is not surprising that investigators pursued work on only one of these messengers to the exclusion of the other. Moreover, substantially less interest has been expressed, at least to the present time, in several other neurotransmitters, hormones, and metabolites which affect feeding and may contribute to satiety.

The end result of a historically circumscribed view of the chemical determinants of satiety is to reinforce the schema of a simple behavioural response, requiring only the mediation of an uncomplicated neurochemical mechanism. It is unlikely that each simple neurochemical model of satiety does justice to the physiological mechanisms which can be brought into play to solve the rat's feeding requirements in a variable complex environment.

The main empirical content of this chapter is concerned with ways in which apparently discrete neurochemical and humoral factors may interact together to determine satiety (the inhibition of feeding) in the rat. The aim is to add an extra degree of complexity to current views of the underlying controls of feeding behaviour. The chapter will conclude with a brief account of the implications of new work and some indications of future research directions.

CCK

Induction of satiety

As well as being found in the central nervous system, CCK is located peripherally in the gastrointestinal tract, pancreas, vagus, and elsewhere (see Chapter 18 of this volume). Gibbs *et al.* (1973*a*) reported that the intraperitoneal (i.p.) administration of partially purified porcine CCK significantly reduced food consumption in rats. They also showed that CCK reduced consumption of a liquid food in food-deprived rats, but not of water in thirsty rats. Thus it appeared that the peptide affected food ingestion relatively specifically, and did not act merely in a general disruptive fashion to impede ingestive responses.

Using human subjects, Kissileff *et al.* (1981) found that intravenous (i.v.) infusion of CCK octapeptide (CCK-8S) significantly reduced the intake of a liquified blend of several foods. In a separate study, infusion of porcine CCK-33 into hungry human subjects reduced their feelings of hunger and depressed appetite normally aroused by the sight, sound, and smell of food preparation (Stacher *et al.* 1979). I.v. infusion of CCK-8S also proved effective in reducing food intake in obese male subjects (Pi-Sunyer *et al.* 1982).

A behavioural index used to identify satiety after a meal has finished is the behavioural satiety sequence (BSS), which consists of a characteristic sequence of non-feeding activities (grooming and movement about the cage) followed by rest and sleep (see Chapter 21 of this volume). Antin *et al.* (1975) showed that injection of partially purified porcine CCK elicited a typical BSS. CCK, released in the gastrointestinal tract as a consequence of initiating a meal, was deemed to act as a satiety signal (Smith and Gibbs 1979). In animals that are sham feeding and hence show a considerable satiety deficit, administration of exogenous CCK brings liquid diet consumption to an end and elicits a normal BSS (Gibbs *et al.* 1973*b*; Antin *et al.* 1975).

This central idea—that exogenous CCK (and, by extension, *endogenous* CCK) produces or enhances a natural satiety state—has been vigorously challenged. The main counter-argument is that administration of exogenous CCK induces malaise, produces aversive effects, and consequently disrupts feeding (Deutsch and Hardy 1977; Swerdlow *et al.* 1983; Moore and Deutsch 1985; Verbalis *et al.* 1986). Since injection of lithium chloride (LiCl) has been used routinely to produce food aversion, it seems appropriate to use it as a standard for comparison with the effects of CCK. Yet Davidson *et al.* (1988) found that CCK-8 did not produce effects like those of LiCl in rats trained with food deprivation and satiation as discriminative signals for shock. These authors concluded that the sensory consequences of the state produced by CCK have little in common with those produced by injection of LiCl. Likewise, Ervin and Teeter (1986) furnished data on feeding and taste aversion learning which clearly distinguished the effects of CCK-8S from those of LiCl. Hence it appears unlikely that all the effects of CCK on feeding responses can be attributed to the kinds of aversive effect and malaise produced by LiCl. In support of an effect of CCK on satiation we can cite the recent data of Corwin *et al.* (1990). These authors trained rats to discriminate between 3 h and 22 h periods of food deprivation. In subsequent substitution tests, rats that were deprived of food for 22 h responded as if they had been deprived for only 3 h after the administration of CCK or if they had been given a preload of sweetened condensed milk. These results indicate that the discriminative stimulus effects of CCK are similar to those produced following ingestion of food. In contrast, injection of LiCl did not produce the same kind of response, and animals rarely responded as if they were only deprived of food for 3 h. Therefore it seems that there is positive evidence to add to earlier work (e.g. Antin *et al.* 1975) that administration of CCK contributes towards an increase in satiety.

Establishing that exogenous CCK may act in a way that is consistent with an increase in satiety provides no direct indication of the normal functional significance of endogenous CCK. Coming to grips with the modes in which endogenous CCK may operate has to a great extent depended on our detailed understanding of the receptors for CCK, and on the development of potent and selective CCK receptor antagonists. Advances in these areas have allowed us to come reasonably close to a basic understanding of the function of endogenous CCK in relation to the control of feeding.

Multiple CCK receptors

Soon after specific binding sites for CCK were identified in rat, mouse, and human brain (Hays *et al.* 1980, 1981; Saito *et al.* 1981; Zarbin *et al.* 1983), it became clear that a distinction could be drawn between CCK receptors in the brain and those found in the pancreas (Innis and Snyder 1980; Sakamoto *et al.* 1984). Two types of CCK receptor were then identified: an A type found in many peripheral locations (e.g. pancreas, vagus, and pyloric sphincter) and a B-type found in the brain. However, further study showed that while the CCK_B receptor is indeed widely distributed in the brain, there are several central locations at which CCK_A receptors are found (Moran *et al.* 1986; Dourish and Hill 1987; Hill *et al.* 1987, 1988, 1990; Moran and McHugh 1988; Chapter 5 of this volume).

Selective CCK receptor antagonists

In the experiments that we have recently conducted, we used two new non-peptide compounds, both of which show high specificity for CCK receptors and which act potently as CCK receptor antagonists. The first of these compounds is devazepide, which shows high affinity for CCK_A receptors, with a good deal of selectivity for this receptor subtype over the CCK_B receptor (Chang and Lotti 1986; Evans *et al.* 1986; Lotti *et al.* 1987). The second is L-365,260, which also proved to be a potent receptor antagonist but with a high degree of selectivity for the CCK_B and gastrin receptors (Bock *et al.* 1989; Lotti and Chang 1989; Chapter 2 of this volume). With these two compounds, devazepide and L-365,260, it becomes possible to address the issues of the functional significance of endogenous CCK, and of the identity of the receptor subtype(s) involved in the response under consideration.

Endogenous CCK and satiety

In the following experiments, the assumption is made that use of selective CCK receptor antagonists will reveal the actions of endogenous CCK, since the antagonists should do little other than block the effects of endogenous CCK. If endogenous CCK does contribute to the development of satiety, then, it is presumed, selective CCK antagonists should retard the development of satiety and possibly increase the level of food intake in a measure of consumption.

Recent reports confirm that devazepide increases food intake in rats and pigs (Dourish *et al.* 1988; Hewson *et al.* 1988; Reidelberger and O'Rourke 1989; Ebenezer *et al.* 1990; see Chapter 21 of this volume). Thus it appears that selective blockade of CCK receptors may be sufficient to stimulate food consumption, and this result provides strong support for the view that endogenous CCK may act to limit food intake (see Chapter 17 of this volume). In further confirmation, Weller *et al.* (1990) have shown that intragastric administration of soybean trypsin inhibitor (STI), which causes release of CCK from the small intenstine, reduced the size of a test meal. Moreover, the effect of STI was reversed by prior treatment with devazepide. This evidence suggests that endogenously released CCK may reduce the size of a meal by acting at CCK receptors which can be blocked by devazepide. Furthermore, in human subjects, devazepide has been reported to reduce ratings of hunger expressed before a test meal (Wolkowitz *et al.* 1990).

Using partially satiated rats, Dourish *et al.* (1989*a*) have recently demonstrated not only that devazepide increases food intake in rats, but also that the selective CCK_B antagonist L-365,260 has a similar effect. Furthermore, both devazepide and L-365,260 proved effective in postponing the onset of satiety. Most interestingly, L-365,260 was the more potent of the two, leading to the suggestion that endogenous CCK causes satiety through an action at brain CCK_B receptors (Dourish *et al.* 1989*a*). This suggestion highlights a significant difference which may separate the satiety-inducing effects of endogenous and exogenous CCK. The effect of exogenous CCK to reduce food consumption is selectively antagonized by devazepide, the CCK_A receptor antagonist, and unaffected by L-365,260 (Dourish *et al.* 1989*b*).

Having presented evidence that endogenous CCK may act in a way indicating that it increases satiety and limits meal size, the question we wish to consider now is whether or not it acts in isolation. Does it exert its influence on feeding responses in a direct and simple manner, or does it contribute effects which interact with other systems to determine the final end result of satiety?

CCK–monoamine interactions

In response to this question, Dourish and I have proposed that CCK interacts with monoamine systems in the brain, and at the present time we place particular emphasis on interactions with a subset of central 5-HT neuronal activity (Cooper and Dourish 1990). Nevertheless, there is evidence that, in determining satiety, CCK may also interact with catecholaminergic systems (noradrenalin and dopamine).

CCK and 5-HTergic controls of feeding

The idea that 5-HT mechanisms are closely involved in determining satiety occupies a central position in our thinking about the neurochemical control of meal termination (Blundell 1977, 1984). The basic postulate is that increased 5-HT neuronal activity within certain pathways (probably projecting to the hypothalamus) engages 'satiety' and exerts a brake upon the further continuation of feeding. This postulate received further persuasive support with the discovery that small doses of selective 5-HT_{1A} agonists (like 8-hydroxy-2-(di-*n*-propylamino) tetralin (8-OH-DPAT)) could induce hyperphagia in satiated rats (Dourish *et al.* 1985). This finding was interpreted in terms of a selective stimulation of somatodendritic 5-HT_{1A} autoreceptors in the raphé nuclei which inhibits 5-HT neuronal firing and thereby reduces any 5-HT-dependent satiety effect (Dourish 1992). Hence, regulation of 5-HT neuronal activity serves to modulate the processes terminating feeding and this, in turn, determines the level of food consumption.

Recent experiments have forged links between 5-HT and CCK in the control of feeding responses. In the first experiment, Stallone *et al.* (1989) found that the 5-HT antagonist metergoline was effective in attenuating the reduction in food intake produced by CCK-8. In contrast, the peripheral 5-HT antagonist, xylamidine, had no effect on the anorectic action of CCK. Therefore Stallone

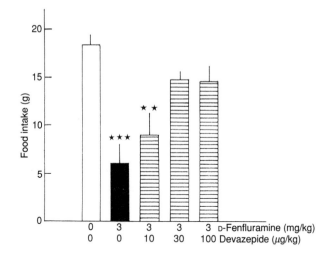

Fig. 23.1. The anorectic effect of D-fenfluramine (3.0 mg/kg) (solid bar) was antagonized by the selective CCK_A receptor antagonist devazepide in a 30 min test of palatable food intake in non-deprived rats. (Reproduced with permission from Cooper *et al.* 1990.)

et al. proposed that exogenous CCK ultimately exerts a satiety effect by altering 5-HT activity in the brain. This formulation does not require the proximal effect of CCK to occur centrally, and indeed Stallone *et al.* favour an action of exogenous CCK at a peripheral site. We have already noted the evidence that the satiety—inducing effect of exogenous CCK appears to depend on stimulation of CCK_A, but not CCK_B, receptors (Dourish *et al.* 1989*b*; Chapter 17 of this volume).

In a second experiment, we reported that the anorectic effect of D-fenfluramine was blocked by devazepide (Cooper *et al.* 1990) (Fig. 23.1), having previously established that D-fenfluramine's effect was mediated by 5-HT$_1$ receptors (Neill and Cooper 1989*a*). A further experiment failed to detect antagonism of D-fenfluramine's anorectic effect by the selective CCK_B gastrin receptor antagonist L-350,260 (Fig. 23.2). These data are consistent with the following scheme: D-fenfluramine acts directly to increase 5-HT neuronal activity, with the anorectic effect then dependent upon 5-HT stimulation of 5-HT$_1$ receptors; 5-HT neuronal activation may, in turn, stimulate endogenous CCK activity which produces a satiety effect through the stimulation of CCK_A receptors.

A further link between CCK and 5-HT has been established experimentally (Cooper *et al.* 1991). Systemic administration of 5-HT brings about a reduction in food intake, and its effect can be blocked by xylamidine (Carruba *et al.* 1986). Our data indicate that the anorectic effect of 5-HT itself can be blocked by devazepide (Fig. 23.3) but not by L-365,260.

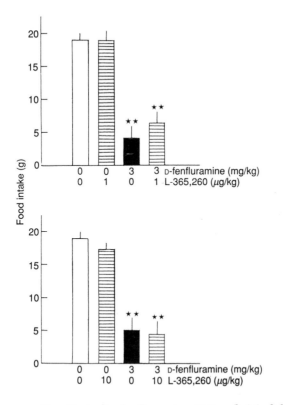

Fig. 23.2. The anorectic effect of D-fenfluramine (3.0 mg/kg) (solid bar) was not antagonized by the selective CCK_B receptor antagonist L-365,260 in a 30 min test of palatable food intake in non-deprived rats. (Data from Cooper *et al.* 1991.)

These results provide a basis for considering the means by which CCK and 5-HT mechanisms may interact to determine satiety. One key assumption, we believe, is that *both elevated 5-HT activity and elevated CCK activity* are required for satiation to be expressed as a reduction in feeding (Cooper and Dourish 1990) (Fig. 23.4). Inhibiting the action of either neurotransmitter will, in turn, affect the capability of the other component to elicit a true satiety state. This explains the result, reported by Stallone *et al.* (1989), that the anorectic effect of exogenous CCK was blocked by metergoline. It also accounts for our results that the anorectic effect of D-fenfluramine, and of systemically administered 5-HT, can be blocked by the selective CCK antagonist devazepide (Cooper *et al.* 1991).

A second element of our model (Fig. 23.4) is that elevated 5-HT will tend to promote CCK activity, and vice versa, suggesting a form of recruitment or mutual co-operativity. One prediction based on this is that CCK should act to enhance central 5-HT activity, and recent evidence provides confirmation of

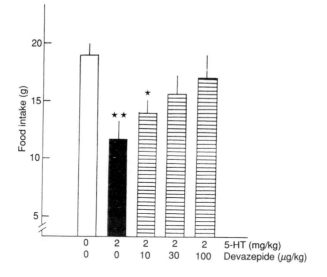

Fig. 23.3. The anorectic effect of systemically administered 5-HT (2 mg/kg) (solid bar) was antagonized by the selective CCK$_A$ receptor antagonist devazepide in a 30 min test of palatable food intake in non-deprived rats. (Data from Cooper *et al.* 1991.)

this form of interaction. Pinnock *et al.* (1990) made intra- and extracellular recordings from dorsal raphé neurons in a slice preparation prepared from rat pons-mesencephalon. A subpopulation of neurons in the dorsal raphé were excited by CCK-8, and this excitatory effect was blocked by devazepide but not by L-365,260. Therefore Pinnock *et al.* concluded that CCK-8 excites a subpopulation of neurons in the dorsal raphé by a CCK$_A$ receptor (see Chapter 5 of this volume). An interaction of the form they describe could provide the basis for part of the mutual co-operativity between 5-HT and CCK that we envisage. The second prediction, that increased 5-HT activity will promote CCK activity, has yet to be confirmed by experimental evidence.

Without going much further into the details of the model shown in Fig. 23.4, it is worth bearing in mind that it allows two separate effects of CCK at its receptors. It does not seem inconceivable that both CCK receptor subtypes (A and B) may be involved in mediating different aspects of the satiety effect of CCK.

In the past, authors have drawn a distinction between the *size* and the *frequency* of meals (Brobeck 1955; Le Magnen 1971). Food consumption could be suppressed either by reducing meal size or by lengthening the interval between meals to reduce their frequency. These two possibilities are reflected in the use of the term *satiety* to refer to either within-meal alterations in feeding or the post-prandial state which can be assessed by, for example, the BSS or

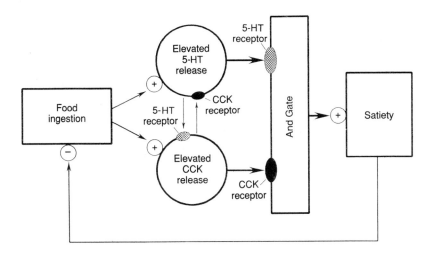

Fig. 23.4. The proposed co-operativity and interdependence of 5-HT and CCK receptor increases in satiety (Cooper and Dourish 1990). The model assumes, first, that following food ingestion, both endogenous 5-HT and CCK are released. Second, the principle of co-operativity implies that, on elevation of 5-HT, activity will tend to increase CCK release and action, and vice versa. Thus there will be a tendency for one system to recruit the other. Third, the interdependence assumption indicates that both 5-HT and CCK actions at their respective receptors are required for the normal development of satiety. Hence pharmacological blockade of either set of receptors will reduce the satiety-inducing effect of either transmitter.

The value of the model may lie in the further research it suggests: (i) the characterization of the receptors mediating the direct and interactive effects of CCK and 5-HT; (ii) the locations for each type of receptor involved in the satiety-promoting effects of the two chemical signals; (iii) the primary eliciting signals for CCK and 5-HT release (are they similar or distinct?); (iv) the behavioural and physiological outputs affected by CCK and 5-HT activity; (v) modulating influences on CCK and 5-HT interactions; (vi) the effects, on the system as a whole, of single or combined pharmacological interventions; (vii) the possibility that the model could be expanded to include other neurotransmitter and hormonal factors involved in the control of satiety, incorporating similar principles of co-operativity and interdependence.

by measurement of the intermeal interval (IMI).

As far as exogenous CCK is concerned, the interesting results obtained by West *et al.* (1984) indicate that it affects only meal size. With administration of CCK, meals are smaller but animals can compensate for the reduction in food intake by adjusting their frequency. If these data provide some insight into the functional significance of endogenous CCK, then we should focus our search for interactions between CCK and 5-HT mechanisms on within-meal satiety, as distinct from the post-prandial phase which determines meal frequency.

Methods incorporating automated data collection and analysis have been in use for a number of years to determine the pattern of meals in rats with free access to food and drink (Clifton 1987). An important aspect of this analysis is that meals can be identified, their size measured, and the intervals between successive meals can be determined. Using this approach, therefore, should provide us with more specific information about the nature of interactions between CCK and 5-HT.

Some years ago, Burton *et al.* (1981) reported the effects of DL-fenfluramine on meal patterns in rats. Reductions in food consumption were related to reductions in meal size and in the rate of eating (cf. Blundell and Latham 1978; Grinker *et al.* 1980). Clifton and Cooper (Chapter 25 of this volume) have recently examined the effects of D-fenfluramine, both alone and in conjunction with the selective CCK_A antagonist devazepide, on food intake and meal patterns in rats. This experiment indicated (i) that D-fenfluramine reduced meal size and (ii) that its effect was attenuated by devazepide. If confirmed, this result implies a 5-HT-dependent mediation of the control of meal size, involving in addition the action of endogenous CCK at CCK receptors. This interpretation would be consistent with the result of West *et al.* (1984), cited above, and suggests that Fig. 23.4 could be more specific and refer to CCK and 5-HT interactions in relation to within-meal satiety determining the size of meals.

This reminds us that, early in the study of CCK's effects on ingestion, it was shown that CCK terminates sham feeding in the gastric-fistulated rat (Gibbs *et al.* 1973*b*). Sham feeding rats exhibit a considerable deficit in within-meal satiety, so that in effect their meals are exceptionally large. Administration of CCK acts to impose satiety, so that the meals are reduced in size and the animals show evidence of a satiety sequence. Effects of drugs, whose actions are related to 5-HT activity, on sham feeding have also been described. Both D-fenfluramine and systemically administered 5-HT reduce sucrose sham feeding in rats (Neill and Cooper 1988, 1989*b*), and it remains to be determined if their effects in the gastric-fistulated rat are reversed by selective CCK receptor antagonists. If they are, it would constitute additional evidence for CCK and 5-HT interactions in relation to the development of within-meal satiety in the rat.

Can these results be related to experimental and clinical data derived from human subjects? According to Liddle *et al.* (1985), fasting levels of plasma CCK are very low, but rise following the ingestion of fat, protein, or amino acids. Plasma CCK levels remain elevated for at least 2 h following a meal. Geracioti and Liddle (1988) investigated a number of bulimic women, and found that not only did they manifest subnormal levels of satiety after a test meal, but also that they exhibit lower levels of plasma CCK compared with controls. The two findings may be related, of course, so that the impairment in subjective experience of satiety could depend upon the lower levels of circulating CCK. Therefore it is interesting that DL- and D-fenfluramine have been reported to have some benefit in treating symptoms of bulimia (Blouin *et al.*

1988; Russell *et al.* 1988; Chapter 27 of this volume). According to our model (Fig. 23.4), this may indicate that increasing 5-HT activity with fenfluramine may have two effects: one to feed directly into the determination of satiety, and a second to enhance CCK activity. This would suggest that bulimic patients receiving fenfluramine may show levels of plasma CCK that have been brought closer to normal.

CCK and noradrenergic controls of feeding

A substantial literature has developed concerning the detailed way in which hypothalamic noradrenaline is involved in the control of feeding responses (Leibowitz 1992). Myers and McCaleb (1981) formulated the hypothesis that CCK may stimulate noradrenergic neurons that are involved in mediating satiety. Thus i.p. injection of CCK caused the release of noradrenaline at medial and rostral hypothalamic sites (Myers and McCaleb 1981). Their idea deserves further scrutiny in the light of the theme, developed in this chapter, that CCK interacts with monoamine systems in determining satiety (Wilson *et al.* 1983; cf. Tsai *et al.* 1984).

CCK and dopaminergic controls of feeding

Lindén (1989) deserves credit for obtaining evidence for a close link between dopaminergic and CCK activity in relation to feeding responses. Her results indicate that the satiety effect of CCK may depend, to some degree, on the central release of dopamine. Thus the dopamine antagonist *cis*-flupentixol blocked not only apomorphine's anorectic effect but also that of CCK-8. Vickroy *et al.* (1988) reported that dopamine efflux in tissue miniprisms prepared from rat nucleus accumbens was facilitated by CCK-8, and that this facilitatory effect may be mediated by CCK_A receptors. Hill *et al.* (1990) have recently described the presence of CCK_A receptors on dopamine cell bodies in primate brain. Hence there may be strong influences of CCK on brain dopamine activity mediated by CCK_A receptors located at the levels of both the cell bodies and the terminals (see Chapters 5 and 35 of this volume for further discussion).

One of the more puzzling features of dopamine's involvement in the control of feeding (Cooper *et al.* 1991*b*) is that not only do dopamine agonists reduce food intake, but reductions are also found following treatment with dopamine antagonists. One possible explanation for the latter type of effect is that dopamine receptor blockade may indirectly bring about the enhancement of CCK activity. In support of this view, we have recently reported that the reduction in food intake produced by the dopamine D_1 receptor antagonist SCH 23390 was blocked by the selective CCK_A antagonist devazepide (Cooper and Barber 1990) (Fig. 23.5). In contrast, the reduction in food intake produced by the D_2 receptor antagonist raclopride was not affected by devazepide. Hence the key to enhanced CCK activity may be the selective blockade of dopamine D_1 receptors (see Chapter 40 of this volume for further discussion

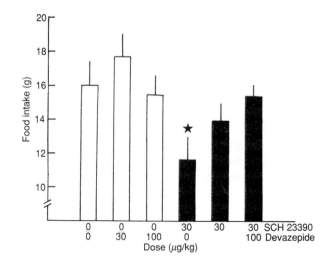

Fig. 23.5. The feeding suppressant action of the selective dopamine D_1 receptor antagonist SCH 23390 (30 μg/kg) was blocked by the selective CCK_A receptor antagonist devazepide (30 and 100 μg/kg). (Reproduced with permission from Cooper and Barber 1990.)

of interactions between CCK and dopamine D_1 receptors). The data also indicate that the reductions in ingestive behaviour produced by D_2 antagonists appear to relate to a CCK-independent mechanism.

Both the D_1 antagonist SCH 23390 and D_2 antagonists (e.g. raclopride, sultopride, sulpiride) have been shown to reduce sucrose sham feeding in the gastric-fistulated rat (Schneider *et al.* 1988; Schneider 1989). The conclusion drawn from these and other results is that both central D_1 and D_2 receptor stimulation are critical to the *reward* of sweet taste in the rat (for details see Schneider 1989). It would be of considerable interest to determine if the effect on sham feeding of either D_1 receptor blockade or D_2 receptor blockade was altered by selective CCK receptor antagonists. If, as we have seen in a real-feeding situation (Cooper and Barber 1990), the effect of SCH 23390 alone was affected by a CCK antagonist, then we would wish to consider distinctly different mechanisms underlying the reductions produced by D_1 and D_2 receptor antagonists respectively. We could imagine that SCH 23390 may activate CCK mechanisms, which in turn would produce a reduction in sham feeding. But a CCK-dependent mechanism such as this might be better interpreted in terms of the imposition of satiety in the sham feeding animal, consistent with the interpretation of the effects of exogenous CCK itself. Clearly, further experimental data are required to address this point.

Implications and future directions

I began this chapter by drawing a simple analogy between the feeding requirements of the laboratory rat and the demands placed on a car and its fuel
requirements under optimal constant conditions of travel. Under such constrained conditions, it is easy to envisage a relatively straightforward means of
managing feeding requirements and thereby to conceive of mechanisms that
involve little complexity. In general, it has been perfectly acceptable amongst
research workers for them not to contemplate or seek greater sophistication in
their concepts of mechanisms of control. My argument, put forward at the start
of this chapter, is that rats (and human beings) evolved to face far greater
demands on their abilities to secure food for themselves, to select and evaluate
food items, to assess the consequences of food ingestion, and to anticipate
future food requirements and ways to meet them. It is the evolved biological
systems that we are trying to understand.

   Smith *et al*. (1990) have recently charted the course 'from the tip of the tongue
to the end of the small intestine' in terms of the succession of afferent information conveyed to the brain as food is ingested, swallowed, and digested. An
array of cranial nerves carry chemosensory and somatosensory fibres, which
are affected sequentially as the food makes its progress from the mouth. These
afferent signals receive initial processing in the hindbrain before being pro-
jected in widespread networks of neurons in the higher levels of the brain.
Clearly, there is a complexity of information processing involved that we are
only just beginning to apprehend. The central theme of the present chapter
deals with the relationships between one neuropeptide, CCK, and several
systems of widely distributed neurons within the brain that utilize monoamines
as neurotransmitters. It is perhaps not too far-fetched to believe that these
systems, interacting with CCK and other neuropeptides, are conveying a variety
of qualitatively distinct forms of information relevant to the feeding process,
the determination of meal size, and the onset of satiety. If this is so, then there
is an experimental challenge of considerable dimensions in trying to characterize these forms of information being conveyed within these several neuronal
systems.

Co-operative systems

Using the model of possible interactions between CCK and 5-HT as an example
(Fig. 23.4), I have tried to depict a co-operative interaction between two
systems. This is a simple case, and no doubt in due course we shall have to
consider co-operative interactions between many more than two systems. In
very general terms, what should we expect from a co-operative combination of
systems? Intuitively, we can understand the increase in the power to process
information that comes with a larger interconnected system comprising several
components. This implies a greater capacity to handle information effectively
and efficiently, to achieve greater degrees of discrimination in the sensory

domain, and to maintain a flexible adaptive form of response in the face of environmental fluctuation and unpredictability. Co-operative systems may have the means of recruiting additional systems to the pool when great demands are placed on their capacity to process and utilize information. Moreover, we can assume that systems involved in the control of feeding can incorporate change through learning, and also make use of models to predict future feeding requirements and to plan food-procurement and utilization strategies.

As it happens, the present chapter places CCK in an important strategic position – in contact, it seems, with several neural systems that are known to be of great significance in determining feeding behaviour. There is no doubt that the new drugs now available to investigators, which act selectively as agonists or antagonists at CCK receptor subtypes, will prove of enormous value in placing CCK's several functions in the broad context of feeding control.

An important corollary of not trying to constrain the study of feeding processes to the very simplest case is, of course, the view that we should be broadening considerably the scope of the experimental paradigms deployed to investigate issues related to feeding. A conservative unadventurous approach will not grant insight into the full complement of properties of the systems that we are engaged in studying.

Clinical implications

Once we relinquish an overly simplistic view of the controls of appetite and satiety, and recognize that we are inevitably confronting a complex set of inter-acting co-operating systems with a great variety of neurochemical features, we can consider the problem of possible pharmacological interventions in a differ-ent light. A single pharmacological intervention cannot simulate, amend, or correct the operations of a complex neural system; the connectivity of bio-logical systems will ensure that drug-induced effects will be widespread and will bring compensatory responses into play. Classically, a single pharmacological treatment is designed to interact with a single receptor to initiate a uniform quantifiable response. We shall have to learn how to employ drugs in more informed ways to achieve useful interventions in complex systems that mediate flexible, adaptive, and heterogeneous responses. Perhaps a first step may be to give greater consideration to the possible uses of drugs in combination, e.g. a 5-HT-stimulating drug with a CCK analogue for treating bulimia, or a 5-HT antagonist in combination with a CCK antagonist for treating anorexic patients. Any success in investigating such combinations will provide the first steps in pharmacological interventions designed to take greater account than we can at the moment of the complex nature of the biological systems that deter-mine the character of the feeding responses that we observe and attempt to change.

Acknowledgement

I am very grateful to Dorothy Trinder for preparing the manuscript.

References

Antin, J., Gibbs, J., Holt, J., Young, R., and Smith, G. P. (1975). Cholecystokinin elicits the complete behavioral sequence of satiety in rats. *J. Comp. Physiol. Psychol.* **89**, 784–90.

Blouin, A. G., Blouin, J. H., Perez, E. L., Bushnick, T., Zuro, C., and Mulder, E. (1988). Treatment of bulimia with fenfluramine and desipramine. *J. Clin. Psychopharm.* **8**, 261–9.

Blundell, J. E. (1977). Is there a role for serotonin (5-hydroxytryptamine) in feeding? *Int. J. Obesity* **1**, 15–42.

Blundell, J. E. (1984). Serotonin and appetite. *Neuropharmacology* **23**, 1537–52.

Blundell, J. E. and Latham, C. J. (1978). Pharmacological manipulation of feeding behaviour: possible influences of serotonin and dopamine on food intake. In *Central mechanisms of anorectic drugs* (ed. S. Garattini and R. Samanin), pp. 83–109. Raven Press, New York.

Bock, M. G., DiPardo, R. M., Evans, B. E., Rittle, K. E., Whitter, W. L., Veber, D. F., Anderson, P. S., and Freidinger, R. M. (1989). Benzodiazepine gastrin and brain cholecystokinin receptor ligands: L-365,260. *J. Med. Chem.* **32**, 13–16.

Brobeck, J. R. (1955). Neural regulation of food intake. *Ann. NY Acad. Sci.* **63**, 44–55.

Burton, M. J., Cooper, S. J., and Popplewell, D. A. (1981). The effect of fenfluramine on the microstructure of feeding and drinking in the rat. *Br. J. Pharmacol.* **72**, 621–33.

Carruba, M. O., Mantegazza, P., Memo, M., Missale, C., Pizzi, M., and Spano, P. F. (1986). Peripheral and central mechanisms of action of serotonergic anorectic drugs. *Appetite* **7** (Suppl.), 105–13.

Chang, R. S. L. and Lotti, V. J. (1986). Biochemical and pharmacological characterization of an extremely potent and selective nonpeptide cholecystokinin antagonist. *Proc. Natl Acad. Sci. USA* **83**, 4923–6.

Clifton, P. G. (1987). Analysis of feeding and drinking patterns. In *Feeding and drinking* (ed. F. M. Toates and N. E. Rowland), pp. 19–35. Elsevier, Amsterdam.

Cooper, S. J. and Barber, D. J. (1990). SCH 23390-induced hypophagia is blocked by the selective CCK-A receptor antagonist devazepide, but not by the CCK-B/gastrin receptor antagonist L-365,260. *Brain Res. Bull.* **24**, 631–3.

Cooper, S. J. and Dourish, C. T. (1990). Multiple cholecystokinin (CCK) receptors and CCK-monoamine interactions are instrumental in the control of feeding. *Physiol. Behav.* **48**, 849–57.

Cooper, S. J., Dourish, C. T., and Barber, D. J. (1990). Reversal of the anorectic effect of (+)-fenfluramine in the rat by the selective cholecystokinin receptor antagonist MK-329. *Br. J. Pharmacol.* **99**, 65–70.

Cooper, S. J., Dourish, C. T., Barber, D. J., and Moseley, S. M. (1991). Reversal of the anorectic effect of systemically-administered 5-hydroxytryptamine by the selective cholecystokinin receptor antagonist devazepide. Submitted for publication.

Cooper, S. J., Rusk, I. N., and Clifton, P. G. (1992). Dopamine, D$_1$ and D$_2$ receptor subtypes, and feeding behaviour. In *The neuropharmacology of appetite* (ed. S. J. Cooper and J. M. Liebman). Oxford University Press, Oxford, in press.

Corwin, R. L., Woolverton, W. L., and Schuster, C. R. (1990). Effects of cholecystokinin d-amphetamine and fenfluramine in rats trained to discriminate 3 from 22 hr of food deprivation. *J. Pharmacol. Exp. Ther.* **253**, 720–8.

Crawley, J. N. (1992). The role of cholecystokinin in feeding behaviour. In *The neuropharmacology of appetite* (ed. S. J. Cooper and J. M. Liebman). Oxford University Press, Oxford, in press.

Davidson, T. L., Flynn, F. W., and Grill, H. J. (1988). Comparison of the interoceptive sensory consequences of CCK, LiCl, and satiety in rats. *Behav. Neurosci.* **102**, 134–40.

Deutsch, J. A. and Hardy, W. T. (1977). Cholecystokinin produces bait shyness in rats. *Nature* **266**, 196.

Dourish, C. T. (1992). The role of multiple serotonin receptors in the control of feeding. In *The neuropharmacology of appetite* (ed. S. J. Cooper and J. M. Liebman), Oxford University Press, Oxford, in press.

Dourish, C. T. and Hill, D. R. (1987). Classification and function of CCK receptors. *Trends Pharmacol. Sci.* **8**, 207–8.

Dourish, C. T., Hutson, P. H., and Curzon, G. (1985). Low doses of the putative serotonin agonist 8-hydroxy-2-(di-n-propylamino) tetralin (8-OH-DPAT) elicit feeding in the rat. *Psychopharmacology* **86**, 197–204.

Dourish, C. T., Coughlan, J., Hawley, D., Clark, M., and Iversen, S. D. (1988). Blockade of CCK-induced hypophagia and prevention of morphine tolerance by the CCK antagonist L-374,718. In *Cholecystokinin antagonists* (ed. R. Y. Wang and R. Schoenfeld), pp. 307–25. Alan R. Liss, New York.

Dourish, C. T., Rycroft, W., and Iversen, S. D. (1989a). Postponement of satiety by blockade of brain cholecystokinin (CCK-B) receptors. *Science* **245**, 1509–11.

Dourish, C. T., Ruckert, A. C., Tattersall, F. D., and Iversen, S. D. (1989b). Evidence that decreased feeding induced by systemic injection of cholecystokinin is mediated by CCK-A receptors. *Eur. J. Pharmacol.* **173**, 233–4.

Ebenezer, I. S., De La Riva, C., and Baldwin, B. A. (1990). Effects of the CCK-receptor antagonist MK-329 on food intake in pigs. *Physiol. Behav.* **47**, 145–8.

Ervin, G. N. and Teeter, M. N. (1986). Cholecystokinin octapeptide and lithium produce different effects on feeding and taste aversion learning. *Physiol. Behav.* **36**, 507–12.

Evans, B. E., Bock, M. G., Rittle, K. E., DiPardo, R. M., Whitter, W. L., Veber, D. F., Anderson, P. S., and Freidinger, R. M. (1986). Design of potent, orally effective nonpeptidal antagonists of the peptide hormone cholecystokinin. *Proc. Natl Acad. Sci. USA* **83**, 4918–22.

Geracioti, T. D., and Liddle, R. A. (1988). Impaired cholecystokinin secretion in bulimia nervosa. *New Engl. J. Med.* **319**, 683–8.

Gibbs, J., Young, R. C., and Smith, G. P. (1973a). Cholecystokinin decreases food intake in rats. *J. Comp. Physiol. Psychol.* **84**, 488–95.

Gibbs, J., Young, R. C., and Smith, G. P. (1973b). Cholecystokinin elicits satiety in rats with open gastric fistulas. *Nature* **245**, 323–5.

Grinker, J. A., Drewnowski, A., Enns, M., and Kissileff, H. (1980). Effects of d-amphetamine and fenfluramine on feeding patterns and activity of obese and lean Zucker rats. *Pharmacol. Biochem. Behav.* **12**, 265–75.

Hays, S. E., Beinfeld, M. C., Jensen, R. I., Goodwin, F. K., and Paul, S. M. (1980). Demonstration of a putative receptor site for cholecystokinin in rat brain. *Neuropeptides* **1**, 53–62.

Hays, S. E., Goodwin, F. K., and Paul, S. M. (1981). Cholecystokinin receptors are decreased in basal ganglia and cerebral cortex of Huntingdon's disease. *Brain Res.* **225**, 452–6.

Hewson, G., Leighton, G. E., Hill, R. J., and Hughes, J. (1988). The cholecystokinin receptor antagonist L-364,718 increases food intake in the rat by attenuation of the action of endogenous cholecystokinin. *Br. J. Pharmacol.* **93**, 79–84.

Hill, D. R., Campbell, N. J., Shaw, T. M., and Woodruff, G. N. (1987).

Autoradiographic localization and biochemical characterization of peripheral type CCK receptors in rats CNS using highly selective nonpeptide CCK antagonists. *J. Neurosci.* **7**, 2967–76.

Hill, D. R., Shaw, T. M., Dourish, C. T., and Woodruff, G. N. (1988). CCK-A receptors in the rat interpeduncular nucleus: evidence for a presynaptic location. *Brain Res.* **454**, 101–5.

Hill, D. R., Shaw, T. M., Graham, W., and Woodruff, G. N. (1990). Autoradiographical detection of cholecystokinin-A receptors in primate brain using ^{125}I-Bolton Hunter CCK-8 and ^{3}H-MK-329. *J. Neurosci.* **10**, 1070–81.

Innis, R. B. and Snyder, S. (1980). Distinct cholecystokinin receptors in brain and pancreas. *Proc. Natl Acad. Sci. USA* **77**, 6917–21.

Kissileff, H. R., Pi-Sunyer, F. X., Thornton, J., and Smith, G. P. (1981). C-terminal octapeptide of cholecystokinin decreases food intake in man. *Am. J. Clin. Nutr.* **34**, 154–60.

Leibowitz, S. F. (1992). Brain monoamines in the control of eating behavior: interaction with peptides, steroids and nutrients. In *The neuropharmacology of appetite* (ed. S. J. Cooper and J. M. Liebman). Oxford University Press, Oxford, in press.

Le Magnen, J. (1971). Advances in studies on the physiological control and regulation of food intake. In *Progress in physiological psychology*, Vol. 4 (ed. E. Stellar and A. N. Epstein), pp. 203–61. Academic Press, New York.

Liddle, R. A., Goldfine, I. D., Rosen, M. S., Taplitz, R. A., and Williams, J. A. (1985). Cholecystokinin bioactivity in human plasma: molecular forms, responses to feeding, and relationship to gall bladder contraction. *J. Clin. Invest.* **75**, 1144–52.

Lindén, A. (1989). Role of cholecystokinin in feeding and lactation. *Acta Physiol. Scand.* **137** (Suppl. 585), 1–49.

Lotti, V. J. and Chang, R. S. L. (1989). A new potent and selective non-peptide gastrin antagonist and brain cholecystokinin receptor (CCK-B) ligand: L-365,260. *Eur. J. Pharmacol.* **162**, 273–80.

Lotti, V. J., Pendleton, R. G., Gould, R. J., Hanson, H. M., Chang, R. S. L., and Clineschmidt, B. V. (1987). *In vivo* pharmacology of L-364,718, a new potent nonpeptide peripheral cholecystokinin antagonist. *J. Pharmacol. Exp. Ther.* **241**, 103–9.

Moore, B. O. and Deutsch, J. A. (1985). An antiemetic is antidotal to the satiety effects of cholecystokinin. *Nature* **315**, 321.

Moran, T. H. and McHugh, P. R. (1988). Anatomical and pharmacological differentiation of pyloric, vagal and brain stem cholecystokinin receptors. In *Cholecystokinin antagonists* (ed. R. Y. Wang and R. Schoenfeld), pp. 117–32. Alan R. Liss, New York.

Moran, T. H., Robinson, P. H., Goldrich, M. S., and McHugh, P. R. (1986). Two brain cholecystokinin receptors: implications for behavioral actions. *Brain Res.* **362**, 175–9.

Myers, R. D. and McCaleb, M. L. (1981). Peripheral and intrahypothalamic cholecystokinin act on the noradrenergic 'feeding circuit' in the rat's diencephalon. *Neuroscience* **6**, 645–55.

Neill, J. C. and Cooper, S. J. (1988). Evidence for serotonergic modulation of sucrose sham feeding in the gastric-fistulated rat *Physiol. Behav.* **44**, 453–9.

Neill, J. C. and Cooper, S. J. (1989*a*). Evidence that D-fenfluramine anorexia is mediated by 5-HT$_1$ receptors. *Psychopharmacology* **97**, 213–18.

Neill, J. C. and Cooper, S. J. (1989*b*). Effects of 5-hydroxytryptamine and

D-fenfluramine on sham feeding and sham drinking in the gastric-fistulated rat. *Physiol. Behav.* **46**, 949–53.

Pinnock, R. D., Woodruff, G. N., and Boden, P. R. (1990). Cholecystokinin excites dorsal raphe neurones via a CCK$_A$ receptor. *Brit. J. Pharmacol.* **100**, 49P.

Pi-Sunyer, X., Kissileff, H. R., Thornton, J., and Smith, G. P. (1982). C-terminal octapeptide of cholecystokinin decreases food intake in obese men. *Physiol. Behav.* **29**, 627–30.

Reidelberger, R. D. and O'Rourke, M. F. (1989). Potent cholecystokinin antagonist L-364,718 stimulates food intake in rats. *Am. J. Physiol.* **257**, R1512–18.

Russell, G. F. M., Checkley, S. A., Feldman, J., and Eisler, I. (1988). A controlled trial of D-fenfluramine in bulimia nervosa. *Clin. Neuropharmacol.* **11** (Suppl. 1), S146–59.

Saito, A., Goldfine, I. D., and Williams, J. A. (1981). Characterization of receptors for cholecystokinin and related peptides in mouse cerebral cortex. *J. Neurochem.* **37**, 483–90.

Sakamoto, C., Williams, J. A., and Goldfine, I. D. (1984). Brain CCK receptors are structurally distinct from pancreas CCK receptors. *Biochem. Biophys. Res. Commun.* **124**, 497–502.

Schneider, L. H. (1989). Orosensory self-stimulation by sucrose involves brain dopaminergic mechanisms. *Ann. NY Acad. Sci.* **575**, 307–19.

Schneider, L. H., Greenberg, D, and Smith, G. P. (1988). Comparison of the effects of selective D-1 and D-2 receptor antagonists on sucrose sham feeding and water sham drinking. *Ann. NY Acad. Sci.* **537**, 534–7.

Smith, G. P. and Gibbs, J. (1979). Postprandial satiety. In *Progress in psychobiology and physiological psychology*, Vol. 8 (ed. J. M. Sprague and A. N. Epstein), pp. 179–242. Academic Press, New York.

Smith, G. P., Greenberg, D., Corp, E., and Gibbs, J. (1990). Afferent information in the control of eating. In *Obesity: towards a molecular approach* (ed. G. Bray, D. Ricquier, and B. Spiegelman), pp. 63–79. Alan R. Liss, New York.

Stacher, G., Bauer, H., and Steinringer, H. (1979). Cholecystokinin decreases appetite and activation evoked by stimuli arising from the preparation of a meal in man. *Physiol. Behav.* **23**, 325–31.

Stallone, D., Nicolaidis, S., and Gibbs, J. (1989). Cholecytokinin-induced anorexia depends on serotoninergic function. *Am. J. Physiol.* **256**, R1138–41.

Stellar, E. (1954). The physiology of motivation. *Psychol. Rev.* **61**, 5–22.

Swerdlow, N. R., van der Kooy, D., Koob, G. F., and Wenger, J. R. (1983). Cholecystokinin produces conditioned place-aversions, not place preferences, in food-deprived rats: evidence against involvement in satiety. *Life Sci.* **32**, 2087–93.

Tsai, S. H., Passaro, E., and Lin, M. T. (1984). Cholecystokinin acts through catecholaminergic mechanisms in the hypothalamus to influence ingestive behaviour in the rat. *Neuropharmacology* **23**, 1351–6.

Verbalis, J. G., McCann, M. J., McHale, C. M., and Stricker, E. M. (1986). Oxytocin secretion in response to cholecystokinin and food intake: differentiation of nausea from satiety. *Science* **232**, 1417–19.

Vickroy, T. W., Bianchi, B. R., Kerwin, J. F., Kopecka, H., and Nadzan, A. M. (1988). Evidence that type A CCK receptors facilitate dopamine efflux in rat brain. *Eur. J. Pharmacol.* **152**, 371–2.

Weller, A., Smith, G. P., and Gibbs, J. (1990). Endogenous cholecystokinin reduces feeding in young rats. *Science* **247**, 1589–91.

West, D. B., Fey, D., and Woods, S. C. (1984). Cholecystokinin persistently suppresses meal size but not food intake in free-feeding rats. *Am. J. Physiol.* **246**, R776–87.

Wilson, M. C., Denson, D., Bedford, J. A., and Hunsinger, R. N. (1983). Pharmacological manipulation of sincalide (CCK-8)-induced suppression of feeding. *Peptides* **4**, 351–7.

Wolkowitz, O. M., Gertz, B., Weingartner, H., Beccaria, L., Thompson, K., and Liddle, R. A. (1990). Hunger in humans induced by MK-329, a specific peripheral-type cholecystokinin receptor antagonist. *Biol. Psychiat.* **28**, 169–73.

Zarbin, M. A., Innis, R. B., Walmsley, J. K., Snyder, S. H., and Kuhar, M. J. (1983). Autoradiographic localization of cholecystokinin receptors in rodent brain. *J. Neurosci.* **3**, 877–906.

24. Selective CCK receptor antagonists and macronutrient consumption in the rat

John Francis and Steven J. Cooper

It has been shown that the selective CCK_A receptor antagonist devazepide reliably blocked the reduction in food intake produced by exogenously administered CCK-8 (Dourish *et al.* 1989*a*). Reidleberger and O'Rourke (1989) demonstrated that devazepide administered alone increased food intake in rats deprived of food for 1 h. Dourish *et al.* (1989*b*) have shown that not only devazepide but also the selective CCK_B receptor antagonist L-365,260 increased food intake in satiated rats. The increases in food intake in these experiments were modest, and rats had to be deprived of food for only a short period of time or presatiated before the effects of the antagonists could be observed.

Liddle *et al.* (1986) showed that neither carbohydrates nor fats, high percentage components of test meals, caused significant CCK release in rats. Only ingestion of a protein meal (casein) caused significant increases in plasma CCK concentration.

Therefore it would seem that ingestion of a protein-rich diet would cause optimum CCK release during feeding. If CCK release is dependent on macronutrients (protein in the rat), then it follows that the effects of selective CCK antagonists on food intake may in turn be dependent on the nutrient composition of the diet. In general it would be expected, on this argument, that increases in intake produced by antagonists should be more closely related to protein ingestion than to either fat or carbohydrate ingestion (but see also Chapter 18 of this volume).

The primary aim of this experiment was to investigate whether exogenously administered CCK-8 would affect the intake of a protein meal and, if it did, to determine its duration of action and the period of its maximal effect. The second aim was to examine further the effect of endogenous CCK on feeding by investigating whether the selective CCK_A antagonist devazepide and the selective CCK_B antagonist L-365,260 have an effect on the intake of a protein meal. The third aim was to compare any effect of CCK-8 and the antagonists on the intake of a protein meal (casein) with the intake of a carbohydrate meal (sucrose) to determine whether the intake of different macronutrients affects the action of exogenous and endogenous CCK in feeding.

Individually housed male hooded rats (300–375 g) were deprived of food for 17 h overnight. They were randomly divided into two groups ($n = 10$–11 per group) and presented on at least four occasions with either a sucrose solution (10 per cent w/v) or a casein solution (9 per cent w/v) depending on the experiment. By the time of testing the amount of solution consumed was generally greater than 18 ml.

Following the overnight period of food deprivation, rats were injected with devazepide (0, 30, 100, or 300 μg/kg s.c.) or L-365,260 (0, 0.1, 1.0, or 10 μg/kg s.c.) 30 min prior to testing or with CCK-8 (0, 2, 4, or 8 μg/kg i.p.) 20 min prior to testing. The rats were then allowed to feed with the solution presented for 60 min. Intake was measured volumetrically at 5 min intervals throughout the test. Consecutive drug test days were separated by at least 48 h.

The data were analysed using one-way repeated ANOVA measures and Dunnett's *t* test.

Casein intake

CCK-8 (2–8 μg/kg) significantly reduced casein solution intake. The maximum reductions occurred after 25 min with a 58 per cent decrease in intake at the 8 μg/kg dose ($F(3, 27) = 9.474$, $p < 0.001$). Individual comparisons indicated that a significant reduction in intake occurred at each of the doses tested (Dunnett's *t* test, $p < 0.05$ and $p < 0.01$). However, after 60 min feeding there were no significant differences between conditions ($F(3, 27) = 1.581$, n.s.) (Fig. 24.1).

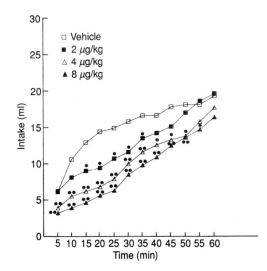

Fig. 24.1. Effects of CCK-8 on the intake of a 9 per cent casein solution in rats deprived of food for 17 h ($n = 10$ per group): $\star$ $p < 0.05$, $\star\star$ $p < 0.01$ compared with vehicle (ANOVA and Dunnett's *t* test).

Devazepide (30–300 μg/kg) produced no significant effect on the intake of casein solution after 60 min ($F(3, 30) = 0.520$, n.s.) or at any point in the test period when the data for the total group of animals were analysed. Nevertheless, additional analyses were carried out when the animals were subdivided into two groups of high and low baseline intake. The low baseline group consumed less than 25 ml of solution per 60 min, and the high baseline group consumed more than 25 ml.

The low baseline intake ($n = 6$) showed a significant increase in intake after 25 min ($F(3, 15) = 3.443$, $p < 0.05$) at the 30 μg/kg and 300 μg/kg dose levels. After 60 min there was a significant increase in casein intake ($F(3, 15) = 0.486$, $p < 0.005$) with a 32.6 per cent increase in casein intake in the 300 μg/kg dose level compared with the vehicle. Individual comparisons indicated a significant increase at each dose used (Table 24.1 and Fig. 24.2). The high baseline intake group ($n = 5$) showed a significant decrease in intake ($F(3, 12) = 3.768$, $p < 0.05$) with a 17.9 per cent decrease in intake at the 30 μg/kg and 100 μg/kg dose levels compared with the vehicle after 60 min. Individual comparisons indicated that this reduction was significant at each drug dose used.

When the group of animals was considered as a whole, L-365,260 (0.1–10 μg/kg) produced an overall effect on the intake of casein solution after 60 min which was significant ($F(3, 30) = 3.035$, $p < 0.05$). However, there were no significant differences between vehicle and drug conditions following comparisons using Dunnett's t test (Table 24.2).

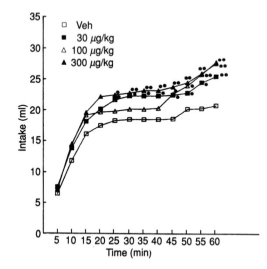

Fig. 24.2. Effects of devazepide on the intake of a 9 per cent casein solution in rats deprived of food for 17 h with low baseline intake (n = 6 per group): * p < 0.05, ** p < 0.01 compared with vehicle (ANOVA and Dunnett's t test).

Table 24.1. The effects of devazepide on the intake of a 9 per cent casein solution in real-feeding rats deprived of food for 17 h

Devazepide (μg/kg)	0	30	100	300
Low baseline (N = 6)				
Mean intake/60 min (ml)	20.83	25.50**	27.83**	27.67**
SEM	± 0.95	± 1.18	± 1.19	± 1.09
Change (%)		+ 22.4	+ 33.6	+ 32.6
High baseline (N = 5)				
Mean intake/60 min (ml)	29.00	23.80*	23.80*	27.67*
SEM	± 1.05	± 2.44	± 1.02	± 1.47
Change (%)		− 17.9	− 17.9	− 4.6

Dunnett's t test, comparisons with vehicle: *$p < 0.05$, **$p < 0.01$

Table 24.2. Effects of L-365,260 on the intake of a 9 per cent casein solution in real-feeding rats deprived of food for 17 h

L-365,260 (μg/kg)	0	0.1	1	10
Mean intake/60 min (n = 11) (ml)	23.83	25.82	22.73	22.27
SEM	± 0.84	+ 1.39	+ 1.06	± 1.05
Change (%)		+ 8.3	− 4.6	− 6.5

Sucrose intake

CCK-8 (2–8 μg/kg) significantly reduced the intake of sucrose solution. The maximum reductions occurred after 20 min with a 50 per cent decrease at the 8 μg/kg dose ($F(3, 27) = 13.716$, $p < 0.001$). Individual comparisons with Dunnett's t test indicated that a significant reduction in intake occurred at each of the doses tested after 60 min.

Neither devazepide (30–300 μg/kg) nor L-365,260 (0.1–10 μg/kg) had any significant effect on sucrose intake ($F(3, 27) = 0.215$ n.s.) and ($F(3, 27) = 0.067$ n.s.) respectively.

Conclusions

First, the data demonstrate that exogenous CCK-8 reduces the intake of both casein and sucrose solutions. This suggests that the effect of exogenous CCK-8 on food intake is unlikely to be macronutrient-specific, although we have yet to examine the effects of exogenous CCK-8 on fat intake. Exogenous CCK-8 reduces the intake of both casein and sucrose solutions within the first 5 min of the start of feeding, reaching its maximal effect at 25 min for casein intake and 20 min for sucrose intake. Exogenous CCK-8 at a dose of 8 μg/kg continues to suppress food intake until 55 min for casein and 60 min for sucrose. Therefore it would seem that exogenous CCK-8 affects the intake of protein and carbohydrate in relatively similar ways.

Devazepide produced a dose-dependent increase in casein intake in the low baseline intake group after 60 min. Devazepide significantly increased the intake 25 min after feeding began; this is also the period when exogenous CCK-8 had its maximal effect. If the effects produced by administration of exogenous CCK-8 are similar to those produced by release of endogenous CCK, then this is when we would expect to see the effects of devazepide. Devazepide produced a significant decrease in casein intake in the high baseline group. However, this effect was not significant until 60 min and may depend more on the late increase in intake in the vehicle condition than on an authentic depression of intake by devazepide. A possible explanation for an effect of devazepide on feeding in the low baseline group and not on feeding in the high baseline group is that low baseline animals may release more endogenous CCK in response to the ingestion of a protein meal.

L-365,260 produced significant differences in casein intake at 60 min. This appears to reflect a modest increase in intake at the $0.1 \mu g/kg$ dose. However, there was no evidence of an increase in intake at the two higher doses of 1.0 and $10 \mu g/kg$. Hence the interpretation of the significant result for L-365,260 is not clear.

These results suggest that the increase in the intake of casein solution may be mediated by antagonism at CCK_A receptors rather than at CCK_B receptors. These data are comparable with those of Dourish et $al.$ (1989a) who found that the decrease in feeding produced by exogenous CCK-8 was mediated by CCK_A receptors. Furthermore, these data suggest that individual differences between rats in terms of the baseline levels of feeding may reflect differences in levels of activity of endogenous CCK in response to protein-containing meals.

Neither devazepide nor L-365,260 had an effect on the intake of the sucrose solution. These data are consistent with results reported by Dourish et $al.$ (1989a). These results, taken together with the increase in casein produced by antagonists, suggest that the increases in intake produced by antagonists are more related to protein ingestion than to carbohydrate ingestion.

In conclusion it would seem that the effect of exogenous CCK-8 on feeding is not macronutrient-specific and also that the use of a protein meal enables the effects of devazepide to be observed in unsatiated rats deprived of food for 17 h. Finally, the increases in casein intake appear to depend on mediation via the CCK_A receptor.

References

Dourish, C. T., Ruckert, A. C., Tattersall, F. D., and Iversen, S. D. (1989a). Evidence that decreased feeding induced by systemic injection of cholecystokinin is mediated by CCK-A receptors. $Eur.$ $J.$ $Pharmacol.$ **174**, 233–4.

Dourish, C. T., Rycroft, W., and Iversen, S. D. (1989*b*). Postponement of satiety by blockade of brain cholecystokinin (CCK-B) receptors. *Science* **245**, 1509–11.

Liddle, R. A., Green, G. A., Conrad, C. K., and Williams, J. A. (1986). Proteins but not amino acid, carbohydrates, or fats stimulate cholecystokinin secretion in the rat. *Am. J. Physiol.* **251**, G243–8.

Reidleberger, R. D. and O'Rourke, M. F. (1989). Potent cholecystokinin antagonist L-364,718 stimulated food intake in rats. *Am. J. Physiol.* **257**, R1512–18.

25. CCK–5-HT interactions influence meal size in the free-feeding rat

Peter G. Clifton and Steven J. Cooper

Food intake in most mammals is behaviourally complex. In the rat larger amounts are eaten at the beginning and end of the night with relatively lower consumption during the day. In addition, food is consumed in 'meals' — periods of relatively intense feeding behaviour occupying 2–5 per cent of the animal's time budget. Changes in food intake may be achieved by modulating meal size, meal frequency, or some combination of both together with readjustment of the usual diurnal intake pattern. The analysis of pharmacologically induced changes in food intake has become widely used in recent years (Clifton 1987).

Chronic meal-contingent infusion of CCK substantially and permanently reduces meal size, although meal frequency increases so that the total food intake is only slightly reduced for the first 2 or 3 days and subsequently is no different from control levels (West *et al.* 1984). The indirect serotonin agonist DL-fenfluramine also reduces food intake, primarily by reducing meal size (Popplewell *et al.* 1981). As tolerance to the anorectic effect of fenfluramine develops and overall food intake increases, meal size remains reduced and meal frequency increases (Barnfield and Clifton, unpublished data). In both cases the reduction in meal size has been taken as evidence of an effect on satiety rather than hunger.

Recent evidence suggests possible interactions between CCK and 5-hydroxytryptamine (5-HT) in the control of feeding. Stallone *et al.* (1989) have shown that the 5-HT antagonist metergoline blocks the anorectic action of CCK-8 in short-term intake tests. They favoured a central site for this interaction in which the antagonist blocked a vagally transmitted peripheral CCK signal. Cooper *et al.* (1990) have demonstrated a rather different interaction. They showed that the CCK antagonist devazepide, at a dose which when given alone had little effect on intake, could block the anorectic effect of D-fenfluramine. Their study also used a simple short-term intake test. Here we report a meal pattern study in which the same drug interaction is examined.

Male rats were habituated to large home cages and a 12 h:12 h light–dark cycle (lights off at 18.00 h). Pellets (45 mg) and water were available *ad libitum*, and intake patterns were recorded using a microprocessor-based system (Clifton 1987). On experimental days the animals were given either 30 μg/kg of

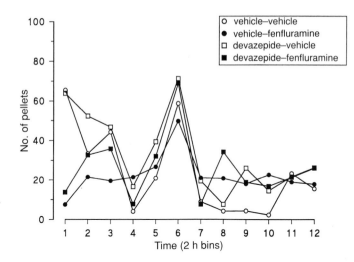

Fig. 25.1. The number of 45 mg pellets consumed in successive 2 h bins after treatment with devazepide (30 μg/kg) and d-fenfluramine (3 mg/kg) or their respective vehicles.

devazepide or vehicle (1 per cent methyl cellulose) at 17.00 h, followed by 3 mg/kg D-fenfluramine or vehicle (water) at 17.30 h. The drugs were gifts from Merck Sharp & Dohme (Dr C. T. Dourish) and Servier (Ms O. Arnaud) respectively. Food and water intake patterns were recorded for the following 24 h. The animals received all four treatment combinations counterbalanced in such a way that a further experimental record was separated from the last fenfluramine treatment by at least 2 days and from the last devazepide treatment by at least 5 days.

Initial analysis of food and water intake in 2 h blocks demonstrated a clear anorexia produced by D-fenfluramine. This anorexia was evident for about 6 h. There was little sign of antagonism by devazepide (Fig. 25.1). This

Table 25.1. Devazepide attenuates the reduction of meal size produced by D-fenfluramine treatment

Condition	Meal size	No. of meals	Latency to feed
Vehicle–vehicle	27.9	9.5	5.1
Devazepide–vehicle	25.1	8.1	1.8
Vehicle–fenfluramine	5.8	9.4	69
Devazepide–fenfluramine	15.7	6.8	109

Drug treatment was 30 μg/kg devazepide or vehicle (1 per cent methyl cellulose) at 17.00 h, followed by 3 mg/kg D-fenfluramine or vehicle (water) at 17.30 h. The data come from meals taken between 18.00 h and 24.00 h. The latency data were transformed to logarithmic values before statistical treatment and the means derive from that analysis.

impression was confirmed by an analysis of variance. There was an interaction between drug (fenfluramine or vehicle) and time ($F(11, 66) = 4.21, p < 0.001$), but effects involving the antagonist fell well below significance. Water intake showed a similar but less marked pattern and the interaction between drug and time was only marginally significant ($F(11, 66) = 1.95, p = 0.049$). The latency to feed was greatly increased by fenfluramine treatment (Table 25.1), giving a highly significant effect in an ANOVA ($F(1, 7) = 37.2, p < 0.001$). Devazepide had no significant impact on this effect of the drug.

Figure 25.2 shows the distribution of interpellet intervals during meals taken while the drug treatment was effective. The vehicle–vehicle group are quite typical of normal rats, eating pellets every 7–8 s during a meal. Devazepide had little effect on this pattern. Treatment with D-fenfluramine completely disrupted and flattened this distribution. Again the CCK_A antagonist produced little or no behavioural antagonism of this effect.

However, at least one feature of the D-fenfluramine anorexia was antagonized by devazepide. Meal size, which was reduced to less than a quarter of the control value by D-fenfluramine, was substantially increased when the animals were pretreated with devazepide (Table 25.1). This led to a highly significant interaction between drug and antagonist treatment in the relevant ANOVA ($F(1, 4) = 18.75, p = 0.012$). Total food intake showed little recovery because of a tendency to take fewer meals.

This set of results, although unanticipated, is clearly compatible with previously published data. At the level of a single meal, or a short test session, the

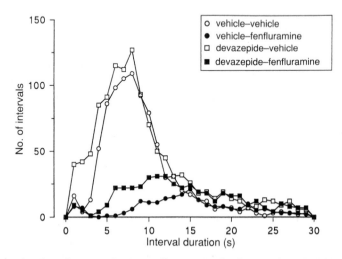

Fig. 25.2. The distribution of inter-pellet intervals during the 6 h following drug treatment with devazepide (30 μg/kg) and D-fenfluramine (3 mg/kg) or their respective vehicles.

CCK_A antagonist devazepide can appreciably reverse the anorexia that results from D-fenfluramine treatment (Cooper *et al.* 1990). However, a second longer-term action of fenfluramine must be invoked to explain the meal-patterning studies reported here. Not only is meal size reduced by fenfluramine, but these small meals lead to longer post-prandial pauses than expected for their size. Only the first action of fenfluramine is antagonized by devazepide. As a result the enhanced meal size when fenfluramine is preceded by devazepide is compensated for by a decrease in meal frequency; total food intake over the longer term shows only a marginal and non-significant recovery. The failure to reverse the increased latency to the first meal is also consistent with this pattern. It is also compatible with the results of West *et al.* (1984), which imply that CCK may contribute to the short-term development of satiety during a meal but does not sustain it from one meal to the next.

Perhaps the CCK_A-receptor-mediated input to serotonergic neurons in the raphé nucleus, which has been identified in neurophysiological studies (Pinnock *et al.* 1990; see Chapter 5 of this volume), signals meal size and is responsible for the termination of a meal. Other serotonergic neurons in the raphé, which lack this input, maintain the state of satiety that persists from one meal to the next as a result of the longer-term consequences of feeding. Although such a hypothesis is speculative, it is open to investigation using currently available pharmacological and behavioural tools.

References

Clifton, P. G. (1987). Analysis of feeding and drinking patterns. In *Methods for the study of feeding and drinking* (ed. F. M. Toates and N. R. Rowland), pp. 19–35. Elsevier, Amsterdam.

Cooper, S. J., Dourish, C. T., and Barber, D. (1990). Reversal of the anorectic effect of (+)-fenfluramine in the rat by the selective cholecystokinin antagonist MK-329. *Br. J. Pharmacol.* **99**. 65–70.

Pinnock, R. D., Woodruff, G. N., and Boden, P. R. (1990). Cholecystokinin excites dorsal raphé neurones via a CCK-A receptor. *Br. J. Pharmacol.* **100**, 49P.

Popplewell, D. A., Burton, M. J., and Cooper, S. J. (1981). The effect of fenfluramine on the microstructure of feeding and drinking in the rat. *Br. J. Pharmacol.* **72**, 621–33.

Stallone, D., Nicolaïdis, S., and Gibbs, J. (1989). Cholecystokinin-induced anorexia depends on serotonergic function. *Am. J. Physiol.* **256**, R1138–49.

West, D. B., Fey, D., and Woods, S. C. (1984). Cholecystokinin persistently suppresses meal size but not food intake in free feeding rats. *Am. J. Physiol.* **246**, R776–87.

26. Effects of CCK receptor blockade on hunger and gastric emptying in humans

Rodger A. Liddle

Introduction

CCK is a classical gastro-intestinal hormone produced by discrete endocrine cells of the upper small intestine. The first biological actions of CCK to be identified were its ability to stimulate gall bladder contraction and pancreatic exocrine secretion. Other actions of CCK that were subsequently noted by injection of the hormone into animals included effects on gastric emptying, bowel motility, and food intake. Studies in several species have indicated that administration of CCK inhibits gastric emptying (Debas *et al.* 1975; Anika 1982; Moran and McHugh 1982; Valenzuela and Defilippi 1981); however, it has been difficult to establish whether these effects were physiological or pharmacological. In some studies, simultaneous measurements of pancreatic secretion (Valenzuela and Defilippi 1981) and gall bladder contraction (Debas *et al.* 1975) were used to estimate physiological CCK levels. Unfortunately these studies have offered conflicting conclusions regarding the role of CCK in gastric emptying.

A major difficulty in establishing the physiological effects of CCK has been the limitation in measuring plasma levels of the hormone. This problem has been circumvented only with the recent development of specific and sensitive assays for CCK (Schlegel *et al.* 1977; Byrnes *et al.* 1981; Calam *et al.* 1982; Schafmayer *et al.* 1982; Chang and Chey 1983; Himeno *et al.* 1983; Jansen and Lamers 1983; Liddle *et al.* 1985; Turkelson *et al.* 1986). The traditional way to prove that a candidate hormone functions in physiological concentrations has been to measure the circulating hormone concentration simultaneously with its target tissue response, and then to reproduce those blood concentrations by infusion of exogenous hormone and determine the effect on the target tissue. With this approach it has been demonstrated that CCK has several hormonal actions, including delaying the gastric emptying of liquids (Liddle *et al.* 1986).

CCK has been localized in neurons of the brain and the peripheral nerves where it probably functions as a neurotransmitter. It is present in very high concentrations in several brain regions thought to mediate sensory, motor, and associational processes (Crawley 1985) and is co-localized with other neuropeptides. The hallmark discovery by Gibbs *et al.* (1973) that intraperitoneal (i.p.) administration of CCK decreased food intake in rats launched numerous investigations examining the satiety effects of CCK. Not only does CCK reduce food intake but it also induces the behavioural sequence of satiety, including grooming and short-term exploration, that is commonly demonstrated by rats after feeding. These satiety-inducing effects of peripherally administered CCK were shown to be abolished by subdiaphragmatic vagotomy (Smith *et al.* 1981). In addition to the effects of peripherally administered CCK, Della-Fera and Baile (1979) found that intracerebroventricular (i.c.v.) administration of CCK also suppressed feeding. A physiological role for CCK in satiety was strongly suggested by the demonstration that CCK antisera injected into the cerebro-spinal fluid (CSF) of sheep delayed the onset of satiety (Della-Fera and Baile 1985).

Thus both central and peripheral effects of CCK on satiety have been demon-strated. The central effects of CCK are probably transmitted via neurally released CCK. However, CCK in the periphery may originate from both endocrine cells of the gut and circulate in the plasma as a true hormone, or it may be released from peripheral nerves and act as a neurotransmitter. Neural CCK could then innervate end-organs such as the stomach or signal satiety to the brain via other neural connections. Moran and McHugh (1982) linked the effects of CCK on satiety with its effects on the stomach when they demon-strated that CCK suppressed food intake in rhesus monkeys by inhibiting gastric emptying. A gastric satiety signal has also been demonstrated in rats (Gibbs and Smith 1986), although satiating effects of the hormone cannot be entirely attributed to effects on the stomach (Moran and McHugh 1988; Chapter 18 of this volume).

The possibility that CCK may regulate both gastric emptying and satiety by either hormonal or neural mechanisms has made it difficult to establish the physiological role of endogenous CCK. These investigations have been further complicated by the discovery of different types of CCK receptors, subclassified as the peripheral (type A) and central (type B) receptors (Moran *et al.* 1986; Dourish and Hill 1987). Although located primarily in the digestive system, peripheral type receptors are also found in the CNS in a species-specific distri-bution (Moran *et al.* 1986; Dourish and Hill 1987; Chapter 5 of this volume). Studies of CCK using infusion of exogenous hormone have not always distin-guished physiological from pharmacological effects. In addition, it has not been possible to determine whether the effects of CCK are central or peripheral in origin. For these reasons an agent that specifically blocks the actions of CCK should be useful for investigating the role of endogenous CCK. Furthermore, the ability to distinguish peripheral from central effects of CCK would depend on the selectivity of the antagonist for A and B type receptors respectively. The recent development of devazepide, a specific antagonist selective for the CCK_A receptor, has provided the opportunity to examine the effects of CCK on gastric emptying and satiety that may be conferred by the actions of endogenous CCK through CCK_A receptors. The initial studies using devazepide to evaluate the effect of CCK receptor blockade on gastric emptying and hunger in humans are summarized below.

Effect of devazepide on gastric emptying

In a double-blind two-period cross-over design, eight normal male volunteers received single doses of placebo or 10 mg devazepide, administered orally, 2 h prior to ingestion of a mixed meal (Liddle *et al.* 1989). This dose of devazepide was chosen because in an earlier study it had been shown to inhibit CCK-stimulated gall bladder contraction completely. The meal was 614 kcal and consisted of two scrambled eggs, two slices of white toasted bread, 8.2 g butter, 75 ml of half-and-half (one-half milk and one-half cream), and 300 ml clear apple juice. This meal contained 66.8 g carbohydrate, 29.4 g fat, and 20 g protein. 500 μCi[^{99m}Tc] sulphur colloid was cooked into the eggs and served as a marker of solid gastric emptying. 100 μCi [^{111}In] diethylenetriamine pentaacetic acid was added to the apple juice and served as a marker of liquid gastric emptying. Determinations of gastric emptying were made by gamma scintigraphy of the abdomen. Plasma CCK levels were measured before and at various times after drug administration by a specific radio-immunoassay, as previously described (Himeno *et al.* 1983).

Basal plasma CCK concentrations were 0.7–1.0 pM and increased with devazepide treatment (Fig. 26.1). Plasma CCK concentrations gradually increased in both placebo- and devazepide-treated subjects beginning 15 min after ingestion of the meal. In subjects receiving devazepide there was a marked accentuation of the post-prandial increase in plasma CCK concentrations. This finding strongly supports the concept of negative feedback regulation of CCK secretion (Green and Lyman 1972; Louie *et al.* 1986; Gomez *et al.* 1988). The basis of this mechanism is that CCK is released from the duodenum in the absence of active pancreatic proteases and bile within the gut lumen. CCK then stimulates the pancreas and gall bladder causing pancreatic juice and bile to flow into the duodenum where it suppresses further CCK release. CCK receptor blockade, as demonstrated with devazepide, inhibits CCK-stimulated pancreatic secretion and gall bladder contraction causing CCK levels to remain elevated.

To confirm the effectiveness of devazepide to antagonize the actions of CCK, gall bladder contraction was measured by abdominal ultrasonography (Fig. 26.2). Gall bladder volumes decreased by 68.4 ± 3.8 per cent within 2 h of eating in placebo-treated subjects. In contrast, devazepide completely inhibited meal-stimulated gall bladder contraction. This inhibition occurred despite higher plasma CCK levels in these subjects.

The effects of devazepide on gastric emptying of both the solid and liquid components of the meal were measured simultaneously. Gastric emptying of the solid component of the meal was based on the rate at which the ^{99m}Tc label

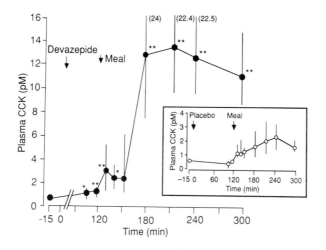

Fig. 26.1. Effect of devazepide on plasma CCK concentrations with ingestion of a mixed meal. Eight subjects received placebo (○) or 10 (●) mg of devazepide orally 2 h before ingestion of a 614 kcal mixed meal. Plasma CCK results are expressed as the median values and the interquartile ranges. All CCK values after ingestion of the meal were significantly greater than premeal values for both treatment groups. Post-prandial CCK levels in subjects receiving devazepide were significantly greater than those observed in placebo-treated subjects. (Reproduced with permission from Liddle *et al.* 1989.)

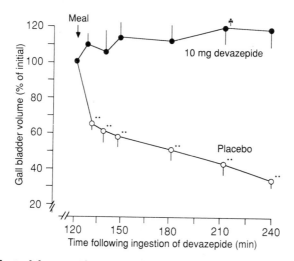

Fig. 26.2. Effect of devazepide on gall bladder volumes after ingestion of a mixed meal. Gall bladder volumes were measured by abdominal ultrasonography in the subjects described in Fig. 26.1. Results are expressed as the percentage of initial (premeal) gall bladder volume. Values are expressed as the mean ± SEM: ✶✶ $p < 0.01$; † $p < 0.05$ within-treatment comparison with premeal value. (Reproduced with permission from Liddle *et al.* 1989.)

disappeared from the area of the stomach (Fig. 26.3). After placebo treatment the solid marker slowly emptied from the stomach with a half-time for emptying of 128 ± 8 min. Devazepide treatment did not modify this emptying pattern. Liquids emptied from the stomach more rapidly, with a half-time for emptying of the ^{111}In marker of 58 ± 10 min (Fig. 26.4). The rate of liquid emptying from the stomach was also unaffected by devazepide.

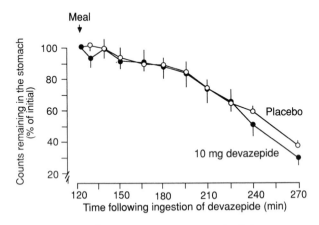

Fig. 26.3. Effect of devazepide on gastric emptying of solids. Gastric emptying was measured in the subjects described in Figs 26.1 and 26.2. The solid component of the meal was labelled with [^{99m}Tc] sulphur colloid. Results are expressed as the mean ± SEM. (Reproduced with permission from Liddle *et al.* 1989.)

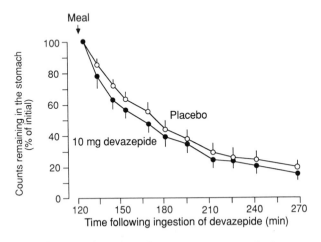

Fig. 26.4. Effect of devazepide on liquid gastric emptying. The liquid component of the mixed meal described in Fig. 26.3 was labelled with [^{111}In]diethylene triamine pentaacetic acid for the simultaneous measurement of liquid gastric emptying. Results are expressed as the mean ± SEM. (Reproduced with permission from Liddle *et al.* 1989.)

Effect of devazepide on hunger

To examine the possibility that endogenous CCK, released after eating, plays a physiological role in regulating appetite, normal subjects were treated with devazepide prior to ingestion of a meal. In the same paradigm as described above, behavioural assessments were measured 15 min prior to, and 90 and 155 min following, ingestion of 10 mg devazepide. Behavioural assessments included visual analogue scale self-ratings of hunger, tiredness, alertness, and fullness. Subjects rated these items by placing a vertical line on a 100 mm horizontal line anchored at the ends with the polar descriptors 'not at all' and 'extremely'.

Hunger ratings were similar in treatment groups prior to devazepide administration. However, by 90 min after treatment, hunger ratings were significantly higher in subjects receiving devazepide (Fig. 26.5). At 155 min after drug administration (30 min after the meal) there was a reduction in hunger in both placebo- and devazepide-treated subjects. There was also a trend towards less reduction in hunger post-prandially with devazepide. Ratings of fullness, tiredness, and alertness were not different, although there was a trend towards reduced fullness in subjects receiving devazepide.

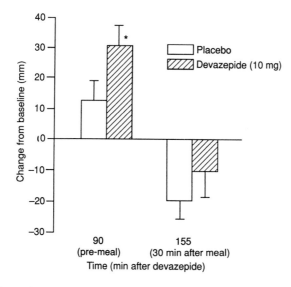

Fig. 26.5. Effect of devazepide on subjective ratings of hunger. In the same eight subjects as described previously, subjective hunger was measured using visual analogue ratings. Results are expressed as changes (in mm) from baseline ratings. Times shown are in relation to the time of devazepide or placebo treatment. A mixed meal was ingested over 5 min beginning at 120 min. Values are expressed as the mean ± SD: * $p < 0.05$. (Modified from Wolkowitz *et al.* 1990.)

It has also been proposed that CCK modulates cognitive function. Exogenous CCK was found to enhance memory retention (Flood *et al.* 1987) and to block experimentally induced amnesia in rats (Katsuura and Itoh 1986). If CCK has similar effects in humans, then we might expect memory to be adversely affected by CCK receptor blockade. To examine this possibility, the same eight volunteers underwent testing of cognitive function during treatment with devazepide (Wolkowitz *et al.* 1990). Tests of attention, free recall, word recognition, knowledge, and automatic processing were performed in a manner that has previously been shown to be sensitive to drug effects in a variety of clinical and normal populations (Roy-Byrne *et al.* 1986; Wolkowitz *et al.* 1987). As shown in Table 26.1, devazepide administration had no significant effect on any of the cognitive parameters assessed (Wolkowitz *et al.* 1990).

Discussion

These studies with devazepide implicate a physiological role for CCK in regulating gall bladder contraction and appetite in humans. However, they do not confirm a physiological role for CCK in gastric emptying. If the hypothesis that endogenous CCK delays gastric emptying were true, then we would expect acceleration of gastric emptying with CCK receptor blockade. Several studies have demonstrated acceleration of gastric emptying in a variety of species using different CCK receptor antagonists. Proglumide, a glutaramic acid derivative and a weak CCK antagonist, has been found to increase gastric emptying in rats (Shillabeer and Davison 1987). Similar results have been obtained with

Table 26.1.   Effect of devazepide on cognitive performance

Test	Changes from baseline		Significance
	Placebo		
90 min after devazepide treatment (but prior to meal)			
Attention	− 0.13	0.25	NS
Free recall	− 0.63	− 1.0	NS
Recognition	− 0.38	− 0.38	NS
Knowledge	0.25	2.5	NS
Automatic processing	− 0.05	− 0.13	NS
155 min after devazepide (30 min after meal)			
Attention	0.13	0.25	NS
Free recall	− 0.13	− 0.88	NS
Recognition	0.25	− 0.38	NS
Knowledge	1.75	5.0	NS
Automatic processing	− 0.12	− 0.03	NS

Cognitive performance tests were conducted on the same subjects as described in Fig. 26.5.
Values represent the number of words.
Adapted from Wolkowitz *et al.* (1990).

devazepide (Decktor *et al.* 1988; Dimaline *et al.* 1988). Loxiglumide, also a glutaramic acid derivative but a more potent CCK antagonist than proglumide, has been found to accelerate the gastric emptying of non-absorbable markers in humans (Meyer *et al.* 1989). In contrast, a similar effect on food-stimulated gastric emptying has not been observed (Corazziari *et al.* 1990; Konturek *et al.* 1990). Our observations with devazepide suggest that endogenous CCK alone does not control gastric emptying. However, it is important to interpret these results with caution since it has been demonstrated that CCK concentrations in the physiological range may delay gastric emptying (Liddle *et al.* 1986). It is possible that other factors compensate for, or obscure, the effects of CCK in the milieu of ingestion of a mixed meal. Alternatively, it is possible that higher doses of devazepide might demonstrate an acceleration of gastric emptying if the CCK receptors involved were different in some way from those mediating gall bladder contraction. Finally, since gastric emptying and CCK release are regulated by various properties of a meal (Chapter 18 of this volume), including nutrient composition, it is also possible that an alteration of gastric emptying with devazepide may be demonstrable with other diets. This latter point may explain the discrepancy between studies using different meals (Meyer *et al.* 1989; Corazziari *et al.* 1990; Konturek *et al.* 1990).

If the hypothesis that endogenous CCK regulates appetite is correct, then CCK receptor antagonists devoid of intrinsic activity should increase appetite. This phenomenon was suggested by studies demonstrating that food intake was increased in rats treated with proglumide (Shillabeer and Davison 1984) and in non-fasted rats treated with devazepide (Dourish *et al.* 1988; Hewson *et al.* 1988). Studies with devazepide suggest that CCK, acting through CCK_A receptors, participates in the regulation of appetite. Dourish and colleagues, in studies using devazepide and L-365,260 (a CCK receptor antagonist selective for CCK_B receptors), found that both antagonists delayed satiety in rats but L-365,260 was 100 times more potent than devazepide, implying that CCK acting through central-type receptors mediates satiety (Dourish *et al.* 1989; Chapter 21 of this volume). Our finding that devazepide increased hunger in humans suggests that endogenous CCK acting through CCK_A receptors also has a physiological role in regulating appetite (Chapters 17 and 19).

Conclusions

The development of specific CCK receptor antagonists provides useful tools for studying the physiology of endogenous CCK. It has been unequivocally demonstrated that CCK receptor blockade inhibits meal-stimulated gall bladder contraction, indicating that CCK is a physiological regulator of this organ. CCK antagonists have been shown to accelerate gastric emptying in some species and in humans under some conditions. However, the observation that devazepide did not modify the emptying of either solids or liquids suggests that

CCK alone was not a major regulator of gastric emptying, at least under the conditions tested.

Stimulation of hunger with devazepide provides strong evidence that endogenous CCK modifies appetite in humans. As CCK antagonists selective for A and B type receptors become available it should be possible to determine whether CCK regulates appetite by central or peripheral mechanisms. The clinical utility of selective CCK antagonists remains to be determined, but it is possible that antagonists could be used to stimulate appetite in such conditions as anorexia nervosa, hypophagic depression, and cancer (Gertz 1988).

References

Anika, M. S. (1982). Effects of cholecystokinin and caerulein on gastric emptying. *Eur. J. Pharmacol.* **85**, 195-9.

Byrnes, D. J., Henderson, L., Borody, T., and Rehfeld, J. F. (1981). Radioimmunoassay of cholecystokinin in human plasma. *Clin. Chim. Acta.* **111**, 81-9.

Calam, J., Ellis, A., and Dockray, G. J. (1982). Identification and measurement of molecular variants of cholecystokinin in duodenal mucosa and plasma. *J. Clin. Invest.* **69**, 218-5.

Chang, T. M. and Chey, W. Y. (1983). Radioimmunoassay of cholecystokinin. *Dig. Dis. Sci.* **28**, 456-68.

Corazziari, E., Ricci, R., Biliotti, D., Bontempo, I., Demedici, A., Pallotta, N., and Torsoli, A. (1990). Oral administration of loxiglumide CCK antagonist inhibits postprandial gall bladder contraction without affecting gastric emptying. *Dig. Dis. Sci.* **35**, 50-4.

Crawley, J. N. (1985). Comparative distribution of cholecystokinin and other neuropeptides: why is this peptide different from all other peptides? *Ann NY Acad. Sci.* **448**, 1-8.

Debas, H. T., Farooq, O., and Grossman, M. I. (1975). Inhibition of gastric emptying is a physiological action of cholecystokinin. *Gastroenterology.* **68**, 1211-17.

Decktor, D. L., Pendleton, R. G., Elnitsky, A. T., Jenkins, A. M., and McDowell, A. P. (1988). Effect of metoclopramide, bethanechol and the cholecystokinin receptor antagonist, L-364,718, on gastric emptying in the rat. *Eur. J. Pharmacol.* **147**, 313-16.

Della-Fera, M. A. and Baile, C. A. (1979). Cholecystokinin octapeptide: continous picomole injections into the cerebral ventricles of sheep suppress feeding. *Science*, **206**, 471-3.

Della-Fera, M. A. and Baile, C. A. (1985). Central nervous system cholecystokinin and the control of feeding behavior in sheep. *Prog. Clin. Biol. Res.* **192**, 115-2.

Dimaline, R., Dockray, G. J., and Green, T. (1988). The action of the cholecystokinin antagonist L-364,718 on gastric emptying in the rat. *J. Physiol.* **396**, 17P.

Dourish, C. T. and Hill, D. R. (1987). Classification and function of CCK receptors. *Trends Pharmacol. Sci.* **8**, 207-8.

Dourish, C. T., Coughlan, J., Hawley, D., Clark, M., and Iversen, S. D. (1988). Blockade of CCK-induced hypophagia and prevention of morphine tolerance by the CCK antagonist L-364,718. In *Cholecystokinin antagonists* (ed. R. Y. Wang and R. Schoenfeld), pp. 307-25. Alan R. Liss, New York.

Dourish, C. T., Rycroft, W., and Iversen, S. D. (1989). Postponement of satiety by blockade of brain cholecystokinin (CCK-B) receptors. *Science.* **245**, 1509-11.

Flood, J. F., Smith, G. E., and Morley, J. E. (1987). Modulation of memory processing by cholecystokinin: dependence on the vagus nerve. *Science*, **236**, 832-4.

Gertz, B. J. (1988). Potential clinical applications of a CCK antagonist. In *Cholecystokinin antagonists* (ed. R. Y. Wang and R. Schoenfeld), pp. 327-42. Alan R. Liss, New York.

Gibbs, J. and Smith, G. P. (1986). Satiety: the roles of peptides from the stomach and the intestine. *Fed. Proc.* **45**, 1391-5.

Gibbs, J., Young, R. C., and Smith, G. P. (1973). Cholecystokinin decreases food intake in rats. *J. Comp. Physiol. Psychol.* **84**, 488-95.

Gomez, G., Upp, J. R., Jr., Lluis, F., Alexander, R. W., Poston, G. J., Greeley, G. H., Jr., and Thompson, J. C. (1988). Regulation of the release of cholecystokinin by bile salts in dogs and humans. *Gastroenterology.* **94**, 1036-46.

Green, G. M. and Lyman, R. L. (1972). Feedback regulation of pancreatic enzyme secretion as a mechanism for trypsin inhibitor-induced hypersecretion in rats. *Proc. Soc. Exp. Biol. Med.* **140**, 6-12.

Hewson, G., Leighton, G. E., Hill, R. G., and Hughes, J. (1988). The cholecystokinin receptor antagonist L364,718 increases food intake in the rat by attenuation of the action of endogenous cholecystokinin. *Br. J. Pharmacol.* **93**, 79-84.

Himeno, S., Tarui, S., Kanayama, S., Kuroshima, T., Shinomura, Y., Hayashi, C., Tateishi, K., Imagawa, K., Hashimura, E., and Hamaoka, T. (1983). Plasma cholecystokinin responses after ingestion of liquid meal and intraduodenal infusion of fat, amino acids, or hydrochloric acid in man: analysis with region specific radioimmunoassay. *Am. J. Gastroenteral.* **78**, 703-7.

Jansen, J. B. M. J. and Lamers, C. B. H. W. (1983). Radioimmunoassay of cholecystokinin in human tissue and plasma. *Clin. Chim. Acta.* **131**, 305-16.

Katsuura, G. and Itoh, S. (1986). Prevention of experimental amnesia by peripherally administered cholecystokinin octapeptide in the rat. *Drug Dev. Res.* **7**, 269-76.

Konturek, S. J., Kwiecien, N., Obtulowicz, W., Kopp, B., Oleksy, J., and Rovati, L. (1990). Cholecystokinin in the inhibition of gastric secretion and gastric emptying in humans. *Digestion.* **45**, 1-8.

Liddle, R. A., Goldfine, I. D., Rosen, M. S., Taplitz, R. A., and Williams, J. A. (1985). Cholecystokinin bioactivity in human plasma: molecular forms, responses to feeding, and relationship to gall bladder contraction. *J. Clin. Invest.* **75**, 1144-52.

Liddle, R. A., Morita, E. T., Conrad, C. K. and Williams, J. A. (1986). Regulation of gastric emptying in humans by cholecystokinin. *J. Clin. Invest.* **77**, 992-6.

Liddle, R. A., Gertz, B. J., Kanayama, S., Beccaria, L., Coker, L. D., Turnbull, T. A., and Morita, E. T. (1989). Effects of a novel cholecystokinin (CCK) receptor antagonist, MK-329, on gall bladder contraction and gastric emptying in humans: implications for the physiology of CCK. *J. Clin. Invest.* **84**, 1220-5.

Louie, D. S., May, P., Miller, P., and Owyang, C. (1986). Cholecystokinin mediates feedback regulation of pancreatic enzyme secretion in rats. *Am. J. Physiol.* **250**, G252-9.

Meyer, B. M., Werth, B. A., Beglinger, C., Hildebrand, P., Jansen, J. B. M. J., Zach, D., Rovati, L. C., and Stalder, G. A. (1989). Role of cholecystokinin in regulation of gastrointestinal motor functions. *Lancet.* **2**, 12-5.

Moran, T. H. and McHugh, P. R. (1982). Cholecystokinin suppresses food intake by inhibiting gastric emptying. *Am. J. Physiol.* **242**, R491-7.

Moran, T. H. and McHugh, P. R. (1988). Gastric and nongastric mechanisms for satiety action of cholecystokinin. *Am. J. Physiol.* **254**, R628-32.

Moran, T. H., Robinson, P. H., Goldrich, M. S., and McHugh, P. R. (1986). Two brain cholecystokinin receptors: implications for behavioral actions. *Brain Res.* **362**, 175–9.

Roy-Byrne, P. P., Weingartner, H., Bierer, I. M., Thompson, K., and Post, R. M. (1986). Effortful and automatic cognitive processes in depression. *Arch. Gen. Psychiatry.* **43**, 265–9.

Schafmayer, A., Werner, M., and Becker, H. D. (1982). Radioimmunological determination of cholecystokinin in tissue extracts. *Digestion.* **24**, 146–54.

Schlegel, W., Raptis, S., Grube, D., and Pfeiffer, E. F. (1977). Estimation of cholecystokinin-pancreozymin (CCK) in human plasma and tissue by a specific radioimmunoassay and the immunohistochemical identification of pancreozymin-producing cells in the duodenum of humans. *Clin. Chim. Acta.* **80**, 305–6.

Shillabeer, G. and Davison, J. S. (1984). The cholecystokinin antagonist, proglumide, increases food intake in the rat. *Regul. Pept.* **8**, 171–6.

Shillabeer, G. and Davison, J. S. (1987). Proglumide, a cholecystokinin antagonist, increases gastric emptying in rats. *Am. J. Physiol.* **252**, R353–60.

Smith, G. P., Jerome, C., Cushin, B. J., Eterno, R. and Simansky, K. J. (1981). Abdominal vagotomy blocks the satiety effect of cholecystokinin in the rat. *Science.* **213**, 1036–7.

Turkelson, C. M., Dale, W. E., Reidelberger, R. and Solomon, T. E. (1986). Development of cholecystokinin radioimmunoassay using synthetic CCK-10 as immunogen. *Regul. Pept.* **15**, 205–17.

Valenzuela, J. E. and Defilippi, C. (1981). Inhibition of gastric emptying in humans by secretin, the octapeptide of cholecystokinin, and intraduodenal fat. *Gastroenterology*, **81**, 898–902.

Wolkowitz, O. M., Weingartner, H., Thompson, K., Pickar, D., Paul, S. M., and Hommer, D. W. (1987). Diazepam-induced amnesia: a neuropharmacological model of an 'organic amnestic syndrome'. *Am. J. Psychiatry.* **144**, 25–9.

Wolkowitz, O. M., Gertz, B., Weingartner, H., Beccaria, L., Thompson, K. and Liddle, R. A. (1990). Hunger in humans induced by MK-329, a specific peripheral-type cholecystokinin receptor antagonist. *Biol. Psychiatry.* **28**, 169–73.

27. CCK, gastric emptying, and satiety in man

P. H. Robinson

Introduction

In primates, gastric emptying is subject to tight physiological control. In human subjects, emptying of liquids, for example, is slowed by an increase in the caloric content, acidity, or tonicity of the meal (Hunt and Knox 1962; Hunt and Stubbs 1975; Brener *et al.* 1983), and by the presence of nutrient in the jejunum (Vidon *et al.* 1989). Liquid meals of fat, protein, and most carbohydrates empty from the stomach of the rhesus monkey at a constant linear rate of calories (McHugh and Moran 1979), while emptying of saline solutions, which proceeds exponentially, is retarded by the infusion of calories into the duodenum (McHugh *et al.* 1982). The presence of a regulated system of control over gastric emptying provides a possible mechanism for caloric control of food intake, as meals of higher caloric density are emptied from the stomach more slowly and therefore provide a more persistent inhibitory influence on feeding (Hunt 1980). If control of gastric emptying can be shown to be an important influence in determining the onset of satiety and of meal size, then the mechanism by which emptying is controlled comes under scrutiny as a possible mediator of satiety. Several candidates exist, and, because of its known satiety and gastric inhibitory actions, CCK may be a physiological control of gastric emptying and hence of satiety and feeding. In some morbid states, gastric emptying and satiety are both abnormal, and changes in CCK function might provide the pathophysiological basis for the observed behavioural and functional disturbances.

In this chapter we shall consider CCK in relation to the role of gastric slowing in the modulation of food intake by nutrients and anorectic drugs in normal primates, including man, and in disorders of eating and weight, namely anorexia nervosa, bulimia nervosa, and morbid obesity.

Role of gastric modulation in the control of food intake

Evidence from a number of sources suggests that gastric distension is an important stimulus which contributes significantly to the termination of a bout of eating. Rhesus monkeys, fitted with in-dwelling gastric cannulae, have been shown to modify their intake of liquid food depending on the volume of stomach contents (Wirth and McHugh 1983). When gastric contents were withdrawn through the tube, the animals consumed more liquid food until the gastric volume before the withdrawal had been re-established. In man, liquid meals of known volume have been infused into the stomachs of normal subjects using nasogastric tubes and the subjects have shown a reasonable ability to judge the volume that has been introduced (Coddington and Bruch 1970). Using a different approach, gastric contents have been followed using gastric scintiscanning while subjects completed visual analogue scales of their hunger, satiety, and other sensations and emotions (Robinson 1989). It was found that gastric contents, represented by the radioactive counts remaining within the gastric outline, were fairly well correlated with visual analogue scale measures of fullness, and inverse correlations between gastric counts and hunger, or urge to eat, were found. The average correlation between these eating-related scales and gastric contents, computed using Spearman's rank coefficient, was around 0.6. Therefore perception of gastric fullness correlates fairly well with gastric contents, providing further evidence that visceral sensory information from the upper gastro-intestinal tract reaches awareness.

A third line of evidence, which is more indirect, concerns the effect on food intake of agents that influence gastric emptying, including both foods and drugs. Infusion of nutrients into the duodenum of the rhesus monkey results in a slowing of gastric emptying proportional to the calories infused (McHugh et al. 1982). Infusion of nutrients into the stomach of the monkey leads to a reduction in subsequent food consumption equivalent to the number of calories that have been infused (McHugh and Moran 1978). This phenomenon has been explained by suggesting that each calorie slows gastric emptying by a certain amount, and that slowed gastric emptying leads to gastric distension during a meal which inhibits feeding to an extent proportional to the caloric load delivered to the intestine. Whether this mechanism extends to human feeding is uncertain. Preloads of different calorie content but similar hedonic qualities have indistinguishable effects on subsequent consumption (Rolls et al. 1988), suggesting that the sensory attributes of food may be more influential than its physiological consequences in controlling human eating behaviour. Fenfluramine is a drug used in the treatment of obesity because of its anorectic action. Until recently, its mode of action has been thought to reside exclusively within the CNS, where it enhances serotonin function (Rowland and Carlton 1986). The drug also inhibits gastric emptying, and work in the rat, the rhesus monkey, and man (Rowland and Carlton 1984; Horowitz et al. 1985; Robinson et al. 1986) has suggested that the gastric inhibitory action of fenfluramine is an important component in its feeding inhibitory action. Indeed, it has been suggested on the basis of rat experiments that fenfluramine anorexia might be mediated, in part, by an action involving CCK, as the CCK_A antagonist devazepide has been found to antagonize fenfluramine anorexia (Cooper et al. 1990; see Chapters 23 and 25 of this volume). In the majority of feeding and gastric emptying experiments published so far, fenfluramine has been used as the racemic mixture (DL-fenfluramine), and recent work (Robinson *et al.*, unpublished results) has suggested that , while D-fenfluramine has a potency several times that of the L-form when tested for anorectic action, the two isomers are approximately equivalent when tested for gastric inhibitory action. This evidence casts doubt on the gastric inhibition theory of fenfluramine action and awaits confirmation.

Role of endogenous CCK in the control of gastric emptying and feeding

Three general approaches to this question have been adopted: measurement of gastric emptying during CCK infusions, measurement of plasma CCK levels during digestion, and observation of the effects of antagonists of CCK on gastric function.

Gastric emptying during CCK infusions

Infusions of CCK in the physiological range result in significant slowing of gastric emptying (Liddle *et al.* 1986; Robinson *et al.* 1988*a*). The change in gastric function was observed scintigraphically by Robinson *et al.* (1988*a*). The outline of the stomach was observed to change substantially in the few minutes following the onset of CCK infusion, with contraction and apparent obliteration of the antral cavity. This change reversed rapidly when the CCK infusion was stopped. Hamilton *et al.* (1976) found that pentagastrin produced a similar contraction of the antral part of the stomach. The antropyloric area was further studied using CCK-33 radio-labelled with Bolton-Hunter reagent and ^{125}I in autoradiographic studies of specimens of human pylorus obtained at operation (Robinson *et al.* 1988*a*). Specific CCK binding was observed in the circular muscle of the pyloric sphincter, providing a structural substrate for the physiological change observed after CCK. Therefore the evidence is that intravenous (i.v.) CCK readily influences gastric emptying of liquid meals. Infusions of CCK have been given to human volunteers to observe the effect on feeding. The findings conflict, although it appears that CCK does reduce food intake when the test meal has been preceded by a preload (Kissileff *et al.* 1982). In other studies, increased food intake has been observed after CCK administration (Sturdevant and Goetz 1976). There is also a suggestion that CCK reduces the stimulating effect of food preparation (Stacher *et al.* 1979). It has been suggested that the inhibitory effects of CCK on feeding can be explained, at least in part, by its slowing of gastric emptying (Moran and McHugh 1982). CCK undoubtedly inhibits gastric emptying, as well as feeding. However, the two effects can be separated (Chapter 18 of this volume). In the rhesus monkey, a threshold dose of CCK (0.2–0.8 μg/kg i.p.) is found to inhibit gastric emptying but not feeding (Moran and McHugh 1982). However, if a preload of saline is infused into the stomach and the same threshold dose of CCK is given, feeding is inhibited. This is good evidence for an interaction between gastric distension and CCK satiety in the rhesus monkey, and is compatible with the proposed mechanism of action of CCK as inhibiting feeding, in part, by reducing the rate of gastric emptying. In man, gastric emptying and food intake have been measured contemporaneously during CCK infusions (Muurahainen *et al.* 1988). Gastric emptying did not correlate with intake during placebo infusions, but there was a significant correlation when CCK was infused, suggesting that CCK may increase the sensitivity of feeding control mechanisms to gastric signals.

Measurement of CCK in plasma during digestion

The detection of CCK in plasma has given rise to a number of problems, in part because of the large number of CCK peptides that have been isolated from human plasma (e.g. CCK-8, CCK-33, and CCK-12) (Izzo *et al.* 1984). However, sensitive bioassay and radio-immunoassay techniques now give reliable and reproducible measures of plasma CCK levels. They suggest that, after a mixed meal, plasma levels of CCK rise from around 15 pg/ml to a peak of around 20 pg/ml (Izzo *et al.* 1984) or from a basal level of 1 pM to a peak of 6.5 pM (Kleibeuker *et al.* 1988). Delayed gastric emptying has been produced with infusions of 12 pmol/kg/h, resulting in a plasma level of 3.4 pM, i.e. well within the physiological range. However, feeding has been inhibited using much higher doses, e.g. 240 pmol/kg/h (Kissileff *et al.* 1982), a dose that would lead to plasma levels far above those observed physiologically. Therefore this evidence suggests that while delayed gastric emptying could be a physiological function of CCK, the reduced feeding observed during CCK infusions would not be expected during normal digestion as a sole result of the CCK secreted by the small intestine.

CCK antagonists and control of gastric emptying and feeding

The most compelling evidence for a role of CCK in the modulation of gastric emptying in man would be the demonstration that it was accelerated by CCK antagonists. Work performed in man has been rather limited, and the results are conflicting. Devazepide, a potent and specific CCK_A antagonist, did not influence gastric emptying of a mixed meal in human subjects (Liddle *et al.* 1989; Chapter 26 of this volume). Loxiglumide, another CCK antagonist, had no effect on gastric emptying in two studies (Corazziari *et al.* 1990; Konturek *et al.* 1990) but did accelerate gastric emptying of radio-opaque markers in another study of gastric emptying (Meyer *et al.* 1989).

Eating and weight disorders, gastric emptying, and CCK

The eating disorders, anorexia nervosa and bulimia nervosa, and morbid obesity are areas of considerable interest for considering the possible role of CCK as an aetiological or exacerbating factor. All have been associated with abnormalities of gastric emptying in some circumstances, and CCK has also received some attention in these disorders.

Anorexia nervosa

In anorexia nervosa, patients, who are generally young women, suffer severe weight loss associated with loss of menstruation and disturbed attitudes to food and weight, namely a morbid fear of fatness and weight gain, and a distortion of body image such that they claim to feel fat even when emaciated (Russell 1983). Abnormalities of satiety have long been recognized (Coddington and Bruch 1970), and many patients feel full for a prolonged period even after small meals. A physiological basis for this symptom was discovered when delayed gastric emptying in anorexia nervosa was found in studies using both nasogastric intubation (Dubois *et al.* 1979) and radio-labelled meals (Robinson *et al.* 1988*b*). The delay in gastric emptying is substantial, with the rate reduced to half or less of control values. In one study (Robinson *et al.* 1988*b*) the delay was only observed in patients who were both underweight and acutely starving. Anorexic patients undergoing refeeding in a special refeeding unit and the same patients who had reached their target weight in the unit all had normal gastric emptying. This suggested that severe acute starvation was the essential causative factor in this group of patients, and an animal study, in which rats were allowed to eat for only 2 h per day, confirmed this impression (Robinson and Stephenson 1990). Gastric emptying, measured using gastric scintigraphy, was grossly delayed in experimental animals compared with freely feeding controls. The possible role of CCK was explored in this animal model. Gastric emptying was measured after intraperitoneal (i.p.) administration of the CCK_A antagonist devazepide. The antagonist did not reverse delayed gastric emptying induced by food restriction. This suggests that abnormal CCK secretion or sensitivity is not a major contributor to delayed gastric emptying in dietary restriction in rats and, by inference, in anorexia nervosa, assuming that the rat model reproduces the essential features of the gastric inhibition observed in anorexia nervosa.

Bulimia nervosa

Bulimia nervosa is a condition linked to anorexia nervosa, but having its own distinctive features. Patients share the excessive concern with weight and body shape, but in addition the condition is characterized by episodes of massive overeating and behaviours aimed at reducing the 'fattening' effects of such overeating including self-induced vomiting, laxative abuse, and exercise abuse (Russell 1979). Most patients with this condition, which is common, affecting about 1 per cent of young women (Fairburn and Beglin 1990), are of normal weight. A patient who suffers typical episodes of bulimia in addition to being grossly underweight and amenorrhoeic, is said to be suffering from both anorexia nervosa and bulimia nervosa. Gastric emptying has been measured in patients of normal weight with bulimia nervosa and found to be normal

(Robinson *et al.* 1988*b*). Patients with bulimia nervosa who were also under-weight and amenorrhoeic, and who were acutely starving, did have delayed gastric emptying. In that group, self-induced vomiting was causing a caloric deficit severe enough to lead to chronic weight loss. In another group, a proportion of patients with bulimia nervosa were found to have delayed gastric empty-ing, indicating that gastric function can be impaired in this disorder. Plasma CCK levels were measured in patients with bulimia nervosa and were found to be low (Geracioti and Liddle 1988). However, gastric emptying was not measured, and the low values might therefore reflect slow gastric emptying in some patients. Despite this, the findings are of great interest, particularly as low satiety ratings were also found. CCK has been administered to patients with bulimia nervosa (Mitchell *et* al. 1986) and found not to prevent episodes of overeating, suggesting that abnormally low CCK secretion, while it may be present and may contribute to overeating in this condition, is not sufficient for its persistence.

Morbid obesity

Gastric emptying was found to be abnormally rapid in a group of patients with massive obesity (Wright *et al.* 1983) and this rate did not reduce to normal in a small subgroup who lost weight to the normal range. This is of substantial interest, as accelerated gastric emptying might prove to be an important aetiological mechanism in obesity. However, the findings have not been con-firmed. CCK, given i.v., inhibited feeding in a group of obese men (Pi-Sunyer *et al.* 1982), just as it did in lean men (Kissileff *et al.* 1982), indicating that CCK, or an orally active analogue, might have a place in the future management of obesity.

Conclusions

There is good evidence of a role for gastric inhibition in the reduced feeding associated with increasing caloric concentration of nutrient, CCK administra-tion, fenfluramine treatment, and anorexia nervosa. CCK may well be an important physiological mediator of gastric emptying, but its effect on feeding can only be obtained using doses in the pharmacological range. Many investiga-tions using the various different CCK antagonists remain to be performed in animal studies, during normal digestion and feeding, and in patients with eating and weight disorders so that the role of CCK under physiological and patho-logical conditions can be established.

References

Brener, W., Hendrix, T. R., and McHugh, P. R. (1983). Regulation of the gastric emp-tying of glucose. *Gastroenterology* **85**, 76–82.

Coddington, R. D. and Bruch, H. (1970). Gastric perceptivity in normal, obese and schizophrenic subjects. *Psychosomatics* **11**, 571–9.

Cooper, S. J., Dourish, C. T., and Barber, D. J. (1990). Reversal of the anorectic effect of (+)-fenfluramine in the rat by the selective cholecystokinin receptor antagonist MK-329. *Br. J. Pharmacol.* **99**, 65–70.

Corazziari, E., Ricci, R., Biliotti, D., Bontempo, I., De-Medici, A., Pallotta, N., and Torsoli, A. (1990). Oral administration of loxiglumide (CCK antagonist) inhibits postprandial gallbladder contraction without affecting gastric emptying. *Dig. Dis. Sci.* **35**, 50–4.

Dubois, A., Gross, H. A., Ebert, M. H., and Castell, D. O. (1979). Altered gastric emptying and secretion in primary anorexia nervosa. *Gastroenterology* **77**, 319–23.

Fairburn, C. G. and Beglin, S. J. (1990). Studies of the epidemiology of bulimia nervosa. *Am. J. Psychiatry* **147**, 401–8.

Geracioti, T. D. and Liddle, R. A. (1988). Impaired cholecystokinin secretion in bulimia nervosa. *New Engl. J. Med.* **319**, 683–8.

Hamilton, S. G., Sheiner, H. J., and Quinlan, M. F. (1976). Continuous monitoring of the effect of pentagastrin on gastric emptying of solid food in man. *Gut* **17**, 273–9.

Horowitz, M., Collins, P. J., Tuckwell, V., Vernon-Roberts, J., and Sherman, D. J. D. (1985). Fenfluramine delays gastric emptying of solid food. *Br. J. Clin. Pharmacol.* **19**, 845–51.

Hunt, J. N. (1980). A possible relation between the regulation of gastric emptying and food intake. *Am. J. Physiol.* **239**, G1–4.

Hunt, J. N. and Knox, M. T. (1962). The regulation of gastric emptying of meals containing citric acid and salts of citric acid. *J. Physiol.* **163**, 34–45.

Hunt, J. N. and Stubbs, D. F. (1975). The volume and energy content of meals as determinants of gastric emptying. *J. Physiol.* **245**, 209–25.

Izzo, R. S., Brugge, W. R., and Praissman, M. (1984). Immunoreactive cholecystokinin in human and rat plasma: correlation of pancreatic secretion in response to CCK. *Regul. Pept.* **9**, 21–34.

Kissileff, H. R., Pi-Sunyer, F. X., Thornton, J., and Smith, G. P. (1982). C-terminal octapeptide of cholecystokinin decreases food intake in man. *Am. J. Clin. Nutr.* **34**, 154–60.

Kleibeuker, J. H., Beekhuis, H., Jansen, J. B., Piers, D. A., and Lamers, C. B. (1988). Cholecystokinin is a physiological hormonal mediator of fat-induced inhibition of gastric emptying in man. *Eur. J. Clin. Invest.* **18**, 173–7.

Konturek, S. J., Kwiecien, N., Obtulowicz, W., Kopp, B., Oleksy, J., and Rovati, L. (1990). Cholecystokinin in the inhibition of gastric secretion and gastric emptying in humans. *Digestion* **45**, 1–8.

Liddle, R. A., Morita, E. T., Conrad, C. K., and Williams, J. A. (1986). Regulation of gastric emptying in humans by cholecystokinin. *J. Clin. Invest.* **77**, 992–6.

Liddle, R. A., Gertz, B. J., Kanayama, S., Beccaria, L., Coker, L. D., Turnbull, T. A., and Morita, E. T. (1989). Effects of a novel cholecystokinin (CCK) receptor antagonist, MK-329, on gallbladder contraction and gastric emptying in humans. Implications for the physiology of CCK. *J. Clin. Invest.* **84**, 1220–5.

McHugh, P. R. and Moran, T. H. (1978). Accuracy of the regulation of caloric ingestion in the rhesus monkey. *Am. J. Physiol.* **235**, R29–34.

McHugh P. R. and Moran, T. H. (1979). Calories and gastric emptying, a regulatory capacity with implications for feeding. *Am. J. Physiol.* **236**, R254–60.

McHugh, P. R., Moran, T. H., and Wirth, J. B. (1982). Postpyloric regulation of gastric emptying in rhesus monkeys. *Am. J. Physiol.* **243**, R408–57.

Meyer, B. M., Werth, B. A., Beglinger, C., Hildebrand, P., Jansen, J. B., Zach, D., Rovati, L. C., and Stalder, G. A. (1989). Role of cholecystokinin in regulation of gastrointestinal motor functions. *Lancet* ii, 12–15.

Mitchell, J. E., Laine, D. E., Morley, J. E., and Levine, A. S. (1986). Naloxone but not CCK-8 may attenuate binge-eating behavior in patients with the bulimia syndrome. *Biol. Psychiatry* 21, 1399–1406.

Moran, T. H. and McHugh, P. R. (1982). Cholecystokinin suppresses food intake by inhibiting gastric emptying. *Am. J. Physiol.* 242, R491–7.

Muurahainen, N., Kissileff, H. R., Derogatis, A. J., and Pi-Sunyer, F. X. (1988). Effects of cholecystokinin-octapeptide (CCK-8) on food intake and gastric emptying in man. *Physiol. Behav.* 44, 645–9.

Pi-Sunyer, X., Kissileff, H. R., Thornton, J., and Smith, G. P. (1982). C-terminal octapeptide of cholecystokinin decreases food intake in obese men. *Physiol. Behav.* 29, 627–30.

Robinson, P. H. (1989). Perceptivity and paraceptivity during measurement of gastric emptying in anorexia and bulimia nervosa. *Br. J. Psychiatry* 154, 400–5.

Robinson, P. H., Stephenson, J. D. (1990). Dietary restriction delays gastric emptying in rats. *Appetite*, 14, 193–201.

Robinson, P. H., Moran, T. H., and McHugh, P. R. (1986). Inhibition of gastric emptying and feeding by fenfluramine. *Am. J. Physiol.* 250, R764–9.

Robinson, P. H., McHugh, P. R., Moran, T. H., and Stephenson, J. D. (1988a). Gastric control of food intake. *J. Psychosom. Res.* 32, 593–606.

Robinson, P. H., Barrett, J., and Clarke, M. (1988b). Determinants of delayed gastric emptying in anorexia nervosa and bulimia nervosa. *Gut* 29, 458–64.

Rolls, B. J., Hetherington, M., and Burley, V. J. (1988). Sensory stimulation and energy density in the development of satiety. *Physiol. Behav.* 44, 727–33.

Rowland, N. and Carlton, J. (1984). Inhibition of gastric emptying by peripheral and central fenfluramine in rats: correlation with anorexia. *Life Sci.* 34, 2495–9.

Rowland, N. E. and Carlton, J. (1986). Neurobiology of an anorectic drug: fenfluramine. *Prog. Neurobiol.* 27, 13–62.

Russell, G. F. M. (1979). Bulimia nervosa: an ominous variant of anorexia nervosa. *Psychol. Med.* 9, 429–48.

Russell, G. F. M. (1983). Anorexia nervosa and bulimia nervosa. In *Handbook of psychiatry*, Vol. 4, *Neuroses and personality disorders* (ed. G. F. M. Russell and L. A. Herson), pp. 285–98. Cambridge University Press, Cambridge.

Sturdevant, R. A. L. and Goetz, H. (1976). Cholecystokinin both stimulates and inhibits human food intake. *Nature* 261, 713–15.

Stacher, G., Bauer, H., and Steinringer, H. (1979). Cholecystokinin decreases appetite and activation evoked by stimuli arising from the preparation of a meal in man. *Physiol. Behav.* 23, 325–31.

Vidon, N., Pfeiffer, A., Chayvialle, J. A., Merite, F., Maurel, M., Franchisseur, C., Huchet, B., and Bernier, J. J. (1989). Effect of jejunal infusion of nutrients on gastrointestinal transit and hormonal response in man. *Gastroenterol. Clin. Biol.* 13, 1042–9.

Wirth, J. B. and McHugh, P. R. (1983). Gastric distension and short-term satiety in the rhesus monkey. *Am. J. Physiol.* 245, R174–80.

Wright, R. A., Krinsky, S., Fleeman, C., Trujillo, J., and Teague, E. (1983). Gastric emptying and obesity. *Gastroenterology* 84, 747–51.

28. Satiation in humans induced by the dipeptide L-aspartyl-L-phenylalanine methyl ester (aspartame): possible involvement of CCK

Peter J. Rogers and John E. Blundell

Post-ingestive suppression of food intake by aspartame

In a series of studies we have used intense sweeteners, such as saccharin, aspartame, and acesulfame-K, as experimental tools to separate the influences of sweetness and caloric dilution on appetite control (e.g. Rogers and Blundell, 1989*b*). However, in addition to their sweet taste, there is the possibility that at least some of these substances may have biologically significant post-ingestive effects (Rogers and Blundell 1989*a*). This was confirmed by studies which showed that oral administration of capsulated aspartame suppresses food intake in humans (Rogers *et al.* 1990). Since the aspartame was ingested without being tasted, the effect must be due to post-ingestive action. Moreover, this anorexic effect of aspartame is both robust and potent. We have replicated the result in six separate experiments to date in which, for example, 200 mg of aspartame produced an acute 10–15 per cent reduction in food intake (Rogers *et al.* 1990 and unpublished data).

Characterization of the post-ingestive anorexic action of aspartame

Table 28.1 shows the results of a study comparing the effects of capsulated aspartame and its constituent amino acids, phenylalanine and aspartic acid, on short-term food intake in 16 male and female undergraduates. Each subject received these three treatments and placebo in a counterbalanced order. On each occasion, the subjects were served a buffet lunch 1 h after ingesting the appropriate capsule and invited to 'eat as much or as little as you like'. Aspartame significantly reduced intake, but there were no reliable differences among the other treatments.

The subjects also completed visual analogue ratings of their hunger, desire to eat, and fullness both before and for 3 h after lunch. Despite the clear reduction in food intake following ingestion of aspartame, there were no significant treatment effects on these subjective ratings of motivation to eat. That is, aspartame did not reduce motivation to eat in anticipation of lunch, and at lunch the same satiating effect was produced by a smaller amount of food—the decrease in motivation to eat during lunch was the same after aspartame as for the other treatments. Furthermore, there were no differences in the rate of recovery of motivation to eat during the post-lunch period. In one previous study (Rogers *et al.* 1990) we detected a weak effect of aspartame on subjective motivation to eat, but this was a much smaller reduction than that obtained with certain other anorexic agents, such as amphetamine, which produced a comparable effect on food intake (Rogers and Blundell 1979). Similarly, a caloric load had a strong suppressive effect on both subjective motivation to eat and food intake (Rogers *et al.* 1988). Taken together, these results indicate that the anorexic action of capsulated aspartame is due primarily to an amplification of the satiating effect of food and not to a reduction in hunger, i.e. an effect on processes involved in the maintenance and termination of eating rather than on processes involved in its initiation (Blundell 1979).

In turn, the above conclusion can help to explain certain paradoxical effects of aspartame-sweetened foods and beverages. For example, we found that consumption of water sweetened with aspartame significantly increased pre-lunch ratings of hunger and desire to eat, but resulted in a small *reduction* ($p < 0.1$) in food intake (Rogers *et al.* 1988). Two effects of aspartame may be operating here: a stimulation of appetite by sweetness (see Rogers and Blundell (1989*a*) for review) leading to an increase in hunger in anticipation of eating, and a post-ingestive inhibitory action amplifying the satiating effect of food consumed during lunch. The results suggest that this second action of aspartame antagonized the stimulatory effect of sweetness during but not before the meal (see Rogers *et al.* (1990) for further discussion).

Mechanisms for the post-ingestive anorexic action of aspartame: possible involvement of CCK

Aspartame appears to be digested in the small intestine to yield methanol, aspartic acid, and phenylalanine, and these products are then absorbed and metabolized as normal dietary constituents (Ranney *et al.* 1976). Intragastrically administered phenylalanine reduces food intake in rats and monkeys by an amount considerably greater than that predicted by simple caloric compensation (Gibbs *et al.* 1976; Anika *et al.* 1977). Although a similar effect has been discounted in humans (Ryan-Harshman *et al.* 1987), re-analysis of those results,* together with at least one other study (Muurahainen *et al.* 1988),

---

* Ryan-Harshman *et al.* (1987) report that neither phenylalanine nor aspartame reduces food intake in humans. However, alanine was used as the 'placebo' treatment, and examination of the data suggests that this was an inappropriate control. Results for four doses of phenylalanine (0.84–10.08 g) are reported. These show a dose-dependent decrease in food intake from 1543 to 1070 kcal. Food intakes after the highest doses (10.08 g) of alanine and aspartame were 1230 kcal

Table 28.1.   The effect of oral administration of capsulated aspartame and its constituent amino acids on food intake in humans

Treatment and dose	Mean food intake ± SE (kcal)
Aspartame (400 mg)	1430 ± 133[a]
L-aspartic acid (200 mg)	1662 ± 133
L-phenylalanine (200 mg)	1730 ± 120
Placebo (400 mg cornflour)	1683 ± 121

[a] ANOVA revealed a significant treatment effect: $F(3, 45) = 6.23$, $p < 0.005$. Paired comparisons showed that intake after aspartame was significantly different from all other treatments: smallest $t(15) = 2.93$, $p < 0.05$.

appears to confirm a suppression of food intake following oral administration of capsulated phenylalanine. In turn, there is strong evidence that phenylalanine is a potent releaser of CCK: gastric and duodenal perfusion with phenylalanine produces marked pancreatic enzyme secretion (Go *et al.* 1970; Meyer and Grossman 1972), and reduces the gastric emptying rate (as well as markedly increasing serum gastrin concentration) (Byrne *et al.* 1977). Together with the evidence demonstrating a role for CCK in satiety (see Chapter 17 of this volume), this suggests that an effect on CCK release mediates the suppression of food intake following administration of phenylalanine.

None the less, the results presented in Table 28.1 show that an individual action of phenylalanine cannot account for the potent anorexic effect of aspartame. Thus 200 mg of phenylalanine failed to reduce food intake, whereas doses of aspartame yielding this and even smaller amounts of phenylalanine produce significant anorexia. Aspartic acid also had no effect on intake (Table 28.1). Furthermore, the involvement of certain other breakdown products of aspartame can probably be discounted. Quantitative analyses suggest that the amount of methanol liberated by such doses of aspartame will not have significant physiological consequences, and similarly the presence of minor breakdown products (e.g. aspartyl diketopiperazine) are unlikely to be important (Pardridge 1986). Therefore it may be that the decrease in intake is due to aspartame itself, or aspartylphenylalanine.

Although phenylalanine reduces food intake only when given at comparatively high doses, there is still strong circumstantial evidence suggesting that

---

and 1124 kcal respectively. Taken together with our findings (Table 28.1), this suggests that aspartame, phenylalanine, and perhaps alanine all suppress food intake; however, aspartame is by far the most potent treatment, being effective at doses as low as 100 mg (Rogers, Lambert, and Blundell, unpublished results). The similar suppression of intake by the highest doses of aspartame and phenylalanine in the studies by Ryan-Harshman *et al.* (1987) may indicate the maximal or asymptotic effect of the treatments. Therefore there would appear to be strong evidence that these substances are anorexic, with the dose–response curve for aspartame to the left of that for phenylalanine.

aspartame-induced anorexia is mediated by CCK. First, as indicated above, aspartame appears to amplify the satiating effect of food, an action entirely consistent with the proposed role of CCK in satiety (see Chapter 17 of this volume). Second, the amino acid sequence of aspartame (Asp–Phe) is the same as that of the C-terminal dipeptide of CCK. Perhaps aspartame (or aspartylphenylalanine), promotes CCK release from the small intestine, thus having a similar though more potent effect than phenylalanine. More remotely, a post-absorptive site of action is possible. There is abundant evidence that small peptides can enter the systemic circulation intact (Gardner 1988; Meisel and Schlimme 1990), and therefore it is conceivable that aspartame acts directly at CCK receptors. Against this is the failure to detect aspartame in plasma after oral administration (Ranney *et al.* 1976), and evidence showing the lack of significant agonist activity at receptor sites of C-terminal CCK fragments less than four amino acids in length (e.g. Rehfeld *et al.* 1980; Cherner *et al.* 1988).

We are currently investigating further the post-ingestive suppression of food intake by aspartame. In this study, aspartame and amino acid treatments will be compared for their effects on appetite and blood CCK levels.

Lack of a substantial anorexic effect of aspartame on food intake in rats

The development of an animal model of aspartame-induced anorexia would greatly facilitate the investigation of this phenomenon. Unfortunately, we have been unable to demonstrate any marked effects on food intake in rats of aspartame administered in the drinking water, by gavage, or by intraperitoneal (i.p.) injection (Table 28.2). The moderate reduction in intake following the largest dose of aspartame given by gavage could be due to phenylalanine (Anika *et al.* 1977), (see above) rather than aspartame or a related dipeptide. On a body weight basis this is 200 times larger than the dose producing an effect of similar magnitude in humans. If these preliminary results are confirmed, it will be important to find the mechanism underlying this species difference. Since most of the studies on the behaviourial and physiological effects of aspartame have been carried out on animals, and in particular on rats, this may explain why the post-ingestive anorexic effect of aspartame had not been reported previously. Finally, it is worth noting that rats and humans also differ in their sensory responses to aspartame; rats appear not to taste aspartame as sweet (Nowlis *et al.* 1980; Sclafani and Abrams 1986).

Therapeutic implications

The above findings add to the growing body of evidence showing that orally ingested peptides can have specific and potent physiological activities (Meisel and Schlimme 1990). Some of these effects may have therapeutic value. Thus the post-ingestive anorexic action of aspartame suggests that it might be

Table 28.2. The effect of aspartame administered by various routes on food intake in rats[a]

Route of administration	Aspartame dose (mg/kg)	Mean intake of wet mash ± SD (g)[b]
I.p. injection	40	14.2 ± 2.5
	20	14.9 ± 2.9
	10	13.4 ± 3.4
	Vehicle	14.5 + 2.3
Oral (drinking water)	180	13.3 ± 2.8
	90	12.6 + 3.3
	36	13.5 ± 4.0
	Vehicle	13.7 ± 1.2
Gavage	312	12.5 + 1.4[c]
	156	13.2 ± 1.3
	78	13.0 ± 3.1
	Vehicle	13.7 ± 1.4

[a] The rats were male Lister hooded rats (mean weight 320 g), eight in each treatment group. The treatments were administered after 18 h food deprivation, 30 min before food was returned and 3 h after the start of the dark phase of the lighting cycle. The 'oral' rats were also water deprived, and they were given access to aspartame solutions (0.2, 0.5, or 1 per cent) for 30 min before both food and water were returned after a further 30 min.
[b] Food intake during first 20 min of access.
[c] $p < 0.05$ (two-tail), versus vehicle paired t test.

exploited as an anti-obesity agent. The possibility that aspartame acts locally on part of the appetite control system to intensify the satiating effect of food makes it an excellent candidate for such a use. Furthermore, aspartame has already undergone extensive safety testing, and as a food additive has been allocated an advisable daily intake of 0–40 mg/kg body weight (FAO–WHO 1981). It remains to be seen, however, whether or not aspartame administered in capsules prior to eating on a regular basis can produce a sustained decrease in food intake and a clinically significant weight loss.

References

Anika, S. M., Houpt, T. R., and Houpt, K. A. (1977). Satiety elicited by cholecystokinin in intact and vagotomised rats. *Physiol. Behav.* **19**, 761–6.

Blundell, J. E. (1979). Hunger, appetite and satiety — constructs in search of identities. In *Nutrition and lifestyles* (ed. M. Turner), pp. 21–42. Applied Science Publishers, Barking.

Byrne, W. J., Walsh, J. H., and Ament, M. E. (1977). The effect of individual amino acids on gastric emptying time in man. *Gastroenterology* **72**, 1035.

Cherner, J. A., Sutcliff, V. E., Grybowski, D. M., Jensen, R. T., and Gardner, J. D. (1988). Functionally distinct receptors for cholecystokinin and gastrin on dispersed chief cells from guinea pig stomach. *Am. J. Physiol.* **254**, G151–5.

FAO–WHO (1981). *Joint FAO–WHO Expert Committee on Food Additives, 25th Report, Technical Report Series No. 669.* WHO, Geneva.

Gardner, M. L. G. (1988). Intestinal absorption of peptides. In *Nutritional modulation of neural function* (ed. J. E. Morley, M. B. Sternman, and J. H. Walsh), pp. 29–38. Academic Press, San Diego, CA.

Gibbs, J., Falasco, J. D., and McHugh, P. R. (1976). Cholecystokinin-decreased food intake in monkeys. *Am. J. Physiol.* **230**, 15–18.

Go, V. L. W., Hofmann, A. F., and Summerskill, W. H. J. (1970). Pancreozymin bioassay in man based on pancreatic enzyme secretion: potency of specific amino acids and other digestive products. *J. Clin. Invest.* **49**, 1558–64.

Meisel, H. and Schlimme, E. (1990). Milk proteins: precursors of bioactive peptides. *Trends Food Sci. Technol.* **1**, 41–3.

Meyer, J. H. and Grossman, M. I. (1972). Comparison of D- and L-phenylalanine as pancreatic stimulants. *Am. J. Physiol.* **222**, 1058–63.

Muurahainen, N. E., Kissileff, H. R., and Pi-Sunyer, F. X. (1988). L-phenylalanine and L-tryptophan reduce food intake when given with a food preload. *Am. J. Clin. Nutr.* **47**, 774.

Nowlis, G. H., Frank, M. E., and Pfaffmann, C. (1980). Specificity of acquired aversions to taste qualities in hamsters and rats. *J. Comp. Physiol. Psychol.* **94**, 932–42.

Pardridge, W. M. (1986). Potential effects of the dipeptide sweetener aspartame on the brain. In *Nutrition and the brain* (ed. R. J. Wurtman and J. J. Wurtman), Vol. 7, pp. 199–241. Raven Press, New York.

Ranney, R. E., Oppermann, J. A., Muldoon, E., and McMahon. (1976). Comparative metabolism of aspartame in experimental animals and man, *J. Toxicol. Environ. Health* **2**, 441–51.

Rehfeld, J. H., Larsson, L-l., Goltermann, N. R., Schwartz, T. W., Holst, J. J., Jensen, S. L., and Morley, J. S. (1980). Neural regulation of pancreatic hormone secretion by the C-terminal tetrapeptide of CCK. *Nature* **270**, 33–8.

Rogers, P. J., and Blundell, J. E. (1979). Effect of anorexic drugs on food intake and the micro-structure of eating in human subjects. *Psychopharmacology (Berlin)* **66**, 159–65.

Rogers, P. J. and Blundell, J. E. (1989a). Evaluation of the influence of intense sweeteners on the short-term control of appetite and caloric intake—a biopsychological approach. In *Progress in sweeteners* (ed. T. H. Grenby), pp. 267–89. Elsevier Applied Science, London.

Rogers, P. J. and Blundell, J. E. (1989b). Separating the actions of sweetness and calories: effects of saccharin and carbohydrates on hunger and food intake in human subjects. *Physiol. Behav.* **45**, 1093–9.

Rogers, P. J., Carlyle, J-A., Hill, A. J., and Blundell, J. E. (1988). Uncoupling sweet taste and calories: Comparison of the effects of glucose and three intense sweeteners on hunger and food intake. *Physiol. Behav.* **43**, 547–52.

Rogers, P. J., Pleming, H. C., and Blundell, J. E. (1990). Aspartame ingested without tasting inhibits hunger and food intake. *Physiol. Behav.* **47**, 1239–43.

Ryan-Harshman, M., Leiter, L. A., and Anderson, G. H. (1987). Phenylalanine and aspartame fail to alter feeding behavior, mood and arousal in men. *Physiol. Behav.* **39**, 247–53.

Sclafani, A. and Abrams, M. (1986). Rats show only a weak preference for the artificial sweetener aspartame. *Physiol. Behav.* **37**, 253–6.

29. CCK-induced activation of oxytocin neurons is blocked by a selective CCK$_A$ receptor antagonist

Gareth Leng, Mitsuko Hamamura, Joanne Bacon, and
Piers C. Emson

In anaesthetized rats, following intravenous (i.v.) administration of CCK, octapeptide, the magnocellular oxytocin neurons of the supra-optic nucleus show a highly reproducible transient excitation lasting 10–15 min and with a mean peak response of about 2 spikes/s above basal firing rates (usually 1–3 spikes/s). This neuronal excitation (Higuchi *et al.* 1991; Renaud *et al.* 1987) results in a significant elevation of plasma oxytocin concentrations, which appears as a peak of approximately 50 pg/ml above basal levels at about 5 min after CCK injection (Verbalis *et al.* 1986*a*, *b*; Blackburn and Leng 1990). The level of oxytocin release is similar in conscious and urethane-anaesthetized rats, suggesting that urethane anaesthesia does not impair the responsiveness of oxytocin neurons to this stimulus. In contrast, the vasopressin cells of the supra-optic nucleus are either transiently inhibited or unaffected following CCK injections, and plasma vasopressin concentrations show little or no change. Hence the pathway activated by CCK injections appears relatively specific.

The pathway mediating CCK-induced oxytocin release

Although the supra-optic nucleus contains CCK-binding sites, which respond by up-regulation following saline drinking (Day *et al.* 1989, Blackburn *et al.* 1990) and which appear to be functional (Jarvis *et al.* 1988), peripherally administered CCK probably does not act at this site. Selective gastric vagotomy eliminates the oxytocin response to CCK (Verbalis *et al.* 1986*b*), and lesions in the region of the area postrema attenuate the response (Carter and Lightman, 1987). It appears likely that peripherally administered CCK acts via the ascending gastric vagus, and probably influences the magnocellular oxytocin system via a direct projection from the region of the nucleus tractus solitarius, which is adjacent to the area postrema and which, unlike the area postrema, is the source of a direct afferent input to the magnocellular oxytocin neurons (Day and Sibbald 1988*a*, *b*; Raby and Renaud 1989). The neurons in the nucleus tractus solitarius that project to the supra-optic nucleus include at least some that belong to the A2 noradrenergic cell group. We have recently shown that CCK injections are followed by increases in extracellular noradrenalin concentrations in both the paraventricular nucleus and the dorsal areas of the supra-optic nucleus, as measured by intracranial microdialysis in anaesthetized rats (Kendrick *et al*. 1991). This release of noradrenaline in the dorsal supra-optic nucleus is accompanied by a lesser release of serotonin, but no release of dopamine.

A CCK$_A$ receptor antagonist blocks CCK-induced oxytocin release

The oxytocin response to i.v. injection of CCK is abolished in the presence of the specific CCK$_A$ receptor antagonist devazepide (1–1000 μg/kg i.v.) but unaffected in the presence of the CCK$_B$ receptor antagonist L-365,260 at similar concentrations (Figs 29.1 and 29.2). Neither antagonist produced a significant change in the basal firing rate of oxytocin neurons or in the basal plasma concentrations of oxytocin in these experiments; hence it appears that the pathway stimulated by CCK is probably not tonically active in urethane-anaesthetized rats. The physiological role of this pathway is not known. Since one of the postulated functions of oxytocin in the rat is to promote sodium excretion, one possibility is that the pathway initiates a reflex natriuresis

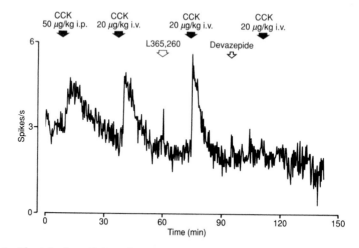

Fig. 29.1. Electrical activity of a putative oxytocin neuron recorded from the supra-optic nucleus of a urethane-anaesthetized rat. The neuron was excited following i.p. and i.v. injections of CCK at the doses and times indicated. Administration of L-365,260 (1 mg/kg i.v.) had no effect upon CCK-induced excitation, but administration of devazepide at the same dose eliminated the neuronal response to CCK. For details of antagonist, vehicle and pretreatments see Fig. 29.2.

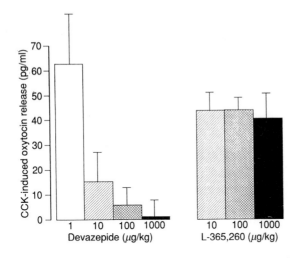

Fig. 29.2. Incremental oxytocin release following i.v. injections of CCK (20 μg/kg) in anaesthetized rats, measured following administration of the specific CCK antagonists L-365,260 and devazepide (drugs supplied courtesy of Merck Sharp & Dohme) at various concentrations. The antagonists were dissolved in propane-1,1-diol and administered i.v. 15 min prior to injection of CCK. Oxytocin was measured in jugular blood samples (0.3 ml) by specific radio-immunoassay (Higuchi *et al.* 1985), and CCK-induced release was calculated as the difference between oxytocin concentration measured in a sample taken 5 min after CCK injection and the basal concentration immediately prior to injection. Bars are means ± SE (n ≥ 9) for each group of rats. CCK-induced oxytocin release was unaffected by the CCK$_B$ antagonist but blocked in a dose-related manner by the CCK$_A$ antagonist at similar concentrations.

following a meal. An alternative interpretation comes from the evidence that, in the rat, oxytocin is a 'stress' hormone, strongly activated by diverse acute stressors but serving no purpose that has yet been clearly identified.

References

Blackburn, R. E. and Leng, G. (1990). Ablation of the region anterior and ventral to the third ventricle (AV3V region) in the rat does not abolish the release of oxytocin in response to systemic cholecytoskinin. *Brain Res.* **508**, 156–60.

Blackburn, R. E., Day, N. C., Leng, G., and Hughes, J. (1990). The effect of antero-ventral third ventricular lesions on the changes in cholecystokinin receptor density in the rat supra-optic nucleus following saline drinking. *J. Neuroendocrinol.* **2**, 323–8.

Carter, D. A. and Lightman S. L. (1987). A role for the area postrema in mediating cholecytoskinin-stimulated oxytocin secretion. *Brain Res.* **435**, 327–30.

Day, N. C., Hall, M. D., and Hughes, J. (1989). Modulation of hypothalamic cholecystokinin receptor density with changes in magnocellular activity: a quantitative autoradiographic study. *Neuroscience* **29**, 371–83.

Day, T. and Sibbald, J. R. (1988a). Direct catecholaminergic projection from nucleus tractus solitarii to supra-optic nucleus. *Brain Res.* **454**, 387–92.

Day, T. and Sibbald, J. R. (1988b). Solitary nucleus excitation of supra-optic vasopressin cells via adrenergic afferents. *Am J. Physiol.* **254**, R711–16.

Higuchi, T., Bicknell, R. J., and Leng, G. (1991). Reduced oxytocin release from the neural lobe of lactating rats is associated with reduced pituitary content and does not reflect reduced excitability of oxytocin neurons. *J. Neuroendocrinol.* **3**, 297–302.

Higuchi, T., Hondo, K., Fukuoka, T., Negoro, H., and Wakabayashi, K. (1985). Release of oxytocin during suckling and parturition in the rat. *J. Endocrinol.* **105**, 339–46.

Jarvis, C. R., Bourque, C. W., and Renaud, L. P. (1988). Cholecystokinin (CCK) depolarises rat supra-optic nucleus (SON) neurosecretory neurons. *Soc. Neurosci. Abstr.* **14**, 145.

Kendrick, K., Leng, G. and Higuchi, T. (1991). Noradrenaline, dopamine and serotonin release in the paraventricular and supra-optic nuclei of the rat in response to cholecystokinin injections *J. Neuroendocrinol.* **3**, 139–44.

Raby, W. N. and Renaud, L. P. (1989). Dorsomedial medulla stimulation activates rat supra-optic oxytocin and vasopressin neurons through different pathways. *J. Physiol.* **417**, 279–94.

Renaud, L. P., Tang, M., McCann, M. J., Stricker, E. M., and Verbalis, J. G. (1987). Cholecystokinin and gastric distension activate oxytocinergic cells in rat hypothalamus. *Am. J. Physiol.* **253**, 661–5.

Verbalis, J. G., McCann, M. J., McHale, C. M., and Stricker, E. M. (1986a). Oxytocin and vasopressin secretion in response to stimuli producing learned taste aversions in rats. *Behav. Neurosci.* **100**, 466–75.

Verbalis, J. G., McCann, M. J., McHale, C. M., and Stricker, E. M. (1986b). Oxytocin secretion in response to cholecystokinin and food: differentiation of nausea from satiety. *Science* **232**, 1417–19.

30. c-*fos* mRNA in oxytocin neurons: selective induction by CCK

Mitsuko Hamamura, Gareth Leng, Piers C. Emson, and Hiroshi Kiyama

The magnocellular neurosecretory system of the rat hypothalamus has long been considered to be an excellent model for the study of mammalian peptidergic neurons. The magnocellular neurons are confined to relatively homogenous nuclei within the hypothalamus — the supra-optic nucleus (SON) and the paraventricular nucleus (PVN) — and the neuronal cell bodies are separated anatomically from the nerve terminals that are present at the site of neurosecretion in the neural lobe of the pituitary (Sofroniew 1983). The outputs of these neurons are the hormones oxytocin and vasopressin, the concentrations of which are readily measurable in the systemic circulation; hence the state of activation of these neuronal pools can be readily assessed *in vivo* as well as *in vitro*. As with all neurons, the secretory activity of magnocellular neurons is governed from minute to minute largely by synaptic events at the cell bodies. However, it is now clear that trans-synaptic activation of neurons also elicits longer-term responses that are correlated with, and may be dependent upon, the induction of new patterns of gene expression. Peptide gene expression in neurons can be modulated by many factors, including membrane electrical activity, neurotrophic growth factors, steroids, and neurotransmitters (e.g. Lightman and Young 1987). One of the most important regulators of peptide gene expression appears to be the proto-oncogene c-*fos* (Morgan and Curran 1989). c-*fos* messenger RNA (mRNA) and its proteins are expressed in neurons *in vitro* following intense depolarization induced by high potassium concentrations (Morgan and Curran 1986) and also following applications of nerve growth factor (Curran and Morgan 1985; Greenberg *et al.* 1985), nicotine (Greenberg *et al.* 1986), barium (Curran and Morgan 1986), and BayK8644 (Morgan and Curran 1986). The c-*fos* gene is activated rapidly and transiently, and such so-called immediate early genes have been proposed to encode regulatory proteins that control the expression of late onset genes, including peptide genes (Sonnenberg *et al.* 1989). In particular, c-*fos* protein forms a heterodimeric transcription factor complex with c-*jun* protein, which is the product of a second immediate early gene, and this heterodimer binds with high affinity and specificity to the AP-1 binding site for transcription factor activity on DNA (Rauscher *et al.* 1988).

The induction of c-*fos* mRNA may involve both calcium-dependent and calcium-independent mechanisms (Sheng and Greenberg 1990). Either calcium entry into neurons via voltage-dependent calcium channels in the cell membrane or increases in the intracellular concentration of cyclic adenosine monophosphate (cAMP) following activation of neuronal membrane receptors will activate a calcium or cAMP response element of the c-*fos* promoter (Sheng *et al.* 1990). In addition, c-*fos* mRNA expression can be induced by protein kinase C through activation of a serum response element of the c-*fos* promoter (Sheng and Greenberg 1990).

We have studied the activation of c-*fos* mRNA expression in the SON and PVN after activation of the oxytocin neurosecretory system by systemic administration of CCK as shown in Fig. 30.1. As described elsewhere in this volume (Chapter 17), CCK induces activation of peripheral CCK receptors on the afferent gastric vagal nerve endings and hence leads to activation of a neuronal pathway involving the nucleus tractus solitarius, terminating in the

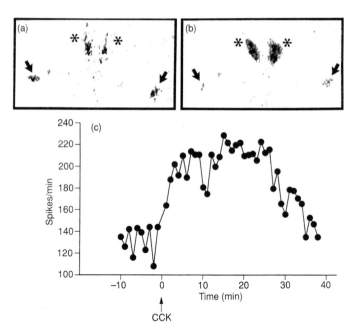

Fig. 30.1. Induction of c-*fos* mRNA expression following systemic CCK injection (CCK-8S, 50 µg/kg i.p.). (a), (b) Localization of the c-*fos* mRNA signal to the SON (arrows) and PVN (asterisks) ((b) is 80 µm caudal to (a)) detected by *in situ* hybridization histochemistry. A ^{35}S-labelled anti-sense oligonucleotide probe specific for the 5′ sequence of rat c-*fos* mRNA was used. Images shown by optical density in hypothalamic sections are derived from rats injected 30 min previously. (c) An increase in the mean firing rates after injection of the same dose of CCK in identified oxytocin neurons in urethane-anaesthetized rats as described in the text.

SON and PVN. Conscious rats were given a single injection of CCK (50 μg/kg i.p.) and were killed 10 or 30 min later. No c-*fos* mRNA signal was detectable in the hypothalamus of uninjected control rats, but in CCK-injected rats strong expression of c-*fos* mRNA was detectable in both the SON and the PVN. We also studied the effects of CCK administration upon the electrical activity of supra-optic neurons: vasopressin neurons were unaffected, but oxytocin neurons showed consistent activation. The activation was surprisingly modest: over the 10 min following CCK injection the mean firing rate of oxytocin neurons was elevated by only 1 spike/s on average. Thus c-*fos* expression in the SON after CCK injection is probably confined to the oxytocin neurons, and apparently accompanies a relatively small change in neuronal excitability.

There is already considerable evidence that c-*fos* mRNA can be expressed in many neurons *in vivo* following intense sustained activation. In the spinal cord, noxious and electrical stimuli induce c-*fos* gene expression in some dorsal horn neurons, and in hippocampal pyramidal neurons c-*fos* mRNA is expressed after administration of convulsants (Hunt *et al.* 1987; Wisden *et al.* 1990). It is known that injection of hypertonic saline produces an intense and sustained activation of both oxytocin and vasopressin neurons in the hypothalamic neurosecretory system (Brimble *et al.* 1978) and also results in c-*fos* mRNA expression in the SON and PVN (Hamamura *et al.* 1991*b*). However, by comparison, systemic CCK injection is a modest, transient, and selective stimulus, and the c-*fos* mRNA activation following CCK points clearly to a role for c-*fos* proteins in physiological circumstances rather than a 'pathological' response to cell injury.

Magnocellular neurosecretory neurons show prolonged and complex action potentials which display a prominent calcium component (Mason and Leng 1984). It is possible that the high sensitivity of the c-*fos* system in magnocellular oxytocin neurons is a consequence of an unusually large calcium entry accompanying electrical activation. On the other hand, in the PVN, systemic administration of CCK also clearly resulted in activation of c-*fos* mRNA expression in parvocellular neurosecretory neurons in the region showing the highest density of mRNA for corticotrophin-releasing factor (CRF) and resulted in an increase in the CRF mRNA signal (Hamamura *et al.* 1991*a*). It is not known whether these neurons also have a significant calcium component to their action potentials.

Thus c-*fos* gene expression appears to be a sensitive indicator of neuronal excitation in the CNS (Sagar *et al.* 1988). We have found that systemic injection of CCK induced c-*fos* mRNA in both the SON and the PVN, which are indispensable neural structures for CCK-induced increases in oxytocin release (Verbalis *et al.* 1986) and (in the case of the PVN) for CCK-induced inhibition of feeding behaviour (Morley, 1987; Chapter 21 of this volume). Whether c-*fos* mRNA expression is induced in all elements of the afferent pathways activated following systemic administration of CCK remains to be established.

References

Brimble, M. J., Dyball, R. E. J., and Forsling, M. L. (1978). Oxytocin release following osmotic activation of oxytocin neurons in the paraventricular and supra-optic nuclei. *J. Physiol.* **278**, 69–78.

Curran, T. and Morgan, J. I. (1985). Superinduction of c-*fos* by nerve growth factor in the presence of peripherally active benzodiazepines. *Science* **229**, 1265–8.

Curran, T. and Morgan, J. I. (1986). Barium modulates c-*fos* expression and post-transcriptional modification. *Proc. Natl Acad. Sci. USA.* **83**, 8521–4.

Greenberg, M. E., Green, L. A., and Ziff, E. B. (1985). Nerve growth factor and epidermal growth factor induce rapid transient changes is proto-oncogene transcription in PC12 cells. *J. Biol. Chem.* **260**, 14101–10.

Greenberg, M. E., Ziff, E. B., and Reene, L. A. (1986). Stimulation of neural acetylcholine receptors induces rapid gene transcription. *Science* **234**, 80–3.

Hamamura, M., Emson, P. C., and Kiyama, K. (1991a). Rapid osmotic activation of c-*fos* mRNA expression in the antero-ventral third ventricular region (AV3V) and subfornical organ of rats. *J. Physiol.* **434**, 90P.

Hamamura, M., Leng, G. Emson, P. C., and Kiyama, H. (1991b). Electrical activation and c-fos mRNA expression in rat neurosecretory neurones after systemic administration of cholecystokinin. *J. Physiol.* **444**, 51–63.

Hunt, S. P., Pini, A., and Evan, G. (1987). Induction of c-*fos*-like protein in spinal cord neurons following sensory stimulation. *Nature* **328**, 632–4.

Lightman, S. L. and Young, W. S., III (1987). Vasopressin, oxytocin, dynorphin, enkephalin and corticotrophin-releasing factor mRNA stimulation in the rat. *J. Physiol.* **394**, 23–9.

Mason, W. T. and Leng, G. (1984). Complex action potential waveform recorded from supra-optic and paraventricular neurons of the rat: evidence for sodium and calcium spike components at different membrane site. *Exp. Brain Res.* **41**, 135–43.

Morgan, J. I. and Curran, T. (1986). Role of ion flux in control of c-*fos* expression. *Nature.* **322**, 550–5.

Morgan, J. I. and Curran, T. (1989) Stimulus-transcription coupling in neurons: role of cellular immediate-early genes. *Trends Neurosci.* **12** 459–62.

Morley, J. E. (1987). Neuropeptide regulation of appetite and weight. *Endocr. Rev.* **8**, 256–87.

Rauscher, F. J., III, Voulalsa, P. J., Franza, B. R., Jr., and Curran, T. (1988). *Fos* and *Jun* bind cooperatively to the AP-1 site: reconstitution *in vitro*. *Genes Dev.* **2**, 1687–99.

Sagar, S. M., Sharp, F. R., and Curran, T. (1988). Expression of c-*fos* protein in brain: metabolic mapping at the cellular level. *Science* **244**, 1328–31.

Sheng, M. and Greenberg, M. E. (1990). The regulation and function of c-*fos* and other immediate early genes in the nervous system. *Neuron* **4**, 477–85.

Sheng, M., McFadden, G., and Greenberg, M. E. (1990). Membrane depolarization and calcium induce c-*fos* transcription factor CREB. *Neuron* **4**, 571–82.

Sofroniew M. V. (1983). Morphology of vasopressin and oxytocin neurons and their central and vascular projections. In *The neurohypophysis: structure, function and control* (ed. B. A. Cross and G. Leng). *Prog. Brain Res.* **60**, 101–14.

Sonnenberg, J. L., Rouscher, F. J., Morgan, J. I., and Curran, T. (1989). Regulation of proenkephalin by *Fos* and *Jun*. *Science*. **246**, 1623–5.

Verbalis, J. G., McCann, M. J., McHale, C. M., and Stricker, E. D. (1986) Oxytocin secretion in response to cholcytokinin and food: differentiation of nausea from satiety. *Science* **232**, 1417–19.

Wisden, W., Errington, M. L., Williams, S., Dunnett, S. B., Waters, C., Hitchcock, D., Evans, G., Bliss, T. V., and Hunt, S. P. (1990). Differential expression of immediate early genes in the hippocampus and spinal cord. *Neuron* **4**, 603–14.

Part IV

CCK–dopamine interactions

31. CCK–dopamine interactions: introduction

Jacqueline N. Crawley

Current information on the coexistence of CCK and dopamine in mid-brain neurons of several species is presented in this part. The mesocorticolimbic dopamine system mediates a variety of spontaneous behaviour, including exploration, feeding, stress responses, drug self-administration, and electrical self-stimulation (Stellar and Stellar 1985; Kalivas and Nemeroff 1988). The nigrostriatal dopamine pathway mediates motor functions including co-ordinated locomotor activities and extrapyramidal stereotypies (DeLong 1990). The tubero-infundibular dopamine tract regulates the release of many pituitary hormones (Meltzer *et al.* 1978). Dopamine receptor antagonists are the primary treatment for schizophrenia (Carlsson 1988), while dopamine precursors are the primary treatment for Parkinson's disease (Calne 1977).

The discovery by Hökfelt and co-workers that two neuropeptides, CCK and neurotensin, coexist with dopamine in mid-brain neurons of the ventral tegmentum and substantia nigra (Hökfelt *et al.* 1986) was an exciting revelation for both basic researchers and clinicians studying dopaminergic mechanisms in neuropsychiatric syndromes. The anatomical distribution of these peptide neuromodulators in discrete subdivisions of the mesocorticolimbic and nigrostriatal pathways is described in detail in Chapter 32.

Functional studies of CCK–dopamine interactions explore the qualitative action(s) of CCK, the site(s) of action of CCK, and the receptor subtypes mediating the actions of CCK, particularly in regions of CCK–dopamine coexistence. Several approaches have been utilized to investigate the effects of exogenously administered CCK on dopaminergic cell bodies, terminals and postsynaptic neurons. Neurophysiological studies have demonstrated excitatory actions of CCK on many central neurons, including substantia nigra, ventral tegmental, and nucleus accumbens cell bodies (Hommer *et al.* 1986). In addition, CCK potentiates the inhibitory effects of dopamine on ventral tegmental and substantia nigra zona compacta cell bodies. The dual effects of CCK in the substantia nigra–ventral tegmental area appear to be differentially mediated by the two CCK receptor subtypes (Hommer *et al.* 1986), while the excitatory effects of CCK in the nucleus accumbens appear to be mediated by the CCK_A receptor subtype (Wang *et al.* 1988). Receptor binding studies have shown reciprocal interactions between CCK receptors and dopamine receptors;

for example, chronic infusion of CCK into the lateral ventricles increased the B_{max} for D_2 receptors in the rat nucleus accumbens and striatum (Dumbrille-Ross and Seeman 1984), while chronic administration of the D_2 antagonist haloperidol increased the B_{max} for CCK in the nucleus accumbens, olfactory tubercle, and cerebral cortex in guinea pig and mouse (Chang *et al.* 1983). As reviewed in Chapter 35, release studies have demonstrated both augmentation and attenuation of the release of dopamine by CCK, reflecting differences in species, CCK concentration, and anatomical region. Behavioural studies, reviewed in the same chapter, have also shown both inhibitory and potentiating effects of CCK on dopamine-mediated behaviour in the mesolimbic pathway which appear to depend on a biphasic dose–response curve for CCK and on the site of administration. For example, nanogram doses of CCK potentiate dopamine-induced and amphetamine-induced hyperlocomotion when micro-injected into the medial posterior nucleus accumbens, while CCK has no effect or inhibits dopamine-induced and amphetamine-induced hyperlocomotion when micro-injected into the lateral anterior nucleus accumbens (Crawley *et al.* 1985; Vaccarino and Rankin 1989). The effects of CCK on the neuro-physiology, release, and behavioural output of dopaminergic neurons are further delineated by CCK receptor subtypes, as described in detail in the following chapters. This complex literature, in which CCK is reported to have different qualitative effects on dopaminergic functions dependent on on the physiological versus pharmacological dose of CCK, on the anatomical site of administration, representing regions of coexistence versus non-coexistence, and on the CCK receptor subtype involved, makes for difficult reading but provides insights into tailoring CCK-based therapeutics to fit highly specific circumstances of dopaminergic dysfunctions.

The ultimate underlying question of whether endogenous CCK contributes to the regulation and dysregulation of dopaminergic activity in the human brain remains to be answered. The newly developed CCK receptor antagonists are excellent tools which now make possible the study of endogenous CCK in neurophysiological, behavioural, receptor binding, and release paradigms. Additional questions raised in the following chapters include the possible differential interactions of CCK receptor subtypes with each of the three identified dopamine receptor subtypes, and interactions of CCK receptor subtypes with opiate receptor subtypes known to modulate dopaminergic functions. Evidence that CCK messenger RNA (mRNA) appears in mid-brain dopamine neurons of neuroleptic-treated schizophrenics (Schalling *et al.* 1989; Chapter 32 of this volume) in greater concentrations than in normal human mid-brain dopamine neurons, where CCK mRNA is often not detectable (Palacios *et al.* 1989; Chapter 33 of this volume), provides a fascinating first glimpse into a peculiarity of endogenous CCK in a human disease state.

There is much evidence to implicate the mesocorticolimbic dopamine pathway in schizophrenia (Chapters 36 and 37 of this volume) and in drug abuse, and the nigrostriatal dopamine pathway is known to be abnormal in

Parkinson's disease and Huntington's chorea (Chapters 38 and 39). These are chronic debilitating health problems in our society, for which better treatments would yield major humanitarian and economic advances. The newly developed agonists and antagonists of CCK_A and CCK_B receptors may provide ligands for *in vivo* imaging in the human brain to visualize the role of endogenous CCK in neuropsychopathologies. Clinicians of the 1990s will be able to test alternative hypotheses that CCK-based agonists or antagonists of the CCK_A or the CCK_B receptor subtype, given alone or in combination with dopaminergic drugs, will provide more efficacious treatments, with greater selectivity and fewer side-effects, than present treatments based solely on the classical neurotransmitter dopamine.

References

Calne, D. B. (1977). Developments in the pharmacology and therapeutics of Parkinsonism. *Ann. Neurol.* **1**, 111-9.

Carlsson, A. (1988). The current status of the dopamine hypothesis of schizophrenia. *Neuropsychopharmacology* **1**, 179-86.

Chang, R. S. L., Lotti, V. J., Martin, G. E., and Chen, T. B. (1983). Increase in brain ^{125}I-cholecystokinin (CCK) receptor binding following chronic haloperidol treatment, intracisternal 6-hydroxydopamine or ventral tegmental lesions. *Life Sci.* **32**, 871-8.

Crawley, J. N., Hommer, D. W., and Skirboll, L. R. (1985). Topographical analysis of nucleus accumbens sites at which cholecystokinin potentiates dopamine-induced hyperlocomotion. *Brain Res.* **355**, 337-41.

DeLong, M. R. (1990). Primate models of movement disorders of basal ganglia origin. *Trends Neurosci.* **13**, 281-5.

Dumbrille-Ross, A. and Seeman, P. (1984). Dopamine receptor elevation by cholecystokinin. *Peptides* **5**, 1207-12.

Hökfelt, T., Holets, V. R., Staines, W., Meister, B., Melander, T., Schalling, M., Schultzberg, M., Freedman, J., Bjorklund, H., Olson, L., Lindh, B., Elfvin, L. G., Lundberg, J. M., Lindgren, J. A., Samuelsson, B., Pernow, B., Terenius, L., Post, C., Everitt, B., and Goldstein, M. (1986). Coexistence of neuronal messengers — an overview. *Prog. Brain Res.* **68**, 33-70.

Hommer, D. W., Stoner, G., Crawley, J. N., Paul, S. M., and Skirboll, L. R. (1986). Cholecystokinin–dopamine coexistence: electrophysiological actions corresponding to cholecystokinin receptor subtype. *J. Neurosci.* **6**, 3039-43.

Kalivas, P. W. and Nemeroff, C. B. (ed.) (1988). *The mesocorticolimbic dopamine system. Ann. NY Acad. Sci.* **537**, 540pp. New York Academy of Sciences, NY.

Meltzer, H. Y., Goode, D. J., and Fang, V. S. (1978). The effect of psychotropic drugs on endocrine function. I. Neuroleptics, precursors, and agonists. In *Psychopharmacology: a generation of progress* (ed. M. A. Lipton, A. DiMascio, and K. F. Killam), pp. 509-29. Raven Press, New York.

Palacios, J. M., Savasta, M., and Mengod, G. (1989). Does cholecystokinin colocalize with dopamine in the human substantia nigra? *Brain Res.* **488**, 369-75.

Schalling, M., Friberg, K., Bird, E., Goldstein, M., Shiffmann, S., Mailleux, P., Vanderhaeghen, J. J., and Hökfelt, T. (1989). Presence of cholecystokinin mRNA in dopamine cells in the ventral mesencephalon of a human with schizophrenia. *Acta Physiol. Scand.* **137**, 467–8.

Stellar, J. R. and Stellar, E. (1985). *The neurobiology of motivation and reward.* Springer-Verlag, New York.

Vaccarino, F. G. and Rankin, J. (1989). Nucleus accumbens cholecystokinin (CCK) can either attenuate or potentiate amphetamine-induced locomotor activity: evidence for rostral–caudal differences in accumbens CCK function. *Behav. Neurosci.* **103**, 831–6.

Wang, R. Y., Kasser, R. J., and Hu, X. T. (1988). Cholecystokinin receptor subtypes in the rat nucleus accumbens. In *Cholecystokinin antagonists* (ed. R. Y. Wang and R. Schoenfeld), pp. 199–216. Alan R. Liss, New York.

32. Immunohistochemical and *in situ* hybridization studies of the coexistence of CCK and dopamine in mesencephalic neurons

T. Hökfelt, M. Schalling, K. Seroogy, P. Frey, J. Walsh, and M. Goldstein

Introduction

CCK was originally isolated by Mutt and Jorpes (1968) from porcine intestine. Vanderhaeghen *et al.* (1975) described CCK/gastrin-like material in the rat brain, and it was then shown to represent mainly the C-terminal octapeptide (CCK-8) (Dockray 1976, 1980; Müller *et al.* 1977; Dockray *et al.* 1978; Rehfeld 1978*a, b*; Beinfeld *et al.* 1980). In fact, in several brain regions CCK-8 is the most abundant neuropeptide (Crawley 1985). In addition to these radio-immunoassay studies, numerous immunohistochemical investigations showing extensive CCK immunoreactive neuron systems in the brain and spinal cord have been published (Innis *et al.* 1979; Larsson and Rehfeld 1979; Lorén *et al.* 1979; Vanderhaeghen *et al.* 1980, 1982; Kiyama *et al.* 1983; Kubota *et al.* 1983; Fuji *et al.* 1985; Hökfelt *et al.* 1985, 1988; Vanderhaeghen 1985). More recently, *in situ* hybridization with radio-labelled probes complementary to CCK messenger RNA (mRNA) has been employed and not only confirmed most results from immunohistochemical studies but also provided interesting new findings (Siegel and Young 1985; Burgunder and Young 1988, 1990; Savasta *et al.* 1988, 1989; Voigt and Uhl 1988; Ingram *et al.* 1989; Lanaud *et al.* 1989; Seroogy *et al.* 1989*b*, 1990; Cortés *et al.* 1991*a, b*; Jayaraman *et al.* 1990; Sutin and Jakobowitz 1990).

It was demonstrated in early studies that CCK-like immunoreactivity (CCK-LI) is present in a subpopulation of dopamine neurons in the ventral mesencephalon (Hökfelt *et al.* 1980*a, b*). These initial studies suggested that this coexistence was most frequent in the ventral tegmental area, i.e. the A10 dopamine cell group according to the nomenclature of Dahlström and Fuxe (1964). Moreover, it could be established that at least one projection site for these neurons was the medial posterior nucleus accumbens (Hökfelt *et al.* 1980*b*). Since then, several other studies have dealt with the dopamine–CCK systems, and it has been shown that this coexistence in the ventral

mesencephalon is more extensive than originally reported (Hökfelt *et al.*
1985, 1986; Seroogy *et al.* 1989*a*). Immunohistochemical studies have also
detailed other aspects of dopamine–CCK coexistence at the mesencephalic
level (Seroogy *et al.* 1988*a*, 1989*a*). Finally, a number of *in situ* hybridization
studies have demonstrated the occurrence of CCK mRNA expressing neurons
in the ventral mesencephalon (Savasta *et al.* 1988, 1989; Voigt and Uhl 1988;
Ingram *et al.* 1989; Lanaud *et al.* 1989; Seroogy *et al.* 1989*b*; Cortés *et al.* 1991*b*;
Jayaraman *et al.* 1990), and some of these studies directly or indirectly support
occurrence of CCK mRNA in dopamine neurons (Savasta *et al.* 1989; Seroogy
et al. 1989*b*).

Most studies on CCK–dopamine coexistence have dealt with the rat, and
information based on histochemical studies in other species is limited. In a
preliminary form, the presence of CCK-L I has been reported in dopamine
neurons in mouse, cat, and monkey (Hökfelt *et al.* 1985, 1986), as well as
in the ventral mesencephalon of an infant brain (Hökfelt *et al.* 1980*a*). We
have recently summarized further studies of different species based on *in situ*
hybridization (Schalling *et al.* 1991).

There is evidence that CCK-8 may act as a neurotransmitter or neuromodulator in many areas of the CNS (Vanderhaeghen and Crawley 1985).
Recently, new CCK antagonists have been developed (Wang and Schoenfeld
1988; Hughes *et al.* 1990; Chapters 2, 3, and 5 of this volume), opening up
possibilities of understanding the functional significance of CCK peptides in
neurons and other systems. It is therefore important to consider the localization
of CCK systems in neuronal and non-neuronal tissues since this will provide a
provisional 'map' of putative sites of action of these drugs. In this chapter we
briefly review the distribution of CCK in relation to mesencephalic dopamine
neurons and their projections, and it will be interesting to see if these systems
are related to the actions of the new CCK antagonists. In a parallel paper
(Hökfelt *et al.* 1991) we focus on some other CCK systems, particularly CCK
in cerebral cortex and spinal cord and CCK/gastrin peptides in sperm.

Aspects of methodology

The analysis of the expression and distribution of CCK in neurons in the ventral
mesencephalon is based on two major methodological approaches: immunohistochemistry, either immunofluorescence (Coons 1958) or various modifications of the peroxidase–antiperoxidase method (Coons 1958; Sternberger
1979), and more recently *in situ* hybridization (Uhl 1986; Schalling 1990;
Young 1990).

Immunohistochemistry

With this approach, double-staining techniques can be applied (Wessendorf
and Elde 1985), employing antisera raised in different species and secondary
antibodies labelled by different fluorophores. Multiple compounds in a single

section can also be studied using elution-restaining techniques, whereby the antibodies are eluted after photography of the first staining pattern. The section is then reincubated with a new primary antibody (Tramu *et al.* 1978). Furthermore, the projections of CCK dopamine neurons have been studied by combining immunohistochemistry with retrograde tracing using fluorescent markers (Hökfelt *et al.* 1983; Skirboll *et al.* 1984).

In our laboratory, the immunohistochemical analysis is based on animals fixed with formalin-containing picric acid (Zamboni and de Martino 1967) by cardiac perfusion. In most cases colchicine is given intraventricularly 24 h before sacrifice. Sections are cut in a cryostat and incubated with a variety of primary antisera (both polyclonal rabbit antisera and mouse monoclonal antibodies). These include sequence-specific antisera directed to different portions of the CCK molecule, thus providing a means of analysing specificity (Hökfelt *et al.* 1988). Dopamine neurons are identified using antisera to tyrosine hydroxylase (TH) (Markey *et al.* 1980), the first enzyme in the catecholamine synthesis. Subsequently, the sections are incubated with secondary antibodies, labelled with fluorescein isothiocyanate (FITC) (green fluorescence) or tetramethylrhodamine isothiocyanate (TRITC) (red fluorescence).

In situ hybridization

The introduction of *in situ* hybridization studies has provided support for earlier views and has added new information. Indeed, CCK mRNA is present in many of the systems previously shown by immunochemistry to contain CCK-LI, thus confirming specificity. Furthermore, mRNA for CCK and other peptides can be visualized in untreated animals, in contrast with immunohistochemistry where in most cases the animals have to be injected with the mitosis inhibitor colchicine (Dahlström 1971) in order to obtain detectable levels of the peptides in cell bodies. The use of *in situ* hybridization seems important, since recent results suggest that colchicine not only blocks axonal transport but can also affect mRNA levels (Cortés *et al.* 1991*b*). *In situ* hybridization is a particularly important approach for mapping CCK-producing neurons in the human brain, where colchicine cannot be applied. Finally, *in situ* hybridization permits analysis of mRNA levels before and after various experimental manipulations and can provide some clues to the 'functional state' and role of peptides and proteins.

Cryostat sections of non-fixed brains from untreated animals (rat, cat, monkey, guinea pig, hamster) were used for *in situ* hybridization. Human brains were also investigated: the ventral mesencephalan of patients diagnosed to be schizophrenic and of controls without this diagnosis were analysed. All the former brains were from patients treated with neuroleptics, whereas the controls had not received this type of drug (for further details see Schalling *et al.* (1991)).

Oligonucleotide probes were used throughout this study. They had been synthesized on an Applied Biosystems DNA synthesizer, and the CCK probe was complementary to mRNA coding for amino acids 89–103 of human CCK (Takahashi *et al.* 1985), which is a region with 100 per cent homology to the rat (Deschenes *et al.* 1984). The TH oligonucleotide probe was complementary to amino acids 40–52 of the human enzyme (Grima *et al.* 1987) and has 98 per cent homology to the rat. All probes were labelled with $\alpha^{35}S$ (dATP) at the 3' end. The *in situ* hybridization was carried out as described in detail by Schalling (1990), and the sections were subsequently exposed to x-ray film and/or dipped in NTB2 nuclear track emulsion, developed, and fixed.

All microscopy sections were analysed using a Nikon Mikrophot-FX microscope equipped for epifluorescence with appropriate filter combinations as well as for light microscopy and bright-field light microscopy.

Dopamine–CCK coexistence in rat

Cell bodies in the ventral mesencephalon

The proportion of dopamine cells in the ventral mesencephalon that contain demonstrable levels of CCK-LI varies in different subgroups of these neurons and in different parts of this brain area (Fig. 32.1). Whereas in the substantia nigra pars lateralis almost all dopamine cells contain CCK-LI, only single cells

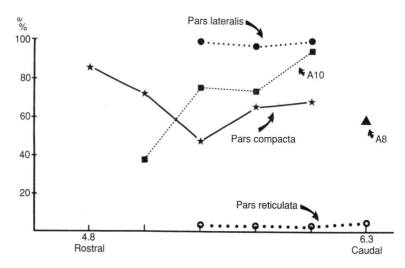

Fig. 32.1. Schematic illustration of the percentage of dopamine neurons containing CCK-LI in various subregions of the rat ventral mesencephalon at different rostro-caudal levels (the most rostral level is approximately 4.8 mm behind the Bregma, and the most caudal point is 6.3 mm behind the Bregma). Sections were analysed at 0.3 mm intervals. Areas analysed were the pars compacta, pars lateralis, and pars reticulata of the substantia nigra as well as the ventral tegmental area (A10 DA cell group) and the A8 DA cell group in the mesencephalic reticular formation. (Reproduced from Hökfelt *et al.* 1986.)

with this coexistence can be observed in the substantia nigra pars reticulata. In the substantia nigra pars compacta the highest percentage (80–90 per cent) is found in the anterior parts and decreases in the caudal direction, but then increases at the most caudal levels. The A10 group shows an increasing incidence of coexistence in the caudal direction. Approximately 60 per cent of the A8 dopamine cells are CCK immunoreactive. There are also some other mid-brain areas which exhibit a moderate proportion of coexistence, including the interfascicular nucleus and the rostral and caudal linear nuclei (Seroogy *et al.* 1989*a*). Coexistence can also be observed in some neurons in the central and ventral periaqueductal grey matter, the supramammilliary region, the peripeduncular region, and the retrorubral field (Seroogy *et al.* 1989*a*).

Recent studies (Savasta *et al.* 1988, 1989; Voigt and Uhl 1988; Ingram *et al.* 1989; Lanaud *et al.* 1989; Seroogy *et al.* 1989*b*; Jayaraman *et al.* 1990) have described the distribution of CCK mRNA in the rat brain (Fig. 32.2(a)), including the ventral mesencephalon (Fig. 32.2(b)), with overlap between CCK mRNA (Fig. 32.2(b)) and TH mRNA (Fig. 32.2(c)). It has also been directly demonstrated, by combining *in situ* hybridization and immunohistochemistry on the same section (Fig. 32.3), that CCK mRNA is present in TH immunoreactive neurons (Seroogy *et al.* 1989*b*). Savasta *et al.* (1989) have shown a close overlap of TH mRNA and CCK mRNA, although in some regions such as the zona reticulata and the medial ventral zona compacta only TH mRNA positive cells were found. Furthermore, Savasta *et al.* demonstrated that destruction of dopamine neurons with the neurotoxin 6-hydroxydopamine (6-OHDA) caused a parallel loss of CCK and TH mRNA expression thoughout the substantia nigra pars compacta and also to a considerable extent in the ventral tegmental area. Finally, the ontogeny of expression of TH and CCK mRNA has been analysed in the rat mesencephalon by Burgunder and Young (1990). They reported that both transcripts appear on embryonic day 13 in the ventral caudal mesencephalon, that there is a synchronous increase in both transcripts during the second half of pregnancy, and that the adult patterns are already established at birth.

Dopamine–CCK projections, terminals, and binding sites

By combining retrograde tracing and double labelling for CCK and TH, it was shown that dopamine–CCK neurons in the ventral tegmental area project to the posterior medial nucleus accumbens (Hökfelt *et al.* 1980*b*). Seroogy *et al.* (1989*b*) have extended these studies, and have demonstrated that dopamine–CCK neurons in the ventral mid-brain project to the caudate-putamen, nucleus accumbens, prefrontal cortex, and amygdala (Fig. 32.4) (see also Seroogy and Fallon 1989).

The question of the localization of the terminal ramifications of the dopamine–CCK neurons (Fig. 32.5) has been addressed in only a few studies (Hökfelt *et al.* 1980*b*, 1988; Seroogy *et al.* 1989*b*). It was observed in early work that there is a strongly CCK immunoreactive network overlapping with TH

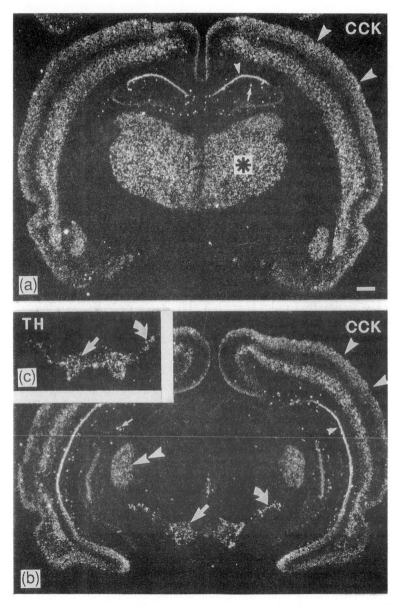

Fig. 32.2. Autoradiograms of frontal sections of rat brain at (a) a rostral and (b, c) a caudal level of rat forebrain after hybridization with a probe complementary to CCK mRNA (a, b) or THmRNA (c). Very dense labelling is seen in cortical layers II, III, V, and VI at both levels (large arrowheads). The thalamus (asterisk), the medial geniculate body (double arrowhead), the substantia nigra (curved arrows), and the ventral tegmental area (straight arrow) and the ventral periaqueductal central grey (Edinger–Westphal nucleus) also show strong hybridization signals. Note the dense labelling of the hippocampus, where the strongest hybridization is seen in CA1 (small arrow head) with little hybridization in CA2 and a medium intensity in CA3. Thin arrows indicate strongly labelled single neurons. Note similar distribution of CCK (b) and TH (c) mRNAs in the substantia nigra-ventral tegmental area (curved and straight arrows.) Scale bar, 400 μm.

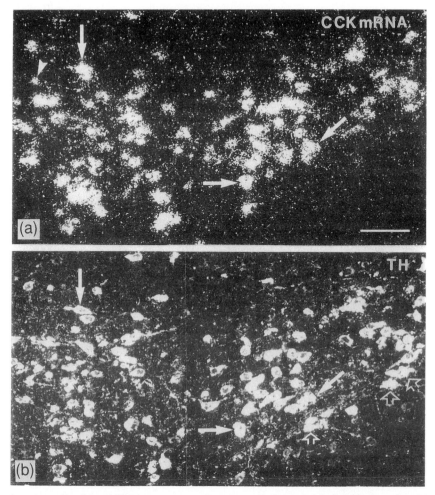

Fig. 32.3. Combined TH immunocytochemistry and *in situ* hybridization for CCK mRNA: (a) immunofluorescence photomicrographs and (b) dark-field autoradiograph of the same section through the substantia nigra pars compacta showing neurons containing (a) TH-LI and also labelled with the probe complementary to (b) CCK mRNA. Arrows indicate examples of double-labelled perikarya. Open arrows in (b) indicate examples of TH immunoreactive soma which do not contain CCK mRNA. The arrowhead in (a) identifies a CCK mRNA labelled cell body which lacks TH-LI. Scale bar, 50 μm.

positive fibres in the posterior medial nucleus accumbens and olfactory tubercle. These strongly CCK immunoreactive nerve terminals disappear to some extent after 6-OHDA treatment, suggesting that they belong to dopamine neurons. Coexistence of CCK-LI and TH-LI in the nucleus accumbens of the rat has in fact been directly demonstrated at the ultrastructural level. Thus,

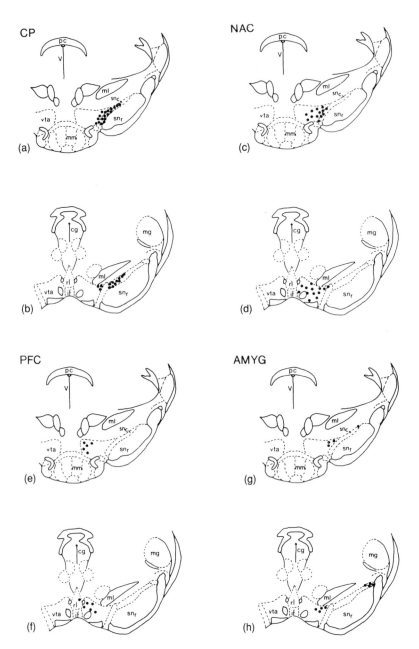

Fig. 32.4. Schematic illustrations of the distribution of triple-labelled neurons, i.e. neurons containing retrogradely transported dye, CCK-LI and TH-LI at ((a), (c), (e), (g)) a rostral and an ((b), (d), (f), (h)) intermediate level of the ventral mesencephalon following dye injections into ((a), (b)) the caudate-putamen (CP), ((c), (d)) the nucleus accumbens (NAC), ((d), (e)) the prefrontal cortex (PFC), and ((g), (h)) the amygdala (AMYG): cg, central grey matter; if, interfascicular nucleus; mg, medial geniculate nucleus; ml, medial lemniscus; mm, medial mammillary nucleus; pc, posterior commissure; rl, rostral linear nucleus; sn_c, substantia nigra pars compacta; sn_r, substantia nigra pars reticulata; V, third ventricle; vta, ventral tegmental area. (Reproduced from Seroogy *et al.* 1989*a*.)

Loopuijt and Van der Kooy (1985) used protein A conjugated to colloidal gold particles of different sizes and demonstrated CCK-LI in large granular vesicles with a diameter of 70–160 nM and TH-LI in the axoplasm of the same nerve endings, most of which did not show pre- or postsynaptic specializations.

There is also a fine weakly fluorescent fibre network in the medial aspects of the caudate nucleus, which also disappears after 6-OHDA treatment (Fig. 32.5). In addition to these coexistence networks, there is a dense weakly fluorescent plexus distributed over the entire striatum (Fig. 32.5) and extending into the nucleus accumbens and olfactory tubercle. Finally, there are patches with even more strongly immunoreactive and very densely packed CCK fibres, located mainly in the medial aspects of the caudate nucleus (Fig. 32.5). Neither of the two latter networks disappear after 6-OH-DA treatment and thus are not related to dopamine–CCK coexistence. In view of the results obtained by Seroogy *et al.* (1989*b*) showing a substantial number of retrogradely labelled cell bodies in the medial pars compacta after Fast Blue injections into the caudate nucleus, it is possible that CCK levels in the dopaminergic nerve endings in the caudate nucleus are too low to be detected with the present sensitivity of the technique. Under all circumstances these findings, taken together, suggest that the ventral tegmental dopamine–CCK system projecting to the posterior medial nucleus accumbens and the olfactory tubercle has considerably higher concentrations of CCK peptide than of other systems in its terminal ramifications. With regard to dopamine–CCK nerve endings in the prefrontal cortex and amygdala, to our knowledge no attempts have been made to establish their exact localization and morphological characteristics.

An interesting question concerns the distribution of binding sites/receptors in the areas innervated by dopamine–CCK neurons. Binding studies with radio-labelled CCK ligands have been carried out on such areas by several groups (Gaudreau *et al.* 1983; Zarbin *et al.* 1983; Van Dijk *et al.* 1984; Sekiguchi and Moroji 1986; Pelaprat *et al.* 1987, 1988). In some of these studies it is particularly clear that the dorsal medial posterior nucleus accumbens has a very high density of binding sites (Sekiguchi and Moroji 1986; Pelaprat *et al.* 1988), i.e an area containing densely packed nerve endings with a pronounced dopamine–CCK coexistence (Hökfelt *et al.* 1980*b*). The results of recent functional studies (Marshall *et al.* 1991) suggest that different CCK receptor subtypes may be present in the regions of nucleus accumbens innervated by CCK–dopamine containing neurons and regions innervated by dopamine neurons that do not contain CCK (see Chapter 5 of this volume).

Coexistence of dopamine with other peptides in the ventral mesencephalon

It has been shown that a population of dopamine neurons in the ventral mesencephalon contains a neurotensin-like peptide (Hökfelt *et al.* 1984). These neurons were localized almost exclusively in the ventral tegmental area, and they represented a small population of all dopamine neurons. The population

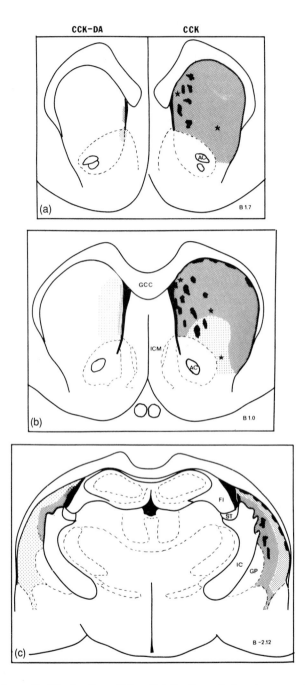

Fig. 32.5. Schematic illustration of the distribution of structures containing CCK only (right-hand side) and CCK–dopamine (left-hand side) structures at three different levels: (a) B1.7; (b) B1.0; (c) B2.12. Right-hand side: dense patches (black) of CCK-LI as well as dense (strong dots) or moderately dense (fine dots) fibre networks and some positive cell bodies (stars) are seen. Left-hand side: dense (strong dots)

was also small compared with the population of dopamine–CCK neurons. It was subsequently shown that more than 90 per cent of the neurotensin (NT) positive prikarya throughout the midbrain also contained CCK-LI (Seroogy *et al.* 1987). In fact, virtually all neurotensin positive cell bodies in these areas, including the ventral tegmental area, the medial substantia nigra pars compacta, the retrorubral field, and the rostral and caudal linear nuclei, contained both NT-LI and TH-LI (Seroogy *et al.* 1988*a*). Recently, the distribution of neurotensin mRNA in the ventral mesencephalon has been studied and described in detail using the *in situ* hybridization technique (Jayaraman *et al.* 1990).

With regard to projections, ventral mesencephalic neurotensin neurons (which also contain dopamine and CCK) have been shown to innervate the nucleus accumbens, prefrontal cortex, and amygdala (Seroogy *et al.* 1987). In fact, nerve terminals containing neurotensin and TH-LI have been demonstrated in prefrontal and limbic cortical areas (Studler *et al.* 1988; von Euler *et al.* 1991). It should be emphasized that the major feature of the well-known interaction between neurotensin and dopamine (Nemeroff 1986; Kitabgi 1989) is related to the dense neurotensin innervation of dopamine cell bodies in both the zona compacta of the substantia nigra and the ventral tegmental area (Jennes *et al.* 1982; Hökfelt *et al.* 1984). This is correlated with a high density of neurotensin binding sites in these areas (Young and Kuhar 1981; Quirion *et al.* 1985; Hervé *et al.* 1986; Szigethy *et al.* 1986; Moyse *et al.* 1987; Dana *et al.* 1989; Schotte and Leysen 1989; Szigethy and Baudet 1989), and direct connections have been demonstrated between neurotensin-positive nerve endings and dopamine neurons in the ventral mesencephalon (Woulfe and Beaudet 1989).

Finally, it should be noted that a small proportion of dopamine neurons in the ventral mesencephalon and the supramammillary region contain vasoactive intestinal polypeptide (VIP)-LI, peptide histidine isoleucine (PHI)-LI, and substance P-LI (Seroogy *et al.* 1988*b*).

Dopamine–CCK coexistence in cat

It has been reported in a preliminary form (Hökfelt *et al.* 1985, 1986) that numerous dopamine neurons in the cat ventral mesencephalon, particularly in the zona compacta of the substantia nigra, contain CCK-LI. Furthermore,

---

or moderately dense (fine dots) fibre networks containing CCK-LI and TH-LI are present mainly at mid and posterior levels of the caudate nucleus with the highest concentrations in the medial aspects. AC, anterior commissure; GCC, genu corpus callosum; ICM, major island of Calleja; FI, fimbria hippocampus; ST, stria terminalis; IC, internal capsule; GP, globus pallidus. (Reproduced from Hökfelt *et al.* 1988.)

injection of the retrogradely transported fluorescent dye Fast Blue into the nucleus caudatus putamen labelled numerous TH- and CCK-positive cell bodies in the pars compacta (Hökfelt *et al.*, unpublished data). Artaud *et al.* (1989) have used biochemical methods to analyse dopamine and CCK levels in the ventral mesencephalon of the cat after 6-OHDA lesions. This treatment resulted in a dramatic decrease in dopamine levels in both the substantia nigra and the ventral tegmental area, whereas CCK levels were markedly reduced in the substantia nigra but to a smaller extent in the ventral tegmental area. No clear decrease in CCK-LI could be seen in the caudate nucleus, whereas dopamine disappeared almost completely. These findings are in agreement with the histochemical observation of the coexistence of dopamine and CCK-LI in the zona compacta of the substantia nigra. In fact, recent double-labelling

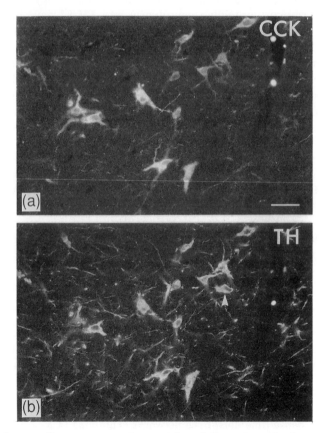

Fig. 32.6. Immunofluorescence micrographs of the cat ventral mesencephalon after processing for double labelling with (a) monoclonal mouse anti-CCK antibodies and (b) rabbit anti-TH antiserum. Virtually all TH-positive cell bodies contain CCK-LI. The arrowhead points to a TH-positive, CCK-negative cell body. Scale bar, 50 μm.

studies have clearly established a high percentage of coexistence of CCK and TH-LI in the cat ventral mesencephalon (Fig. 32.6), and overlapping distribution patterns of CCK mRNA and TH mRNA in this area have also been demonstrated (Schalling *et al.* 1991).

The surprising lack of effect of the 6-OHDA lesion on the striatal CCK levels may reflect that the amount of peptide transported into the axons and terminal ramifications of the nigrostriatal dopamine–CCK pathway is small (see above). However, several other explanations can also be found (Artaud *et al.* 1989). The source of neostriatal CCK in the cat has been analysed using histochemical techniques by Adams and Fisher (1990). After injection of retrograde tracer into the nucleus caudatus putamen they found labelled CCK-positive neurons in the ventral tegmental area, the compacta, and lateral divisions of the substantia nigra and the retrorubral area, further supporting the existence of ascending mesencephalic striatal CCK projections, presumably also containing dopamine. In addition, they observed CCK inputs from thalamus as well as local striatal non-spiny CCK immunoreactive neurons in both the caudate and the putamen.

Dopamine–CCK coexistence in monkey

So far, only limited information is available on the coexistence of dopamine and CCK in the monkey ventral mesencephalon. *In situ* hybridization reveals a marked overlap between CCK mRNA and TH mRNA, particularly at the mid and posterior levels (Schalling *et al.* 1991). Double-labelling immunofluorescence analysis has directly demonstrated the coexistence of CCK-LI and TH-LI (Schalling *et al.* 1991), which is particularly extensive in the mid-line areas although overlapping distribution patterns were also observed in the medial zona compacta. TH and CCK immunoreactive nerve endings with a similar distribution have been described in the nucleus accumbens of monkey (Hökfelt *et al.* 1986). Therefore it seems likely that at least a population of dopamine neurons projecting to forebrain areas exhibit coexistence between dopamine and CCK.

CCK–dopamine coexistence in man

It has been reported that cells containing both CCK-LI and TH-LI could be observed in human mesencephalon (Hökfelt *et al.* 1980*a*). The brain analysed was from an infant, and the CCK-LI was weak. We have subsequently tried to analyse adult brains using immunohistochemistry, but with this. approach no definitely CCK-positive neurons could be observed in the ventral mesencephalon (Hökfelt *et al.*, unpublished data). This problem has recently been studied using *in situ* hybridization, which should be a more favourable method since, as discussed above, colchicine treatment is not required. Thus Palacios *et al.* (1989) reported that dopaminergic cell bodies in the human substantia nigra did not show detectable amounts of CCK mRNA but that low levels could be seen in the nucleus paranigralis. In another study Schalling *et al.* (1989, 1991) analysed 10 post-mortem human brains, five of which were from neuroleptic-treated patients diagnosed to be schizophrenic. In all patients, a strong and reproducible hybridization signal was observed with the probe complementary to TH mRNA. Furthermore, in all patients diagnosed to be schizophrenic, CCK mRNA could be observed in the substantia nigra. A par-

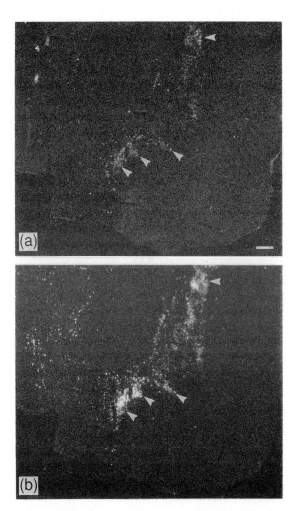

Fig. 32.7. Autoradiographs of two adjacent sections of human ventral mesencephalon of a patient with the diagnosis of schizophrenia showing a good overlap between the distribution of (a) CCK mRNA-positive cells and (b) positive cells (arrowheads) in the zona compacta of the substantia nigra. However, TH mRNA-positive cells are more numerous and more intensively labelled. Scale bar, 1 mm.

ticularly strong signal was seen in the youngest patient studied (36 years old), with CCK mRNA positive cells in the entire zona compacta and a high degree of overlap with the TH mRNA positive cells (Fig. 32.7). In the age-matched control brain a CCK signal was only observed in the mid-line area of the ventral mesencephalon (the paranigral area). In the remaining four control brains a weak but distinct CCK mRNA signal was found in two cases. At the cellular level CCK mRNA could be seen overlying many cells with melanin pigment. However, there were also pigment-containing cells without grain accumulations and CCK mRNA-positive neurons without apparent pigment. In a few cases it was possible to process CCK mRNA autoradiographs for TH immunostaining and a coexistence could be established in some cells (Schalling *et al.* 1991).

The present findings suggest that, in addition to paranigral cells (cf. Palacios *et al.* 1989), CCK mRNA can also be observed in some dopamine cells in the zona compacta in brains from a control group. However, it seems likely that the CCK mRNA levels in normal subjects are low, sometimes above and sometimes below the detection limit of our technique. In contrast, in schizophrenic patients treated with neuroleptics a CCK mRNA signal could consistently be observed, suggesting that CCK mRNA levels are indeed higher in these subjects. Whether this is related to the disease as such or to the treatment with neuroleptics or to other unknown factors is at present unclear. It would therefore be interesting to study brains from patients who had received neuroleptic treatment for disorders other than psychosis. Furthermore, if, as indicated here, CCK mRNA is in fact upregulated, and if this is resulting in increased peptide levels in the nerve endings and increased CCK release, then it would be interesting to find out what the effects of this CCK are. Are they of significance for the symptomatology either in positive or negative direction, or perhaps for side-effects? Such questions could perhaps be answered by administration of the type of drugs, i.e. CCK antagonists, discussed in this book. Further careful studies, including quantitative analysis, are needed to understand better the occurrence and significance of CCK mRNA in dopamine neurons in the human ventral mesencephalon, particularly in patients with schizophrenia (Wang *et al.* 1984; Chapter 36 of this volume).

CCK–dopamine coexistence in guinea pig and hamster

Early immunohistochemical studies suggested that CCK-LI is not present in guinea pig (Hökfelt *et al.* 1985, 1986). This species has now been reinvestigated using *in situ* hybridization. So far no signal has been seen in the ventral tegmental area/substantia nigra using the oligonucleotide probe complementary to CCK mRNA (Fig. 32.8(a)). In contrast, a very strong signal was observed over the cortical areas and the hippocampus (Fig. 32.8(a)). A strong hybridization was seen in the ventral mesencephalon with the probe complementary to TH mRNA (Fig. 32.8(b)). Similar findings were obtained in the ventral mesencephalon of the hamster (Fig. 32.9). In view of the results obtained on

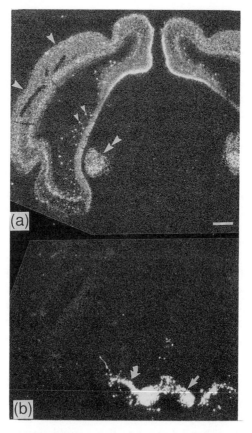

Fig. 32.8. Autoradiographs of two pairs of adjacent frontal sections of guinea pig mesencephalon and cortex. A strong TH mRNA message can be seen in the zona compacta of the substantia nigra (curved arrow) and the ventral tegmental area (arrow) (b), but no corresponding signal is seen for CCK mRNA (a). Note the strong message for CCK in the cortex (large arrowheads) and hippocampus (small arrowheads) (a). Double arrowheads in (a) indicate the medial geniculate body. Scale bar, 1 mm.

human brains with an apparent increase in CCK mRNA levels in patients treated with neuroleptics (see above), some hamsters and guinea pigs were treated for 6 weeks with an oil emulsion preparation of the neuroleptic flupenthixol (guinea pigs received 40 mg every 2 weeks and hamsters 10 mg every 2 weeks). No hybridization signal for CCK mRNA was seen in the ventral mesencephalon of these animals. These results suggest that dopamine neurons may not synthesize CCK in guinea pig and hamster, and that true species differences exist between, on the one hand, these two species and, on the other hand, rat, cat, monkey, and presumably man. Species differences between

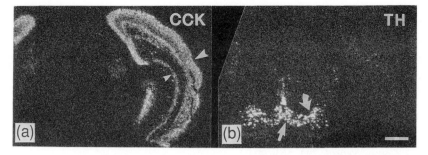

Fig. 32.9. Autoradiographs of two pairs of adjacent frontal sections of hamster mesencephalon and cortex. A strong TH mRNA message can be seen in the zona compacta of the substantia nigra (curved arrow) and the ventral tegmental area (arrow) (b), but no corresponding signal is seen for CCK mRNA (a). Note the strong message for CCK in the cortex (large arrowheads) and hippocampus (small arrowheads) (a). Scale bar, 1 mm.

CCK distribution patterns have also been observed when comparing hamster and rat forebrain (Miceli *et al.* 1987).

Concluding remarks

In this chapter we have reviewed the evidence for the coexistence of dopamine and the neuropeptide CCK in neurons in the ventral mesencephalon projecting to various forebrain areas, and particular emphasis has been put on the situation in different species. Dopamine and CCK seem to coexist in rat, cat, monkey, and presumably man. In man, CCK mRNA was easily detectable in brains from neuroleptic-treated patients diagnosed to be schizophrenic. No evidence for dopamine–CCK coexistence was obtained in hamster and guinea pig, suggesting that true species differences may be present. Moreover, certain species differences were observed with regard to the coexistence patterns. Thus, in rat and monkey the dopamine–CCK coexistence was particularly pronounced in mid-line areas, particularly with regard to to the peptide levels. In cat a somewhat reversed situation was encountered with an extensive coexistence in the dopamine neurons in the zona compacta, but a lower incidence in the mid-line areas. The amount of peptide produced and transported into the axons and terminal ramifications seems to vary considerably between various groups of mesencephalic dopamine cells in a single species, and also between different species. The significance of these findings still remains unclear, but numerous interactions between CCK and dopamine have been described both at the level of the mesencephalon and in various forebrain areas to which the dopamine–CCk neurons project. However, we shall not deal with these aspects here but refer the reader to other chapters in this volume (see Chapters 35, 36, and 40).

Acknowledgements

This study was supported by grants from the Swedish MRC (04X-2887), National Institute of Mental Health (MH-43230), and Marianne och Marcus Wallenbergs Stiftelse.

References

Adams, C. E. and Fisher, R. S. (1990). Sources of neostriatal cholecystokinin in the cat. *J. Comp. Neurol.* **292**, 563–74.

Artaud, F., Baruch, P., Stutzmann, J. M., Saffroy, M., Godehue, G., Barbeito, L., Hervé, D., Studler, J. M., Glowinski, J., and Chérmany, A. (1989). Cholecystokinin: corelease with dopamine from nigrostriatal neurons in the cat. *Eur. J. Neurosci.* **1**, 162–71.

Beinfeld, M. G., Meyer, D. K., and Brownstein, M. (1980). Cholecystokinin octapeptides in the rat hypothalamohypophyseal system. *Nature* **288**, 376–8.

Burgunder, J.-M. and Young, W. S, III (1988). The distribution of thalamic projection neurons containing cholecystokinin messenger RNA, using *in situ* hybridization histochemistry and retrograde labeling. *Mol. Brain Res.* **4**, 179–89.

Burgunder, J.-M. and Young, W. S. (1990). Ontogeny of tyrosine hydroxylase and cholecystokinin gene expression in the rat mesencephalon. *Dev. Brain Res.* **52**, 85–93.

Coons, A. H. (1958). Fluorescent antibody methods. In *General cytochemical methods* (ed. J. F. Danielli), pp. 399–422. Academic Press, New York.

Cortés, R., Arvidsson, U., Schalling, M., and Ceccatelli, S. (1991*a*). Studies on mRNA for cholecystokinin, calcitonin gene-related peptide and choline acetyltransferase in the lower brain stem, spinal cord and dorsal root ganglia of rat and guinea pig with special reference to motoneurons. *J. Chem. Neuroanat.*, in the press.

Cortés, R., Ceccatelli, S., Schalling, M. and Hökfelt, T. (1991*b*). Differential effects of intracerebroventricular colchicine adminstration on the expression of mRNA for neuropeptides and neurotransmitter enzymes, with special emphasis on galanin: an *in situ* hybridization study. *Synapse*, in the press.

Crawley, J. N. (1985). Comparative distribution of cholecystokinin and other neuropeptides. Why is this peptide different from all other peptides? In *Neuronal cholecystokin* (ed. J. J. Vanderhaeghen and J. Crawley). *Ann. NY Acad. Sci.* **448**, 1–8.

Dahlström, A. (1971). Effects of vinblastine and colchicine on monoamine containing neurons of the rat with special regard to the axoplasmic transport of amine granules. *Acta Neuropathol.* **5** (Suppl.), 226–37.

Dahlström, A. and Fuxe, K. (1964). Evidence of the existence of monoamine-containing neurons in the central nervous system. I. Demonstration of monoamines in the cell bodies of brain stem neurons. *Acta Physiol. Scand.* **62** (Suppl. 232), 1–55.

Dana, C., Vial, M., Leonard, K., Beauregard, A., Kitabgi, P., Vincent, J.-P., Rostène, W., and Beaudet, A. (1989). Electron microscopic localization of neurotensin binding sites in the midbrain tegmentum of the rat. I. Ventral tegmental area and interfascicular nucleus. *J. Neurosci.* **9**, 2247–57.

Deschenes, R. J., Lorenz, L. J., Haun, R. S., Roos, B. A., Collier, K. J., and Dixon, J. E. (1984). Cloning and sequence analysis of cDNA encoding rat preprocholecystokinin. *Proc. Natl Acad. Sci. USA* **81**, 726–30.

Dockray, G. J. (1976). Immunochemical evidence of cholecystokinin-like peptides in brain. *Nature* **264**, 568–70.

Dockray, G. J. (1980). Cholecystokinins in rat cerebral cortex: identification, purification and characterization by immunochemical methods. *Brain Res.* **188**, 155–65.

Dockray, G. J., Gregory, R. A., Hutchinson, J. B., Harris, J. J., and Runswick, M. J. (1978). Isolation, structure and biological activity of two cholecystokinin octapeptides from sheep brain. *Nature* **274**, 711–13.

Euler G., von Meister, B., Hökfelt, T., Eneroth, P., and Fuxe, K. (1991). Intraventricular injection of neurotensin reduces dopamine D-2 agonist binding in rat forebrain and intermediate lobe of the pituitary gland. Relationship to serum hormone levels and nerve terminal coexistence. *Brain Res.*, in the press.

Fuji, K., Senba, E., Fujii, S., Nomura, I., Wu, J.-Y., Ueda, Y., and Tohyama, M. (1985). Distribution, ontogeny and projections of cholecystokinin-8, vasoactive intestinal polypeptide and γ-aminobutyrate-containing neuron systems in the rat spinal cord: an immunohistochemical analysis. *Neuroscience* **14**, 881–94.

Gaudreau, P., Quirion, R., St-Pierre, S., and Pert, C. B. (1983). Characterization and visualization of cholecystokinin receptors in rat brain using (^{3}H)pentagastrin. *Peptides* **4**, 755–62.

Grima, B., Lamouroux, A., Blanot, F., Faucon-Biguet, N., and Mallet, J. (1985). Complete coding sequence of rat tyrosine hydroxylase mRNA. *Proc. Natl Acad. Sci. USA* **82**, 617–21.

Hervé, D., Tassin, J. P., Studler, J. M, Dana, C., Kitabgi, P., Vincent, J.-P., Glowinski, J., and Rostène, W. (1986). Dopaminergic control of ^{125}I-labeled neurotensin binding site density in corticolimbic structures of the rat brain. *Proc. Natl Acad. Sci. USA* **83**, 6203–7.

Hökfelt, T., Rehfeld, J. F., Skirboll, L., Ivemark, B., Goldstein, M., and Markey, K. (1980a). Evidence for coexistence of dopamine and CCK in mesolimbic neurons. *Nature* **285**, 476–8.

Hökfelt, T., Skirboll, L., Rehfeld, J. F., Goldstein, M., Markey, K., and Dann, O. (1980b). A subpopulation of mesencephalic dopamine neurons projecting to limbic areas containing cholecystokinin-like peptide: evidence from immunohistochemistry combined with retrograde tracing. *Neuroscience* **5**, 2093–124.

Hökfelt, T., Skagerberg, G., Skirboll, L., and Björklund, A. (1983). Combination of retrograde tracing and neurotransmitter histochemistry. In *Handbook of chemical neuroanatomy*, Vol. 1, *Methods in chemical neuroanatomy* (ed. A. Björklund and T. Hökfelt), pp. 228–85, Elsevier, Amsterdam.

Hökfelt, T., Everitt, B. J., Theodorsson-Norheim, E., and Goldstein, M. (1984). Occurrence of neurotensin like immunoreactivity in subpopulations of hypothalamic, mesencephalic and medullary catecholamine neurons. *J. Comp. Neurol.* **222**, 543–59.

Hökfelt, T., Skirboll, L., Everitt, B. J., Meister, B., Brownstein, M., Jacobs, T., Faden, A., Kuga, S., Goldstein, M., Markstein, R., Dockray, G., and Rehfeld, J. (1985). Distribution of cholecystokinin-like immunoreactivity in the nervous system with special reference to coexistence with classical neurotransmitters and other neuropeptides. *Ann. NY Acad. Sci.* **448**, 255–74.

Hökfelt, T., Holets, V., Staines, W., Meister, B., Melander, T., Schalling, M., Schultzberg, M., Freedman, J., Björklund, H., Olson, L., Lindh, B., Elfvin, L.-G., Lundberg, J. M., Lindgren, J. Å., Samuelsson, B., Pernow, B., Terenius, L., Post, C., Everitt, B., and Goldstein, M. (1986). Coexistence of neuronal messengers—an overview. In *Progress in brain research*, Vol. 68. (ed. T. Hökfelt, K. Fuxe and B. Pernow), pp. 33–70. Elsevier, Amsterdam.

Hökfelt, T., Herrera-Marschitz, M., Seroogy, K., Ju, G., Staines, W. A., Holets, V.,

Schalling, M., Ungerstedt, U., Post, C., Rehfeld, J. F., Frey, P., Fischer, J., Dockray, G., Hamaoka, T., Walsh, J. H., and Goldstein, M. (1988). Immunohistochemical studies on cholecystokinin (CCK)-immunoreactive neurons in the rat using sequence specific antisera and with special reference to the caudate nucleus and primary sensory neurons. *J. Chem. Neuroanat.* **1**, 11–52.

Hökfelt, T., Cortés, R., Schalling, M., Ceccatelli, S., Pelto-Huikko, M., Persson, H., and Villar, M. J. (1991). Distribution patterns of CCK and CCK mRNA in some neuronal and non-neuronal tissue. *Neuropeptides* **19** (Suppl.), in press.

Hughes, J., Boden, P., Costall, B., Domeney, A., Kelly, E., Horwell, D. C., Hunter, J. C., Pinnock, R. D., and Woodruff, G. N. (1990). Development of a class of selective cholecystokinin type B receptor antagonists having potent anxiolytic activity. *Proc. Natl Acad. Sci. USA* **87**, 6728–32.

Ingram, S. M., Krause, R. G., II, Baldino, F., Jr, Skeen, L. C., and Lewis, M. E. (1989). Neuronal localization of cholecystokinin mRNA in the rat brain by *in situ* hybridization histochemistry. *J. Comp. Neurol.* **287**, 260–72.

Innis, R. N., Correa, F. M. T., Uhl, G. R., Schneider, B., and Snyder, S. H. (1979). Cholecystokinin octapeptide-like immunoreactivity: histochemical localization in rat brain. *Proc. Natl Acad. Sci. USA* **76**, 521–5.

Jayaraman, A., Nishimori, T., Dobner, P., and Uhl, G. R. (1990). Cholecystokinin and neurotensin mRNAs are differently expressed in subnuclei of the ventral tegmental area. *J. Comp. Neurol.* **296**, 291–302.

Jennes, L., Stumpf, W. E., and Kalivas, P. W. (1982). Neurotensin: topographical distribution in rat brain by immunohistochemiastry. *J. Comp. Neurol.* **210**, 211–24.

Kitabgi, P. (1989). Neurotensin modulates dopamine neurotransmission at levels along brain dopaminergic pathways. *Neurochem. Int.* **14**, 111–19.

Kiyama, H., Shiosaka, S., Kubota, Y., Cho, H. J., Takagi, H., Tateishi, K., Hashimura, E., Hamaoka, T., and Tohyama, M. (1983). Ontogeny of cholecystokinin-8 containing neuron system of the rat: an immunohistochemical analysis – II. Lower brain stem. *Neuroscience* **10**, 1341–59.

Kubota, Y., Inagaki, S., Shiosaka, S., Cho, H. J., Tateishi, K., Hashimura, E., Hamaoka, T., and Tohyama, M. (1983). The distribution of cholecystokinin octapeptide-like structures in the lower brain stem of the rat: an immunohisto-chemical analysis. *Neuroscience* **9**, 587–604.

Lanaud, P., Popovici, T., Normand, E., Lemoine, C., Bloch, B., and Roques, B. P. (1989). Distribution of CCK mRNA in particular regions (hippocampus, periaqueductal grey and thalamus) of the rat by *in situ* hybridization. *Neurosci. Lett.* **104**, 38–42.

Larsson, L.-I. and Rehfeld, J. F. (1979). Localization and molecular heterogeneity of cholecystokinin in the central and peripheral nervous system. *Brain Res.* **165**, 201–18.

Loopuijt, L. D. and Van der Kooy, D. (1985). Simultaneous ultrastructural localization of cholecystokinin- and tyrosine hyroxylase-like immunoreactivity in nerve fibers of the rat nucleus accumbens. *Neurosci. Lett.* **56**, 329–34.

Lorén, I., Alumets, J., Håkanson, R., and Sundler, F. (1979). Distribution of gastrin and CCK-like peptides. *Histochemistry* **59**, 249–57.

Markey, K. A., Kondo, S., Shenkman, L., and Goldstein, M. (1980). Purification and characterization of tyrosine hydroxylase from a clonal pheochromocytoma cell line. *Mol. Pharmacol.* **17**, 79–85.

Marshall, F. H., Barnes, S., Woodruff, G. N., Hughes, J., and Hunter, J. C. (1991). Cholecystokinin modulates the release of dopamine from the anterior and posterior nucleus accumbens by two different mechanisms. *J. Neurochem.* **56**, 917–22.

Miceli, M. O., Van der Kooy, D., Post, C. A., Della-Fera, M. A., and Baile, C. A. (1987). Differential distributions of cholecystokinin in hamster and rat forebrain. *Brain Res.* **402**, 318–30.

Moyse, E., Rostène, W., Vial, M., Leonard, K., Mazella, J., Kitabgi, P., Vincent, J. P., and Beaudet, A. (1987). Regional distribution of neurotensin binding sites in rat brain: a film and light microscopic radioautographic study using monoiodo ^{125}I-Tyr^3-neurotensin. *Neuroscience* **22**, 527–36.

Müller, J. E., Straus, E. and Yalow, R. W. (1977). Cholecystokinin and its COOH-terminal octapeptide in the pig brain. *Proc. Natl Acad. Sci. USA* **74**, 3035–7.

Mutt, V. and Jorpes, J. E. (1968). Structure of porcine cholecystokinin-pancreozymin. I. Cleavage with thrombin and trypsin. *Eur. J. Biochem.* **6**, 156–62.

Nemeroff, C. B. (1986). The interaction of neurotensin with dopaminergic pathways in the central nervous system: basic neurobiology and implications for the pathogenesis and treatment of schizophrenia. *Psychoneuroendocrinology* **11**, 15–37.

Palacios, J. M., Savasta, M. and Mengod, G. (1989). Does cholecystokinin colocalize with dopamine in the human substantia nigra? *Brain Res.* **488**, 369–75.

Pelaprat, D., Broer, Y., Studler, J. M., Peschanski, M., Tassin, J. P., Glowinski, J., Rostene, W., and Roques, B. P. (1987). Autoradiography of CCK receptors in the rat brain using (^{3}H) Boc (Nle$^{28, 31}$) CCK$_{27-33}$ and (^{125}I) Bolton-Hunter CCK$_8$. Functional significance of subregional distributions. *Neurochem. Int.* **10**, 495–508.

Pelaprat, D., Dusart, I., and Peschanski, M. (1988). Postnatal development of cholecystokinin (CCK) binding sites in the rat forebrain and midbrain: an autoradiographic study. *Develop. Brain Res.* **44**, 119–32.

Quirion, R., Chiueh, C C., Everist, H. E., and Pert, A. (1985). Comparative localization of neurotensin receptors on nigrostriatal and mesolimbic dopaminergic terminals. *Brain Res.* **327**, 385–9.

Rehfeld, J. F. (1978a). Immunochemical studies on cholecystokinin. I. Development of sequence-specific radioimmunoassays for porcine triacontatriapeptide cholecystokinin. *J. Biol. Chem.* **253**, 4016–21.

Rehfeld, J. F. (1978b). Immunochemical studies on cholecystokinin. II. Distribution and molecular heterogeneity in the central nervous system and small intestine of man and hog. *J. Biol. Chem.* **253**, 4022–30.

Savasta, M., Palacios, J. M., and Mengod, G. (1988). Regional localization of the mRNA encoding for the neuropeptide cholecystokinin in the rat brain studied by *in situ* hybridization. *Neurosci. Lett.* **93**, 132–8.

Savasta, M., Ruberte, E., Palacios, J. M., and Mengod, G. (1989). The colocalization of cholecystokinin and tyrosine hydroxylase mRNA in mesencephalic dopaminergic neurons in the rat brain examined by *in situ* hybridization. *Neuroscience* **29**, 363–9.

Schalling, M. (1990). *In situ* hybridization studies on regulatory molecules in neural and endocrine tissues with special reference to expression of coexisting peptides. M. D. Thesis, Karolinska Institutet, Stockholm.

Schalling, M., Friberg, K., Bird, E., Goldstein, M., Schiffmann, S., Mailleux, P., Vanderhaeghen, J. J., and Hökfelt, T. (1989). Presence of cholecystokinin mRNA in dopamine cells in the ventral mesencephalon of a human with schizophrenia. *Acta Physiol. Scand.* **137**, 467–8.

Schalling, M., Friberg, K., Seroogy, K., Riederer, P., Bird, E., Schiffmann, S., Mailleux, P., Vanderhaeghen, J.-J., Kuga, S., Goldstein, M., Kitahama, K., Luppi, P. H., Jouvet, M., and Hökfelt, T. (1991). Analysis of expression of cholecystokinin in dopamine cells in the ventral mesencephalon of several species and in humans with schizophrenia. *Proc. Natl Acad. Sci.*, in the press.

Schotte, A. and Leysen, J. E. (1989). Autoradiographic evidence for the localization of high affinity neurotensin binding sites on dopaminergic nerve terminals in the nigrostriatal and mesolimbic pathways in rat brain. *J. Chem. Neuroanat.* **2**, 253–7.

Sekiguchi, R. and Moroji, T. (1986). A comparative study on the characterization and distribution of cholecystokinin binding sites among the rat, mouse and guinea pig brain. *Brain Res.* **99**, 271–81.

Seroogy, K. B. and Fallon, J. H. (1989). Forebrain projections from cholecystokinin like-immunoreactive neurons in the rat midbrain. *J. Comp. Neurol.* **279**, 415–35.

Seroogy, K. B., Mehta, A., and Fallon, J. H. (1987). Neurotensin and cholecystokinin coexistence within neurons of the ventral mesencephalon: projections to forebrain. *Exp. Brain Res.* **68**, 277–89.

Seroogy, K., Ceccatelli, S., Schalling, M., Hökfelt, T., Frey, P., Walsh, J., Dockray, G., Brown, J., Buchan, A., and Goldstein, M. (1988a). A subpopulation of dopaminergic neurons in rat ventral mesencephalon contains both neurotensin and cholecystokinin *Brain Res.* **455**, 88–98.

Seroogy, K., Tsuruo, Y., Hökfelt, T., Walsh, J., Fahrenkrug, J., Emson, P. C., and Goldstein, M. (1988b). Further analysis of presence of peptides in dopamine neurons. *Exp. Brain Res.* **72**, 523–34.

Seroogy, K. B., Dangaran, K., Lim, S., Haycock, J. W., and Fallon, J. H. (1989a). Ventral mesencephalic neurons containing both cholecystokinin- and tyrosine hydroxylase-like immunoreactivities project to forebrain regions. *J. Comp. Neurol.* **279**, 397–414.

Seroogy, K., Schalling, M., Brené, S., Dagerlind, Å., Chai, S. Y., Hökfelt, T., Persson, H., Brownstein, M., Huan, R., Dixon, J., Filer, D., Schlessinger, D., and Goldstein, M. (1989b). Cholecystokinin and tyrosine hydroxylase messenger RNAs in neurons of rat mesencephalon: peptide/monoamine coexistence studied using *in situ* hybridization combined with immunohistochemistry. *Exp. Brain Res.* **74**, 149–62.

Seroogy, K. B., Mohapatra, N. K., Lund, P. K., Réthelyi, M., McGehee, D. S., and Perl, E. R. (1990). Species-specific expression of cholecystokinin messenger RNA in rodent dorsal root ganglia. *Mol. Brain Res.* **7**, 171–6.

Siegel, R. E. and Young, W. S., III (1985). Detection of preprocholecystokinin and preproenkephalin A mRNAs in rat brain by hybridization histochemistry using complementary RNA probes. *Neuropeptides* **6**, 573–80.

Skirboll, L., Hökfelt, T., Norell, O., Phillipson, O. Kuypers, H. G. J. M., Bentivoglio, M., Catsman-Berrevoets, C. E., Visser, T. J., Steinbusch, H., Verhofstad, A., Cuello, A. C., Goldstein, M., and Brownstein, M. (1984). A method for specific transmitter identification of retrogradely labeled neurons: Immunofluorescence combined with fluorescence tracing. *Brain Res. Rev.* **8**, 99–127.

Sternberger, L. A. (1979). *Immunohistochemistry* (2nd edn). Wiley, New York.

Studler, J.-M., Kitabgi, P., Tramu, G., Hervé, D., Glowinski, J., and Tassin, J.-P. (1988). Extensive co-localization of neurotensin with dopamine in rat meso-cortico-frontal dopaminergic neurons. *Neuropeptides* **11**, 95–100.

Sutin, E. L. and Jacobowitz, D. M. (1990). Detection of CCK mRNA in the motor nucleus of the rat trigeminal nerve with *in situ* hybridization histochemistry. *Mol. Brain Res.* **8**, 63–8.

Szigethy, E. and Beaudet, A. (1989). Correspondence between high affinity ^{125}I-neurotensin binding sites and dopaminergic neurons in the rat substantia nigra and ventral tegmental area: a combined radioautographic and immunohistochemical light microscopic study. *J. Comp. Neurol.* **279**, 128–37.

Szigethy, E., Kitabgi, P., and Beaudet, A. (1986). Localization of neurotensin binding sites to dopaminergic and putative cholinergic neurons in rat central nervous system. *Soc. Neurosci. Abstr.* **12**, 1003.

Takahashi, Y., Kato, K., Hayashizaki, Y., Wakabayashi, T., Ohtsuka, E., Atsuki, S., Ikehara, M., and Matsubara, K. (1985). Molecular cloning of the human cholecystokinin gene by use of a synthetic probe containing deoxyinosine. *Proc. Natl Acad. Sci. USA* **82**, 1931–5.

Tramu, G., Pillez, A., and Leonardelli, J. (1978). An efficient method of antibody elution of the successive or simultaneous location of two antigens by immunocytochemistry. *J. Histochem. Cytochem.* **26**, 322–4.

Uhl, G. R. (1986). *In situ hybridization in brain*. Plenum, New York.

Vanderhaeghen, J. J. (1985). Neuronal cholecystokinin. In *Handbook of chemical neuroanatomy*, Vol. 4, *GABA and neuropeptides in the CNS* (ed. A. Björklund and T. Hökfelt), pp. 406–35. Elsevier, Amsterdam.

Vanderhaeghen, J.-J. and Crawley, J. N. (ed.) (1985). Neuronal cholecystokinin. *Ann. NY Acad. Sci.* **448**.

Vanderhaeghen, J.-J., Signeau, J. C., and Gepts, W. (1975). New peptide in the vertebrate CNS reacting with antigastrin antibodies. *Nature* **257**, 604–5.

Vanderhaeghen, J.-J., Lostra, F., De Mey, J., and Gilles, C. (1980). Immunohistochemical localization of cholecystokinin- and gastrin-like peptides in the brain and hypophysis of the rat. *Proc. Natl Acad. Sci. USA* **77**, 1190–4.

Vanderhaeghen, J. J., Deschepper, C. Lostra, F., Vierendeels, G., and Schoenen, J. (1982). Immunohistochemical evidence for cholecystokinin-like peptides in neuronal cell bodies of the rat spinal cord. *Cell Tiss. Res.* **223**, 463–7.

Van Dijk, A., Richards, J. G., Trzeciak, A., Gillessen, D., and Möhler, H. (1984). Cholecystokinin receptors: biochemical demonstration and autoradiographical localization in the rat brain and pancreas using (^{3}H) cholecystokinin$_8$, as radioligand. *J. Neurosci.* **4**, 1021–33.

Voigt, M. M. and Uhl, G. R. (1988). Preprocholecystokinin mRNA in rat brain: regional expression includes thalamus. *Mol. Brain Res.* **4**, 247–53.

Wang, R. Y. and Schoenfeld, R. (eds) (1988). *Cholecystokinin antagonists*. Alan R. Liss, New York.

Wang, R. Y., White, F. J., and Voigt, M. M. (1984). Cholecystokinin, dopamine and schizophrenia. *Trends Pharmacol. Sci.* **5**, 436–8.

Wessendorf, M. W. and Elde, R. P. (1985). Characterization of an immunofluorescence technique for the demonstration of coexisting neurotransmitters within nerve fibers and terminals. *J. Histochem. Cytochem.* **3**, 984–94.

Woulfe, J. and Beaudet, A. (1989). Immunocytochemical evidence for direct connections between neurotensin-containing axons and dopaminergic neurons in the rat ventral midbrain tegmentum. *Brain Res.* **479**, 402–6.

Young, W. S., III, (1990). *In situ* hybridization histochemistry. In *Handbook of chemical neuroanatomy*, Vol. 8, *Analysis of neuronal microcircuits and synaptic interactions* (ed. A. Björklund, T. Hökfelt, F. G. Wouterlood, and A. N. van den Pol), pp. 481–512. Elsevier, Amsterdam.

Young, W. S., III and Kuhar, M. J. (1981). Neurotensin receptor localization by light microscopic autoradiography in rat brain. *Brain Res.* **206**, 273–85.

Zamboni, L. and de Martino, C. (1967). Buffered picric-acid formaldehyde: a new rapid fixative for electron-microscopy. *J. Cell. Biol.* **35**, 148A.

Zarbin, M. A., Innis, R. B., Wamsley, J. K., Snyder, S. H., and Kuhar, M. J. (1983). Autoradiographic localization of cholecystokinin receptors in rodent brain. *J. Neurosci.* **3**, 877–906.

33. A study of the 'coexpression' of CCK and tyrosine hydroxylase messenger RNAs in the rat substantia nigra

H. Kiyama, E. McGowan, and P. C. Emson

Developments in the technique of *in situ* hybridization using enzyme-labelled or biotin-labelled DNA probes have enabled us to utilize a combination of isotopic and non-isotopic *in situ* procedures to demonstrate the presence of two messenger RNAs (mRNAs) in one cell, i.e. 'coexpression' studies (Kiyama and Emson 1990). The study of 'coexpression' is a logical development of many of the earlier studies of coexistence, where double-labelling techniques using antibodies visualized two or more immunoreactivities in a single cell (Hökfelt *et al.* 1980*a*). The problem with methods used to demonstrate coexistence was that they depended on the use of antibodies which recognized, to variable extents, epitopes related to the desired antigen. Thus in studies of coexistence possible cross-reactivity of the antibodies used may confound the study and the application of drugs such as colchicine is often required to increase the neuronal content of the antigen. As colchicine itself can induce changes in neuropeptide gene expression, the use of this drug in neuroanatomical studies needs to be carefully controlled (Kiyama and Emson 1991) as it may result in false positives in coexpression studies.

It was because of these potential problems with present coexistence studies that we developed a technique that enabled us to use two oligonucleotide probes (one radio-labelled and the other non-isotopically labelled) to visualize two mRNAs in a single neuron. This technique is much less likely than antibody studies to suffer from problems of cross-reactivity because probe sequences and

Fig. 33.1. Simultaneous demonstration of CCK and TH mRNA in substantia nigra neurons.

(a) Low magnification of a double-labelled section. At this magnification, only the TH mRNA signal (alkaline phosphatase stain) can be seen with high contrast and resolution. SNC, substantia nigra pars compacta; SNL, subtantia nigra pars lateralis; SNR, substantia nigra pars reticulata; vmSNC, ventromedial part of SNC; VTA, ventral tegmental area. Scale bar, 400 μm.

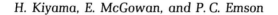
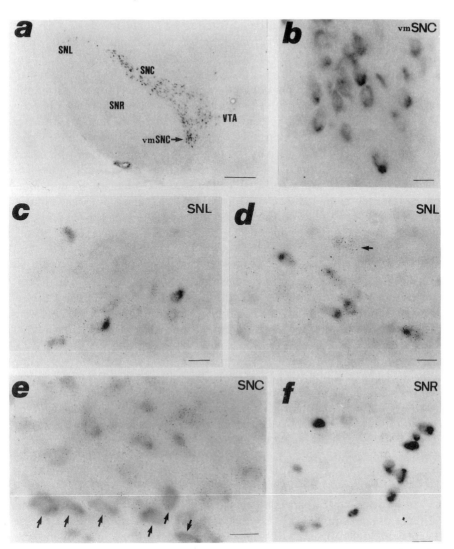

(b) Higher magnification of vmSNC (see arrow in (a)). In this ventromedial edge of substantia nigra pars compacta, only TH mRNA positive cells predominate and no CCK mRNA positive signal (silver grains) was apparent. Scale bar, 30 μm.

(c), (d) In contrast with (b), almost all the TH mRNA positive cells have a CCK mRNA signal in SNL, with an occasional single-labelled CCK mRNA positive cell (arrow) observed.

(e) In substantia nigra pars compacta proper (SNC), 60–70 per cent of TH mRNA positive cells also contain CCK mRNA. The cells which express only TH mRNA are predominantly localized in the most ventral tier (arrows). Scale bar, 30 μm.

(f) In the substantia nigra pars reticulata (SNR) the majority of cells are single-labelled and express only TH mRNA. In this region only a few TH/CCK mRNA double-labelled cells are found per brain. Scale bar, 30 μm.

conditions of hybridization can be adjusted to be highly stringent. Also, the mRNAs are visualized on fresh frozen cryostat cut sections from rats which have received no drugs such as colchicine to block axonal transport. The example we chose to investigate was the coexistence of the neuropeptide CCK with tyrosine hydroxylase (TH) in a population of dopamine cells in rat substantia nigra (Hökfelt *et al.* 1980*b*, *c*). This example of coexistence aroused interest because of the possibility that release of the neuropeptide CCK might influence dopamine release from nerve terminals in which CCK and dopamine coexist, and modulation of CCK activity in this system may provide alternative approaches to the treatment of conditions such as Parkinson's disease or schizophrenia which are associated with functional changes in dopamine activity (Hökfelt *et al.* 1980*b*, *c*; Chapters 36, 37, 38, and 39 of this volume).

The two probes used for this study were a radio-labelled ^{35}S-antisense oligodeoxynucleotide probe specific for CCK mRNA (48 mer) and a 30 mer antisense enzyme-labelled sequence specific for TH mRNA (Burgunder and Young 1988; Kiyama *et al.* 1990, 1991). The two probes were hybridized together on the same tissue section and washed to high stringency. Then the colour reaction was carried out for visualization of the enzyme-labelled probe. Once the colour reaction (TH mRNA signal) was clearly visible, sections were washed and dipped in autoradiography emulsion. After a suitable incubation time in the dark (4–6 weeks) the slides were developed and the sites of CCK mRNA expression were detected by the presence of silver grains corresponding to the presence of ^{35}S-labelled CCK probe bound to CCK mRNA.

Examination of the autoradiographs revealed that, as expected, many of the dopamine cells in rat substantia nigra contained two mRNA signals, a coloured enzyme product (TH mRNA) and silver grains corresponding to the site of CCK mRNA expression (Fig. 33.1). However, the method reveals that not all dopamine cells contained CCK mRNA: those in the pars reticulata contained only a TH mRNA signal, whilst those in the dorsal tier of the substantia nigra pars compacta contained both TH and CCK mRNAs. Detailed examination of the various zones of the pars compacta/pars reticulata revealed that there was a distinct organization of cells containing both TH and CCK mRNA. These 'coexpressing' cells appeared to belong to those populations of dopamine cells projecting to the striatal matrix, whilst the cells containing only TH mRNA are believed to project to the patch compartment of the striatum (Gerfen *et al.* 1987) (Fig. 33.2). If this is the case, it suggests that functional interactions between CCK and dopamine occur preferentially in the matrix compartment of the striatum, and that the expression of CCK mRNA marks a distinct sub-population of dopamine cells.

Acknowledgements

H. K. was supported by the Japanese Society for the Promotion of Science and The Royal Society, and E. Mc G. is a Medical Research Council Student.

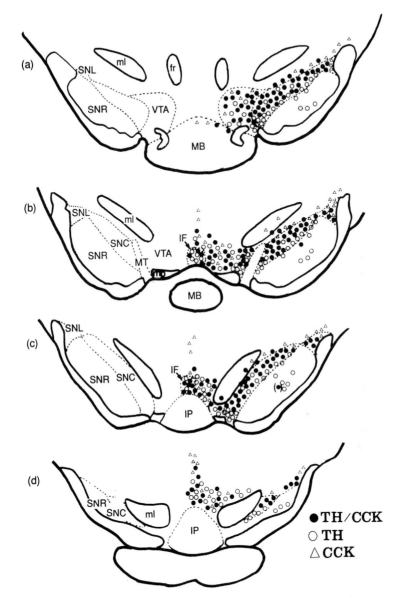

Fig. 33.2. Schematic map showing the distribution of the two mRNA signals (CCK and TH) in the rat substantia nigra and ventral tegmental area: (●), areas where double-labelled cells (CCK/TH mRNA cells) predominate; ○, single-labelled cells (TH mRNA); △, single-labelled cells (CCK mRNA). FR, fasciculus retroflexus; IF, interfascicular nucleus; IP, interpeduncular nucleus; MB, mammillary body, ml, medial lemniscus; MT, medial terminal nucleus accessory optic tract; SNC, substantia nigra pars compacta; SNL, substantia nigra pars lateralis; SNR, substantia nigra pars reticulata; VTA, ventral tegmental area.

References

Burgunder, J. M. and Young, W. S., III (1988). The distribution of thalamic projection neurons containing cholecystokinin messenger RNA, using *in situ* hybridization histochemistry and retrograde labelling. *Mol. Brain Res.* **4**, 179–89.

Gerfen, C. R., Herkenham, M., and Thibault, J. (1987). The neostriatal mosaic: II patch- and matrix-directed mesostriatal dopaminergic and non-dopaminergic systems. *J. Neurosci.* **7**, 3915–34.

Hökfelt, T., Johansson, O., Ljungdahl, A., Lundberg, J. M., and Schultzberg, M. (1980a). Peptidergic neurons. *Nature* **284**, 515–21.

Hökfelt, T., Rehfeld, J. F., Skirboll, L., Invemark, B., Goldstein, M., and Markey, K. (1980b). Evidence for the co-existence of dopamine and cholecystokinin in mesolimbic neurons. *Nature* **285**, 476–8.

Hökfelt, T., Skirboll, L., Rehfeld, J. F., Goldstein, M., Markey, K., and Dann, O. (1980c). A subpopulation of mesencephalic dopamine neurons projecting to limbic areas contain a cholecystokinin-like peptide: evidence from immunohisto-chemistry combined with retrograde tracing. *Neuroscience* **5**, 2093–124.

Kiyama, H. and Emson, P. C. (1990). Evidence for the co-expression of oxytocin and vasopressin mRNAs in magnocellular neurosecretory cells: Simultaneous demonstration of two neurohypophysin mRNA's by hybridization histochemistry. *J. Neuroendocrinol.* **2**, 257–9.

Kiyama, H. and Emson, P. C. (1991). Colchicine induced expression of proneurotensin mRNA in rat striatum and hypothalamus. *Mol. Brain Res.*, **9**, 353–7.

Kiyama, H., Emson, P. C., Ruth, J., and Morgan, C. (1990). Sensitive non-radioactive *in situ* hybridization histochemistry: demonstration of tyrosine hydroxylase gene expression in rat brain and adrenal. *Mol. Brain Res.* **7**, 213–19.

Kiyama, H., McGowan, E. M., and Emson, P. C. (1991). Co-expression of cholecystokinin mRNA and tyrosine hydroxylase mRNA in populations of rat substantia nigra cells; a study using a combined radioactive and non-radioactive *in situ* hybridisation procedure. *Mol. Brain Res.*, **9**, 87–93.

34. Does hippocampal CCK influence dopaminergic input to the nucleus accumbens? Some anatomical observations

S. Totterdell and A. D. Smith

The peptide CCK is widely distributed in the brain, where it is thought to act as a neurotransmitter/neuromodulator (Crawley, 1988). Concentrations are high in the cerebral cortex and also in the hippocampus (Beinfeld, 1983), where it is considered to be located in intrinsic and projection neurons (Handelman *et al*. 1983). Levels of CCK are reported to be decreased post mortem in the hippocampus and amygdala of patients diagnosed as schizophrenic (Roberts *et al*. 1983).

Neurons in the hippocampus project to the nucleus accumbens (Kelley and Domesick 1982), an area which also receives dopaminergic input from the ventral tegmental area (VTA). There is considerable evidence that CCK coexists with dopamine in at least some of the VTA neurons which project to the nucleus accumbens (Hökfelt *et al*. 1980), and this has led to studies of the interactions between CCK and dopamine release in the nucleus accumbens. However, it might also be possible for CCK in the hippocampus to influence dopamine activity in the nucleus accumbens. Therefore we investigated whether the neurons which project from the hippocampus to the nucleus accumbens receive input from intrinsic hippocampal CCK neurons, and whether this hippocampal input forms contacts with the same neurons in the nucleus accumbens that receive input from the VTA.

Stereotaxic injections of the retrograde tracer WGA-HRP were made into the nucleus accumbens of anaesthetized rats. After 48–68 h survival, the rats were fixed by perfusion (Totterdell and Smith 1986). The brains were removed and washed thoroughly in 0.1 M phosphate buffer, and 70 μm sections of the hippocampus were cut using a vibrating microtome. The sections were reacted using diaminobenzidine (DAB) as the chromagen to reveal the retrogradely transported WGA-HRP. Some of these hippocampal sections were then incubated with an antibody to CCK (Dockray *et al*. 1981) and the immunolabelled profiles were revealed by the PAP method. After treatment with 1 per cent osmium tetroxide, about half of these sections were further processed by the single-section Golgi procedure (Freund and Somogyi 1983) and impregnated neurons were then 'gold-toned' (Fairen *et al.* 1977). Finally, all sections were dehydrated and mounted in resin for light microscopic examination. Areas of interest were selected for further study in the electron microscope.

In the second part of the study, the hippocampal input to the nucleus accumbens was labelled in a number of ways. An electrolytic lesion of the fimbria-fornix, made to interrupt the pathway, resulted in degeneration of boutons of hippocampal origin in the nucleus accumbens. In other rats the anterograde tracer, *Phaseolus vulgaris* leucoagglutinin (PHA-L) was iontophoresed into the ventral subiculum (+5 mA, 7 s on, 7 s off), and the PHA-L-containing fibres in the nucleus accumbens were revealed using an antibody to PHA-L and the PAP or avidin–biotin method, with DAB as the chromagen, The anterograde tracer biocytin, which can be pressure-injected and requires shorter survival times than PHA-L, was also used to demonstrate these fibres. After fixation by perfusion, sections taken through the nucleus accumbens were first reacted to reveal the anterograde tracer (for PHA-L or biocytin) and then the dopaminergic input was demonstrated using an antibody to tyrosine hydroxylase (TH) (Van den Pol *et al.* 1984). In combinations where the anterograde tracer had been revealed with DAB, TH was demonstrated with benzidine dihydrochloride (BDHC). Various methods such as detergents, partial enzyme digestion, and freeze–thawing were used to enhance penetration of the immunoreagents into the tissue.

Again some of the sections were processed by the single-section Golgi method, and all were dehydrated and mounted on slides in resin for light microscopy. Selected areas were then studied in the electron microscope.

Light microscopy in the hippocampus

Neurons retrogradely labelled from the nucleus accumbens were found mainly in the pyramidal layer of the ventral subiculum. Large WGA-HRP injections also labelled neurons in the dorsal subiculum and CA1. The retrogradely labelled neurons had characteristic dense granules in their somata and proximal dendrites (Fig. 34.1) and the cell bodies were generally pyramidal in appearance, although a very small number of non-pyramidal neurons were found to contain these granules (Totterdell and Hayes 1987). Golgi impregnation of some retrogradely labelled neurons confirmed their pyramidal morphology, with branching basal dendrites and a single main apical dendrite (Fig. 34.1).

CCK immunocytochemistry resulted in the labelling of cells in all hippocampal layers. These CCK neurons were generally bipolar and most frequent in the stratum radiatum and the stratum moleculare-lacunosum. CCK-immunoreactive fibres were found to form pericellular nets around the somata in the pyramidal cell layer.

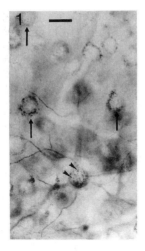

Fig. 34.1. Light micrograph of a Golgi-impregnated gold-toned pyramidal projection neuron in the subiculum which contains granular HRP reaction product (arrowheads). Other neurons which also project to the nucleus accumbens, but which are not Golgi-impregnated, are present (arrows). Scale bar, 20 μm.

Electron microscopy in the hippocampus

Neuronal cell bodies containing characteristic electron-dense HRP granules (Fig. 34.2) had the typical ultrastructural features of pyramidal neurons such as a smooth nuclear envelope, no nuclear inclusions, and no boutons in asymmetric synaptic contact with the soma.

CCK-immunoreactive boutons contained generally round synaptic vesicles, coated with reaction product (Fig. 34.2). These boutons were elongated or rounded, and often contained mitochondria. They were found to make symmetrical synaptic contacts with the somata (Fig. 34.2), proximal dendrites, and somatic spines of neurons that project to the nucleus accumbens.

Light microscopy in the nucleus accumbens

The anterograde tracers PHA-L and biocytin labelled fibres in the medial nucleus accumbens which were fine and varicose. The distribution of this hippocampal input overlapped with that of the dopaminergic input, demonstrated with an antibody to TH and presumably originating in the VTA. In material in which both hippocampal and dopaminergic fibres were labelled, these could be distinguished by the use of DAB or BDHC as chromagens. The anterograde label in fibres reacted with DAB is brown and amorphous, whereas the TH-immunoreactive fibres reacted with BDHC are blue–black and crystalline.

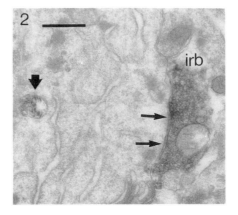

Fig. 34.2. A CCK-immunoreactive bouton (irb) is in symmetrical synaptic contact (arrows) with a soma in the pyramidal layer of the subiculum containing HRP (broad arrow) retrogradely transported from the nucleus accumbens. Scale bar, 0.5 μm.

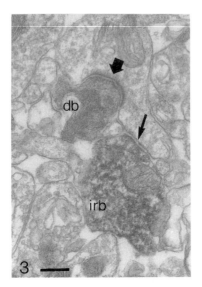

Fig. 34.3. A bouton of hippocampal origin, degenerating following a fimbria-fornix lesion (db), is in asymmetrical synaptic contact (broad arrow) with a spine in the nucleus accumbens which is also contacted by a TH-immunoreactive bouton (irb) making a symmetrical synapse (arrow). Scale bar, 0.2 μm.

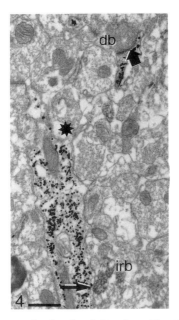

Fig. 34.4. Electron micrograph of a dendrite of an identified spiny neuron in the nucleus accumbens containing the dense particles typical of a Golgi-impregnated gold-toned neuron. Note the emergence of a spine (star). A bouton of hippocampal origin, degenerating after a fimbria-fornix lesion (db), is in asymmetrical synaptic contact (broad arrow) with the head of a spine. Another bouton, immunoreactive for TH (irb), is in symmetrical synaptic contact (arrow) with dendritic shaft. Scale bar, 0.5 μm.

Electron microscopy in the nucleus accumbens

The hippocampal input to the nucleus accumbens, whether identified by degeneration as a result of a fimbria-fornix lesion or by the anterograde transport of PHA-L or biocytin, terminated in boutons which were in asymmetric synaptic contact principally with spines (Figs 34.3 and 34.4). The TH-immunoreactive boutons demonstrated with BDHC contain irregular electron-dense crystals which often cause local disruption of the ultrastructure, obscuring synaptic contacts. Boutons immunoreactive for TH, demonstrated with DAB, contain vesicles coated with the typical amorphous deposit (Figs 34.3 and 34.4), and were seen in symmetrical synaptic contact with dendritic shafts (65 per cent) (Fig. 34.4) or spines (34 per cent) (Fig. 34.3).

Examples of a structure receiving both hippocampal and dopaminergic inputs were found. In all, ten examples of convergence were found in six blocks, sectioned for electron microscopy, taken from three rats. In some cases the two boutons were in synaptic contact with the same profile (Fig. 34.3). However, in order to increase the likelihood of identifying examples of convergence,

Golgi-impregnated neurons in the nucleus accumbens, where both degenerating boutons and TH-immunoreactive boutons had been located, were sectioned for electron microscopy. In a section from an identified medium-sized spiny neuron (Fig. 34.4), it can be seen that a TH-immunoreactive bouton is in symmetrical synaptic contact with a dendritic shaft from which arises a spine, also Golgi-impregnated, which is contacted by a degenerating bouton in asymmetrical synaptic contact.

Conclusions

These studies demonstrate that the hippocampal input to the nucleus accumbens originates principally from the pyramidal neurons of the ventral subiculum. Larger injections, which may have involved the lateral septum, resulted in labelling of CA1 cells. These subicular projection neurons receive input from CCK-immunoreactive boutons, possibly arising from intrinsic neurons (Nunzi *et al.* 1985) which are in symmetrical synaptic contact with somata and proximal dendrites. It was not possible in our material to examine whether there is CCK-immunoreactive input to more distal dendrites of projection neurons.

A very small number of non-pyramidal neurons were found to project to the nucleus accumbens, but these neurons appeared not to be CCK-immunoreactive (Totterdell and Hayes 1987).

It was of interest that all the CCK-immunoreactive boutons found in this study were in symmetrical synaptic contact with their targets. CCK is reported to be excitatory in its action (Dodd and Kelly 1981), and 'excitatory synapses' are generally thought to be asymmetrical. It may be that the coexistence of CCK with γ-aminobutyrate (GABA) (Somogyi *et al.* 1984) explains this discrepancy. CCK-immunoreactive boutons in the nucleus accumbens are reported to form mainly asymmetrical synapses (Baali-Cherif *et al.* 1984).

In the nucleus accumbens an antibody to TH was used to demonstrate dopaminergic input, presumably from the VTA (Hökfelt *et al.* 1980). TH is the rate-limiting enzyme in the synthetic pathway for both dopamine and noradrenaline, but staining patterns with antibodies to dopamine itself and dopamine β-hydroxylase, the enzyme which converts dopamine to noradrenaline, would suggest that the TH immunoreactivity here coincides with dopamine-containing profiles (Totterdell and Smith 1989). TH-immunoreactive boutons were seen to make only symmetrical synaptic contacts, 65 per cent with dendritic shafts. The observation that 90 per cent of CCK-immunoreactive boutons in the nucleus accumbens participate in asymmetrical synaptic contacts (Baali-Cherif *et al.* 1984), mainly with spines, would suggest that the majority of the boutons examined in that study did not also contain dopamine. However, not all the CCK in the nucleus accumbens originates in the A10 area.

The different methods used to label the hippocampal input to the nucleus accumbens identify similar populations of boutons, found largely in asymmterical synaptic contact with spines. This observation supports the use of a fimbria-fornix lesion to identify the hippocampal input to the nucleus accumbens. The tracer biocytin would seem to have some advantages over PHA-L in this pathway, since it is transported more quickly and does not rely on an antibody for detection, resulting in fewer problems with penetration and better ultrastructural preservation.

The problem of immunocytochemically labelling two populations of boutons in the same tissue has been addressed by using two different chromagens, DAB and BDHC. It is possible to distinguish between these two end-products in the electron microscope, but the coarser BDHC crystals compromise the ultrastructure. Consequently, examples of convergence have so far come from experiments combining a fimbria-fornix lesion with TH immunocytochemistry. These examples are still quite difficult to locate since both the lesion, which may be incomplete, and the TH labelling, which is often restricted by poor penetration, result in an underestimation of the respective inputs. The use of Golgi impregnation increases the chance of identifying convergence, but places additional constraints on the processing of the tissue.

There are several possible sites for the interaction between CCK and dopamine. Many studies have focused on their coexistence and corelease (Studler et al. 1984) in the VTA–nucleus accumbens pathway. CCK would seem to be excitatory in the nucleus accumbens but acts as a modulator of dopamine activity in a dose- and location-specific way (Crawley 1988). Dopaminergic pathways are also reported from the VTA to various limbic areas, including the hippocampus (Campana et al. 1989) where this input could interact with the intrinsic CCK. Finally, it is suggested that dopamine release in the striatum is regulated by the glutamatergic corticostriatal fibres, either presynaptically (Chéramy et al. 1987) or postsynaptically (Freund et al. 1985). Similarly, dopamine release and activity in the nucleus accumbens could be influenced by input to the accumbens from neurons in the hippocampus, which might themselves be modulated by intrinsic CCK input.

Neither of the first two alternatives have yet been demonstrated at the synaptic level. However, we have provided anatomical evidence to support the third hypothesis at the level of the synapse. Our observations could also provide an anatomical explanation for the effectiveness of drugs that act on the dopamine system to treat a disorder such as schizophrenia, which may arise from abnormal transmission in the limbic system.

References

Baali-Cherif, H., Arluison, M., and Tramu, G. (1984). Cholecystokinin-like immunoreactivity in the rat nucleus accumbens. An ultrastructural study. *Neurosci. Lett.* **49**, 331–5.

Beinfeld, M. C. (1983). Cholecystokinin in the central nervous system: a minireview. *Neuropeptides* **3**, 411–27.

Campana, E., Gasbarri, A., and Pacitti, C. (1989). The mesolimbic dopaminergic projection to the hippocampal formation in the rat: distribution and organization of nucleus ventralis tegmenti efferents. *Behav. Pharmacol.* **1**, S211.

Chéramy, A., Romo, R., and Glowinski, J. (1987). Role of corticostriatal glutamatergic neurons in the presynaptic control of dopamine release. In *Neurotransmitter interactions in the basal ganglia* (eds M. Sandler, C. Feuerstein, and B. Scatton), pp. 133–41. Raven Press, New York.

Crawley, J. N. (1988). Neuronal cholecystokinin. In *ISI atlas of science: pharmacology*, Vol. 2, pp. 84–90.

Dockray, G. J., Williams, R. G., and Zhu, W.-Y. (1981). Development of region specific antisera for the C-terminal tetrapeptide of gastrin/cholecystokinin and their use in studies of immunoreactive forms of cholecystokinin in rat brain. *Neurochem. Int.* **3**, 281–8.

Dodd, J. and Kelly, J. S. (1981). The actions of cholecystokinin and related peptides on pyramidal neurones of the mammalian hippocampus. *Brain Res.* **205**, 337–50.

Fairén, A., Peters, A., and Saldanha, J. (1977). A new procedure for examining Golgi impregnated neurons by light and electron microscopy. *J. Neurocytol.* **6**, 311–37.

Freund, T. F. and Somogyi, P. (1983). The section Golgi impregnation procedure. 1. Description of the method and its combination with histochemistry after intracellular iontophoresis or retrograde transport of horseradish peroxidase. *Neuroscience* **9**, 463–74.

Freund, T. F., Powell, J. F., and Smith, A. D. (1985), Tyrosine hydroxylase-immunoreactive boutons in synaptic contact with identified striatonigral neurons, with particular reference to dendritic spines. *Neuroscience* **13**, 1189–1215.

Handelmann, G. E., Beinfeld, M. C., O'Donohue, T. L., Nelson, J. B., and Brenneman, D. E. (1983). Extra-hippocampal projections of CCK neurons of the hippocampus and subiculum. *Peptides* **4**, 331–4.

Hökfelt, T., Skirboll, L., Rehfeld, J. F., Goldstein, M., Markey, K., and Dann, O. (1980). A subpopulation of mesencephalic dopamine neurons projecting to limbic areas contains a cholecystokinin-like peptide: evidence from immunohistochemistry combined with retrograde tracing. *Neuroscience* **5**, 2093–124.

Kelley, A. E. and Domesick, V. B. (1982). The distribution of the projection from the hippocampal formation to the nucleus accumbens in the rat: an anterograde- and retrograde- horseradish peroxidase study. *Neuroscience* **7**, 2321–35.

Nunzi, M. G., Gorio, A., Milan, F., Freund, T. F., Somogyi P., and Smith, A. D. (1985). Cholecystokinin-immunoreactive cells form symmetrical synaptic contacts with pyramidal and nonpyramidal neurons in the hippocampus. *J. Comp. Neurol.* **237**, 485–505.

Roberts, G. W., Ferrier, N., Lee, Y., Crow, T. J., Johnstone, E. C., Owens, D. G. C., Bacarese-Hamilton, A. J., McGregor, G., O'Shaughnessy, D., Polak, J. M., and Bloom, S. R. (1983). Peptides, the limbic lobe and schizophrenia. *Brain Res.* **288**, 199–211.

Somogyi, P., Hodgson, A. J., Smith, A. D., Nunzi, M., Gorio, A., and Wu, J.-Y. (1984). Different populations of GABAergic neurons in the visual cortex and hippocampus of the cat contain somatostatin- or cholecystokinin-like immunoreactivity. *J. Neurosci.* **4**, 2590–603.

Studler, J. M., Reibaud, M., Tramu, G., Blanc, G., Glowinski, J., and Tassin, J. P. (1984). Pharmacological study on the mixed CCK8/DA meso-nucleus accumbens pathway: evidence for the existence of storage sites containing the two transmitters. *Brain Res.* **298**, 91–7.

Totterdell, S. and Hayes, L. (1987). Non-pyramidal hippocampal projection neurons: a light and electron miscroscopic study. *J. Neurocytol.* **16**, 477–85.

Totterdell, S. and Smith, A. D. (1986). Cholecystokinin-immunoreactive boutons in synaptic contact with hippocampal pyramidal neurons that project to the nucleus accumbens. *Neuroscience* **19**, 181–92.

Totterdell, S. and Smith, A. D. (1989). Convergence of hippocampal and dopaminergic input onto identified neurons in the nucleus accumbens of the rat. *J. Chem. Neuroanat.* **2**, 285–98.

Van den Pol, A. N., Herbst, R., and Powell, J. F. (1984). Tyrosine hydroxylase-immunoreactive neurons of the hypothalamus: a light and electron microscopic study. *Neuroscience* **13**, 1117–56.

35. Functional analyses of the coexistence of CCK and dopamine

Jacqueline N. Crawley, Susan M. Fiske, John R. Evers, Mark C. Austin, and Margery C. Beinfeld

Introduction

The discovery of the peptide CCK in porcine gut (Mutt 1980) and rat brain (Vanderhaegen *et al.* 1975), was quickly followed by the discovery of CCK-like immunoreactivity in mid-brain dopamine (DA) neurons (Hökfelt *et al.* 1980*a*). Immunohistochemical and *in situ* hybridization studies have confirmed the presence of CCK in dopaminergic neurons of the ventral tegmentum and substantia nigra of rats, cats, and primates (Hökfelt *et al.* 1980*a, b*; Freeman and Chiodo 1988; Ingram *et al.* 1989; Savasta *et al.* 1989; Seroogy *et al.* 1989*a, b*; Burgunder and Young 1990; Cottingham *et al.* 1990; Jayaraman *et al.* 1990; see Chapter 32 of this volume). Retrograde tracing studies and lesion techniques have identified the major projections of this CCK–DA coexistence as terminating in the medial caudal nucleus accumbens, the medial olfactory tubercle, and the medial septum, with a minor projection to the caudate nucleus (Hökfelt *et al.* 1980*b*; Marley *et al.* 1982; Fallon *et al.* 1983; Studler *et al.* 1981; Seroogy *et al.* 1989*a*; Adams and Fisher 1990). Demonstration of the anatomical distribution of CCK in the mesolimbic DA pathway sparked the intriguing hypothesis that CCK contributes to the functional activity of these dopaminergic brain regions, which have been implicated in the pathophysiology and treatment of schizophrenia.

The first functional examination of the CCK–DA coexistence took the form of neurophysiological studies of the effects of CCK on neuronal firing rates of mid-brain DA neurons in the rat. Skirboll and co-workers found that micro-iontophoresis of CCK increased the firing rate of ventral tegmental and substantia nigra neurons (Skirboll *et al.* 1981; Hommer *et al.* 1985). This excitatory effect of CCK was blocked by the CCK receptor antagonist proglumide (Chiodo *et al.* 1987). When given in combination with DA, or the dopaminergic agonist apomorphine, CCK facilitated the inhibitory effects of DA or apomorphine, potentiating the reduction in firing rates induced by the dopaminergic drugs (Hommer and Skirboll 1983; Hommer *et al.* 1986; Brodie and Dunwiddie 1987; Freeman and Bunney 1987; Freeman and Chiodo 1988; Stittsworth and Mueller 1990). The ability of CCK to increase the firing rate of mesolimbic neurons, and to potentiate the DA-induced decrease in their firing rate, was specific to regions of the ventral tegmentum and substantia nigra zona compacta that have been reported to contain coexisting DA–CCK (Hommer *et al.* 1986).

Behavioural studies

The second approach to investigating the functional importance of CCK in the mesolimbic pathway was to use behavioural measures. Simple rodent paradigms, including exploratory locomotion and stereotyped sniffing, which are well-established measures of dopaminergic activity of the mesolimbic and nigrostriatal pathways respectively, were employed to test the ability of CCK to mimic, block, or modulate dopaminergically induced behaviour. High (microgram) doses of CCK were found to inhibit spontaneous locomotion, rearing, and grooming (Britton *et al.* 1989; see Chapter 22 of this volume) and to inhibit amphetamine-induced rearing behaviour (Schneider *et al.* 1983). Lower (nanogram) doses of CCK micro-injected into the nucleus accumbens have been reported to potentiate DA-induced hyperlocomotion (Crawley *et al.* 1985*a*), to inhibit amphetamine-induced hyperlocomotion (Weiss *et al.* 1988), to potentiate apomorphine-induced stereotypy (Crawley *et al.* 1985*a*; Weiss *et al.* 1988), to antagonize apomorphine-induced hypolocomotion (Van Ree *et al.* 1983), and to decrease spontaneous locomotion and rearing (Katsuura *et al.* 1985; Daugé *et al.* 1989*a*, *b*). In addition, the CCK antagonist proglumide produced a shift to the left in the dose-response curve for the effects of haloperidol on stereotyped behaviours (Csernansky *et al.* 1987).

These conflicting reports of both facilitatory and inhibitory effects of CCK on DA-mediated behaviour can be resolved if a number of factors, including the motivational state of the rat, the dose of CCK, and the region of the nucleus accumbens, are taken into account. Daugé *et al.* (1989*a*) found that CCK alone decreased exploratory locomotion and rearing when administered into the nucleus accumbens of rats not habituated to the test environment, but had no effect when administered to habituated rats. Their results suggested that the motivational level, or the level of spontaneous exploration, may determine basal dopaminergic activity and the subsequent response to CCK. Crawley and co-workers tested a wide range of doses of CCK, alone and with DA, and analysed topographical regions within the nucleus accumbens. CCK micro-injected into the medial posterior nucleus accumbens did not mimic or block DA-induced hyperlocomotion over a dose range of 20 pg–4 μg, but potentiated DA-induced hyperlocomotion when given in combination with DA and apomorphine-induced stereotypy when given in combination with apomorphine at CCK doses of 20 pg–200 ng (Crawley *et al.* 1985*a*). When micro-injected into the caudate nucleus, CCK did not significantly affect spontaneous locomotion or stereotypy, or apomorphine-induced stereotypy (Crawley *et al.* 1985*a*). Topographical analysis of the actions of CCK within subregions of the nucleus accumbens on potentiating DA-mediated behaviour supported the interpretation that CCK influences dopaminergic function primarily at sites of the terminals of the CCK–DA coexistence. When micro-injected into the medial posterior nucleus accumbens, CCK strongly potentiated DA-induced hyper-locomotion. When micro-injected into the anterior nucleus accumbens or into the lateral nucleus accumbens, CCK had no effect or partially inhibited DA-induced hyperlocomotion (Crawley *et al.* 1985*b*). While CCK is found in the anterior nucleus accumbens and caudate nucleus in high concentrations, the majority of CCK-containing terminals in these regions appear to originate from cell bodies in the cerebral cortex (e.g. the claustrum) (Meyer *et al.* 1982; Hökfelt *et al.* 1985). These behavioural findings are consistent with a biochemical study showing that in the rat, CCK potentiated DA-induced adenylate cyclase activity in the posterior nucleus accumbens, while CCK inhibited DA-induced adenylate cyclase activity in the anterior nucleus accumbens (Studler *et al.* 1986) and the caudate nucleus (Chapter 40 of this volume). Subsequent behavioural studies have similarly found differential responses to CCK in the anterior versus posterior nucleus accumbens. CCK potentiated self-stimulation when micro-injected into the medial posterior nucleus accumbens, but inhibited it when micro-injected into the anterior nucleus accumbens of the rat (De Witte *et al.* 1987). Similarly, the CCK antagonist proglumide antagonized intracranial self-stimulation of the ventral tegmental area when administered into the caudal nucleus accumbens, but potentiated it when administered into the rostral nucleus accumbens (Vaccarino and Vaccarino 1989). In addition, nanogram doses of CCK potentiated amphetamine-induced hyperlocomotion when CCK was micro-injected into the caudal nucleus accumbens, but inhibited it when micro-injected into the anterior nucleus accumbens (Vaccarino and Rankin 1989). Taken together, these findings suggest that the nucleus accumbens of the rat consists of rostral and caudal subregions that are distinguishable by dif-ferential responses to CCK, and which correspond to the pattern of innervation by the dopaminergic ventral tegmental neurons which do and do not contain CCK.

The ability of CCK to potentiate DA-induced hyperlocomotion in the nucleus accumbens was not mimicked by unsulphated CCK-8 (CCK-8US) or by CCK-4 (Crawley *et al.* 1986). The ability of CCK to potentiate DA-induced hyperlocomotion in the nucleus accumbens was blocked by proglumide, an antagonist of the peripheral type CCK receptor (Crawley *et al.* 1986).

This pharmacological profile for the behavioural effects of CCK resembles that for the binding characteristics of the CCK receptor found in the gut. Our behavioural findings were the first suggestion of the existence of the peripheral type CCK receptor CCK_A in the nucleus accumbens of the rat. Subsequent behavioural analyses (Daugé *et al.* 1989*b*, 1990), a receptor binding study (Rovati and Makovec 1988), and release studies (Vickroy *et al.* 1988; Marshall *et al.* 1990) have inferentially supported the presence of peripheral type CCK receptors in the nucleus accumbens. However, direct autoradiographic binding studies using radio-labelled CCK or [³H] devazepide, a ligand selective for the peripheral type receptor, have not visualized detectable quantities of peripheral type CCK receptors in the nucleus accumbens (Moran *et al.* 1986; Hill *et al.* 1987, 1990; Peleprat *et al.* 1987).

Analogous studies of the behavioural effects of CCK have been conducted in the cell body region of the ventral tegmentum. Nanogram doses of CCK alone had no effect on exploratory locomotion, but potentiated the inhibitory effects of DA (Crawley 1989). Topographical analysis showed that CCK potentiated DA-induced hypolocomotion when micro-injected into the medial and caudal regions of the ventral tegmental area (VTA), but had no effect when micro-injected into the rostral VTA or regions outside the VTA (Crawley 1989). Pharmacological analysis showed that CCK-8US and CCK-4 also potentiated DA-induced hypolocomotion in the VTA, and that the effects of CCK-8US were not blocked by the peripheral type CCK receptor antagonist proglumide at doses which blocked the behavioural effects of CCK in the nucleus accumbens (Crawley 1989). Thus, a peripheral receptor subtype appears to mediate the behavioural effects of CCK in the nucleus accumbens, while a central receptor subtype appears to mediate the behavioural effects of CCK in the VTA.

The development of highly selective potent antagonists of both the peripheral and the central subtypes of the CCK receptor make possible a more complete analysis of the receptor subtypes mediating the behavioural effects of CCK in both the cell body and the terminal region of the CCK–DA coexistence. Devazepide is a potent non-peptide antagonist selective for the peripheral type CCK receptor (Chang and Lotti 1986; Chapter 2 of this volume). The benzodiazepine-like structure of this compound has been shown to penetrate the blood–brain barrier relatively efficiently (Pullen and Hodgson 1987). L-365,260, an analogue of devazepide is a potent non-peptide antagonist selective for the brain type CCK receptor (Lotti and Chang 1989; Chapter 2 of this volume). Therefore the behavioural paradigms described above were employed to test the ability of devazepide and L-365,260 to block CCK potentiation of DA-mediated locomotion in the nucleus accumbens and VTA of the rat.

Male Sprague–Dawley rats, weighing 250 g at the start of the experiment, were stereotaxically implanted with bilateral 24-gauge hypodermic stainless steel guide cannulae under chloral hydrate anaesthesia. Co-ordinates for the medial posterior nucleus accumbens were 1.2 mm rostral and 1.3 mm lateral to bregma, and 5.7 mm ventral to the surface of the skull; co-ordinates for the central VTA were 5.8 mm caudal and 0.5 mm lateral to bregma, and 6.5 mm ventral to the surface of the skull. A week after surgery, rats were micro-injected with 0.2 μl of treatment A, followed 5 min later by 0.2 μl of treatment B. Treatment A was vehicle (2.5 per cent ethanol, 2.5 per cent propylene glycol, 95 per cent saline), devazepide (10 ng in vehicle), or L-365,260 (10 ng in vehicle) (compounds kindly supplied by Dr R. M. Freidinger, Merck Sharp & Dohme,

West Point, PA). Treatment B was 0.9 per cent physiological saline (SAL), DA (10 μg into the nucleus accumbens, 2.5 μg into the VTA in 1 per cent ascorbate and saline), or DA + CCK (10 ng into the nucleus accumbens, 200 ng into the VTA) (DA was obtained from the Sigma Chemical Co., St. Louis, MO, and CCK from Bachem Bioscience Inc., Philadelphia, PA). Each micro-injection was performed over a 1 min period, plus 30 s before removal of the injection tube and replacement of the stylet closure, using a 31 gauge injection tube which extended 2 mm below the ventral tip of the guide cannula. Immediately after micro-injection, the rat was placed in the centre of a Digiscan optical animal activity monitor (Omnitech Electronics Inc., Columbus, OH) for 15 min. Ambulatory locomotor activity was recorded for statistical analysis by one-way ANOVA, followed by Newman–Keuls analysis of differences between individual means. Each animal was used twice, with at least a week between uses and with different treatments randomized over the two uses. Each drug treatment group consisted of at least six rats.

As shown in Fig. 35.1, in the medial posterior nucleus accumbens DA significantly increased ambulatory locomotion and CCK potentiated DA-induced hyperlocomotion after pretreatment with vehicle (ANOVA $F\,(2, 17) = 52.9$, $p < 0.001$; $p < 0.01$ for SAL versus DA, $p < 0.05$ for DA versus DA + CCK). Devazepide blocked the ability of CCK to potentiate DA-induced hyperlocomotion, while having no effect alone on locomotion in saline-treated rats and no effect alone on DA-induced hyperlocomotion (ANOVA $F\,(2, 19) = 5.38$, $p < 0.02$; $p < 0.05$ SAL versus DA, no significant difference between DA versus DA + CCK). L-365,260 did not change normal locomotion, DA-induced hyperlocomotion, or CCK potentiation of DA-induced hyperlocomotion in the nucleus accumbens (ANOVA $F\,(2, 19) = 15.05$, $p < 0.001$; $p < 0.01$ for SAL versus DA, $p < 0.05$ for DA versus DA + CCK). In the medial VTA, DA inhibited locomotion and CCK potentiated DA-induced hypolocomotion after treatment with vehicle ($F\,(2, 51) = 28.7$, $p < 0.001$; $p < 0.01$ SAL versus DA, $p < 0.05$ DA versus DA + CCK). Devazepide did not significantly affect normal locomotion, DA-induced hypolocomotion, or CCK potentiation of DA-induced hypolocomotion (ANOVA $F\,(2, 13) = 4.35$, $p < 0.05$; $p < 0.05$ SAL versus DA + CCK). L-365,260 blocked CCK potentiation of DA-induced hypolocomotion (ANOVA $F\,(2, 21) = 7.07$, $p < 0.01$; $p < 0.01$ SAL versus DA, $p < 0.05$ DA versus DA + CCK).

These behavioural results suggest that devazepide (10 ng) can block the behavioural effects of CCK (10 ng) in the nucleus accumbens, and that L-365,260 (10 ng) can block the behavioural effects of CCK (200 ng) in the VTA. The specificities of the peripheral type antagonist for the nucleus accumbens and of the central type antagonist for the VTA are consistent with the pharmacological profile previously obtained for CCK agonists at these two sites. The trend for L-365,260 to inhibit exploratory locomotion when given alone, and with DA, is interesting in view of the excitatory neurophysiological effects of CCK at CCK$_B$ receptors throughout the brain (Hommer *et al.* 1985) and will be the subject of further investigations.

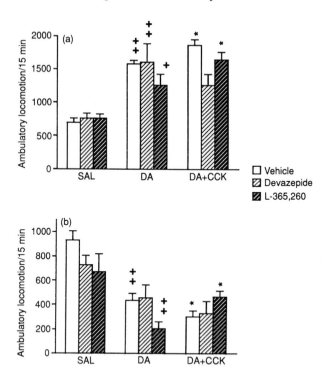

Fig. 35.1. Behavioural analysis of CCK antagonists in the rat nucleus accumbens and ventral tegmental area.

(a) Saline (SAL) (0.2 μl), DA (10 μg), or DA (10 μg) + CCK-8S (10 ng) was bilaterally micro-injected into the medial posterior nucleus accumbens 5 min after microinjection of vehicle (0.2 μl), devazepide (10 ng), or L-365,260 (10 ng) into the same site, and immediately before testing for exploratory locomotor activity in a Digiscan activity monitor. CCK potentiated DA-induced hyperlocomotion after vehicle pretreatment, as previously reported in rats without vehicle pretreatment (Crawley *et al.* 1985*a*, *b*). Devazepide blocked the CCK potentiation of DA-induced hyperlocomotion. L-365,260 was without effect in the nucleus accumbens.

(b) SAL, (0.2 μl), DA (2.5 μg), or DA (2.5 μg) + CCK (200 ng) was bilaterally micro-injected into the medial VTA 5 min after micro-injection of vehicle (0.2 μl), devazepide (10 ng), or L-365,260 (10 ng) into the same site, and immediately before testing for exploratory locomotor activity in a Digiscan activity monitor. CCK potentiated DA-induced hypolocomotion after vehicle pretreatment, as previously reported in rats without vehicle pretreatment (Crawley 1989). L-365,260 blocked the CCK potentiation of DA-induced hypolocomotion. Devazepide had no effect in the VTA.

Data are expressed as mean ± SEM ($n = 6$–8 per data point): $+ p < 0.05$, $+ + p < 0.01$ as compared with SAL; $\star p < 0.05$, compared with DA.

374 Functional analyses of the coexistence of CCK and dopamine

Release studies

A fundamental question in neuropeptide investigations is whether the peptide transmitter is released under physiologically relevant conditions. In the case of coexistence, the multiple transmitters may be differentially released (Bartfai *et al.* 1988). Does one transmitter influence the release of the other coexisting transmitter? Several release studies have addressed these issues for the mesolimbic coexistence of CCK and DA.

In vitro slice perfusion, *in vivo* push–pull cannula perfusion, *in vivo* microdialysis, and *in vivo* voltammetry techniques have been used to assess the role of CCK in regulating the release of DA. In slices from rat nucleus accumbens, Vickroy *et al.* (1988) found that CCK produced a concentration-dependent facilitation of potassium-evoked [³H]DA overflow, maximal at 100 nM CCK-8. The ability of CCK to increase dopamine release was blocked by CCK$_A$ receptor subtype antagonists, but not by CCK$_B$ receptor subtype antagonists (Voigt *et al.* 1986; Vickroy *et al.* 1988; Marshall *et al.* 1990). In slices from cat caudate nucleus, CCK inhibited electrically evoked [³H] DA release, maximally at 10^{-1} M CCK-8S, while CCK-8US was inactive, again suggesting a CCK$_A$ receptor subtype (Markstein and Hökfelt 1984).

Earlier post-mortem studies, measuring DA turnover in mice and rats, found that *in vivo* intraventricular administration of CCK produced a decrease in basal and amphetamine-stimulated DA release in the caudate nucleus, prefrontal cortex, and olfactory tubercle, an effect that showed the pharmacological profile of the CCK$_B$ receptor subtype (Fuxe *et al.* 1980; Altar *et al.* 1988; Altar and Boyar 1989). Recent *in vivo* studies using high speed electrochemical recordings of local DA concentrations in the nucleus accumbens of the anaethestized rat showed that CCK produced a rapid transient increase in DA release, followed by a prolonged decrease (Lane *et al.* 1986; Blaha *et al.* 1987; Gerhardt *et al.* 1989). Using *in vivo* microdialysis, Ruggeri *et al.* (1987) found that CCK administered in the perfusion solution significantly increased the concentration of DA in the perfusate obtained from the microdialysis probe in the anterior, but not the posterior, nucleus accumbens. Also using *in vivo* microdialysis, Laitinen *et al.* (1990) found that CCK micro-injected into the cell body region in the VTA had no significant effect on the release of DA in the nucleus accumbens.

These somewhat conflicting results can be interpreted as showing that CCK facilitates DA release when the *in vivo* sampling time is short or with *in vitro* slice preparations, but has no effect or inhibits DA release when the sampling time is longer or when strong activators such as electrical stimulation or amphetamine are used to evoke high levels of DA release. The short time course of several minutes, detectable with electrochemical probes or in tissue slices, would be consistent with the time course of the behavioural effects of CCK, facilitating DA effects on motor behaviour. Since microdialysis and post mortem studies generally employ sampling times of 20 min or more, a small or short-lived effect of a peptide on transmitter release may not be detectable. The effects of CCK on DA release may also be species-specific and anatomically selective. The rapid facilitation of DA release by CCK in the nucleus accumbens of the rat appears to be mediated by CCK_A receptors, while the slower inhibition of DA release by CCK may be through the CCK_B receptors.

Slice preparations, *in vivo* microdialysis, and *in vivo* push–pull perfusion have been used to investigate the release of endogenous CCK and the ability of DA to modulate the release of CCK from terminal regions of the CCK–DA coexistence. In slices from the rat caudate-putamen, DA and D_2-selective agonists increased CCK release, while the D_2 antagonist haloperidol inhibited CCK release (Meyer and Krauss 1983; Conzelmann *et al.* 1984). Although microdialysis is difficult for CCK, because of the very low recovery of the peptide through the plastic dialysis membrane (De Mesquita *et al.* 1990), in one study CCK has been detected in microdialysate perfusate from rat frontal cortex, showing a significant increase in CCK-like immunoreactivity after treatment with the D_{-2} antagonist sulpiride (Takita *et al.* 1989).

In an elegant push–pull perfusion study in cat caudate nucleus and SN, where CCK–DA coexistence has been demonstrated (Hökfelt *et al.* 1985), simultaneous release of DA and CCK from the caudate was detected after activation of the SN by α-methyl-*p*-tyrosine (Artaud *et al.* 1989). Application of DA and a D_{-1} agonist into the caudate stimulated CCK release (Artaud *et al.* 1989). Application of CCK to the SN significantly increased the firing rate of dopaminergic SN neurons and stimulated ^{3}H[DA] release in the caudate (Artaud *et al.* 1989).

Similarly, to begin to investigate the release of CCK in the medial posterior nucleus accumbens of the rat *in vivo*, we tested microdialysis and push–pull perfusion techniques in the medial posterior nucleus accumbens of rats anaesthetized with chloral hydrate. A 2 mm microdialysis probe, or a push–pull cannula (Carnegie Medicin, Bioanalytical Systems, West Lafayette, IN), was stereotaxically placed at 1.6 mm rostral to bregma, 1.4 mm lateral to bregma, and 8.5 mm ventral to the surface of the skull. Ringer's solution was perfused at a rate of 1 μl/min for microdialysis, or 5 μl/min for push–pull perfusion, with samples taken at 30 min intervals. Aliquots of 120 μl per sample were assayed for CCK concentration by radio-immunoassay as previously described (De Mesquita *et al.* 1990). Assay sensitivity was approximately 0.15 pg, with perfusion solution blanks reading zero. *In vitro* recoveries of CCK through six microdialysis probes were 0.2, 1.1, 0.5, 0.2, 1.8, and 2.4 per cent, detecting 10^{-6} M CCK-8S standard. Basal CCK concentrations were detectable in only a small number of rats tested. *In vitro* recoveries of CCK through four push–pull cannulae were 80, 126, 110, and 24 per cent, detecting 10^{-9} M CCK-8S standard. Basal CCK was detectable in 15 out of 19 animals subjected to push–pull perfusion. As shown in Fig. 35.2, substitution of 55 mM K^+ for 3 mM K^+, with a concomitant decrease in perfusate concentration of Na^+, produced a large increase in CCK concentration in two out of three rats and

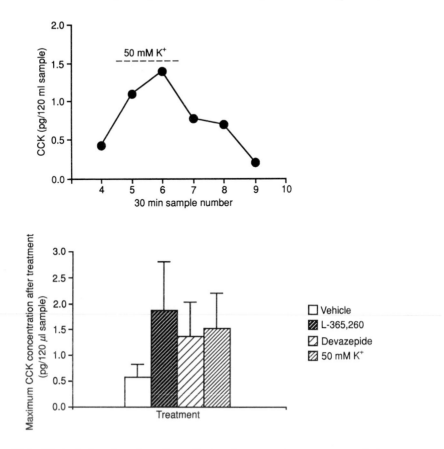

Fig. 35.2. Pilot study of push–pull perfusion for quantitating extracellular CCK concentrations in the medial posterior nucleus accumbens of the anaesthetized rat. Radio-immunoassay of 30 min perfusate samples detected basal CCK concentrations in the range of 0.2–0.8 pg/120 μl aliquot. 50 mM potassium, perfused into the nucleus accumbens over a period of 1 h (during 30 min sampling periods 5 and 6), produced an increase in CCK concentrations during the perfusion period (means of the three animals tested). Pilot studies of devazepide (50 μg/kg i.p.), L-365,260 (50 μg/kg i.p.), or vehicle (1 ml/kg), injected at the beginning of sampling period 5, have been conducted in a small number of animals (n = 4 per group). Considerable variability was seen between animals within each treatment group, particularly with respect to the time course of changes in CCK concentration after treatment. Preliminary data, presented in the bottom panel as mean $\pm$ SEM for the maximum concentration of CCK reached after each treatment, suggest that CCK antagonists may increase extracellular CCK in the medial posterior nucleus accumbens.

a small increase in CCK concentration in one out of three rats. Intraperitoneal (i.p.) administration of devazepide (50 μg/kg) produced a large increase in CCK concentration in two out of four rats. I.p. administration of L-365,260, (50 μg/kg) produced a large increase in CCK concentration in three out of four rats. I.p. administration of vehicle (2.5 per cent ethanol, 2.5 per cent propylene glycol, 95 per cent saline, 1 ml/kg) produced a small increase in CCK concentration in two out of four rats. The increase in CCK concentration occurred during the 1 h perfusion period with high potassium at 90–150 min after the single injection of devazepide and at 30–150 min after the single injection of L-365,260. Variability in magnitude and time course of changes in CCK concentrations after each of these treatments precluded statistical analyses for the small number of animals tested in this pilot experiment. However, these very preliminary data, summarized in Fig. 35.2, suggest that extracellular CCK can be measured in the nucleus accumbens of the rat using push–pull perfusion, that the concentration of CCK is stimulated by local potassium administration, and that L-365,260 may increase extracellular CCK concentration in the nucleus accumbens to a degree similar to that seen after potassium stimulation. These initial results hold promise as an approach to providing conclusive answers to functional questions about the release of endogenous CCK from a terminal region of the CCK–DA coexistence.

In situ hybridization studies

Another major approach to investigating the functional significance of CCK in the mesolimbic pathway is to test whether synthesis of CCK is altered during treatments which activate the mesolimbic dopaminergic neurons. Synthesis of neuropeptides is difficult to measure *in vivo*. Specific enzymes which process the cleavage of CCK-8 from prepro CCK, and which process the metabolism of CCK to smaller inactive fragments, have not been definitively identified. No pharmacological agents exist which specifically inhibit the synthesis or metabolism of CCK. Given the lack of pharmacological tools for assessing the turnover of the peptide, another approach to quantitating the rate of synthesis of new peptide is to measure messenger RNA (mRNA) concentrations for CCK. *In situ* hybridization provides a quantitative approach to assessing CCK mRNA concentrations in the substantia nigra–ventral tegmental area (SN–VTA) in rat brain (Ingram *et al.* 1989; Savasta *et al.* Seroogy *et al.* 1989*a, b*; 1989; Burgunder and Young 1990; Jayaraman *et al.* 1990). Quantitation of mRNA for CCK and for tyrosine hydroxylase (TH), the rate-limiting enzyme in DA synthesis, in adjacent sections through the VTA of rat brain gives a measure of relative rates of synthesis of CCK and DA. Experiments can be performed to quantitate mRNA for CCK and for TH after pharmacological, neurophysiological, or behavioural treatments known to change the activity of ventral tegmental neurons. These studies can address the alternative hypotheses that peptide transmitters are synthesized rapidly to maintain releasable peptide concentrations during periods of high neuronal activity, or that synthesis occurs slowly to replace depleted peptide transmitter gradually. The question of coexisting transmitters influencing each other's synthesis can also be addressed in these studies.

Pharmacological tools for activating the ventral tegmental neurons include DA receptor antagonists such as haloperidol, stimulant drugs such as amphetamine, and depleting agents such as reserpine. Haloperidol is a D_2 receptor blocker which is widely used as an antipsychotic. Rats given haloperidol show an immediate increase in neuronal firing rate in ventral tegmental and substantia nigra neurons, followed by a gradual decrease in firing rate of these mid-brain DA neurons after weeks of haloperidol treatment (Bunney and Grace 1978). To test the possibility that acute and chronic haloperidol would alter mRNA for both TH and CCK, rats were subcutaneously administered 2 mg/kg of haloperidol or vehicle daily, and sacrificed at time points of 3 days or 19 days for *in situ* hybridization and quantitation of TH and CCK mRNA (Cottingham *et al.* 1990). When ^{35}S-labelled oligonucleotide probes to TH and to CCK and the image analysis software Image 1.06 (W. Rasband, NIMH, Bethesda, MD) were used, no significant differences compared with vehicle controls were detected in either TH or CCK mRNA concentrations after haloperidol treatments at these time points (Cottingham *et al.* 1990).

The availability of CCK receptor antagonists selective for the CCK_A and CCK_B subtypes of CCK receptors makes possible an investigation of the role of endogenous CCK in regulating ventral tegmental and substantia nigra neurotransmitter synthesis activity using *in situ* hybridization of CCK and TH mRNA as a measure of synthesis. Male Sprague–Dawley rats were administered

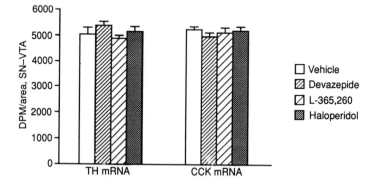

Fig. 35.3. *In situ* hybridization analysis of TH mRNA and CCK mRNA concentration in the SN–VTA of the rat brain. No significant differences were detected between treatment groups for either TH mRNA or CCK mRNA 5 h after i.p. treatment with vehicle (2.5 per cent ethanol, 2.5 per cent propylene glycol, 95 per cent saline), devazepide (50 μg/kg), L-365,260 (50 μg/kg), or haloperidol (2 mg/kg).

devazepide (50 μg/kg i.p.), L-365,260 (50 μg/kg i.p.), haloperidol (2 mg/kg i.p.), or vehicle (i.p.), and were sacrificed 5 h later for CCK and TH mRNA *in situ* hybridization. [^{35}S]dATP radio-labelled oligonucleotides for TH (complementary to bases 1441–1488 of rat TH mRNA, as described by Cottingham *et al.* (1990) and for CCK (complementary to bases 315–362 of rat CCK mRNA, as described by Cottingham *et al.* (1990) were used to hybridize the frozen 12 μm sections taken at the level of the central region of the SN–VTA. Washed sections were exposed to film and quantitated for the amount of radioactivity, using radioactive brain paste standards and an updated version of the Image 1.26 software as previously described (Cottingham *et al.* 1990). Data were analysed using one-way ANOVA. As illustrated in Fig. 35.3, no significant differences were found between vehicle-treated controls, devazepide-treated rats, L-365,260-treated rats, and haloperidol-treated rats for TH mRNA (F (3, 22) = 1.284, not significant) or CCK mRNA (F (3, 22) = 0.547, not significant).

The results of these negative experiments are open to several interpretations.

1. Technically, *in situ* hybridization may not be sufficiently sensitive to detect small but functionally important changes in CCK or TH mRNA concentrations in mid-brain DA neurons. The doses and time points used in the present studies may not have been optimal for detecting changes in mRNA. Additionally, single-cell analysis of concentration of mRNA per neuron, rather than regional analysis of concentration of SN–VTA as in the present studies, may be necessary if subregions within the SN–VTA respond differently to pharmacological treatments.

2. Conceptually, new mRNA for TH and CCK may not be the best measure of new transmitter availability. TH activity may be regulated by cofactor concentrations rather than by new enzyme synthesis. CCK concentrations may be regulated at the level of the enzymes for cleavage from the preproCCK, or at the level of the enzymes for degradation of CCK-8.

3. Functionally, it might be concluded that blockade of CCK receptors does not significantly change neurotransmitter synthesis, suggesting that endogenous CCK does not exert strong tonic input to mid-brain dopaminergic neurons.

Conclusions

A great many studies of exogenously applied CCK have demonstrated the functional activity of this peptide in physiological and behavioural events mediated by the mesolimbic pathway, where CCK coexists with DA in a large proportion of neurons. The crucial question to answer is whether, in fact, endogenous CCK contributes to the activity of this pathway. The new potent long-lasting, subtype-selective antagonists of CCK receptors now enable us to address this question. The behavioural studies reported above suggest that devazepide and

L-365,260, administered centrally, can block the central effects of CCK on at least one type of behavioural paradigm. Further experiments are needed to test more conclusively the ability of either antagonist to influence exploratory locomotion when administered alone or in combination with DA, which would indicate a contribution of endogenous mesolimbic CCK to exploratory locomotion. The mRNA studies reported above detected no effect of haloperidol, devazepide, or L-365,260 on mRNA for CCK or TH in the substantia nigra or VTA at a time point of 5 h after treatment. These negative findings may be due to technical limitations of the *in situ* hybridization technique, or may reflect a lack of neurotransmitter receptor input for regulating the synthesis of mesolimbic neurotransmitters. The very preliminary release studies described above show that extracellular CCK can be measured in the nucleus accumbens using push–pull perfusion and radio-immunoassay. Data from the first pilot study suggest that endogenous CCK is released by potassium stimulation, and that endogenous CCK may be released by antagonists of CCK receptors. Taken together, these first studies open the question of the role of endogenous CCK in regulating mesolimbic function by demonstrating the usefulness of the new highly selective CCK receptor antagonists for investigations of mesolimbic CCK in a variety of targeted functional studies.

Acknowledgements

CCK release studies were funded in part by NS 18667 to M.C.B.

References

Adams, C. E. and Fisher, R. S. (1990). Sources of neostriatal cholecystokinin in the cat. *J. Comp. Neurol.* **292**, 563–74.

Altar, C. A. and Boyar, W. C. (1989). Brain CCK-B receptors mediate the suppression of dopamine release by cholecystokinin. *Brain Res.* **483**, 321–6.

Altar, C. A., Boyar, W. C., Oei, E., and Wood, P. L. (1988). Cholecystokinin attenuates basal and drug-induced increases in limbic and striatal dopamine release. *Brain Res.* **460**, 76–82.

Artaud, F., Baruch, P., Stutzmann, J. M., Saffroy, M., Godeheu, G., Barbeito, L., Herve, D., Studler, J. M., Glowinski, J., and Cheramy, A. (1989). Cholecystokinin: corelease with dopamine from nigrostriatal neurons in the cat. *Eur. J. Neurosci.* **1**, 162–71.

Bartfai, T., Iverfeldt, K., Fisone, G., and Serfözö, P. (1988). Regulation of the release of coexisting neurotransmitters. *Ann. Rev. Pharmacol. Toxicol.* **28**, 285–310.

Blaha, C. D., Phillips, A. G., and Lane, R. F. (1987). Reversal by cholecystokinin of apomorphine-induced inhibition of dopamine release in the nucleus accumbens of the rat. *Reg. Pept.* **17**, 301–10.

Britton, D. R., Yahiro, L., Cullen, M. J., Kerwin, J. F., Kopecka, H., and Nadzan, A. M. (1989). Centrally administered CCK suppresses activity in mice by a peripheral-type CCK receptor. *Pharmacol. Biochem. Behav.* **34**, 779–83.

Brodie, M. S. and Dunwiddie, T. V. (1987). Cholecystokinin potentiates dopamine inhibition of mesencephalic dopamine neurons in vitro. *Brain Res.* **425**, 106–13.

Bunney, B. S. and Grace, A. A. (1978). Acute and chronic haloperidol treatment: comparison of effects on nigral dopaminergic cell activity. *Life Sci.* **23**, 1715–28.

Burgunder, J. M. and Young, W. S. (1990). Ontogeny of tyrosine hydroxylase and cholecystokinin gene expression in the rat mesencephalon. *Dev. Brain Res.* **52**, 85–93.

Chang, R. S. L. and Lotti, V. J. (1986). Biochemical and pharmacological characterization of an extremely potent and selective nonpeptide cholecystokinin antagonist. *Proc. Natl Acad. Sci. USA* **83**, 4923–6.

Chiodo, L. A., Freeman, A. S., and Bunney, B. S. (1987). Electrophysiological studies on the specificity of the cholecystokinin antagonist proglumide. *Brain Res.* **410**, 205–11.

Conzelmann, U., Holland, A., and Meyer, D. K. (1984). Effects of selective dopamine D_2-receptor agonists on the release of cholecystokinin-like immunoreactivity from rat neostriatum. *Eur. J. Pharmacol.* **101**, 119–25.

Cottingham, S. L., Pickar, D., Shimotake, T. K., Montpied, P., Paul, S. M., and Crawley, J. N. (1990). Tyrosine hydroxylase and cholecystokinin mRNA levels in the substantia nigra, ventral tegmental area, and locus ceruleus are unaffected by acute and chronic haloperidol administration. *Cell. Mol. Neurobiol.* **10**, 41–50.

Crawley, J. N. (1989). Microinjection of cholecystokinin into the rat ventral tegmental area potentiates dopamine-induced hypolocomotion. *Synapse* **3**, 346–55.

Crawley, J. N., Stivers, J. A., Blumstein, L. K., and Paul, S. M. (1985a). Cholecystokinin potentiates dopamine-mediated behaviors: Evidence for modulation specific to a site of coexistence. *J. Neurosci.* **5**, 1972–83.

Crawley, J. N., Hommer, D. W., and Skirboll, L. K. (1985b). Topographical analysis of nucleus accumbens sites at which cholecystokinin potentiates dopamine-induced hyperlocomotion. *Brain Res.* **355**, 337–41.

Crawley, J. N., Stivers, J. A., Hommer, D. W., Skirboll, L. R., and Paul, S. M. (1986). Antagonists of central and peripheral behavioral actions of cholecystokinin octapeptide. *J. Pharmacol. Exp. Ther.* **236**, 320–30.

Csernansky, J. G., Glick, S., and Mellentin, J. (1987). Differential effects of proglumide on mesolimbic and nigrostriatal dopamine function. *Psychopharmacology* **91**, 440–4.

Daugé, V., Dor, A., Feger, J., and Roques, B. P. (1989a). The behavioral effects of CCK8 injected into the medial nucleus accumbens are dependent on the motivational state of the rat. *Eur. J. Pharmacol.* **163**, 25–32.

Daugé, V., Steimes, P., Derrien, M., Beau, N., Roques, B. P., and Feger, J. (1989b). CCK8 effects on motivational and emotional states of rats involve CCK_A receptors of the postero-median part of the nucleus accumbens. *Pharmacol. Biochem. Behav.* **34**, 157–63.

Daugé, V., Bohme, G. A., Crawley, J. N., Durieux, C., Stutzmann, J. M., Feger, J., Blanchard, J. C., and Roques, B. P. (1990). Investigation of behavioral and electrophysiological responses induced by selective stimulation of CCK_B receptors by using a new highly potent CCK analog: BC 264. *Synapse* **6**, 73–80.

De Mesquita, S., Beinfeld, M. C., and Crawley, J. N. (1990). Microdialysis as an approach to quantitate the release of neuropeptides. *Prog. Neuro-psychopharmacol. Biol. Psychiat.*, **14**, S5–S15.

De Witte, P., Heibreder, C., Roques, B., and Vanderhaeghen, J. J. (1987). Opposite effects of cholecystokinin octapeptide (CCK-8) and tetrapeptide (CCK-4) after injection into the caudal part of the nucleus accumbens or into its rostral part and the cerebral ventricles. *Neurochem. Int.* **10**, 473–9.

Fallon, J. H., Hicks, R., and Loughlin, S. E. (1983). The origin of cholecystokinin terminals in the basal forebrain of the rat: evidence from immunofluorescence and retrograde tracing. *Neurosci. Lett.* **37**, 29–35.

Freeman, A. S. and Bunney, B. S. (1987). Activity of A9 and A10 dopaminergic neurons in unrestrained rats: further characterization and effects of apomorphine and cholecystokinin. *Brain Res.* **405**, 46–55.

Freeman, A. S. and Chiodo, L. A. (1988). Electrophysiological effects of cholecystokinin octapeptide on identified rat nigrostriatal dopaminergic neurons. *Brain Res.* **436**, 266–74.

Fuxe, K., Andersson, K., Locatelli, V., Agnati, L. F., Hökfelt, T., Skirboll, L., and Mutt, V. (1980). Cholecystokinin peptides produce marked reduction of dopamine turnover in discrete areas in the rat brain following intraventricular injection. *Eur. J. Pharmacol.* **67**, 329–31.

Gerhardt, G. A., Friedemann, M., Brodie, M. S., Vickroy, T. W., Gratton, A. P., Hoffer, B. J., and Rose, G. M. (1989). The effects of cholecystokinin (CCK-8) on dopamine-containing nerve terminals in the caudate nucleus and nucleus accumbens of the anesthetized rat: an in vivo electrochemical study. *Brain Res.* **499**, 157–63.

Hill, D. R., Campbell, N. J., Shaw, T. M., and Woodruff, G. N. (1987). Autoradiographic localization and biochemical characterization of peripheral type CCK receptors in rat CNS using highly selective non-peptide CCK antagonists. *J. Neurosci.* **7**, 2967–76.

Hill, D. R., Shaw, T. M., Graham, W., and Woodruff, G. N. (1990). Autoradiographical detection of cholecystokinin-A receptors in primate brain using ^{125}I-Bolton Hunter CCK-8 and ^{3}H-MK-329. *J. Neurosci.* **10**, 1070–81.

Hökfelt, T., Rehfeld, J. F., Skirboll, L., Ivemark, B., Goldstein, M., and Markey, K. (1980*a*). Evidence for coexistence of dopamine and CCK in mesolimbic neurons. *Nature* **285**, 476–8.

Hökfelt, T., Skirboll, L., Rehfeld, J. F., Goldstein, M., Markey, K., and Dann, O. (1980*b*). A subpopulation of mesencephalic dopamine neurons projecting to limbic areas contains a cholecystokinin-like peptide: evidence from immunohistochemistry combined with retrograde tracing. *Neuroscience* **5**, 2093–124.

Hökfelt, T., Skirboll, L., Everitt, B., Meister, B., Brownstein, M., Jacobs, T., Faden, A., Kuga, S., Goldstein, M., Markstein, R., Dockray, G., and Rehfeld, J. (1985). Distribution of cholecystokinin-like immunoreactivity in the nervous system. In *Neuronal cholecystokinin* (ed. J. J. Vanderhaeghen and J. N. Crawley). *Ann. NY Acad. Sci.* **448**, 255–74.

Hommer, D. W. and Skirboll, L. R. (1983). Cholecystokinin-like peptides potentiate apomorphine-induced inhibition of dopamine neurons. *Eur. J. Pharmacol.* **91**, 151–2.

Hommer, D. W., Palkovits, M., Crawley, J. N., Paul, S. M., and Skirboll, L. R. (1985). Cholecystokinin-induced excitation in the substantia nigra: evidence for peripheral and central components. *J. Neurosci.* **5**, 1387–92.

Hommer, D. W., Stoner, G., Crawley, J. N., Paul, S. M., and Skirboll, L. R. (1986). Cholecystokinin–dopamine coexistence: electrophysiological actions corresponding to cholecystokinin receptor subtype. *J. Neurosci.* **6**, 3039–43.

Ingram, S. M., Krause, R. G., Baldino, F., Skeen, L. C., and Lewis, M. E. (1989). Neuronal localization of cholecystokinin mRNA in the rat brain by using *in situ* hybridization histochemistry. *J. Comp. Neurol.* **287**, 260–72.

Jayaraman, A., Nishimori, T., Dobner, P., and Uhl, G. R. (1990). Cholecystokinin and neurotensin mRNAs are differentially expressed in subnuclei of the ventral tegmental area. *J. Comp. Neurol.* **296**, 291–302.

Katsuura, G., Itoh, S., and Hsiao, S. (1985). Specificity of nucleus accumbens to activities related to cholecystokinin in rats. *Peptides* **6**, 91–6.

Laitinen, K., Crawley, J. N., Mefford, I. N., and De Witte, Ph. (1990). Neurotensin and cholecystokinin microinjected into the ventral tegmental area modulate microdialysate concentrations of dopamine and metabolites in the posterior nucleus accumbens. *Brain Res.* **523**, 342–6.

Lane, R. F., Blaha, C. D., and Phillips, A. G. (1986). *In vivo* electrochemical analysis of cholecystokinin-induced inhibition of dopamine release in the nucleus accumbens. *Brain Res.* **397**, 200–4.

Lotti, V. J. and Chang, R. S. L. (1989). A new potent and selective non-peptide gastrin antagonist and brain cholecystokinin receptor (CCK-B) ligand: L-365,260. *Eur. J. Pharmacol.* **1162**, 273–80.

Markstein, R. and Hökfelt, T. (1984). Effect of cholecystokinin-octapeptide on dopamine release from slices of cat caudate nucleus. *J. Neurosci.* **4**, 570–5.

Marley, P. D., Emson, P. C., and Rehfeld, J. F. (1982). Effect of 6-hydroxydopamine lesions of the medial forebrain bundle on the distribution of cholecystokinin in rat forebrain. *Brain Res.* **252**, 382–5.

Marshall, F. H., Barnes, S., Pinnock, R. D., and Hughes, J. (1990). Characterization of cholecystokinin octapeptide-stimulated endogenous dopamine release from rat nucleus accumbens *in vitro*. *Br. J. Pharmacol.* **99**, 845–8.

Meyer, D. K. and Krauss, J. (1983). Dopamine modulates cholecystokinin release in neostriatum. *Nature* **301**, 338–40.

Meyer, D. K., Beinfeld, M. C., Oertel, W. H., and Brownstein, M. J. (1982). Origin of the cholecystokinin-containing fibres in the rat caudatoputamen. *Science* **215**, 187–8.

Moran, T. H., Robinson, P. H., Goldrich, M. S., and McHugh, P. R. (1986). Two brain cholecystokinin receptors: implications for behavioral actions. *Brain Res.* **362**, 175–9.

Mutt, V. (1980). Cholecystokinin: isolation, structure, and function. In *Gastrointestinal hormones* (ed. G. B. J. Glass), pp. 169–221. Raven Press, New York.

Peleprat, D., Broer, Y., Studler, J. M., Peschanski, M., Tassin, J. P., Glowinski, J., Rostene W., and Roques, B. P. (1987). Autoradiography of CCK receptors in the rat brain using $[^3H]Boc[Nle^{28,31}]CCK_{27-33}$ and $[^{125}I]$Bolton-Hunter CCK_8. Functional significance of subregional distributions. *Neurochem. Int.* **87**, 495–508.

Pullen, R. G. L. and Hodgson, O. J. (1987). Penetration of diazepam and the non-peptide CCK antagonist, L-364,718, into rat brain. *J. Pharm. Pharmacol.* **39**, 863–4.

Rovati, L. C. and Makovec, F. (1988). New pentanoic acid derivatives with potent CCK antagonistic properties: different activity on the periphery vs. central nervous system. In *Cholecystokinin antagonists* (ed. R. Y. Wang and R. Schoenfeld), pp. 1–11. Alan R. Liss, New York.

Ruggeri, M., Ungerstedt, U., Agnati, L. F., Mutt, V., Harfstrand, A., and Fuxe, K. (1987). Effects of cholecystokinin peptides and neurotensin on dopamine release and metabolism in the rostral and caudal part of the nucleus accumbens using intracerebral dialysis in the anethestized rat. *Neurochem. Int.* **10**, 509–20.

Savasta, M., Ruberte, E., Palacios, J. N., and Mengod, G. (1989). The colocalization of cholecystokinin and tyrosine hydroxylase mRNAs in mesencephalic dopaminergic neurons in the rat brain examined by *in situ* hybridization. *Neuroscience* **29**, 363–9.

Schneider, L. H., Alpert, J. E., and Iversen, S. D. (1983). CCK-8 modulation of mesolimbic dopamine: antagonism of amphetamine-stimulated behaviors. *Peptides* **4**, 749–53.

Seroogy, K. B., Dangaran, K., Lim, S., Haycock, J. W., and Fallon, J. F. (1989a).

Ventral mesencephalic neurons containing both cholecystokinin- and tyrosine hydroxylase-like immunoreactivities project to forebrain regions. *J. Comp. Neurol.* **279**, 397–414.

Seroogy, K., Schalling, M., Brene, S., Dagerlind, A., Chai, S. Y., Hökfelt, T., Persson, H., Brownstein, M., Huan, R., Dixon, J., Filer, D., Schlessinger, D., and Goldstein, M. (1989*b*). Cholecystokinin and tyrosine hydroxylase messenger RNAs in neurons of rat mesencephalon: peptide/monoamine coexistence studies using in situ hybridization combined with immunocytochemistry. *Exp. Brain Res.* **74**, 149–62.

Skirboll, L. R., Grace, A. A., Hommer, D. W., Rehfeld, J., Goldstein, M., Hökfelt, T., and Bunney, B. S. (1981). Peptide-monoamine coexistence: studies of the actions of cholecystokinin-like peptide on the electrical activity of midbrain dopamine neurons. *Neuroscience* **6**, 2111–24.

Stittsworth, J. D. and Mueller, A. L. (1990). Cholecystokinin octapeptide potentiates the inhibitory response mediated by D_2 dopamine receptors in slices of the ventral tegmental area of the brain in the rat. *Neuropharmacology* **29**, 119–27.

Studler, J. M., Simon, H, Cesselin, F., Legrand, J. C., Glowinski, J., and Tassin, J. P. (1981). Biochemical investigation on the localization of cholecystokinin octapeptide in dopaminergic neurons originating from the ventral tegmental area of the rat. *Neuropeptides* **2**, 131–9.

Studler, J. M., Reibaud, M., Herve, D., Blanc, G., Glowinski, J., and Tassin, J. P. (1986). Opposite effects of sulphated cholecystokinin on DA-sensitive adenylate cyclase in two areas of the rat nucleus accumbens. *Eur. J. Pharmacol.* **126**, 125–8.

Takita, M., Tsuruta, T., Oh-hashi, Y., and Kato, T. (1989). *In vivo* release of cholecystokinin-like immunoreactivity in rat frontal cortex under freely moving conditions. *Neurosci. Lett.* **100**, 249–53.

Vaccarino, F. J. and Rankin, J. (1989). Nucleus accumbens cholecystokinin (CCK) can either attenuate or potentiate amphetamine-induced locomotor activity: evidence for rostral–caudal differences in accumbens CCK function. *Behav. Neurosci.* **103**, 831–6.

Vaccarino, F. J. and Vaccarino, A. L. (1989). Antagonism of cholecystokinin function in the rostral and caudal nucleus accumbens: differential effects on brain stimulation reward. *Neurosci. Lett.* **97**, 151–6.

Vanderhaeghen, J. J., Signeau, J. C., and Gepts, W. (1975). New peptide in the vertebrate CNS reacting with antigastrin antibodies. *Nature* **257**, 604–5.

Van Ree, J. M., Gadfori, O., and DeWeid, D. (1983). In rats, the behavioral profile of CCK-8 related peptides resembles that of antipsychotic agents. *Eur. J. Pharmacol.* **93**, 63–78.

Vickroy, T. W., Bianchi, B. R., Kerwin, J. F., Kopecka, H., and Nadzan, A. M. (1988). Evidence that type A CCK receptors facilitate dopamine efflux in rat brain. *Eur. J. Pharmacol.* **152**, 371–2.

Voigt, M., Wang, R. Y., and Westfall, R. C. (1986). Cholecystokinin octapeptides alter the release of endogenous dopamine from the rat nucleus accumbens *in vitro*. *J. Pharmacol. Exp. Ther.* **237**, 147–53.

Weiss, F., Tanzer, D. J., and Ettenberg, A. (1988). Opposite actions of CCK-8 on amphetamine-induced hyperlocomotion and stereotypy following intracerebroventricular and intra-accumbens injections in rats. *Pharmacol. Biochem. Behav.* **30**, 309–17.

36. CCK–dopamine interactions: implications for treatment of schizophrenia

N. M. J. Rupniak

There has been great interest in the possible role of CCK in the pathophysiology and therapy of diseases in which an abnormality of cerebral dopamine function is implicated. Most attention has focused on schizophrenia as a target for novel drug development. In this chapter I shall collate evidence from studies in animals and man to assess the ways in which the CCK system might be manipulated by novel drugs to treat psychosis and neuroleptic-induced movement disorders.

Psychosis

Mesolimbic CCK–dopamine system as a site for antipsychotic activity

There has been speculation that the intimately shared mesolimbic peptide–monoamine pathway might be selectively manipulated to develop a novel antipsychotic therapy. Indeed, the ability of neuroleptic drugs to alter mesolimbic CCK function selectively may contribute to the way in which these agents exert antipsychotic activity.

There are no known specific screens for antipsychotic activity in animals. Instead, preclinical screens attempt to demonstrate a neuroleptic-like profile based on that of dopamine receptor antagonists. Blockade of dopamine receptors in the mesolimbic system is often considered an important site for the antipsychotic effects of neuroleptic drugs (e.g. Crow *et al.* 1975). Interference with the *nigrostriatal* dopamine system is not considered essential for antipsychotic activity but is held responsible for induction of movement disorders and is therefore regarded as undesirable. This explanation for the antipsychotic effects of neuroleptic drugs is based largely on experience with atypical agents like clozapine and sulpiride which exhibit selective 'mesolimbic' profiles in animal models (Costall and Naylor 1976; White and Wang 1983) and have a lower propensity to cause extrapyramidal side-effects in man. The notion that this represents an ideal profile for novel antipsychotic drugs receives widespread support, but is questionable on several grounds. First, there is no conclusive neuropathological evidence for an abnormality in mesolimbic dopamine function in schizophrenia or for maintained dopamine receptor blockade in the mesolimbic system after months of neuroleptic treatment (see the review by Rupniak *et al.* 1983). Second, the non-selective pharmacological profile of clozapine and its relative lack of potency as a dopamine antagonist (Hauser and Closse 1978; Burki 1980; Leysen 1982) cast serious doubt on the proposal that its antipsychotic activity is attributable to an interaction with dopamine receptors. Finally, the predictive utility of 'selective mesolimbic' screens for novel antipsychotics is challenged by the recent report of extra-pyramidal disturbances, including tardive dyskinesia, in patients treated with sulpiride (Achiron *et al.* 1990).

Despite these uncertainties about the predictive utility of preclinical screens for antipsychotic agents, they do provide compelling evidence for a modulatory role of CCK on mesolimbic dopamine function and for an alteration in the mesolimbic CCK system as a consequence of chronic treatment with antipsychotic drugs. These data will be reviewed in the following sections and discussed in relation to clinical studies.

Neuroleptic-like effects of CCK in rodents

Numerous studies report a neuroleptic-like profile of CCK analogues in animals. However, many conflicting reports exist, some claiming *inhibition* and others *potentiation* of cerebral dopamine function. These discrepancies appear at least partly attributable to differences in CCK–dopamine interactions between and within tissues which may be mediated via distinct populations of CCK receptors.

CCK–dopamine interactions have been most extensively examined in the nucleus accumbens (Table 36.1). CCK innervation of this structure is heterogeneous, with the anterolateral region being derived from non-dopaminergic neurons unlike the posteromedial portion which receives CCK/dopamine neurons (Studler *et al.* 1985). Opposite effects of CCK on dopamine function have been obtained following micro-iontophoretic application in these regions.

In general a *facilitation* of dopamine-mediated responses is observed following local injections of sulphated CCK octapeptide (CCK-8S) into the *posteromedial* nucleus accumbens, whilst *neuroleptic*-like effects are observed in the *anterolateral* portion. Recent studies measuring potassium-evoked dopamine release in brain slices indicate that the facilitatory effects of CCK in posterior accumbens tissue slices are mediated via stimulation of CCK_A receptors; conversely the inhibitory effects of CCK in the anterior accumbens and striatal tissue appear to be mediated by CCK_B receptors (Altar and Boyar 1989; Marshall *et al.* 1991). The localization of CCK_A and CCK_B receptors may also differ with respect to presynaptic versus postsynaptic sites. Given these opposing effects within the nucleus accumbens, it seems improbable that a non-selective CCK agonist should necessarily exhibit antipsychotic activity following systemic administration. However, we might predict such a profile

Table 36.1. Rostral–caudal differences in CCK–dopamine interactions in the nucleus accumbens of rodents

| Parameter | Effect of micro-injection of CCK-8S | | |
	Posteromedial	Anterolateral	Reference
Dopamine content	None	Increased	Ruggeri et al. (1987)
Potassium-evoked dopamine release	Increased	Decreased	Marshall et al. (1991)
Dopamine-stimulated cAMP formation	Potentiated	Attenuated	Studler et al. (1986)
Dopamine-induced hyperlocomotion	Potentiated	None	Crawley et al. (1985a)
Ampthetamine-induced locomotor stimulation	Potentiated	Attenuated	Vaccarino and Rankin (1989)
Intracranial self-stimulation	Potentiated	Inhibited	De Witte et al. (1987); Vaccarino and Koob (1984)

for a selective CCK_B receptor agonist, or possibly a CCK_A receptor antagonist (see Chapter 35 of this volume), or both.

In contrast with the results for the nucleus accumbens, focal injections of CCK-8S into the caudate nucleus appear to exert little influence on dopamine-mediated behaviour such as hyperlocomotion and stereotypy (Van Ree 1983; Crawley et al. 1985b). However, it might be quite misleading to assume that systemic therapy with a CCK agonist would not influence *striatal* dopamine function, since Vasar et al. (1990) observed an increase in striatal D_2 receptor numbers following repeated treatment with caerulein in mice. This phenomenon of striatal dopamine receptor supersensitivity is thought to be responsible for the development of tardive dyskinesia following chronic administration of classical neuroleptic drugs.

Evidence for an alteration in mesolimbic CCK function by chronic treatment with antipsychotic drugs

The antipsychotic effects of neuroleptic drugs take several weeks to develop in schizophrenia (Johnstone et al. 1978). Therefore the effects of these drugs on brain function may best be examined following repeated administration in animals. Several studies indicate that subchronic treatment with neuroleptic drugs causes selective changes in mesolimbic and mesocortical, but *not* striatal, CCK systems. The specificity of this effect may be attributable to the coexistence of CCK with dopamine in mesolimbic and forebrain pathways which is not present in the nigrostriatal dopamine system (Hökfelt et al. 1980).

The most reproducible effect on CCK function of treatment with haloperidol for several weeks is an increase of up to twofold in [^{125}I]-CCK binding in the nucleus accumbens, olfactory tubercle, and frontal and cingulate cortex (Chang *et al.* 1983; Fukamauchi *et al.* 1987; Debonnel *et al.* 1990). Whilst full Scatchard analysis was not performed in all tissues, the increase in cortical binding at least appears to be attributable to an increase in the number of specific binding sites (Chang *et al.* 1983; Fukamauchi *et al.* 1987). In marked contrast *no* changes in CCK binding were detected in the hippocampus or striatum, areas which are rich in CCK but where neuronal colocalization with dopamine is absent.

Receptor proliferation is usually considered to indicate supersensitivity as a consequence of denervation or chronic pharmacological blockade. Consistent with this interpretation, the development of functional supersensitivity to iontophoretic application of CCK-8S in the nucleus accumbens of rats treated for 3–5 weeks with haloperidol (Debonnel *et al.* 1990) might be consistent with decreased release of CCK, perhaps owing to a depolarization block of mesolimbic neurons (White and Wang 1983). However, CCK turnover cannot be assessed directly, and the possible effects of repeated neuroleptic treatment on CCK content in different brain regions are more controversial. In one study CCK levels were elevated in mesolimbic and striatal tissue from rats treated for up to 4 months with haloperidol or other neuroleptics including clozapine; however, CCK levels in the cortex were initially reduced and then returned to normal values (Frey 1983). These changes in mesolimbic and striatal CCK content were not replicated in subsequent studies by Gysling and Beinfeld (1984) and by Radke *et al.* (1989). However, the decrease in cortical CCK content was confirmed by Radke *et al.* who also reported an increase in CCK levels in the substantia nigra and ventral tegmental area. This latter effect may indicate an increase in CCK synthesis in neurons projecting to forebrain structures.

Neuropathological and therapeutic findings in schizophrenia

The case for an involvement of CCK in the antipsychotic activity of neuroleptic drugs would be considerably strengthened by the demonstration of a convincing abnormality of brain CCK function in schizophrenia. Few studies have been conducted to date and the status remains inconclusive. Most workers have examined CCK concentrations in different brain structures, but it is not known whether changes in this parameter reflect synthesis or release. Kleinman *et al.* (1983) observed no changes in CCK levels in the nucleus accumbens, caudate nucleus, hippocampus, frontal cortex, or temporal cortex of paranoid schizophrenics, but did report an increase in CCK levels in the amygdala of these patients. However, a second group of investigators reported *decreases* in CCK content in temporal lobe structures (cortex, amygdala, and hippocampus) in schizophrenics with *negative* symptoms (Ferrier *et al.* 1983; Roberts *et al.* 1983). The patients examined by Kleinman *et al.* had not been categorized as type I or type II, and this may explain the different findings in these studies. The possibility that an abnormality of limbic CCK function exists in at least a subgroup of schizophrenics requires further investigation, but would be particularly significant for the development of CCK-based therapies since type II symptoms typically respond poorly to neuroleptic drugs. A second post mortem finding of interest was of a reduction in CCK binding in hippocampal and frontal cortex tissue (Farmery *et al.* 1985). Although the effects of medication on these parameters cannot be excluded, they would not be predicted from the effects of chronic neuroleptic exposure in rodents.

CCK or caerulein have been examined for antipsychotic activity in over 500 patients over the last 10 years. The clinical literature on this subject is controversial: some studies report marked improvements in psychosis, while others demonstrate no beneficial effect. In a recent critical appraisal of these studies, Montgomery and Green (1988) point to serious methodological flaws which in their opinion make it impossible to draw any conclusions about the antipsychotic efficacy of CCK.

1. Many studies were not placebo-controlled.

2. The most consistent (and mainly positive) trials of CCK or caerulein were not blind. When this design was used, the induction of gastro-intestinal side-effects would be likely to unblind even a well-controlled study

3. Most studies included chronic schizophrenics, a proportion of whom are likely to be poor responders even using conventional medication

4. Most patients were receiving concurrent neuroleptic medication which could readily mask any antipsychotic effects of the test substance

5. Many investigations examined acute treatment only. A period of at least 4 weeks would be required for conclusive evaluation

6. The number of patients examined in the majority of studies is too small to hope to demonstrate a statistically reliable effect (comparison groups of fewer than 10 people).

The major confounding factor in these investigations is the need for systemic administration of agents whose penetration into the brain is questionable and the use of single or narrow dose ranges limited by gastro-intestinal side-effects.

Tardive dyskinesia

Uncontrollable adventitious flexions and extensions of the musculature of the limbs, trunk, and face (tardive dyskinesias) often emerge as a disturbing consequence of chronic therapy with antipsychotic drugs (Marsden *et al.* 1975). However, it should be noted that a significant proportion of elderly people also develop spontaneous orofacial dyskinesias, especially following institutionalization (Klawans and Barr 1982).

Preliminary evidence from several unrelated sources suggests that CCK agonist therapy may be beneficial in the control of tardive dyskinesia. A possible neuropathological link between CCK and chorea is indicated by the substantial decrease in CCK-like immunoreactivity in the substantia nigra and globus pallidus (but not the caudate, putamen, or frontal cortex) of patients with Huntington's chorea (Emson *et al.* 1980). On the strength of the neuroleptic-like profile of CCK analogues in preclinical screens and the known ability of neuroleptic drugs to suppress chorea in man, Nishikawa *et al.* (1985, 1988) examined the antidyskinetic effect of caerulein in small numbers of chronic schizophrenics receiving concomitant neuroleptic medication. The response of individuals varied considerably, but dramatic and surprisingly long-lasting improvements in tardive dyskinesias (but interestingly not psychosis) were observed in a few patients. The duration of benefit from treatment with caerulein is difficult to reconcile with its relatively short plasma half-life, and therefore the possibility of spontaneous remission cannot be excluded. There have been no reports of antidyskinetic effects of CCK analogues from other clinical groups, which might have been expected in view of the large number of schizophrenics examined for possible antipsychotic activity. However, preliminary evidence from animal models gives further support to a role for CCK agonists in the control of antidyskinetic activity. Whilst there are no satisfactory rodent models of tardive dyskinesia, CCK-8S was able to suppress oral movements associated with chronic exposure to fluphenazine following intraperitoneal administration in rats (Stoessl *et al.* 1989). Similarly, CCK-8S blocked oral movements and the efflux of cyclic adenosine monophosphate (cAMP) in the striatum induced by the D_1 dopamine agonist SKF 38393 (Chapter 40 of this volume). Finally, we recently reported the ability of peripherally administered CCK-8S to block chorea induced by L-dopa in parkinsonian squirrel monkeys. This effect was accompanied by only a slight (non-significant) attenuation of locomotor stimulation (Boyce *et al.* 1990*a*), indicating that CCK agonists might be able to improve the therapeutic window of antipsychotic drugs in man. At present the sites (peripheral or central) and mechanisms (A or B receptor-mediated) for these antidyskinetic effects are not known. As in the case of neuroleptic drugs, there may be a risk that chronic exposure to CCK agonists might also *induce* extrapyramidal disturbances.

Implications for drug design

The most significant evidence supporting a role for CCK agonist therapy in schizophrenia is based on animal screens demonstrating a neuroleptic-like profile. Compelling as this evidence may be, it remains a fact that the predictive value of these models to detect novel antipsychotic agents is not known. Much of the appeal of a CCK-based therapy for schizophrenia was based on speculation about the coexistence of CCK and dopamine in mesolimbic neurons and its functional significance. Some findings in rodents lead to the prediction that in order to produce a neuroleptic-like selective blockade of this pathway a CCK_A receptor antagonist would be required. Such agents are available (e.g.

devazepide), but evidence to support a neuroleptic-like profile following systemic administration in animal models is not yet available. We obtained no evidence for a neuroleptic-like effect of devazepide in primates (Boyce *et al.* 1990*b*). Moreover, other data exist which give rise to quite opposite predictions about the effects of CCK_A antagonists. Alternatively, a selective CCK_B agonist would be expected to mimic the effects of neuroleptic drugs on dopamine function. However, there are several hazards associated with this approach. First, based on current knowledge it is difficult to see how actions on basal ganglia structures (such as the striatum), and hence possible motor disturbances, could be avoided. CCK_B agonist therapy might inevitably induce troublesome gastro-intestinal side-effects (abdominal pain, nausea) and, perhaps more seriously, panic attack (De Montigny 1989; see Chapter 11 of this volume). The benefits of a CCK-agonist-based therapy over conventional treatment would need to be impressive to overcome these concerns. There are insufficient tools available at present to warrant further clinical trials in schizophrenia. The development of selective agonists with good brain penetration would be a prerequisite to reducing the risk of peripheral side-effects. Clinical trials to evaluate the role of CCK receptors in anxiety (see Chapter 11 of this volume) will take priority over assessing whether schizophrenia should be re-examined.

References

Achiron, A., Zoldan, Y., and Melamed, E. (1990). Tardive dyskinesia induced by sulpiride. *Clin. Neuropharmacol.* **13**, 248–52.

Altar, C. A. and Boyar, W. C. (1989). Brain CCK-B receptors mediate the suppression of dopamine release by cholecystokinin. *Brain Res.* **483**, 321–6.

Burki, H. R. (1980). Inhibition of ^{3}H-clozapine binding in rat brain after oral administration of neuroleptics. *Life Sci.* **26**, 2187–93.

Boyce, S., Rupniak, N. M. J., Steventon, M., and Iversen, S. D. (1990*a*). CCK-8S inhibits L-dopa-induced dyskinesias in parkinsonian squirrel monkeys. *Neurology* **40**, 717–8.

Boyce, S., Rupniak, N. M. J., Tye, S., Steventon, S. J., and Iversen, S. D. (1990*b*). Modulatory role for CCK-B antagonists in Parkinson's disease. *Clin. Neuropharmacol.* **13**, 339–47.

Chang, R. S. L., Lotti, V. J., Martin, G. E., and Chen, T. B. (1983). Increase in brain ^{125}I-cholecystokinin (CCK) receptor binding following chronic haloperidol treatment, intracisternal 6-hydroxydopamine or ventral tegmental lesions. *Life Sci.* **32**, 871–8.

Costall, B. and Naylor, R. J. (1976). Antagonism of the hyperactivity induced by dopamine applied intracerebrally to the nucleus accumbens septi by typical neuroleptics and by clozapine, sulpiride and thioridazine. *Eur. J. Pharmacol.* **35**, 161–8.

Crawley, J. N., Hommer, D. W., and Skirboll, L. R. (1985*a*). Topographical analysis of nucleus accumbens sites at which cholecystokinin potentiates dopamine-induced hyperlocomotion. *Brain Res.* **355**, 337–41.

Crawley, J. N., Stivers, J. A., Blumstein, L. K., and Paul, S. M. (1985*b*).

Cholecystokinin potentiates dopamine-mediated behaviours: evidence for modulation specific to a site of coexistence. *J. Neurosci.* **5**, 1972–83.

Crow, T. J., Deakin, J. F. W., and Longden, A. (1975). Do anti-psychotic drugs act by dopamine receptor blockade in the nucleus accumbens? *Br. J. Pharmacol.* **55**, 295P.

Debonnel, G., Gaudreau, P., Quirion, R., and De Montigny, C. (1990). Effects of long-term haloperidol treatment on the responsiveness of accumbens neurons to cholecystokinin and dopamine: electrophysiological and radioligand binding studies in the rat. *J. Neurosci.* **10**, 469–78.

De Montigny, C. (1989). Cholecystokinin tetrapeptide induces panic-like attack in healthy volunteers. *Arch. Gen. Psychiat.* **46**, 511–17.

De Witte, P., Heibreder, C., Roques, B., and Vanderhaeghen, J. J. (1987). Opposite effects of cholecystokinin octapeptide (CCK-8) and tetrapeptide (CCK-4) after injection into the caudal part of the nucleus accumbens or into the rostral part and the cerebral ventricles. *Neurochem. Int.* **10**, 473–9.

Emson, P. C., Rehfeld, J. F., Langevin, H., and Rossor, M. (1980). Reduction in cholecystokinin-like immunoreactivity in the basal ganglia in Huntington's disease. *Brain Res.* **198**, 497–500.

Farmery, S. M., Owen, F., Poulter, M., and Crow, T. J. (1985). Reduced high affinity cholecystokinin binding in hippocampus and frontal cortex of schizophrenic patients. *Life Sci.* **36**, 473–7.

Ferrier, I. N., Roberts, G. W., Crow, T. J., Johnstone, E. C., Owens, D. G. C., Lee, Y. C., O'Shaughnessy, D., Adrian, T. E., Polak, J. M., and Bloom, S. R. (1983). Reduced cholecystokinin-like and somatostatin-like immunoreactivity in limbic lobe is associated with negative symptoms in schizophrenia. *Life Sci.* **33**, 475–82.

Frey, P. (1983). Cholecystokinin octapeptide levels in rat brain are changed after subchronic neuroleptic treatment. *Eur. J. Pharmacol.* **95**, 87–92.

Fukamauchi, F., Yoshikawa, T., Kaneno, S., Shibuya, H., and Takahashi, R. (1987). The chronic administration of dopamine antagonists and methamphetamine changed the [^{3}H]-cholecystokinin-8 binding sites in the rat frontal cortex. *Neuropeptides* **10**, 221–5.

Gysling, K. and Beinfeld, M. C. (1984). Failure of chronic haloperidol treatment to alter levels of cholecystokinin in the rat brain striatum and olfactory tubercle–nucleus accumbens area. *Neuropeptides* **4**, 421–3.

Hökfelt, T., Skirboll, L., Rehfeld, J. F., Goldstein, M., Markey, K., and Dann, O. (1980). A subpopulation of mesencephalic dopamine neurons projecting to limbic areas contains a cholecystokinin-like peptide: evidence from immunohistochemistry combined with retrograde tracing. *Neuroscience* **5**, 2093–124.

Hauser, D. and Closse, A. (1978). ^{3}H-clozapine binding to rat brain membranes. *Life Sci.* **23**, 557–62.

Johnstone, E. C., Crow, T. J., Frith, C. D., Carney, M. W. P., and Price, J. S. (1978). Mechanism of the antipsychotic effect in the treatment of acute schizophrenia. *Lancet* i, 848–51.

Klawans, H. L. and Barr, A. (1982). Prevalence of spontaneous lingual-facial-buccal dyskinesia in the elderly. *Neurology* **32**, 558–9.

Kleinman, J. E., Ladarola, M., Govoni, S., Hong, J., Gillin, J. C., and Wyatt, R. J. (1983). Postmortem measurements of neuropeptides in human brain. *Psychopharmacol. Bull.* **19**, 375–7.

Leysen, J. E. (1982). Review on neuroleptic receptors; specificity and multiplicity of in vitro binding relates to pharmacological activity. In *Clinical pharmacology in psychiatry: Neuroleptic and antidepressant research* (ed. E. Usdu, S. Dahe, L. F. Gran, and O. Lingjaercle), pp. 35–47. Macmillan, Basingstoke.

Marsden, C. D., Tarsy, D., Baldessarini, R. J. (1975). Spontaneous and drug-induced movement disorders in psychotic patients. In *Psychiatric aspects of neurologic disease* (ed. D. F. Benson and D. Blumer), pp. 219–66. Grune and Stratton, New York.

Marshall, F. H., Barnes, S., Woodruff, G. N., Hughes, J., and Hunter, J. C. (1991). Cholecystokinin modulates the release of dopamine from the anterior and posterior nucleus accumbens by two different mechanisms. *J. Neurochem.* **56**, 917–22.

Montgomery, S. A. and Green, M. C. D. (1988). The use of cholecystokinin in schizophrenia: a review. *Psychol. Med.* **18**, 593–603.

Nishikawa, T., Tanaka, M., Koga, I., and Uchida, Y. (1985). Biphasic and long-lasting effect of ceruletide on tardive dyskinesia. *Psychopharmacology* **86**, 43–4.

Nishikawa, T., Tanaka, M., Tsuda, A., Koga, I., and Uchida, Y. (1988). Treatment of tardive dyskinesia with cerelutide. *Prog. Neuro-Psychopharmacol. Biol. Psychiat.* **12**, 803–12.

Radke, J. M., MacLennan, A. J., Beinfeld, M. C., Bissette, G., Nemeroff, C. B., Vincent, S. R., and Fibiger, H. C. (1989). Effects of short- and long-term haloperidol administration and withdrawal on regional brain cholecystokinin and neurotensin concentrations in the rat. *Brain Res.* **480**, 178–83.

Roberts, G. W., Ferrier, I. N., Lee, Y., Crow, T. J., Johnstone, E. C., Owens, D. G. C., Bacarese-Hamilton, A. J., McGregor, G., O'Shaughnessey, D., Polak, J. M., and Bloom, S. R. (1983). Peptides, the limbic lobe and schizophrenia. *Brain Res.* **288**, 199–211.

Ruggeri, M., Understedt, U., Agnati, L. F., Mutt, V., Harfstrand, A., and Fuxe, K. (1987). Effects of cholecystokinin peptides and neurotensin on dopamine release and metabolism in the rostral and caudal part of the nucleus accumbens using intracerebral dialysis in the anaesthetised rat. *Neurochem. Int.* **10**, 509–20.

Rupniak, N. M. J., Jenner, P., and Marsden, C. D. (1983). Long-term neuroleptic treatment and the status of the dopamine hypothesis of schizophrenia. In *Theory in psychopharmacology*, Vol. 2 (ed. S. J. Cooper), pp. 196–237. Academic Press, London.

Stoessl, A. J., Dourish, C. T., and Iversen, S. D. (1989). Chronic neuroleptic-induced mouth movements in the rat: suppression by CCK and selective dopamine D1 and D2 receptor antagonists. *Psychopharmacology* **98**, 372–9.

Studler, J. M., Reibaud, M., Tramu, G., Blanc, G., Glowinski, J., and Tassin, J. P. (1985). Distinct properties of cholecystokinin-8 and mixed dopamine-cholecystokinin-8 neurons innervating the nucleus accumbens. *Ann. NY Acad. Sci.* **448**, 306–14.

Studler, J. M., Reibaud, M., Herve, D., Blanc, G., Glowinski, J., and Tassin, J. P. (1986). Opposite effects of sulfated cholecystokinin on DA-sensitive adenylate cyclase in two areas of the rat nucleus accumbens. *Eur. J. Pharmacol.* **126**, 125–8.

Vaccarino, F. J. and Koob, G. F. (1984). Microinjections of nanogram amounts of sulfated cholecystokinin octapeptide (CCK-8) into the rat nucleus accumbens attenuates brain stimulation reward. *Neurosci. Lett.* **52**, 61–6.

Vaccarino, F. J. and Rankin, J. (1989). Nucleus accumbens cholecystokinin (CCK) can either attenuate or potentiate amphetamine-induced locomotor activity: evidence for rostal-caudal differences in accumbens CCK function. *Behav. Neurosci.* **103**, 831–6.

Van Ree, J. M., Gaffori, O., and De Wied, D. (1983). In rats, the behavioral profile of CCK-8 related peptides resembles that of antipsychotic agents. *Eur. J. Pharmacol.* **93**, 63–78.

Vasar, E., Allikmets, L., Soosaar, A., and Lang, A. (1990). Similar behavioural and biochemical effects of long-term haloperidol and caerulein treatment in albino mice. *Pharmacol. Biochem. Behav.* **35**, 855–9.

White, F. J. and Wang, R. Y. (1983). Differential effects of classical and atypical antipsychotic drugs on A9 and A10 dopamine neurons. *Science* **221**, 1054–7.

37. CCK: its role in dopamine-related disorders

M. A. Kuiper, G. J. van Kamp, and E. Ch. Wolters

Introduction

The peptide now recognized as CCK was originally identified as the compound responsible for causing contraction of the gall bladder and stimulation of pancreatic enzyme secretion (Jorpes and Mutt 1966). However, CCK is also synthesized in the CNS (Dockray *et al.* 1977). The brain peptide differs from the gastric peptide and consists primarily of the biologically active sulphated carboxyl-terminal octapeptide (CCK-8S). It is distributed throughout the whole brain with the exception of the cerebellum. In the cerebral cortex and hippocampus CCK exceeds the level of any other neuropeptide (Crawley 1985*a*), and there is abundant evidence that CCK and related peptides exert neurotransmitter/neuromodulator roles in the CNS (see Chapters 1 and 4 of this volume).

Hökfelt *et al.* (1980) demonstrated colocalization of dopamine (DA) and CCK in mesolimbic neurons projecting from the ventral tegmental area of the mid-brain and innervating the posteriomedian nucleus accumbens. This initiated research on CCK in relation to DA-related disorders, e.g. Parkinson's disease, schizophrenia, and tardive dyskinesia. In Parkinson's disease, progressive loss of nigrostriatal DA neurons leads to striatal DA deficiency and this correlates (after a threshold of about 80 per cent loss) with the severity of parkinsonian disability. Schizophrenia is a disorder believed to be related to hyperactivity of the dopaminergic system, which is treated with DA receptor antagonists. Tardive dyskinesia is one of the (often irreversible) complications of prolonged treatment with dopamine antagonists.

CCK receptors

Two receptor subtypes can be distinguished: CCK_A, which is found not only peripherally but also in the CNS (substantia nigra, caudate nucleus, ventral tegmental area), and CCK_B, which is the main CCK receptor in the cortex and hippocampus (Moran *et al.* 1986; Hill *et al.* 1990). It has been proposed on the basis of *in vivo* and *in vitro* results, that CCK_A receptors mediate an excitatory effect on DA release while CCK_B receptors suppress DA release (Crawley

1985*b*; Vaccarino and Rankin 1989; Marshall *et al.* 1990; Chapter 35 of this volume).

Brain dopamine receptors are also heterogenous, and to date two types have been distinguished: the D_1 and D_2 receptors (Kebabian and Calne 1979). There is evidence that both receptor types may be located pre- and postsynaptically.

Recent evidence suggests that there is a functional interaction between CCK receptors and D_1 receptors (Chapter 40 of this volume). An intramembrane interaction between CCK receptors and D_2 autoreceptors, and possibly D_2 postsynaptic receptors, has also been suggested (Tanganelli *et al.* 1990).

Cerebrospinal fluid studies

CCK has been measured in the cerebrospinal fluid (CSF) of patients with neurological and psychiatric disorders, but the results have been contradictory. Gerner and Yamada (1982) and Tamminga *et al.* (1986) failed to find a statistically significant reduction of CSF levels of CCK in untreated schizophrenics, whereas Verbanck *et al.* (1984) reported such a reduction. Lotstra *et al.* (1985) reported a decrease in CCK levels in the CSF of patients with Parkinson's disease.

Autopsy studies

Examination of post mortem brains of schizophrenics has produced conflicting results (Montgomery and Green 1988). A neurochemical study of post-mortem human parkinsonian brains (Studler *et al.* 1982) has demonstrated decreased CCK concentrations within the substantia nigra.

Electrophysiological and biochemical studies

In experimental electrophysiological studies, CCK is found both to excite and inhibit DA neurons. CCK is able to activate DA neurons in areas where DA and CCK coexist (Wang 1988) and to potentiate the inhibition of DA neurons induced by low doses of apomorphine (Hommer and Skirboll 1983). CCK can reduce DA turnover in part of the caudate nucleus and in the anterior part of the nucleus accumbens where CCK and DA are not costored (Fuxe *et al.* 1980). CCK can also potentiate the efflux of cyclic adenosine monophosphate (cAMP) in neurons which contain CCK and DA, and decrease the efflux of cAMP in neurons containing DA but not CCK (Studler *et al.* 1985). It has been suggested that CCK might regulate DA release via a direct presynaptic action on receptors which display a pharmacological profile that is similar to that of CCK_A receptors (Vickroy *et al.* 1988).

Parkinson's disease

Animal experiments give rise to expectations that CCK and CCK_B antagonists may play a modulatory role in L-dopa-induced (hyper)locomotion. CCK administered systemically markedly inhibited L-dopa-induced dyskinesias in monkeys treated with 1-methyl-4-phenyl-1,2,3,6-tetrahydropyridine (MPTP) (parkinsonian monkeys) (Boyce *et al.* 1990*a*). Administration of a CCK_B receptor antagonist failed to stimulate a locomotor response in parkinsonian monkeys. However, given together with dopamine agonists, it caused a potentiation of the locomotor stimulation. A selective CCK_A antagonist did not show this effect (Boyce *et al.* 1990*b*) (see Chapter 39 of this volume for further discussion).

Although there is lack of unanimity concerning CCK and DA interactions, the CCK decapeptide analogue caerulein has been tested in the clinical management of parkinsonian patients stabilized on L-dopa therapy (Bruno *et al.* 1985). Despite substantially elevated plasma caerulein levels, no improvement in neurological status was observed. This may be because CCK does not easily penetrate the brain at dose levels that do not induce gastro-intestinal side-effects.

Schizophrenia

Several years ago, it was suggested that CCK could be used in the treatment of psychiatric disorders resulting from excessive DA release (Fuxe *et al.* 1980; Hökfelt *et al.* 1980). Subsequently some CCK analogues were used to treat schizophrenic patients and some positive results were reported (Bloom *et al.* 1983; Lotstra *et al.* 1984; Albus *et al.* 1986). However, the alleged antipsychotic action of CCK could not be confirmed in double-blind studies.

In most of the studies CCK was used as an adjuvant to standard neuroleptic treatment. As a rule the number of patients involved was too small to be able to demonstrate efficacy above that of the active compound. Limited efficacy was reported in open studies, but not in placebo-controlled double-blind studies (Lotstra *et al.* 1984; Albus *et al.* 1986). To date, neither the efficacy nor, for that matter, the non-efficacy of CCK in the treatment of schizophrenia has been established (Montgomery and Green 1988).

The non-selective CCK antagonist proglumide has also been tested in the treatment of schizophrenia, but without positive results (Hicks *et al.* 1989).

Tardive dyskinesia

Tardive dyskinesia may be induced by long-term administration of neuroleptics and resembles L-dopa-induced dyskinesias. This effect is thought to be the consequence of postsynaptic compensation in response to chronic receptor blockade which results in an increase receptor sensitivity to DA. Results have been obtained which indicate that a relative increase of D_1 activity, as well as altered CCK function, may contribute to the pathogenesis of tardive dyskinesia (Stoessl *et al.* 1989) (see Chapter 38 of this volume for further discussion). In humans, caerulein has been reported to decrease the severity of neuroleptic-induced tardive dyskinesia (Nishikawa *et al.* 1988). To our knowledge, this result has not been confirmed.

Conclusions

Owing to its pharmacological properties (poor entry into the CNS, gastro-intestinal toxicity), CCK is not likely to be useful in the management of DA-related disorders. Only one study in patients with tardive dyskinesia reported positive results with the CCK analogue caerulein. However, non-peptide CCK antagonists have been shown to be effective tools in some preclinical studies, from which therapeutic agents for neuropsychiatric diseases may be developed.

References

Albus, M., von Gellhorn, K., Munch, U., Naber, D., and Ackenheil, M. (1986). A double-blind study with ceruletide in chronic schizophrenic patients. *Psychiat. Res.* **19**, 1-7.

Bloom, D. M., Nair, N. P. V., and Schwartz, G. (1983). CCK-8 in the treatment of chronic schizophrenia. *Psychopharmacol. Bull.* **19**, 361-3.

Boyce, S., Rupniak, N. M. J., Steventon, M. J., and Iversen, S. D. (1990*a*). CCK-8S inhibits L-dopa-induced dyskinesias in parkinsonian squirrel monkeys. *Neurology* **40**, 717-18.

Boyce, S., Rupniak, N. M. J., Tye, S., Steventon, M. J., and Iversen, S. D. (1990*b*). Modulatory role for CCK-B antagonists in Parkinson's disease. *Clin. Neurophar-macol.* **13**, 339-47.

Bruno, G., Ruggieri, S., Chase, T. N., Bakker, K., and Tamminga, C. A. (1985). Caerulein treatment of Parkinson's disease. *Clin. Neuropharmacol.* **8**, 266-70.

Crawley J. N. (1985*a*). Comparative distribution of cholecystokinin and other neuropeptides. *Ann. NY Acad. Sci.* **448**, 1-8.

Crawley J. N. (1985*b*) Behavioural evidence for cholecystokinin modulation of dopamine in the mesolimbic pathways. *Prog. Clin. Biol. Res.* **192**, 131-8.

Dockray, G. J., Gregory, R. A., Hutchinson, J. B., Harris, J. L., and Runswick, M. J. (1977). Isolation, structure and biological activity of two cholecystokinin octapep-tides from sheep brain. *Nature* **270**, 359-61.

Fuxe, K., Andersson, K., Lacatelli, V., Agnati, L. F., Hökfelt, T., Skirboll, L., and Mutt, V. (1980). Cholecystokinin peptides produce marked reduction of dopamine turnover in discrete areas in the rat brain following intraventricular injection. *Eur. J. Pharmacol.* **67**, 325-31.

Gerner, R. H. and Yamada, T. (1982). Altered neuropeptide concentrations in cerebrospinal fluid of psychiatric patients. *Brain Res.* **238**, 298-302.

Hicks, P. B., Vinogradov, S., Riney, S. J., Su, K., and Csernansky, J. G. (1989). A preliminary dose-ranging trial of proglumide for the treatment of refractory schizophrenics. *J. Clin. Psychopharmacol.* **9**, 209-12.

Hill, D. R., Shaw, T. M., Graham, W., and Woodruff, G. N. (1990). Autoradiographical detection of cholecystokinin-A receptors in primate brain using ^{125}I Bolton Hunter CCK-8 and ^{3}H-MK-329. *J. Neurosci.* **10**, 1070–81.

Hökfelt, T., Rehfeld, J. F., Skirboll, L. R., Ivemark, B., Goldstein, M., and Markey K. (1980). Evidence for coexistence of dopamine and CCK in meso-limbic neurons. *Nature* **285**, 476–8.

Hommer, D. W. and Skirboll, L. R. (1983). Cholecystokinin-like peptides potentiate apomorphine-induced inhibition of dopamine neurons. *Eur. J. Pharmacol.* **91**, 151–2.

Jorpes, E. and Mutt, V. (1966). Cholecystokinin and pancreozymin: one single hormone? *Acta Physiol. Scand.* **66**, 196–202.

Kebabian, J. N. and Calne, D. B. (1979). Multiple receptors for dopamine. *Nature* **277**, 93–6.

Lotstra, F., Verbanck, P., Mendlewicz, J., and Vanderhaeghen, J. J. (1984). No evidence of antipsychotic effect of caerulein in schizophrenic patients free of neuroleptics: a double-blind crossover study, *Biol. Psychiat.* **19**, 877–82.

Lotstra, F., Verbanck, P., Gilles, C., Mendlewicz, J., and Vanderhaeghen, J. J. (1985). Reduced cholecystokinin levels in cerebrospinal fluid of parkinsonian and schizophrenic patients. Effect of ceruletide in schizophrenia. *Ann. NY Acad. Sci.* **448**, 507–17.

Marshall, F. H., Barnes, S., Pinnock, R. D., and Hughes, J. (1990). Characterization of cholecystokinin octapeptide-stimulated endogenous dopamine release from rat nucleus accumbens *in vitro*. *Br. J. Pharmacol.* **99**, 845–8.

Montgomery, S. A. and Green, M. C. (1988). The use of cholecystokinin in schizophrenia: a review. *Psychol. Med.* **18**, 593–603.

Moran, T. H., Robinson, P. H., Goldrich, M. S. and McHugh, P. R. (1986). Two brain cholecystokinin receptors: Implication for behavioural actions. *Brain Res.* **362**, 175–9.

Nishikawa, T., Tanaka, M., Tsuda, A., Koga, I., and Uchida, Y. (1988). Treatment of tardive dyskinesia with ceruletide. *Prog. Neuropsychopharmacol. Biol. Psych.* **12**, 803–12.

Stoessl, A. J., Dourish, C. T., and Iversen, S. D. (1989). Chronic neuroleptic-induced mouth movements in the rat: suppression by CCK and selective dopamine D1 and D2 receptor antagonists. *Psychopharmacology* **98**, 372–9.

Studler, J. M., Javoy-Agid, F., Cesselin, F., Legrand, J. C., and Agid, Y. (1982). CCK-8-immunoreactivity distribution in human brain: selective decrease in the substantia nigra from parkinsonian patients. *Brain Res.* **243**, 176–9.

Studler, J. M., Reibaud, M., Tramu, G., Blanc, G., Glowinski, J., and Tassin, J. P. (1985). Distinct properties of cholecystokinin-8 and mixed dopamine–cholecystokinin-8 neurons innervating the nucleus accumbens. *Ann. NY Acad. Sci.* **448**, 306–14.

Tamminga, C. A., Littman, R. L., Alphs, L. D., Chase, T. N., Tuaker, G. K., and Wagman, A. M. (1986). Neuronal cholecystokinin and schizpohrenia: pathogenic and therapeutic studies. *Psychopharmacology* **88**, 387–91.

Tanganelli, S., Fuxe, K., von Euler, G., Agnati, L. F., Ferraro, L., and Ungerstedt, U. (1990). Involvement of cholecystokinin receptors in the control of striatal dopamine receptors. *Naunya Schmiedebergs Arch. Pharmacol.* **342**, 300–4.

Vaccarino, F. J. and Rankin, J. (1989). Nucleus accumbens cholecystokinin (CCK) can either attenuate or potentiate amphetamine-induced locomotor activity: evidence for rostral–caudal differences in accumbens CCK function. *Behav. Neurosci.* **103**, 831–6.

Verbanck, P. M. P., Lotstra, F., Gilles, C., Linkowsli, P., Mendlewicz, J., and Vanderhaegen, J. J. (1984). Reduced cholecystokinin immunoreactivity in the cerebrospinal fluid of patients with psychiatric disorders. *Life Sci.* **34**, 67–72.

Vickroy, T. W., Bianchi, B. R., Kerwin, J. F., Jr., Kopecka, H., and Nadzan, A. M. (1988). Evidence that type A CCK receptors facilitate dopamine efflux in rat brain. *Eur. J. Pharmacol.* **152**, 371–2.

Wang, R. Y. (1988). Cholecystokinin, dopamine, and schizophrenia: recent progress and current problems. *Ann. NY Acad. Sci.* **537**, 362–79.

38. CCK in experimental models of dyskinesia

A. Jon Stoessl and E. Szczutkowski

Introduction

Dyskinesias are involuntary movements which are usually rapid and relatively stereotyped. A number of different aetiologies are recognized, and a relative excess of dopaminergic activity is frequently thought to underly the movements. Tardive dyskinesia refers to those movements attributed to long-term use of neuroleptic medications, and it affects approximately 20 per cent of those individuals so exposed. Based on observations of enhanced dopamine receptor binding (Burt *et al.* 1977) and behavioural supersensitivity to dopamine agonists following chronic neuroleptic use (Tarsy and Baldessarini 1973), it has long been assumed that dopamine receptor upregulation is the critical underlying abnormality. That this is an oversimplification has been well argued by Fibiger and Lloyd (1984), and is further supported by recent electrophysiological (Jiang *et al.* 1990) and molecular biological (Van Tol *et al.* 1990) data. We were interested in the possible role of CCK in tardive dyskinesia for a variety of reasons:

1. Chronic neuroleptics alter brain levels of CCK (Frey 1983; Fukamauchi *et al.* 1987*a*) and possibly also CCK binding (Chang *et al.* 1983; Fukamauchi *et al.* 1987*b*).

2. There are complex biochemical, electrophysiological, and behavioural interactions between CCK and dopamine.

3. Preliminary clinical reports indicated a beneficial effect of ceruletide in a few patients with tardive dyskinesia (Nishikawa *et al.* 1986).

As a model of tardive dyskinesia, we have used the development of vacuous chewing mouth movements (VCMs) in rats, as previously described by others (Iversen *et al.* 1980; Waddington *et al.* 1983).

Another disorder with great clinical impact is the dyskinesia induced by L-dopa in patients with Parkinson's disease. This troublesome side-effect frequently limits the therapy required in order to maintain adequate mobility. We have therefore looked at the effects of CCK on behavioural responses to L-dopa in a rodent model of parkinsonism.

Neuroleptic-induced mouth movements

This work has been described in detail elsewhere (Stoessl *et al.* 1989). Male Sprague–Dawley rats (initial weight 250 g) received fluphenazine decanoate (25 mg/kg) or its vehicle (sesame oil, 1 ml/kg) intramuscularly (i.m.) every 3–4 weeks for 18 weeks. Observations commenced approximately 4 weeks following

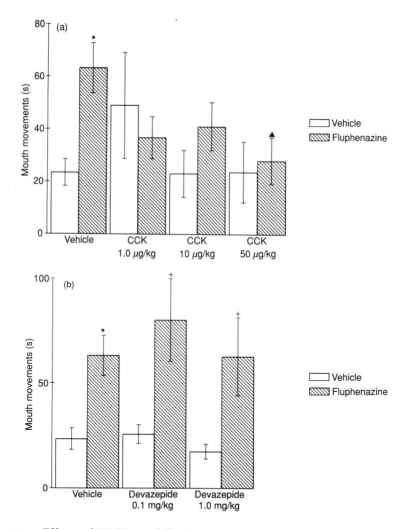

Fig. 38.1. Effects of (a) CCK and (b) devazepide on VCMs (mean ± SEM duration (s) in 15 min) induced by chronic i.m. fluphenazine: * significantly different from vehicle/vehicle; ▲ significantly different from fluphenazine/vehicle; + significantly different from comparable vehicle/drug group. (Reproduced with permission from Stoessl *et al.* 1989.)

the final (sixth) injection. Animals were habituated to Perspex boxes with wire grid floors and a mirror on the rear wall for at least 60 min. Sulphated CCK octapeptide (CCK-8S) (1, 10, 50 μg/kg) or its vehicle (0.9 per cent saline) was administered intraperitoneally (i.p.) and, starting 15 min later, the duration of VCMs was continuously recorded for 15 min using a keyboard linked to a microcomputer. The CCK_A antagonist devazepide was suspended in 0.5 per cent methylcellulose and administered i.p. 30 min prior to observation.

There was a 2.5-fold increase in VCMs in animals chronically treated with fluphenazine compared with the vehicle control group ($p < 0.001$). This increase was suppressed to control levels by CCK ($p < 0.04$, two-way ANOVA) (Fig. 38.1(a)), but devazepide had no effect (Fig. 38.1(b)).

6-hydroxydopamine lesions

Following pretreatment with pargyline (50 mg/kg i.p.) and desipramine (20 mg/kg i.p.), male Sprague–Dawley rats (200–250 g) were anaesthetized with ketamine (60 mg/kg i.p.) and xylazine (15 mg/kg i.p.), and cannulae were stereotaxically placed bilaterally in the medial forebrain bundle. 6-Hydroxydopamine (6-OHDA) (6 μg per side, in 1.5 μl of 0.9 per cent saline containing 0.05 per cent ascorbic acid) was infused over 2 min and the cannulae were kept in position for another 2 min to allow diffusion away from the tip. Bilateral guide cannulae were then implanted 1.5 mm above the lateral ventricles and fixed to the skull using jeweller's screws and dental acrylic. Sham lesions were performed by bilateral infusion of the ascorbate–saline vehicle into the medial forebrain bundle, but otherwise animals were treated in an identical fashion. Post-operatively, nutrition and hydration were maintained in the lesioned animals by gavage feeding until the pre-operative weight was regained.

Approximately 2 weeks after surgery, animals were habituated to Perspex observation boxes for at least 60 min. Carbidopa (25 mg/kg i.p.) was administered, followed 60 min later by L-dopa (0, 10, 30, 50 mg/kg i.p.) and another 15 min later by bilateral intracerebroventricular infusion of 25 ng CCK-8S or its vehicle (2.5 μl 0.9 per cent saline) per side via injection cannulae inserted 1.5 mm below the implanted guide cannulae. Starting 15 min after the infusion of CCK-8S, behavioural responses were recorded for 2 min out of every 8 min block for a total of 10 blocks using a microcomputer programmed to record frequency and duration of a variety of behavioural response categories. Whenever possible, each animal received all four treatments of L-dopa separated by at least 48 h, with the dose varied according to a Latin square design. On each occasion, the animal received the same pretreatment (CCK or saline).

Data were analysed by multifactorial ANOVA (LESION × CCK × DOSE of L-dopa). Because of mortality among the lesioned animals, many animals did not receive all doses of L-dopa and a repeated measures analysis could not be performed.

In sham-operated animals L-dopa induced an increase in locomotion which was attenuated by CCK (Fig. 38.2(a)). Baseline locomotion was actually higher in the lesioned animals (possibly because of increased irritability), but locomotion was completely suppressed at higher doses of L-dopa, secondary to a stereotypic self-injurious grooming response (see below). CCK converted the response pattern to that seen in sham-operated, saline-treated controls ($p = 0.05$, main effect DOSE; $p = 0.012$, LESION × CCK × DOSE interaction; all other terms non-significant).

Rearing was not significantly affected by any of the treatment variables. L-dopa increased sniffing in sham-operated animals, an effect which was if anything potentiated by CCK. The highest dose of L-dopa resulted in suppression of sniffing in the lesioned group, an effect which was reversed by CCK, resulting in a dose–response profile similar to the sham/vehicle animals (Fig. 38.2(b)) ($p = 0.001$, main effect DOSE; $p = 0.014$, main effect CCK; $p = 0.02$, LESION × CCK interaction; $p = 0.03$, LESION × DOSE interaction; $p = 0.01$, CCK × DOSE interaction; $p = 0.04$, LESION × CCK × DOSE interaction).

In intact animals, there was no grooming response to L-dopa. In the lesioned animals, L-dopa induced a marked stereotypic 'grooming' response in which the animals bit their forepaws or trunk, frequently to the point of bleeding. This behaviour was substantially attenuated by CCK (Fig. 37.2(c)) ($p < 0.001$, main effect LESION; $p < 0.01$, main effect CCK; $p = 0.001$, main effect DOSE; $p < 0.01$, LESION × CCK interaction; $p = 0.001$, LESION × DOSE interaction.

Comments

The findings presented here indicate that CCK suppresses behavioural responses that may serve as experimental models of dyskinesia, and suggest that abnormalities of CCK transmission may play a pivotal role in the genesis of these disorders. VCMs in rats induced by chronic neuroleptic treatment are controversial as a model of tardive dyskinesia. Some investigators feel that this response is more equivalent to the development of acute neuroleptic-induced dystonia in humans (Rupniak *et al.* 1986). Nevertheless, other features are reminiscent of tardive dyskinesia (Waddington and Molloy 1987), including suppression of VCMs by an increased dose of neuroleptics (Stoessl *et al.* 1989).

An adequate rodent model of L-dopa-induced dyskinesias has not been identified. However, if dyskinesias are considered to be an unwanted, counterproductive, and relatively stereotyped (within an individual) side-effect of dopaminergic stimulation, then the L-dopa-induced stereotypic self-injurious grooming exhibited by rats with 6-OHDA lesions can clearly be regarded as analogous. Furthermore, suppression of this behaviour in rodents correlates well with the blockade of L-dopa-induced dyskinesias in MPTP-lesioned primates reported by Boyce *et al.* (1990) (see Chapter 39 of this volume).

Other investigators have reported blockade of dopaminergic behaviours in

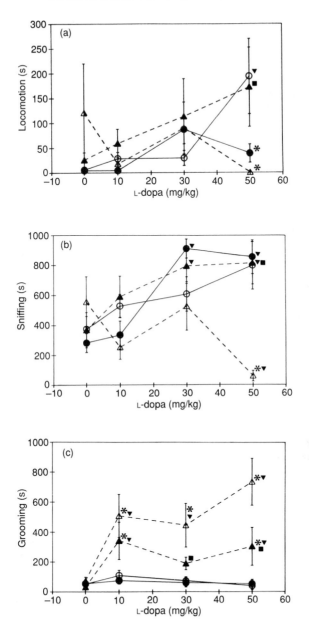

Fig. 38.2. Effects of CCK on (a) L-dopa-induced locomotion, (b) sniffing, and (c) grooming. All values are the mean ± SEM (n = 6–10 animals per group) duration (s) scored from a total of 10 blocks of 2 min every 8 min (○ sham/vehicle, ● sham/CCK; △ lesion/vehicle; ▲ lesion/CCK): ⋆ significantly different from sham/vehicle; ▼ significantly different from 0 mg L-dopa; ■ significantly different from lesion/vehicle.

'sensitized' rodents. Thus, Zetler (1985) found that CCK or ceruletide administered subcutaneously inhibited apomorphine-induced stereotyped gnawing or cage-climbing in mice pretreated with scopolamine or teflutixol. Weiss *et al.* (1989) found that intra-accumbens CCK-8S attenuated the supersensitive locomotor response to apomorphine in rats with bilateral 6-OHDA lesions of the accumbens, or rats treated with subchronic flupenthixol. Stereotypies were not addressed in this study, and the dose response to CCK was markedly biphasic, making the results difficult to interpret. Moroji and Hagino (1986) found blockade of amphetamine or apomorphine-induced hyperactivity by ceruletide in mice, but no dose-related effect on stereotypy.

In contrast, Crawley and co-workers (Crawley *et al.* 1984; Blumstein *et al.* 1987; see Chapter 35 of this volume) have found potentiation of apomorphine-induced stereotypy (and dopamine-induced hyperlocomotion) when low doses of CCK are infused directly into the posterior nucleus accumbens. Differences in their results may relate to the low dose of CCK employed, location within the posterior accumbens, use of a direct postsynaptic agonist, and of course their use of intact animals.

The mechanism of action of CCK in these paradigms is unclear. CCK is thought to attenuate the release of dopamine within the striatum and nucleus accumbens (Markstein and Hökfelt 1984; Voigt and Wang 1984), possibly via a process of depolarization blockade (Lane *et al.* 1986). Within the posterior nucleus accumbens, however, CCK *increases* the *basal* release of dopamine, while K^+-evoked release is attenuated in both the anterior and posterior nucleus accumbens (Voigt *et al.* 1986). Thus some of the behavioural effects observed may relate to presynaptic inhibition of dopamine release by CCK. However, this would not explain a selective inhibition of stereotyped grooming by CCK in 6-OHDA-lesioned animals while locomotion was preserved, unless the effect was indeed selective within the striatum (thought to mediate stereotyped behaviours (Kelly *et al.* 1975)) versus the nucleus accumbens (responsible for locomotion).

Alternatively, the effects of CCK may be mediated postsynaptically, as would be suggested by the electrophysiological observations of Wang and Hu (1986). In this case, behavioural selectivity for the antidopaminergic effects of CCK may be explained either by regional selectivity (striatum versus nucleus accumbens) or possibly by an inhibitory interaction with D_1 receptors. This might account for the suppression of VCMs and grooming, both of which appear to be dependent upon stimulation of D_1 dopamine receptors, while locomotion, which is predominantly mediated by D_2 receptors, is spared. It is interesting to note with regard to this hypothesis that CCK-8S inhibits VCMs and the efflux of cyclic adenosine monophosphate (cAMP) induced by the D_1 agonist SKF 38393 (see Chapter 40 of this volume).

Regardless of mechanism, these results suggest that abnormalities of CCK transmission may play a critical role in the development of dyskinesias, and that further investigation of these mechanisms is warranted. Preliminary clinical studies (Nishikawa *et al.* 1986) indicate that there may indeed be a therapeutic role for CCK-like analogues in the treatment of these disorders.

Acknowledgements

This work was supported by a grant from the Medical Research Council of Canada. The computer software was developed by Dr M. T. Martin-Iverson. A. J. S. is a Career Scientist of the Ontario Ministry of Health.

References

Blumstein, L. K., Crawley, J. N., Davis, L. G. and Baldino, F., Jr (1987). Neuropeptide modulation of apomorphine-induced stereotyped behaviour. *Brain Res.* **404**, 293–300.

Boyce, S., Rupniak, N. M. J., Steventon, M. J., and Iversen, S. D. (1990). CCK-8S inhibits L-dopa-induced dyskinesias in parkinsonian squirrel monkeys. *Neurology* **40**, 717–18.

Burt, D. R., Creese, I., and Snyder, S. H. (1977). Antischizophrenic drugs: chronic treatment elevates dopamine receptor binding in brain. *Science* **196**, 326–8.

Chang, R. S. L., Lotti, V. J., Martin, G. E., and Chen, T. B. (1983). Increase in brain ^{125}I-cholecystokinin (CCK) receptor binding following chronic haloperidol treatment, intracisternal 6-hydroxydopamine or ventral tegmental lesions. *Life Sci.* **32**, 871–8.

Crawley, J. N., Hommer, D. W., and Skirboll, L. R. (1984). Behavioural and neurophysiological evidence for a facilitatory interaction between co-existing transmitters: cholecystokinin and dopamine. *Neurochem. Int.* **6**, 755–60.

Fibiger, H. C. and Lloyd, K. G. (1984). Neurobiological substrates of tardive dyskinesia: the GABA hypothesis. *Trends Neurosci.* **7**, 462–4.

Frey, P. (1983). Cholecystokinin octapeptide levels in rat brain are changed after subchronic neuroleptic treatment. *Eur. J. Pharmacol.* **95**, 87–92.

Fukamauchi, F., Yoshikawa, T., Kaneno, S., Shibuya, H., and Takahashi, R. (1987*a*). Dopaminergic agents affected neuronal transmission of cholecystokinin in the rat brain. *Neuropeptides* **10**, 207–20.

Fukamauchi, F., Yoshikawa, T., Kaneno, S., Shibuya, H., and Takahashi, R. (1987*b*) The chronic administration of dopamine antagonists and methamphetamine changed the [^{3}H]-cholecystokinin-8 binding-sites in the rat frontal cortex. *Neuropeptides* **10**, 221–5.

Iversen, S. D., Howells, R. B., and Hughes, R. P. (1980) Behavioural consequences of long-term treatment with neuroleptic drugs. *Adv. Biochem. Psychopharmacol.* **24**, 305–13.

Jiang, L. H., Kassar, R. J., Altar, C. A., and Wang, R. Y. (1990). One year of continuous treatment with haloperidol or clozapine fails to induce a hypersensitive response of caudate putamen neurons to dopamine D1 and D2 receptor agonists. *J. Pharmacol. Exp. Therp.* **253**, 1198–1205.

Kelly, P. H., Seviour, P. W., and Iversen, S. D. (1975). Amphetamine and apomorphine responses in the rat following 6-OHDA lesions of the nucleus accumbens septi and corpus striatum. *Brain Res.* **94**, 507–22.

Lane, R. F., Blaha, C. D., and Phillips, A. G. (1986) *In vivo* electrochemical analysis of cholecystokinin-induced inhibition of dopamine release in the nucleus accumbens. *Brain Res.* **397**, 200-4.

Markstein, R. and Hökfelt, T. (1984). Effect of cholecystokinin-octapeptide on dopamine release from slices of cat caudate nucleus. *J. Neurosci.* **4**, 570-5.

Moroji, T. and Hagino, Y. (1986) A behavioural pharmacological study on CCK-8 related peptides in mice. *Neuropeptides* **8**, 273-86.

Nishikawa, T., Tanaka, M., Tsuda, A., Kuwahara, H., Koga, I., and Uchida, Y. (1986). Effect of ceruletide on tardive dyskinesia: a pilot study of quantitative computer analyses on electromyogram and microvibration. *Psychopharmacology* **90**, 5-8.

Rupniak, N. M. J., Jenner, P., and Marsden, C. D. (1986). Acute dystonia induced by neuroleptic drugs. *Psychopharmacology* **88**, 403-19.

Stoessl, A. J., Dourish, C. T., and Iversen, S. D. (1989). Chronic neuroleptic-induced mouth movements in the rat: suppression by CCK and selective dopamine D1 and D2 receptor antagonists. *Psychopharmacology* **98**, 372-9.

Tarsy, D. and Baldessarini, R. J. (1973). Pharmacologically induced behavioural supersensitivity to apomorphine. *Nature New Biol.* **245**, 262-3.

Van Tol, H. H. M., Riva, M., Civelli, O., and Creese, I. (1990). Lack of effect of chronic dopamine receptor blockade on D2 dopamine receptor mRNA level. *Neurosci. Lett.* **111**, 303-8.

Voigt, M. M. and Wang, R. Y. (1984). *In vivo* release of dopamine in the nucleus accumbens of the rat: modulation by cholecystokinin. *Brain Res.* **296**, 189-94.

Voigt, M., Wang, R. Y., and Westfall, T. C. (1986). Cholecystokinin octapeptides alter the release of endogenous dopamine from the rat nucleus accumbens *in vitro*. *J. Pharmacol. Exp. Ther.* **237**, 147-53.

Waddington, J. L. and Molloy, A. G. (1987). The status of late-onset vacuous chewing/perioral movements during long-term neuroleptic treatment in rodents: tardive dyskinesia or dystonia? *Psychopharmacology* **91**, 136-7.

Waddington, J. L., Cross, A. J., Gamble, S. J., and Bourne, R. C. (1983). Spontaneous orofacial dyskinesia and dopaminergic function in rats after 6 months of neuroleptic treatment. *Science* **220**, 530-2.

Wang, R. Y. and Hu, X.-T. (1986). Does cholecystokinin potentiate dopamine action in the nucleus accumbens? *Brain Res.* **380**, 363-7.

Weiss, F., Ettenberg, A., and Koob, G. F. (1989). CCK-8 injected into the nucleus accumbens attenuates the supersensitive locomotor response to apomorphine in 6-OHDA and chronic-neuroleptic treated rats. *Psychopharmacology* **99**, 409-15.

Zetler, G. (1985). Antistereotypic effects of cholecystokinin octapeptide (CCK-8), ceruletide and related peptides on apomorphine-induced gnawing in sensitized mice. *Neuropharmacology* **24**, 251-9.

39. CCK and movement disorders

Susan Boyce and Nadia M. J. Rupniak

Basal ganglia diseases seriously compromise an individual's ability to perform activities of daily living, and many limitations are associated with current treatments. Movement disorders range from akinetic/hypokinetic syndromes such as Parkinson's disease, where movement is extremely impoverished, to hyperkinetic conditions involving excessive and uncontrollable movements as in Huntington's chorea and drug-induced dyskinesias. The profound effect of dopamine receptor agonists and antagonists on movement clearly implicate abnormalities of dopamine function in aberrant motor control. The clinical similarity between drug-induced and spontaneous movement disorders further suggests a common pathophysiology. The close association of CCK with dopamine in the mesolimbic and basal ganglia structures suggests that it may also be involved in the regulation of motor activity. In this chapter we shall discuss the functional interactions of CCK and dopamine, the potential use of CCK agonists and antagonists in the treatment of movement disorders, and the possible problems associated with such therapies.

Antidyskinetic properties of CCK analogues

Dyskinesias encompass a variety of abnormal involuntary movements including chorea, dystonia, myoclonus, tics, athetosis, stereotypy, and hemiballismus. These may be of spontaneous origin or the result of chronic treatments with drugs such as L-dopa or neuroleptics which cause dopaminergic overactivity. No effective antidyskinetic agents exist that are appropriate for routine clinical use. Although dopamine receptor antagonists and depleting agents have been successfully employed to suppress abnormal movements, this has been at the expense of impairing motor function acutely and increasing the likelihood of dyskinesias if given chronically (Klawans and Weiner 1974; Marsden *et al.* 1975). Moreover, dopamine-depleting drugs may cause depression and postural hypotension (Snaith and Warren 1974). These side-effects are likely to be mechanism-based and might not be expected of drugs which do not interact directly with catecholamine neurons.

The discovery that CCK immunoreactivity is reduced in the substantia nigra in Parkinson's disease (Studler *et al.* 1982) and Huntington's chorea (Emson *et al.* 1980), may implicate CCK in the pathogenesis of these movement disorders. However, it is difficult to reconcile a loss of CCK with both hypo-kinetic and hyperkinetic syndromes. The additional loss of CCK in globus pallidus in Huntington's chorea may account for this difference. Therefore CCK agonists may be of value in restoring CCK function and modulating dopamine activity in these and other movement disorders.

Following systemic administration, CCK has consistently been found to exert neuroleptic-like effects (see Chapter 36 of this volume). Consistent with this observation, there have been several reports of the beneficial effects of ceruletide in movement disorders including tardive dyskinesia (Chapter 37 of this volume), Huntington's chorea (Nakanishi *et al.* 1988), dystonia, and action tremor (Hirayama *et al.* 1988). Regrettably, many of these studies cannot be assessed adequately. In contrast, in one double-blind cross-over study, ceruletide failed to exert an antidyskinetic effect in parkinsonian patients (Bruno *et al.* 1985). The effectiveness of CCK agonists as antidyskinetic agents therefore remains inconclusive. For these reasons we examined in more detail the ability of CCK to inhibit dyskinesias induced by L-dopa in parkinsonian primates.

We employed squirrel monkeys previously rendered parkinsonian follow-ing treatment with the nigral neurotoxin 1-methyl-4-phenyl-1, 2, 3, 6-tetra-hydropyridine (MPTP). In response to L-dopa treatment the subjects had developed dyskinesias which closely resembled those observed in Parkinson's

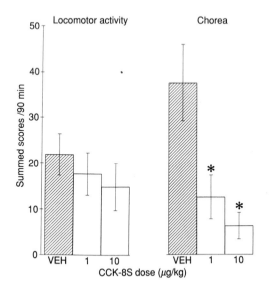

Fig. 39.1. Effect of CCK-8S on locomotor activity and dyskinesias induced by L-dopa (15 mg/kg p.o.) in MPTP-treated squirrel monkeys. Values are the mean (± SEM) obtained from seven animals (ANOVA: locomotor activity, $F_{(2, 12)} = 12.58$, $p = 0.001$). * $p < 0.05$ compared with vehicle treatment (Dunnett's t test).

disease (Boyce *et al.* 1990*a*). Administration of L-dopa (15 mg/kg p.o.) induced marked locomotor stimulation accompanied by severe and debilitating chorea. When sulphated CCK octapeptide (CCK-8S) (1 or 10 μg/kg i.p.) was administered with L-dopa, a dose-related inhibition in dyskinesias was observed but the locomotor response was unaffected (Fig. 39.1) (Boyce *et al.* 1990*b*). All animals improved after treatment with CCK, and a complete antagonism of chorea was observed in some cases. The duration of this effect was short-lived (less than 90 min), as would be expected from the short plasma half-life of CCK. Our findings clearly support clinical evidence for an antidyskinetic effect of CCK. Similarly, Stoessl *et al.* (1989) (see also Chapter 38 of this volume) reported that CCK-8S suppressed neuroleptic-induced mouth movements in rodents, which are regarded by some as a model of tardive dyskinesia.

The mechanism mediating the antidyskinetic effects of CCK-8S is unknown but may involve modulation of central D_1 receptor function. In separate studies we have speculated that chorea may be mediated via stimulation of D_1 receptors (Boyce *et al.* 1990*c*). Similarly, in rodents CCK-8S inhibited perioral movements and striatal cyclic adenosine monophosphate (cAMP) efflux induced by the partial D_1 receptor agonist SKF 38393 (see Chapter 40 of this volume).

Implications for therapy

Unlike conventional antidyskinetic agents, CCK agonists may control overactivity syndromes without aggravating parkinsonism. A second potential advantage of CCK agonists resides in their putative antipsychotic activity (Frey 1983). If this were substantiated, CCK agonists might also be effective in the primary treatment of psychosis in patients with neuroleptic-induced tardive dyskinesia. An antipsychotic effect would also be of great value in controlling dopamine agonist-induced psychosis in Parkinson's disease (Cotzias *et al.* 1969).

Despite these possible advantages, the use of CCK and peptide analogues is not without complications. Rapid metabolism, poor bioavailability, and limited access to the CNS (Oldendorf 1981) could seriously reduce clinical efficacy. Although these problems might be overcome by developing non-peptide analogues of CCK, other actions associated with stimulation of CCK receptors could undermine clinical use of CCK agonists. These include panic attacks (De Montigny 1989; Bradwejn *et al.* 1990; Chapter 11 of this volume), changes in nociception (Faris *et al.* 1983; Chapters 43 and 44 of this volume), and gastro-intestinal side-effects (Bruno *et al.* 1985).

Modulatory role for CCK antagonists in Parkinson's disease

In the preceding section we considered the use of CCK agonists in dopamine overactivity syndromes. In contrast with these conditions, Parkinson's disease is a dopamine deficiency syndrome. Replacement therapy with L-dopa becomes progressively complicated by swings in neurological status caused by short-lived benefit from each dose and by fluctuations unrelated to the timing of medication ('wearing-off' and 'on–off' effects) (Fahn 1974; Marsden and Parkes 1977). The need for more intense dosage regimes to overcome these difficulties may aggravate peak-dose dyskinesias (Barbeau *et al.* 1970) and induce psychosis (Cotzias *et al.* 1969). Therefore a treatment that augmented the antiparkinsonian effects of dopamine agonists would be particularly attractive. Since CCK agonists *inhibit* dopamine function, then CCK antagonists would be expected to enhance dopamine activity. This hypothesis leads to two predictions: first CCK antagonists might themselves stimulate locomotor activity, and second they might potentiate the effects of dopamine agonists.

Do CCK antagonists possess antiparkinsonian activity?

We attempted to address this question by investigating whether CCK antagonists would stimulate activity in MPTP-treated monkeys. Treatment with either the CCK_A receptor antagonist devazepide (1–100 μg/kg i.p.) or the CCK_B receptor antagonist L-365,260 (1–100 μg/kg i.p.) did not affect locomotor activity (Boyce *et al.* 1990*d*). Similarly, in numerous rodent studies CCK antagonists failed to alter locomotor activity (Koshla and Crawley 1988; Soar *et al.* 1989; O'Neill *et al.* 1991; see Chapter 22 of this volume).

Do CCK antagonists augment the effects of dopamine agonists?

CCK antagonists appear not to possess antiparkinsonian activity when given alone. However, when L-365,260 (1 or 10 μg/kg i.p.) was given as a 1 h

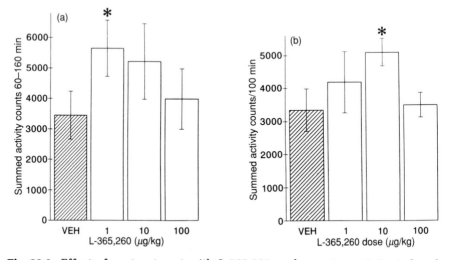

Fig. 39.2. Effect of pretreatment with L-365,260 on locomotor activity induced by (a) L-dopa (20 mg/kg p.o.) and (b) (+)-PHNO (2.5 μg/kg s.c.) in MPTP-treated squirrel monkeys. Values are the mean ($\pm$ SEM) obtained from six or seven animals (ANOVA: L-dopa, $F(3, 18) = 3.66$, $p = 0.032$; (+)-PHNO, $F(3, 15) = 3.31$, $p = 0.049$). $\star$ $p < 0.05$ compared with vehicle treatment (Dunnett's t test).

pretreatment to L-dopa or the selective D_2 agonist (+)-4-propyl-9-hydroxynaphthoxazine (+)-PHNO) a marked potentiation (50–60 per cent) in locomotor stimulation was observed (Fig. 39.2). In contrast, pretreatment with devazepide (1–100 μg/kg i.p.) failed to modify the motor effects of dopamine agonists (Fig. 39.3).

The modulatory effects of CCK on dopamine-mediated locomotor activity might be attributed to alterations in those mesolimbic neurons in which CCK and dopamine are colocalized (Hökfelt *et al.* 1980). Direct injection of CCK into the posterior nucleus accumbens (where CCK and dopamine coexist) *enhanced* the locomotor stimulatory effects of dopamine agonists in rodents (Crawley *et al.* 1985), a finding which apparently contradicts our results. However, an *inhibitory* action of CCK was reported following direct injection into the anterior nucleus accumbens (Vaccarino and Rankin 1989), a region which is innervated independently by CCK and dopamine neurons (Seroogy and Fallon 1989). More recently, Marshall *et al.* (1991) have demonstrated that different subclasses of CCK receptors are responsible for these opposing effects on dopamine function. Thus, stimulation of CCK_A receptors facilitated dopamine release in the nucleus accumbens, whereas an inhibition was observed following activation of CCK_B receptors. The contradictory findings in behavioural experiments probably depend upon the particular population of CCK receptors examined.

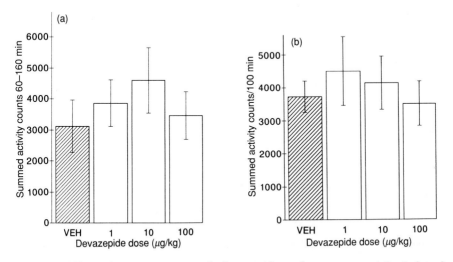

Fig. 39.3. Effect of pretreatment with devazepide on locomotor activity induced by (a) L-dopa (20 mg/kg p.o.) and (b) (+)-PHNO (2.5 μg/kg s.c.) in MPTP-treated squirrel monkeys. Values are the mean (± SEM) obtained from six or seven animals (ANOVA: L-dopa $F(3, 15) = 1.24$, $p = 0.330$; (+)-PHNO, $F(3, 18) = 0.75$, $p = 0.537$).

Implications for therapy

CCK$_B$ receptor antagonists may be useful adjuncts to existing anti-parkinsonian drugs. Nevertheless, several problems may complicate routine clinical use of CCK antagonists. In our study the efficacy of L-365,260 was restricted to only one CCK antagonist–dopamine agonist dose combination. Dopamine agonists themselves have a relatively narrow therapeutic window and require careful titration in individual patients (Stoessl *et al.* 1985). The compounding difficulties of titrating the doses of both CCK antagonists and dopamine agonists may prevent their routine use in Parkinson's disease. It should be noted that in our study the dose range of L-365,260 examined was very wide and that a 10-fold difference separated each dose. Further studies are required to establish the active dose range of this compound more accurately. The potential benefits of such treatments, no matter how small the improvement or how difficult the initial dose titration may be, could prove invaluable to patients with a brittle neurological status.

Even if this objection is overcome, many issues remain unresolved. Since we argued earlier that *stimulation* of CCK receptors might have antidyskinetic and antipsychotic effects, CCK antagonists could be prodyskinetic and induce or exacerbate psychosis. However, these problems may be overcome once the complexities of the CCK–dopamine interactions are unravelled. At present it is not known precisely how the different CCK and dopamine subtypes interact in influencing movement. The facilitatory effect of CCK$_B$ antagonists on locomotion appears to involve an interaction with D$_2$ dopamine receptors, as judged from the effects with the selective D$_2$ agonist (+)-PHNO. Stimulation of D$_2$ receptors appears essential for antiparkinsonian activity (Schachter *et al.* 1980). Conversely, the antidyskinetic effects of CCK-8S may involve D$_1$ receptors; the CCK receptor mediating this response is unknown. If these mechanisms can be dissociated, such complications might be avoided.

References

Barbeau, A., Mars, H., Gillo-Joffroy, L., and Arsenault, A. (1970). A proposed classification of dopa-induced dyskinesias. In *L-dopa and parkinsonism* (ed. A. Barbeau and F. H. McDowell), pp. 118–23. Davis, Philadelphia, PA.

Boyce, S., Rupniak, N. M. J., Steventon, M. J., and Iversen, S. D. (1990a). Characterisation of dyskinesias induced by L-dopa in MPTP-treated squirrel monkeys. *Psychopharmacology* **102**, 21–7.

Boyce, S., Rupniak, N. M. J., Steventon, M. J., and Iversen, S. D. (1990b). CCK-8S inhibits L-dopa-induced dyskinesias in parkinsonian squirrel monkeys. *Neurology* **40**, 717–8.

Boyce, S., Rupniak, N. M. J., Steventon, M. J., and Iversen, S. D. (1990c). Differential effects of D$_1$ and D$_2$ agonists in MPTP-treated primates: functional implications for Parkinson's disease. *Neurology* **40**, 997–1033.

Boyce, S., Rupniak, N. M. J., Tye, S. J., Steventon, M. J., and Iversen, S. D. (1990d).

Modulatory role for CCK-B antagonists in Parkinson's disease. *Clin. Neurophar-macol.* **13**, 339–47.

Bradwejn, J., Koszycki, D. and Meterissian, G. (1990). Cholecystokinin-tetrapeptide induces panic attacks in patients with panic disorder. *Can. J. Psychiat.* **35**, 83–5.

Bruno, G., Ruggieri, S., Chase, T. N., Bakker, K., and Tamminga, C. A. (1985). Caerulein treatment in Parkinson's disease. *Clin. Neuropharmacol.* **80**, 266–70.

Cotzias, G. P., Papavasiliou, P. S., and Gellene, R. (1969). Modification of parkinsonism — chronic treatment with L-dopa. *New Engl. J. Med.* **280**, 337–45.

Crawley, J. N., Stivers, J. A., Blumstein, L. K., and Paul, S. M. (1985). Chol-ecystokinin potentiates dopamine mediated behaviours: evidence for modulation specific to a site of coexistence. *J. Neurosci.* **5**, 1972–83.

De Montigny, C. (1989). Cholecystokinin-tetrapeptide induces panic-like attacks in healthy volunteers. *Arch. Gen. Psychiat.* **46**, 511–7.

Emson, P. C., Rehfeld, J. F., Langevin, H., and Rossor, M. (1980). Reduction in cholecystokinin-like immunoreactivity in the basal ganglia in Huntington's disease. *Brain Res.* **198**, 497–500.

Fahn, S. (1974). 'On–off' phenomenon with L-dopa therapy in parkinsonism. *Neurology.* **24**, 431–41.

Faris, P. L., Kominsaruk, B. R., Watkins, L. R., and Mayer, D. J. (1983). Evidence for the neuropeptide cholecystokinin as an antagonist of opiate analgesia. *Science* **219**, 310–2.

Frey, P. (1983). Cholecystokinin octapeptide levels in rat brain are changed after chronic neuroleptic treatment. *Eur. J. Pharmacol.* **95**, 87–92.

Hirayama, K., Yamada, T., Tokumaru, Y., and Katayama, K. (1988). Clinical effects of ceruletide on dystonia, chorea and action tremor. In *Neuropeptides: basis and clinics*, Vol. 62 (ed. I. Sobue), pp. 374–80. Ministry of Health and Welfare, Nagoya.

Hökfelt, T., Rehfeld, J. F., Skirboll, L., Ivemark, B., Goldstein M., and Markey, K. (1980). Evidence for coexistence of dopamine and CCK in mesolimbic neurons. *Nature* **285**, 476–8.

Klawans, H. L. and Weiner, W. J. (1974). Attempted use of haloperidol in the treatment of L-dopa-induced dyskinesias. *J. Neurol. Neurosurg. Psychiat.* **37**, 427–30.

Koshla, S. and Crawley, J. N. (1988). Potency of L-364,718 as an antagonist of the behavioural effects of peripherally administered cholecystokinin. *Life Sci.* **42**, 153–9.

Marsden, C. D. and Parkes, J. D. (1977). Success and problems of long-term L-dopa therapy in Parkinson's disease. *Lancet* **i**, 345–9.

Marsden, C. D., Tarsy, D., and Baldessarini, R. J. (1975). Spontaneous and drug-induced movement disorders in psychotic patients. In *Psychiatric aspects of neuro-leptic disease* (ed. D. F. Benson and D. Blumer), pp. 219–66. Grune and Stratton, New York.

Marshall, F. H., Barnes, S., Hunter, J. C., and Hughes, J. (1991). Cholecystokinin modulates the release of dopamine from the anterior and posterior nucleus accumbens by two different mechanisms. *J. Neurochem.* **56**, 917–22.

Nakanishi, T., Naito, Y., Komatsuzaki, Y., Kanazawa, I. (1988). Clinical trial of cholecystokinin-octapeptide-like agent (ceruletide) on Huntington's disease. In *Neuropeptides: basis and clinics*, Vol. 62 (ed. I. Sobue), pp. 381–5. Ministry of Health and Welfare, Nagoya.

Oldendorf, W. H. (1981). Blood–brain barrier permeability to peptides: pitfalls in measurement. *Peptides* **2** (Suppl. 2), 109–11.

O'Neill, M. F., Dourish, C. T., and Iversen, S. D. (1991). Hypolocomotion induced by peripheral or central injection of CCK in the mouse is blocked by the CCK-A receptor antagonist devazepide but not by the CCK-B receptor antagonist L-365,260. *Eur. J. Pharmacol.* 93, 203–8.

Schachter, M., Bedard, P., Belsono, A. G., Jenner, P., Marsden, C. D., Price, P., Parkes, J. D., Keenan, J., Smith, B., Rosenthaler, J., Horowski, R., and Dorow, R. (1980). The role of D_1 and D_2 receptors. *Nature* 286, 157–9.

Seroogy, K. M. and Fallon, J. H. (1989). Forebrain projections from cholecystokinin-immunoreactive neurons in the rat brain. *J. Comp. Neurol.* 279, 415–35.

Snaith, R. P. and Warren, H. de B. (1974). Treatment of Huntington's chorea with tetrabenazine. *Lancet* i, 418.

Soar, J., Hewson, G., Leighton, G. E., Hill, R. G., and Hughes, J. (1989). L-364,718 antagonises the cholecystokinin-induced suppression in locomotor activity. *Pharmacol. Biochem. Behav.* 33, 637–40.

Stoessl, A. J., Mak, E., and Calne, D. B. (1985) (+)-4-propyl-9-hydroxynaphthoxazine (PHNO), a new dopamimetic in treatment of Parkinsonism. *Lancet.* ii, 1330–1.

Stoessl, A. J., Dourish, C. T., and Iversen, S. D. (1989). Chronic neuroleptic-induced mouth movements in the rat: suppression by CCK and selective dopamine D_1 and D_2 receptor antagonists. *Psychopharmacology* 98, 372–9.

Studler, J. M., Javoy-Agid, F., Cesselin, F., Legrand, J. C., and Agid, Y. (1982). CCK-8 immunoreactivity distribution in human brain: selective decrease in the substantia nigra from parkinsonian patients. *Brain Res.* 243, 176–9.

Studler, J. M., Reibald, M., Herve, D., Blanc, G. D., Glowinski, J. and Tassin, J. P. (1986). Opposite effects of sulphated cholecystokinin on dopamine-sensitive adenylate cyclase in two areas of the rat nucleus accumbens. *Eur. J. Pharmacol.* 126, 125–8.

Vaccarino, E. I. and Rankin, I. (1989). Nucleus accumbens cholecystokinin (CCK) can either attenuate or potentiate amphetamine-induced locomotor activity: evidence for rostral-caudal differences in accumbens CCK function. *Behav. Neurosci.* 103, 831–6.

40. Behavioural and neurochemical evidence for an interaction of CCK with D_1 dopamine receptors

C. T. Dourish, P. H. Hutson, S. J. Kitchener, M. F. O'Neill, and N. Suman-Chauhan

Introduction

The neuropeptide CCK has complex effects on brain dopaminergic neurotransmission, and both enhancement and inhibition of dopamine-mediated responses have been reported after treatment with CCK (Voigt and Wang 1984; Crawley *et al.* 1985). CCK has been shown to coexist with dopamine in a subpopulation of mesolimbic neurons that project from the ventral tegmental area of the mid-brain and innervate the posteromedian nucleus accumbens (Hökfelt *et al.* 1980). It has been proposed that CCK specifically enhances dopamine-mediated behaviour in brain regions that are innervated by CCK/dopamine-containing neurons (Crawley *et al.* 1984, 1985). Thus injection of CCK into the posteromedian nucleus accumbens enhances the stimulant action of the dopamine agonist apomorphine injected into the same site (Crawley *et al.* 1985). In contrast, injection of CCK into the anterolateral region of the nucleus accumbens, which is innervated by dopamine neurons that do not contain CCK, blocks dopamine-mediated behaviour (Vaccarino and Rankin 1989).

CCK receptors have been classified into two types, CCK_A and CCK_B receptors (Moran *et al.* 1986; Dourish and Hill 1987). Both subtypes are found in the brain, although initial studies using autoradiographical techniques suggested that CCK_A receptors were confined to a few regions of the brainstem (Moran *et al.* 1986; Hill *et al.* 1987). More recently, however, functional evidence has indicated the presence of CCK_A receptors in the nucleus accumbens (Vickroy *et al.* 1988). Furthermore, on the basis of *in vitro* results it has been proposed that CCK_A receptors mediate an excitatory effect of CCK on dopamine release in the posteromedian nucleus accumbens, whereas CCK_B receptors mediate an inhibitory effect of CCK on dopamine release in the anterolateral nucleus accumbens (Marshall *et al.* 1990). Brain dopamine receptors are also known to be heterogenous and have been classified into D_1 and D_2 subtypes (Kebabian and Calne 1979). Although it is known that CCK modifies the behavioural and neurochemical responses to mixed D_1–D_2 dopamine receptor agonists such as dopamine and apomorphine (Voigt and Wang 1984; Crawley *et al.* 1985), the effects of CCK on the responses induced by a selective D_1 dopamine receptor agonist have not been examined. D_1 dopamine receptor stimulation produces a characteristic pattern of behaviour and increases the formation of the cyclic nucleotide 3'5'-monophosphate (cAMP) by stimulating the enzyme adenylate cyclase (Waddington and O'Boyle 1989). In the present study, we examined the effects of CCK on the behavioural responses and the increase in cAMP overflow *in vivo* induced by the selective D_1 dopamine receptor agonist SKF 38393.

Behavioural and neurochemical studies

For the behavioural studies, male Sprague–Dawley rats were habituated to Perspex observation boxes for 2.5 h. In the first experiment, groups of rats (n = 7–8 per group) were injected subcutaneously (s.c.) with (±)-SKF 38393 at doses of 10, 20, 40, or 80 mg/kg or saline vehicle. Beginning 75 min later, the animals were observed for a period of 15 min, and the frequency and duration of sniffing, grooming, and mouth movements were recorded by an observer using a keypad interfaced to a CUBE microcomputer. In the second experiment, rats were injected intraperitoneally (i.p.) with either saline vehicle or sulphated CCK octapeptide (CCK-8S), at doses of 0.001, 0.01, 0.1, 1.0, or 10.0 µg/kg, 10 min prior to injection of vehicle or 20 mg/kg of SKF 38393. Beginning 75 min after injection of SKF 38393, animals were observed for a 15 min period and their behaviour was recorded as described above.

For the neurochemical studies, male Sprague–Dawley rats were anaesthetized with chloral hydrate (400 mg/kg i.p.) and implanted with a dialysis probe in the striatum. The probe was perfused with Krebs Ringer at a rate of 2 µl/min for a period of 1 h before serial samples were collected at 20 min intervals for 3 h. These samples were frozen at $-70\,°C$ for later analysis for cAMP content. Samples were analysed for cAMP content by radio-immunoassay (Amersham) using the high sensitivity acetylation method (see Hutson and Suman-Chauhan (1990) for full details). In the first experiment SKF 38393, SCH 23390, and sulpiride were dissolved in water as a 10 mM stock solution, diluted with Krebs Ringer, and infused through the dialysis probe. In the second experiment, rats were injected i.p. with either saline vehicle or CCK 10 min prior to intrastriatal infusion of vehicle or 10 µM SKF 38393.

SKF 38393 significantly increased the frequency and duration of sniffing ($F(4, 34) = 3.19$, $p < 0.03$), grooming ($F(4, 34) = 4.63$, $p < 0.005$), and mouth movements ($F(4, 34) = 5.10, p < 0.003$) during the 15 min test period. The dose–response curve for each type of behaviour is illustrated in Fig. 40.1. It is apparent that the dose–response curves for sniffing and mouth movements are bell-shaped. On the basis of these data, a dose of 20 mg/kg of SKF 38393 (the minimum dose to induce all three behaviours) was chosen for the CCK

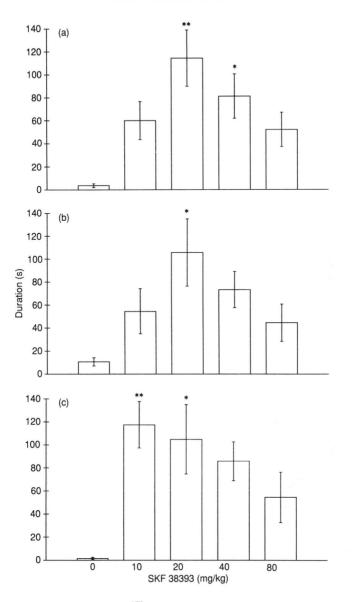

Fig. 40.1. Effects of s.c. injection of SKF 38393 on (a) mouth movements, (b) sniffing, and (c) grooming in rats. Data are expressed as mean ± SEM (n = 7–8 rats per group): ⋆ $p < 0.05$, ⋆⋆ $p < 0.01$ compared with vehicle (ANOVA and Tukey test).

interaction studies. CCK at doses of 0.01–1.0 μg/kg attenuated the mouth movement response induced by SKF 38393 (ANOVA, CCK main effect: $F(5, 142) = 2.45$, $p < 0.04$) (Fig. 40.2), whereas a higher dose of CCK (10.0 μg/kg) was ineffective. Furthermore, CCK at doses of 1.0 and 10.0 μg/kg attenuated sniffing induced by SKF 38393 (ANOVA, CCK main effect:

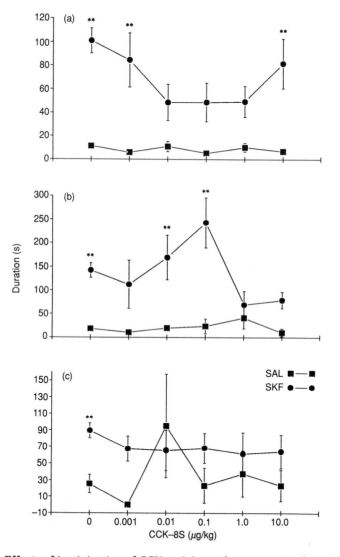

Fig. 40.2. Effects of i.p. injection of CCK on (a) mouth movements, (b) sniffing, and (c) grooming induced by s.c. injection of 20 mg/kg of SKF 38393. Data are expressed as mean ± SEM (n = 8 or more rats per group): ∗∗ p < 0.01 compared with control group (SAL/SAL) (ANOVA and Tukey test).

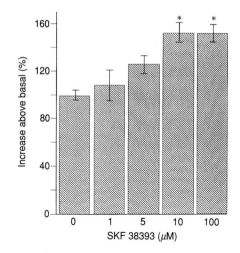

Fig. 40.3. Effect of intrastriatal infusion of SKF 38393 on striatal cAMP efflux *in vivo*. Data are expressed as mean ± SEM percentage increase above basal cAMP values which were 14.5 ± 1.3 fmol/40 μl ($n = 20$), ($n = 3-6$ rats per group): $\ast$ $p < 0.05$ compared with basal values (ANOVA and Tukey test).

$F(5, 142) = 2.70$, $p < 0.03$) (Fig. 40.2). In contrast, CCK had no significant effect on grooming induced by SKF 38393.

Striatal cAMP efflux was relatively stable over a period of 3 h in the absence of drug treatment (basal values approximately 20 fmol/40 μl). As absolute values for basal striatal cAMP efflux varied between animals but were relatively stable within animals, results from drug studies were expressed as the percentage increase above basal. SKF 38393 caused a concentration-dependent

Table 40.1. Effects of SCH 23390 and sulpiride on the efflux of striatal cAMP by SKF 38393

Pretreatment			Treatment		Striatal cAMP efflux (% of basal)	
			SKF 38393	(10 μM)	152.9 ± 8.0	(6)
SCH 23390	(1 μM)	+	SKF 38393	(10 μM)	112.9 ± 5.5	(6)**
SCH 23390	(10 μM)				108.3 ± 4.9	(5)**
SCH 23390	(10 μM)	+	SKF 38393	(10 μM)	118.8 ± 10.0	(5)*
SCH 23390	(100 μM)				106.0 ± 3.7	(5)**
SCH 23390	(100 μM)	+	SKF 38393	(10 μM)	100.5 ± 5.3	(5)**
Sulpiride	(10 μM)				96.9 ± 3.0	(3)** +
Sulpiride	(10 μM)	+	SKF 38393	(10 μM)	138.4 ± 4.5	(3)ns

Values are means ± SEM; number of rats in brackets.
   ns not significant; *$p < 0.05$, **$p < 0.01$ compared with SKF 38393 (10 μM); + $p < 0.05$ compared with rats treated with SKF 38393 + sulpiride using ANOVA and the Tukey test.
   Basal cAMP values were 21.7 ± 1.5 fmol/40 μl ($n = 38$).

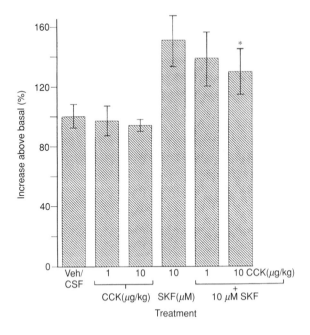

Fig. 40.4. Effect of i.p. injection of CCK on the increase of striatal cAMP efflux induced by SKF 38393. Data are expressed as mean ± SEM percentage increase above basal cAMP values ($n = 4$–11 rats per group): $\ast$ $p < 0.05$ compared with SKF 38393 alone (ANOVA and Tukey test).

increase in the overflow of cAMP in striatal dialysates, which achieved statistical significance at concentrations of 10 and 100 μM (see Fig. 40.3). Pretreatment with the D_1 dopamine antagonist SCH 23390 (1, 10, and 100 μM), but not the D_2 dopamine antagonist sulpiride (10 μM), blocked the stimulation of cAMP efflux induced by SKF 38393 (10 μM). Neither SCH 23390 nor sulpiride had any intrinsic effect on striatal cAMP efflux (Table 40.1). CCK (1 and 10 μg/kg) had no intrinsic effect on cAMP levels but attenuated the increase induced by 10 μM of SKF 38393 (Fig. 40.4).

Conclusions

These results provide the first behavioural evidence for an interaction between CCK and D_1 dopamine receptors. The most striking behavioural effect observed was the attenuation of mouth movements, induced by the selective D_1 dopamine receptor agonist SKF 38393, by a wide range of doses of CCK. As CCK was effective in blocking the mouth movement response at doses as low as 0.1 μg/kg i.p. it appears possible that the initial site of action of the peptide may be in the periphery. Thus recent evidence suggests that peripherally administered CCK may modulate dopaminergic function by activation of vagal afferent fibres (Hamamura *et al.* 1989). Higher doses of CCK (1 and 10 μg/kg) also attenuated sniffing induced by SKF 38393. Surprisingly, 10 μg/kg of CCK failed to block the mouth movement response, a finding for which there is no apparent explanation at present.

The present results are consistent with the observation that CCK suppresses mouth movements induced by withdrawal from chronic neuroleptic treatment (Stoessl *et al.* 1989). It has been proposed that the mouth movement response induced by chronic neuroleptic treatment is similar to that induced by SKF 38393 (Rosengarten *et al.* 1983, 1986). Comparisons have also been drawn between the mouth movements induced by these drug treatments in rodents and the dyskinesias observed in patients following chronic neuroleptic treatment (Waddington *et al.* 1986; Stoessl *et al.* 1989). If the mouth movement response induced by SKF 38393 is a valid model of dyskinesia in humans, it may be predicted that a CCK agonist would be of benefit in disorders such as tardive dyskinesia. Therefore it is relevant that preliminary data from an open study in schizophrenic patients with tardive dyskinesia suggest a beneficial effect of the CCK analogue ceruletide (Nishikawa *et al.* 1986).

In neurochemical studies, we observed that intrastriatal infusion of SKF 38393 produced a concentration-dependent increase in striatal cAMP efflux measured *in vivo* by intracerebral dialysis. The increase in striatal cAMP efflux induced by SKF 38393 was blocked by the selective D_1 dopamine receptor antagonist SCH 23390 but not by the selective D_2 dopamine receptor antagonist sulpiride, thus implicating D_1 dopamine receptors in mediating the response. The increase in striatal cAMP overflow induced by SKF 38393 was attenuated by 1 and 10 μg/kg of CCK. The doses of CCK required to block the neurochemical response to SKF 38393 were somewhat higher than those required to block the mouth movement response but the same as those required to block the sniffing response induced by the drug. This may be an indication that the region of the striatum perfused in the dialysis studies is close to the site of action for the sniffing response induced by SKF 38393, but remote from that involved in the mouth movement response induced by the drug. Previous studies that examined the effect of CCK on the formation of cAMP in the presence of dopamine reported a complex pattern of results that were dependent on the brain region studied. Thus it was reported that CCK failed to modulate basal or dopamine-stimulated adenylate cyclase activity (Morency *et al.* 1988). The contrast between this result and the present findings may be explained by the different conditions used in the two studies (i.e. *in vivo* versus *in vitro*). This also illustrates the utility of *in vivo* neurochemical methods for examining neurotransmitter interactions. Studler *et al.* (1986) found that CCK potentiated the formation of cAMP in the posterior nucleus accumbens (which contains neurons in which dopamine and CCK coexist) but attenuated cAMP formation in the anterior nucleus accumbens (in which most dopamine neurons do not contain CCK). This dissociation correlates with behavioural findings showing that CCK potentiates dopamine-mediated behaviour when injected into the posterior nucleus accumbens but blocks dopaminergic behaviour when injected into anterior nucleus accumbens (Crawley *et al.* 1985; Vaccarino and Rankin 1989). In the present study, we observed that i.p. injection of CCK blocked the behavioural and neurochemical responses to a selective D_1 dopamine agonist. This suggests that the inhibitory effect of CCK on cAMP formation observed by Studler *et al.* (1986) involves an interaction with D_1 dopamine receptors. Furthermore, the present data, when considered together with previous findings, suggest that CCK may interact with D_1 dopamine receptors located on dopamine neurons that do not contain CCK or on post-synaptic target cells which are in synaptic contact with dopamine neurons that do not contain CCK. Our results also suggest that such CCK–D_1 receptor interactions may occur in the striatum (in addition to the anterior nucleus accumbens) which is innervated by dopamine neurons, the majority of which do not contain CCK. Interestingly, Meyer *et al.* (1984) have reported that D_1 dopamine receptor stimulation induced by SKF 38393 reduces CCK release in the striatum. This antagonistic relationship between CCK and D_1 dopamine receptors in the striatum is in accordance with the present results. Previous findings have indicated that injection of CCK into the striatum has no effect on dopamine-mediated behaviour (Crawley *et al.* 1985). In light of the present findings it would be of interest to re-examine this issue using selective D_1 and D_2 dopamine agonists.

References

Crawley, J. N., Hommer, D. W., and Skirboll, L. (1984). Behavioural and neuro-physiological evidence for a facilitatory interaction between co-existing neurotrans-mitters. *Neurochem. Int.* **6**, 755–60.

Crawley, J. N., Stivers, J. A., Blumstein, L. K., and Paul, S. M. (1985). Cholecy-stokinin potentiates dopamine-mediated behaviours: evidence for modulation spe-cific to a site of coexistence. *J. Neurosci.* **5**, 1972–83.

Dourish, C. T. and Hill, D. R. (1987). Classification and function of CCK receptors. *Trends Pharmacol. Sci.* **8**, 207–8.

Hamamura, T., Kazahaya, Y., and Otsuki, S. (1989). Ceruletide suppresses endogenous dopamine release via vagal afferent system, studied by *in vivo* intracerebral dialysis. *Brain Res.* **483**, 78–83.

Hill, D. R., Campbell, N. J., Shaw, T. M., and Woodruff, G. N. (1987). Autoradio-graphic localization and biochemical characterization of peripheral type CCK recep-tors in rat CNS using highly selective non-peptide CCK antagonists. *J. Neurosci.* **7**, 2967–76.

Hökfelt, T., Rehfeld, J. F., Skirboll, L., Ivemark, B., Goldstein, M., and Markey, K. (1980). Evidence for coexistence of dopamine and CCK in meso-limbic neurons. *Nature* **285**, 476–8.

Hutson, P. H. and Suman-Chauhan, N. (1990). Activation of postsynaptic striatal dopamine receptors, monitored by efflux of cAMP *in vivo*. *Neuropharmacology* **29**, 1011–16.

Kebabian, J. N. and Calne, D. B. (1979). Multiple receptors for dopamine. *Nature* **277**, 93–6.

Marshall, F. H., Barnes, S., Hunter, J. C., and Hughes, J. (1990). Cholecystokinin modulates the release of dopamine from the anterior and posterior nucleus accumbens by two different mechanisms. *Br. J. Pharmacol.* **100**, 349P.

Meyer, D. K., Holland, A., and Conzelmann, U. (1984). Dopamine D_1 receptor stimulation reduces neostriatal cholecystokinin release. *Eur. J. Pharmacol.* **104**, 387–8.

Moran, T. H., Robinson, P. H., Goldrich, M. S., and McHugh, P. R. (1986). Two brain cholecystokinin receptors: implications for behavioural actions. *Brain Res.* **362**, 175–9.

Morency, M. A., Ross, G. M., Kajiura, J. S., and Mishra, R. K. (1988). Sulfated cholecystokinin octapeptide (CCK-8) failed to modulate basal or dopamine-stimulated adenylate cyclase activity in rat striatum. *Prog. Neuropsychopharmacol. Biol. Psychiat.* **12**, 331–6.

Nishikawa, T., Tanaka, M., Tsuda, A., Kuwahara, H., Koga, I., and Uchida, Y. (1986). Effect of ceruletide on tardive dyskinesia: a pilot study of quantitative computer analyses on electromyogram and microvibration. *Psychopharmacology* **90**, 5–8.

Rosengarten, H., Schweitzer, J. W., and Freidhoff, A. J. (1983). Induction of oral dyskinesias in naive rats by D1 receptor stimulation. *Life Sci.* **33**, 2479–82.

Rosengarten, H., Schweitzer, J. W., and Freidhoff, A. J. (1986). Selective dopamine D2 receptor reduction enhances a D1 mediated oral dyskinesia in rats. *Life Sci.* **39**, 29–35.

Stoessl, A. J., Dourish, C. T., and Iversen, S. D. (1989). Chronic neuroleptic-induced mouth movements in the rat: suppression by CCK and selective dopamine D1 and D2 receptor antagonists. *Psychopharmacology* **98**, 372–9.

Studler, J. M., Reibaud, M., Herve, D., Blanc, G., Glowinski, J., and Tassin, J. P. (1986). Opposite effects of sulfated cholecystokinin octapeptide on DA-sensitive adenylate cyclase in two areas of rat nucleus accumbens. *Eur. J. Pharmacol.* **126**, 125–8.

Vaccarino, F. J. and Rankin, J. (1989). Nucleus accumbens cholecystokinin (CCK) can either attenuate or potentiate amphetamine-induced locomotor activity: evidence for rostral–caudal differences in accumbens CCK function. *Behav. Neurosci.* **103**, 831–6.

Vickroy, T. W., Bianchi, B. R., Kerwin, J. F., Kopecka, H., and Nadzan, A. M. (1988). Evidence that type A CCK receptors facilitate dopamine efflux in rat brain. *Eur. J. Pharmacol.* **152**, 371–2.

Voigt, M. M. and Wang, R. Y. (1984). *In vivo* release of dopamine in the nucleus accumbens of the rat: modulation by cholecystokinin. *Brain Res.* **296**, 189–194.

Waddington, J. L. and O'Boyle, K. M. (1989). Drugs acting on brain dopamine receptors: a conceptual re-evaluation five years after the first selective D-1 antagonist. *Pharmacol. Ther.* **43**, 1–52.

Waddington, J. L., Youssef, H. A., O'Boyle, K. M., and Molloy, A. G. (1986). A reappraisal of abnormal, involuntary movements (tardive dyskinesia) in schizophrenia and other disorders: animal models and alternative hypotheses. In *The neurobiology of dopamine systems* (ed. W. Winlow and R. Markstein), pp. 266–86. Manchester University Press, Manchester.

41. Involvement of central dopamine and CCK_A receptors in the apomorphine-induced hypotensive response in the anaesthetized rat

Gillian Sturman

Apomorphine has been shown to relieve parkinsonian rigidity in humans, and this is thought to be due to direct stimulation of dopamine receptors in the basal ganglia (Andén *et al.* 1967). After intravenous (i.v.) administration of apomorphine, there is a transient fall in blood pressure. Whether this effect is due to the involvement of the central or peripheral dopamine receptors is under debate (Finch and Haeusler 1973; Mugabo *et al.* 1983). Hökfelt *et al.* (1980) have reported that dopaminergic neurons in the mid-brain contain CCK. This peptide has been shown to facilitate dopamine-induced hyperlocomotion and apomorphine-induced stereotypy in the rat (Crawley *et al.* 1984), suggesting that it increases the neuronal responsiveness to dopamine agonists. With the development of devazepide, a highly potent CCK_A receptor antagonist (Chapter 2 of this volume), and L-365,260, a selective CCK_B receptor antagonist (Lotti and Chang 1989; Chapter 2 of this volume), the possible involvement of CCK and its receptors in the apomorphine-induced hypotensive response in the anaesthetized rat can be investigated.

Male Wistar rats (250–350 g) anaesthetized with urethane (1.0–1.25 g/kg i.p.) were used in the tests. The systemic blood pressure was measured via a cannulated carotid or femoral artery using a Bell & Howell physiological pressure transducer and a Lectromed two-channel pen recorder. All substances used were dissolved in isotonic saline, in volumes not greater than 1 ml/kg, and injected into either a jugular or a femoral vein at a rate of 0.1 ml/s, with a subsequent 0.1 ml saline injection. Apomorphine (10–30 nmol/kg), sulphated CCK octapeptide (CCK-8S) (0.5–5 nmol/kg), and dopamine (0.1–30 μmol/kg) produced immediate dose-dependent alterations in blood pressure. After establishing standard dose–response curves in each rat, the antagonists were injected intravenously (i.v.) at least 10 min before apomorphine, CCK, and dopamine administration was repeated. Each antagonist was tested in a minimum of 10 rats. In some anaesthetized rats, bilateral vagotomy or pithing was performed instead of antagonist administration.

The results obtained with various dopamine antagonists showed that

Table 41.1. Median depressor responses in anaesthetized rats after i.v. injection of apomorphine before and after various dopamine antagonists administered at least 10 min previously

	Dose (μmol/kg)	Response (mmHg)	Range (mmHg)
Control	—	20	14–29
Domperidone	0.2	18	14–21
Haloperidol	5.3	7	0–11
Metoclopromide	0.9	3	0–5
R (+)-sulpiride	6.1	5	3–8

10 rats per group.
The doses of antagonist used were shown previously to block the pressor effects of i.v. administered dopamine in each rat.

haloperidol (5 μmol/kg), R (+)-sulpiride (6 μmol/kg), and metoclopromide (0.9 μmol/kg), which are able to cross the blood–brain barrier, significantly shifted the dose–response curves for apomorphine-induced hypotension to the right, whilst this did not occur with domperidone (0.2 μmol/kg) which only penetrates into the CNS with difficulty (Table 41.1). The hypotensive response induced by apomorphine was also abolished after pithing, bilateral vagotomy, or administration of the cholinergic antagonists atropine sulphate (5 μmol/kg) or hexamethonium (50 μmol/kg).

Injection of CCK-8S (0.5–5 nmol/kg i.v.) produced a transient hypotensive response similar to that produced by apomorphine. Administration of selective CCK antagonists, at doses which abolished the depressor response induced by CCK, showed that the apomorphine-induced hypotensive response was reduced by devazepide (2 μmol/kg), which is selective for the CCK_A receptor, but not by L-365,260 (3 μmol/kg), which is selective for the CCK_B receptor (Table 41.2).

Finch and Haeusler (1973) failed to show apomorphine antagonism with

Table 41.2. Median responses in anaesthetized rats after i.v. injection of apomorphine or CCK-8S before and after i.v. administration of L-365,260 or devazepide

	Response (range) (mmHg)		
Depresor agent	Control	+ devazepide (2 μmol/kg)	+ L-365,260 (3 μmol/kg)
CCK-8S (2.5 nmol/kg)	22 (20–30)	3 (0–8)	3.5 (0–10)
Apomorphine HCl (30 nmol/kg)	7.5 (5–10)	0	10 (8–20)
Apomorphine HCl (100 nmol/kg)	20 (10–25)	5 (0–10)	20 (17–35)
Apomorphine HCl (300 nmol/kg)	33 (20–40)	18 (15–25)	35 (30–40)

12 rats per group.

haloperidol or sulpiride, but our findings are in agreement with those of Finch and Hersom (1976) and of Petitjean *et al.* (1984) who suggested that central or spinal dopamine receptors respectively were involved in the apomorphine-induced hypotensive response. Apomorphine has been shown to be a full agonist of the D$_2$ receptor but a partial agonist–antagonist of the D$_1$ receptor (Goldberg *et al.* 1978). In this work, the use of cholinergic antagonists, pithing, or bilateral vagotomy all abolished the apomorphine-induced hypotensive response, indicating that this effect must involve a central component mediated via the vagus nerve.

The role of brain CCK$_A$ and CCK$_B$ receptors is still being established but CCK has been shown both to enhance and to inhibit central dopamine-mediated responses (Chapter 35 of this volume). Since this work has shown that devazepide, but not L-365,260, inhibited the apomorphine-induced hypotensive response in the rat, CCK$_A$ receptors are likely to be involved. Recent evidence has suggested that peripherally administered CCK may modulate dopaminergic function by activation of vagal afferent fibres (Hamamura *et al.* 1989). Thus, the CCK$_A$ receptors implicated in the results presented here may be located peripherally. Our finding that vagotomy abolishes the apomorphine-induced hypotensive response in the rat provides further evidence that a peripheral CCK$_A$ receptor component mediating a central dopaminergic action is involved in the apomorphine-induced hypotensive response of the rat.

References

Andén, N. E., Rubenson, A., Fuxe, K., and Hökfelt, T. (1967). Evidence for dopamine receptor stimulation by apomorphine. *J. Pharm. Pharmacol.* **19**, 627–9.

Crawley, J. N., Hommer, D. W., and Skirboll, L. R. (1984). Behavioural and neurophysiological evidence for a facilatory interaction between coexisting transmitters: cholecystokinin and dopamine. *Neurochem. Int.* **6**, 755–60.

Finch, L. and Haeusler, G. (1973). The cardiovascular effects of apomorphine in the anaesthetised rat. *Eur. J. Pharmacol.* **21**, 264–70.

Finch, L. and Hersom, A. (1976). Studies on the centrally mediated cardiovascular effects of apomorphine in the anaesthetised rat. *Br. J. Pharmacol.* **56**, 366P.

Goldberg, L. I., Volkman, P. H., and Kohli, J. D. (1978). A comparison of the vascular dopamine receptor with other dopamine receptors. *Ann. Rev. Pharmacol.* **18**, 57–79.

Hamamura, T., Kazahaya, Y., and Otsuki, S. (1989). Ceruletide suppresses endogenous dopamine release via vagal afferent system, studied by *in vivo* intracerebral dialysis. *Brain Res.* **483**, 78–83.

Hökfelt, T., Rehfelt, J. F., Skirboll, L., Ivemark, B., Goldstein, M., and Markey, K. (1980). Evidence for coexistence of dopamine and CCK in mesolimbic neurones. *Nature* **285**, 476–8.

Lotti, V. J. and Chang, R. S. L. (1989). A new potent and selective non-peptide gastrin antagonist and brain cholecystokinin receptor (CCK-B) ligand: L-365,260. *Eur. J. Pharmacol.* **162**, 273–80.

Mugabo, P., Buylaert, W., and Bogaert, M. (1983). Cardiovascular effects of apomorphine in the rat. *Arch. Int. Pharmacodyn.* **262**, 324–5.
Petitjean, P., Mouchet, P., Pellissier, G., Manier, M., Feuerstein, C., and Demenge,
P. (1984). Cardiovascular effects in the rat of intrathecal injections of apomorphine
at the thoracic spinal cord level. *Eur. J. Pharmacol.* **105**, 355–9.

Part V

CCK and opioid analgesia

42. CCK and opioid analgesia: introduction

Susan D. Iversen

There has been a growing awareness that CCK and the opioids have opposite physiological effects on food intake, motor behaviour, and the responsiveness of animals to painful stimuli. The experimental findings in support of this statement have been reviewed in a number of the chapters in this volume. Furthermore, it appears likely that the CCK and opioid systems are not functionally independent. Drugs which increase or decrease CCK function modify the response to morphine, and a number of anatomical and neuropharmacological studies have demonstrated a functional interaction between CCK and opioids at various levels of the neural axis. It is in the area of pain perception that these findings have immediate clinical significance since morphine and related opioids are so widely used as analgesics. Opioid analgesics have many disadvantages: beyond the serious problem of tolerance and dependence are serious acute side-effects, including respiratory depression and constipation.

The observation that CCK at physiological doses blocks opioid analgesia (Itoh *et al.* 1982; Faris *et al.* 1983) is now well confirmed, with the important corollary that CCK antagonist drugs which act on the CNS could be important adjuncts to morphine in the treatment of pain. The interaction between CCK and opioids in relation to food intake and motor behaviour has been less intensively investigated to date. It would be interesting to discover more about the anatomical relationships of CCK and opioid substrates in the circuitry mediating the control of food intake and motor behaviour, since diseases involving these systems may well offer additional targets for CCK antagonist or agonist drugs. It should not be assumed, however, that all the effects of CCK in the CNS involve an interaction with opioid systems. For example, there is no evidence at present that the anxiogenic effect of CCK-4 and the anxiolytic effect of CCK antagonists involve opioid systems (De Montigny 1989).

Chapter 44 reviews the evidence that CCK antagonists enhance morphine analgesia and prevent the rapid development of tolerance to the analgesic effect of morphine. It is notable that the doses of CCK antagonists active in analgesic models are considerably higher than those reported by Dourish and his colleagues to be effective in anxiety and feeding models (see Chapter 21). The basis of the differing potency in these models is unclear, but subtypes of, for example, CCK_B receptors cannot be excluded. However, morphine dependence and the subsequent naloxone-induced withdrawal symptoms are clearly not attenuated by CCK antagonists. Kellstein and Mayer (Chapter 43) have provided a review of CCK and analgesia. The emphasis in their chapter and the concordance of their views with those of Dourish are striking. So at least we can say that there is agreement about the interesting questions that need to be asked. Han (Chapter 45) reviews this field with special reference to electro-acupuncture (EA) in rat and rabbit as a method for inducing endogenous release of opioids in the CNS. Exogenously applied CCK or CCK antagonists modulate EA-induced analgesia in exactly the same way they do when exogenously applied morphine is used to induce analgesia. Dickenson *et al.* (Chapter 46) describe their recent electrophysiological work on the definition of opioid receptor subtypes which may interact with CCK mechanisms in the processing of pain stimuli in the spinal cord of the rat.

The observation that CCK antagonists potentiate morphine analgesia is now well established in rodents, but more work in primates is required first because primate results are likely to translate to man and second because CCK receptor subtypes in the areas of spinal cord implicated in CCK–opioid pain control differ in rodents and monkeys (Chapter 5). A limited number of primate models have been developed for evaluating the response to an acute painful stimulus. One, in the squirrel monkey, involves withdrawal of the tail from warm water, and another, in the baboon, is dolorimetry, in which a fine stimulatory electrode is implanted around the tooth pulp. The current is gradually increased until the monkey indicates with a lick/chew response that the stimulus has been felt. This current level is recorded as a threshold response. In both tests the CCK_A selective antagonist devazepide is reported to enhance the analgesic response to morphine (see Chapter 44 for tail withdrawal and Chapter 50 for dolorimetry).

Wiesenfeld-Hallin *et al.* (1990) have recently reported that PD 134308, another selective CCK_B antagonist, enhances the analgesic and depressive effect of morphine on the spinal nociceptive flexor reflex in rats. During the discussion period, Kelly commented that according to the model proposed by Dourish (Chapter 44) L-365,260 might be expected to show intrinsic analgesic activity. Limited results on this interesting question are available at present.

Dourish *et al.* (1990) had shown earlier than the CCK_A antagonist devazepide is devoid of intrinsic analgesic activity in the squirrel monkey tail-withdrawal model. More recently, O'Neill *et al.* (1990; Chapter 51 of this volume) have found that the selective CCK_B antagonist L-365,260 has intrinsic analgesic activity in that model. A key question now is to determine if the analgesic properties of CCK_B antagonists are naloxone-sensitive or functionally independent of the opioid substrate. Limited data are available, and Wiesenfeld-Hallin *et al.* (1990) have reported that in the rat the effects of PD 134308, and of coadministered PD 134308 and galanin to depress the flexor reflex in spinalized rats, were naloxone-reversible.

It was commented during the discussion that most of the results on morphine enhancement and intrinsic analgesia have been obtained with models of nociception at the spinal cord level. The modulation of supraspinal CCK or CCK–opioid mechanisms remain to be explored in behavioural situations. In response Dourish agreed, but commented that in the studies by Wiesenfeld-Hallin *et al.* (1990) modest analgesic activity of PD 134308 had been demonstrated using the hot-plate, an analgesic test considered to involve supraspinal levels of processing. When considering control of nociceptive responses at the supraspinal levels, interaction of CCK with substrates other than opioids should not be ignored (see Chapters 44 and 45). Comment was made that pain responses in some situations involve refined sensory–motor co-ordination. Dopamine pathways are known to be involved in such responses, and it is possible that CCK–DA interactions of the kind described in Chapter 35 may be functionally important in the antinociceptive effects of CCK antagonists.

Further experimental evidence in rabbits (head withdrawal from thermal stimulus) and rats (tail flick), demonstrating a functional antagonism between *exogenously* administered CCK and morphine, is provided in Chapter 45. It is interesting, as Dourish commented in the discussion, that CCK-8S was more active than CCK-8US in blocking the effect of morphine. The relative potency of these agonists suggests involvement of CCK_A receptors rather than CCK_B receptors. However, by contrast, in studies with CCK antagonists the greater potency of selective CCK_B compounds suggests involvement of the CCK_B receptor. Han agreed that this was puzzling and unexplained.

Han's main interest, however, has been to study *endogenous* interactions between CCK and opioids, and he has developed EA as a means of stimulating the release of endogenous opioids and inducing analgesia. Han stressed in the discussion that such stimulation activates Aδ fibres and not C fibres in the dorsal root. EA induces analgesia which can be blocked with intracerebroventricular (i.c.v.) or intrathecal CCK. Continuous EA for 6 h induces tolerance to the initial analgesic effect which, in discussion, Han said was not due to sensory adaptation. Infusion of CCK antisera (i.c.v.) blocks the development of this tolerance.

The CNS site of these interactions has also been explored, and Han showed that neurons of the intralaminar nuclei of the thalamus, especially the nucleus parafascicularis, respond with either excitation or inhibition to noxious stimuli. Systemically administered morphine suppresses the excitatory effects and enhances the inhibitory effects, and both effects are blocked by i.c.v. CCK. Furthermore, chronic morphine resulted in tolerance to the neurophysiological responses to the acute effects of the opioid. These interactions between CCK and opioids at the neurophysiological level were replicated in the same model using EA instead of systemic morphine, and the effects of EA were blocked by CCK and potentiated by CCK antisera.

Neurophysiological models, as described in Chapters 45 and 46, have a great advantage in studying peptide interactions in relation to pain, since in an anaesthetized preparation motoric, conditioning, and stress-induced analgesia can be excluded as explanations for the results. The anatomical, synaptic, and molecular bases of the CCK–opioid interaction in relation to pain continue to provide a challenge at both the experimental and the theoretical level. Dourish reviews the existing evidence in these three areas in the final sections of Chapter 44 and dares to draw a model which accommodates many of the findings. Models are proposed to generate experiments and to be challenged, which I am sure will be the fate of this one.

However, Han and Dickenson have already provided some clues to the unravelling of these questions. In behavioural and binding studies Han found that CCK blocks analgesia induced by selective μ and κ agonists but not that induced by δ agonists, and similarly inhibits binding to μ and κ but not to δ opioid receptors. Dickenson presented results on a rat spinal cord model, recording from nociceptive neurons, and provided support for the involvement of μ but not δ receptors, although the role of κ receptors had not been investigated. Han and Dickenson also provided evidence on the crucial question of whether the CCK–opioid interaction is at the spinal or the supraspinal level, or more probably both levels, of the neural axis.

Dickenson showed that intrathecal application of CCK prevents the inhibition of dorsal horn neurons by the μ receptor agonist DAGOL, and Wiesenfeld-Hallin *et al.* (1990) demonstrated in a spinal reflex model that the effect of intrathecal morphine was blocked by CCK and enhanced by a CCK antagonist.

In models of neurotransmitter release, Benoliel *et al.* (Chapter 47) and Rodriguez and Sacristan (Chapter 48) show *in vivo* that opiates can modulate release of CCK from the spinal cord. Both agree that μ and δ receptors are involved, with a suggestion that κ receptors may not have a direct modulatory role but influence the role of μ receptors in this neuronal circuitry. However, the supraspinal mechanisms deserve more attention. Han's results on thalamic neurons suggest that, at this level of the pain processing pathway, CCK–opioid interactions similar to those observed in the spinal cord are also operating.

The potential use of CCK antagonists as opioid adjuncts or as analgesics in their own right in the treatment of pain in man has hardly been explored. As mentioned by Dourish in the concluding section of Chapter 44, the evidence obtained using the non-selective and relatively weak antagonist proglumide in man is equivocal. The availability of more selective and potent CCK antagonists now provides the opportunity to test the hypothesis that CCK acts as an endogenous anti-opioid. However, as Kellstein and Mayer (Chapter 43) point out, there are some important findings which argue for caution: first the bell-shaped dose–response curves for CCK antagonist enhancement of opioid analgesia, and second their own findings of unpredictable dose relationships for CCK antagonism of opioid tolerance and, most important, of compensatory re-regulation of CCK and/or opioid systems accompanying *chronic* blockade of CCK receptors.

There are further important unanswered questions which will need to be addressed before the potential therapeutic use of CCK antagonists in pain control can be evaluated. The results of animal experiments suggest that CCK antagonists may be of value as adjuncts to morphine therapy, provided that the enhancement effect does not show tolerance. More interesting, perhaps, is the possibility that the compounds will show intrinsic analgesic activity in man and be valuable not only as a replacement for morphine but also, as Dickenson pointed out, in the treatment of the considerable number of patients who suffer from morphine-resistant pain. Some of Han's findings are relevant to this issue. He reported that not all people or all rats and rabbits respond to EA. However, rat non-responders can be transformed into responders by treatment with the CCK antagonist devazepide. This confirms Han's earlier findings (Han *et al.* 1985) that non-responder EA rats become responders after treatment with antiserum to CCK-8.

The general discussion which followed the analgesia session reiterated some fundamental questions about the CNS effects of CCK antagonists, which had been noted in earlier discussions of feeding, anxiety, and CCK–dopamine interactions in basal ganglia. The CCK antagonists, particularly the CCK_B compounds, show extraordinary potency in a number of models, raising the question of central versus peripheral sites of action. The bell-shaped dose-response curves noted by a number of investigators are difficult to explain in terms of interaction with a single receptor site. However, it may be that the explanation lies in the realm of response competition in behaviour. Antagonism of CCK receptors in one functional circuit may induce a clear behavioural effect but at higher doses many recruit other CCK-sensitive brain systems with response consequences which attenuate the original behaviour. Finally, on a number of occasions discussion focused on the involvement of CCK receptor subtypes in the behavioural effects observed. CCK_A versus CCK_B can be addressed in relative potency terms with the available CCK antagonists if equal bioavailability, pharmacokinetics, brain penetration, and metabolism can be assumed. Much comparative work remains to be done with the available antagonists. More pressing is the question of whether there are just two CCK receptors in body and brain. The exciting but daunting prospect is that there are multiple subtypes of CCK receptor. The challenge of designing selective drugs for these targets and a definition of their functional significance will keep us all busy until the next CCK meeting and beyond.

References

De Montigny, C. (1989). Cholecystokinin tetrapeptide induces panic-like attacks in healthy volunteers: preliminary findings. *Arch. Gen. Psychiat.* **46**, 511–17.
Dourish, C. T., O'Neill, M. F., Schaffer, L. W., Siegl, P. K. S., and Iversen, S. D. (1990). The cholecystokinin receptor antagonist devazepide enhances morphine-induced analgesia but not morphine-induced respiratory depression in the squirrel monkey. *J. Pharmacol. Exp. Ther.* **255**, 1158–65.
Faris, P. L., Komisaruk, B. R., Watkins, L. R., and Mayer, D. J. (1983). Evidence for

the neuropeptide cholecystokinin as an antagonist of opiate analgesia. *Science* **219**, 310–12.

Han, J. S., Ding, X. Z., and Fan, S. G. (1985). Is cholecystokinin octapeptide (CCK-8) a candidate for endogenous anti-opioid substrate? *Neuropeptides* **5**, 399–402.

Itoh, S., Katsuura, G., and Maeda, Y. (1982). Caerulein and cholecystokinin suppress β-endorphin-induced analgesia in the rat. *Eur. J. Pharmacol.* **80**, 421–5.

O'Neill, M. F., Dourish, C. T., Tye, S. J., and Iversen, S. D. (1990). Blockade of CCK-B receptors by L-365,260 induces analgesia in the squirrel monkey. *Brain Res.* **534**, 287–90.

Wiesenfeld-Hallin, Z., Xu, X-J., Hughes, J., Horwell, D. C., and Hökfelt, T. (1990). PD 134308, a selective antagonist of cholecystokinin type B receptor, enhances the analgesic effect of morphine and synergistically interacts with intrathecal galanin to depress spinal nociceptive reflexes. *Proc. Natl Acad. Sci. USA* **87**, 7105–9.

43. CCK and opioid analgesia

David E. Kellstein and David J. Mayer

Introduction

CCK is the most abundant gastro-intestinal peptide in the CNS (Beinfeld 1983; Crawley 1985), where it may function as a neurotransmitter and/ or a neuromodulator (Morley 1982; Baber *et al*. 1989). Although several forms of CCK with amino acid chains of different lengths have been isolated, the sulphated octapeptide variant predominates within the mammalian CNS (Rehfeld 1978; Larssen and Rehfeld 1979) and exhibits the greatest potency (Mutt 1982). Therefore, in this chapter CCK will refer to this form of the peptide.

Anatomical studies strongly suggest a functional interaction between CCK and opioids. The CNS distribution of CCK closely parallels that of endogenous opioids within the brain (e.g. nucleus raphé magnus and periaqueductal grey (Beinfeld and Palkovits 1982), and hypothalamus, basal ganglia, and dorsomedial thalamus (Stengaard-Pedersen and Larsson 1982)) and the spinal cord (e.g. the substantia gelatinosa; Stengaard-Pedersen and Larsson 1981). Further, CCK and enkephalins have been colocalized within neurons in nociceptive areas of the brainstem and thalamus (Gall *et al*. 1987). Accordingly, the brain and spinal cord distribution of CCK binding sites is very similar to that of opioid receptors (Saito *et al*. 1980; Zarbin *et al*. 1983; Van Dijk *et al*. 1984; Hill *et al*. 1988).

Results of functional studies indicate that CCK and opioid systems generally interact in an antagonistic manner. For example, CCK and opioids have opposite effects on satiety (Della-Fera and Baile 1979; Morley and Levine 1983), respiration (Gillis *et al*. 1983; Jensen *et al*. 1983), gut motility (Zetler 1979), locomotion (Schnur *et al*. 1986), and thermoregulation (Kapas *et al*. 1989). In addition, there is considerable evidence to suggest that CCK functions as an endogenous antagonist of opioid analgesia and may also be involved in opioid tolerance. The widespread use of opioid analgesics in the management of chronic pain, and the adverse effects and tolerance development associated with such use, make the study of CCK antagonism of opioid analgesia clinically important. In this chapter we review developments in this field and suggest future directions for investigation.

CCK–opioid interactions in analgesia

Shortly after the demonstration that intraventricular administration of CCK inhibited β-endorphin-induced analgesia, as assessed with the rat hot-plate assay (Itoh *et al*. 1982), Faris *et al*. (1983) found that systemic pretreatment with CCK significantly attenuated both the peak and duration of antinociception elicited by systemic morphine in the rat tail-flick model. In the same study, CCK administered either systemically or intrathecally antagonized endogenous opioid analgesia induced by front-paw footshock; peptide pretreatment also attenuated analgesia classically conditioned to such footshock. Since both types of analgesia are naloxone-reversible (Watkins *et al*. 1982; Watkins and Mayer 1982), these findings suggest that CCK also antagonizes antinociception evoked by release of endogenous opioids. Further, the lack of effect of (i) CCK on non-opioid analgesia induced by hind-paw footshock (Watkins *et al*. 1982; Faris *et al*. 1983) and (ii) unsulphated CCK on analgesia induced by front-paw shock (Faris *et al*. 1983) or intraventricular β-endorphin (Itoh *et al*. 1982) indicate that CCK antagonism is selective to opioid-mediated analgesia and that this effect is specific to the natural sulphated form of CCK.

Subsequent studies confirmed and extended these initial findings. CCK also attenuated opioid analgesia induced by restraint stress (Dourish *et al*. 1988*a*) or electro-acupuncture analgesia (Han *et al*. 1986; Chapter 45 of this volume). An important point in all these studies is that CCK alone did not elicit hyperalgesia, indicating that its anti-opioid effect is not attributable to a non-specific activation of nociceptive pathways.

However, this anti-analgesic effect of CCK on opioids contrasts with studies reporting an antinociceptive effect of CCK (Zetler 1980; Jurna and Zetler 1981; Barbaz *et al*. 1986; Hill *et al*. 1987*b*; Pittaway *et al*. 1987). Interestingly, in these studies naloxone attenuated, while enkephalinase inhibitors enhanced, CCK-induced analgesia, suggesting that CCK-induced analgesia is mediated via activation of opioid pathways. Thus the interaction between CCK and opioids is apparently bidirectional, since not only does activation of opioid circuitry apparently activate CCK pathways but CCK administration appears to elicit release of endogenous opioids.

The apparent contradiction between the findings demonstrating an analgesic effect of CCK and those which suggest an anti-opioid effect of this peptide may be attributable to differences in methodology. CCK consistently inhibited opioid effects in the thermal tail-flick test, whereas CCK-induced analgesia has been demonstrated in other pain assays (e.g. chemical-induced writhing, hot-plate test, mechanical paw pressure) but not generally in the tail-flick test (Barbaz *et al*. 1986; Pittaway *et al*. 1987). However, since CCK antagonism of opioid analgesia has also been demonstrated in the hot-plate and paw pressure tests (Itoh *et al*. 1982; O'Neill *et al*. 1989), and CCK-induced antinociception has been demonstrated in the tail-flick test (Jurna and Zetler 1981), a better explanation might be a biphasic dose response because CCK induces analgesia at doses approximately 10-fold higher (Zetler 1980) than those which inhibit opioid antinociception (Faris *et al.* 1983). As suggested by Baber *et al.* (1989), the latter observation indicates that analgesia induced by high doses of CCK may be a 'pharmacological' effect, whereas antagonism of opioid antinociception by small amounts of the peptide may reflect the physiological action of CCK.

Whereas the experiments summarized above demonstrate that exogenous CCK antagonized opioid analgesia, the involvement of endogenous peptide is supported by the finding that treatment with CCK antibodies enhanced morphine analgesia (Faris *et al.* 1984). In agreement, studies using the CCK antagonist proglumide have consistently demonstrated enhancement of opioid analgesia, and have also supported an anti-opioid effect of endogenous CCK. Thus initial studies by Watkins *et al.* (1984, 1985*a, b*) using the rat tail-flick assay demonstrated proglumide enhancement of antinociception induced by front-paw footshock, by intrathecal (D-Ala^2)-methionine enkephalinamide, and by systemic, intrathecal, or supraspinal morphine. Further, proglumide itself was devoid of analgesic activity (Watkins *et al.* 1984, 1985*a, b*) and did not enhance non-opioid analgesia, i.e. that induced by intrathecal norepinephrine or hind-paw footshock (Watkins *et al.* 1984). These latter findings concur with those obtained using exogenous CCK, and provide evidence that CCK pathways are not tonically involved in nociception but rather are selectively activated by administration of exogenous opioids, or release of endogenous opioids, and function to return pain perception to normal.

However, there is some controversy as to whether proglumide enhances opioid analgesia via binding to the CCK$_B$ receptor subtype which predominates in rat brain and spinal cord (Moran *et al.* 1986; Hill *et al.* 1988). In mouse brain homogenates (rats were not used because of the large amounts of non-specific binding), proglumide has lower binding affinities (IC$_{50}$ values in the millimolar range (Makovec *et al.* 1985; Vigna *et al.* 1985)) than the approximate maximal concentrations (in the micromolar range) available after CNS administration of doses which enhance opioid effects (Watkins *et al.* 1984, 1985*a, b*; Kellstein and Mayer 1990). However, the results of mouse brain binding studies *in vitro* may not accurately reflect the anti-CCK activity of proglumide *in vivo*. In addition to the disruption of neural tissue and dilution of receptors inherent in binding assays, species differences in both the distribution and selective affinities of central CCK receptors have been demonstrated (Vigna *et al.* 1985; Williams *et al.* 1986; Hill *et al.* 1987*a*, 1988). Further, electrophysiological studies (Chiodo and Bunney 1983; Willetts *et al.* 1985) and behavioural studies (Crawley *et al.* 1986; Gillis *et al.* 1986) have consistently demonstrated that proglumide inhibits the effect of CCK, suggesting that the antagonist does bind to CCK receptors in the CNS.

Subsequent experiments demonstrated that the CCK antagonists benzotript

(Watkins *et al.* 1985*b*), lorglumide (Dourish *et al.* 1988*a*; Kellstein and Mayer 1990), devazepide (Dourish *et al.* 1988*a, b*; O'Neill *et al.* 1989), L-365,031 and L-365,260 (Dourish *et al.* 1988*a*, 1990) also enhance opioid analgesia, suggesting that opioid enhancement is characteristic of CCK antagonists in general. CCK antagonism of opioid analgesia is apparently mediated by the CCK_B receptor, since L-365,260 (selective for CCK_B receptors) was more potent than devazepide and L-365,031 (selective for CCK_A receptors) in both enhancing morphine analgesia and blocking development of opioid tolerance

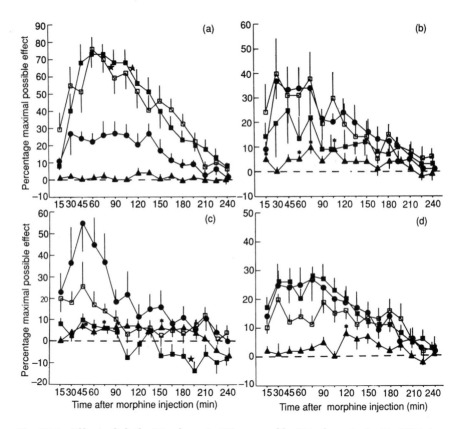

Fig. 43.1. Effect of daily (■, days 1–22) or weekly (□, days 1, 8, 15, 22) intrathecal proglumide treatment on intrathecal morphine analgesia. Proglumide (20 ng total dose) or saline (●, 1 μl total volume) was injected 10 min and immediately before morphine (1 μg); proglumide was also injected before saline (▲) on (a) day 1, (b) day 8, (c) day 22, and (d) day 36. Daily injections were terminated after day 22. ★ $p < 0.01$ for comparison of proglumide + morphine groups with saline + morphine using ANOVA of area under the curve values: ⋆ $p < 0.05$ for comparison of test latencies of the proglumide + saline group with pretreatment latencies using *t* tests. Error bars indicate the SEM ($n = 7$–8). (Reproduced with permission from Kellstein and Mayer 1990.)

(Dourish *et al.* 1990). Interestingly, bell-shaped dose-response curves have been consistently reported for enhancement of opioid analgesia by all CCK antagonists when multiple doses were employed. The reason for this phenomenon is unknown, but could be attributable to a partial agonist effect of these agents at higher doses.

In human studies of experimental pain (Price *et al.* 1985) or clinical pain (Lavigne *et al.* 1989), proglumide facilitated morphine analgesia without exacerbating adverse effects, suggesting that CCK blockade may be a useful

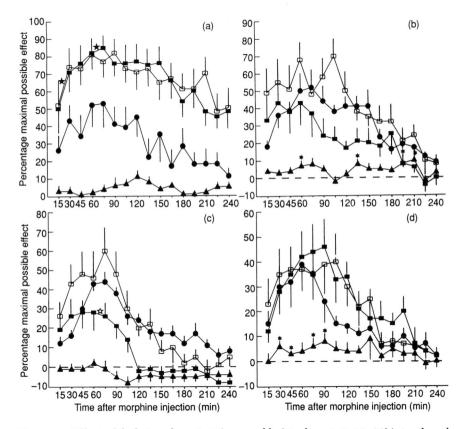

Fig. 43.2. Effect of daily (■ , days 1–22) or weekly (□ , days 1, 8, 15, 22) intrathecal lorglumide treatment on intrathecal morphine analgesia. Lorglumide (7 ng total dose) or saline (●, 1 μl total volume) was injected 10 min and immediately before morphine (1 μg); lorglumide was also injected before saline (▲) on (a) day 1, (b) day 8, (c) day 22, and (d) day 36. Daily injections were terminated after day 22. ★ $p < 0.01$, ⋆ $p < 0.05$ for comparison of lorglumide + morphine groups with saline + morphine using ANOVA of area under the curve values; ⋆ $p < 0.05$ for comparison of test latencies of lorglumide + saline group with pretreatment latencies using *t* tests. Errors bars indicate SEM ($n = 8$). (Reproduced with permission from Kellstein and Mayer 1990.)

adjunct to opioids in the management of pain (see Chapter 44 of this volume). In a recent study (Kellstein and Mayer 1990), however, we found that, although acute pretreatment with CCK antagonists enhanced intrathecal morphine analgesia, chronic (22 day) administration of proglumide (Fig. 43.1) or lorglumide (Fig. 43.2) not only caused loss of the initial enhancement but actually diminished normal levels of opioid antinociception. These results suggest that (i) compensatory alterations in CCK–opioid interactions occur during chronic CCK blockade, and (ii) caution should be employed in any chronic clinical use of CCK antagonists for the treatment of pain.

Involvement of CCK in opioid tolerance

In addition to the evidence summarized above, which suggests acute CCK modulation of opioid pathways, studies using CCK antagonists also support the participation of CCK in the development of opioid tolerance. Thus systemic coadministration of proglumide (Tang *et al.* 1984; Watkins *et al.* 1984; Panerai *et al.* 1987), benzotript (Panerai *et al.* 1987), devazepide (Dourish *et al.* 1988*b*), and L-365,031 or L-365,260 (Dourish *et al.* 1990) with morphine attenuated and/or reversed tolerance. The additional observations that (i) rats exposed to a long-term morphine tolerance regimen required a larger dose of proglumide to restore opioid analgesia than those receiving short-term administration (Kinscheck *et al.* 1983) and (ii) cerebrospinal fluid (CSF) levels of CCK (as determined with radio-immunoassay) were progressively higher in rats receiving saline, morphine acutely, or morphine in short- or long-term tolerance para-

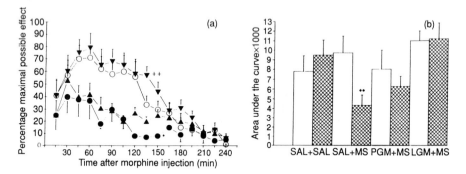

Fig. 43.3. Attenuation of opioid tolerance by intrathecal coadministration of 64 ng of proglumide (PGM) or 70 ng of lorglumide (LGM) with a low (1 μg) dose of morphine (MS) from day 2 to day 7. (a) Analgesia induced by 1 μg of MS on day 8 in animals receiving saline (SAL) ($\bigcirc$), SAL + MS ($\bullet$), PGM + MS ($\blacktriangle$), or LGM + MS ($\blacktriangledown$) from day 2 to day 7: $\star$ $p < 0.05$ compared with SAL; $+ +$ $p < 0.01$ compared with SAL + MS using ANOVA of area under the curve values. (b) Within-group comparisons of antinociception induced by 1 μg of MS on day 1 (open bars) to day 8 (hatched bars): $\star\star$ $p < 0.01$ compared with day 1 using ANOVA. Error bars indicate SEM ($n = 8$).

digms (Watkins *et al.* 1985*c*) suggest that repeated opioid administration elicits a compensatory progressively greater synthesis and/or release of CCK which may underlie the development of morphine tolerance.

In contrast, CCK does not appear to play a role in opioid dependence, since the number or severity of abstinence signs associated with naloxone-precipitated morphine withdrawal was not altered by benzotript, proglumide (Panerai *et al.* 1987), devazepide (Dourish *et al.* 1988*b*), or L-365,260 (Dourish *et al.* 1990).

We recently found that intrathecal coinjection of proglumide (64 ng) or lorglumide (70 ng) with morphine prevented the development of tolerance to the analgesia produced by 1 μg of morphine given intrathecally (Fig. 43.3), and that higher doses of these CCK antagonists (1280 ng and 1400 ng respectively) were required to block the tolerance evoked by a larger dose (10 μg) of morphine (Fig. 43.4). The lower doses of CCK antagonists were ineffective against the tolerance induced by the higher morphine dose (Fig. 43.5) (Kellstein and Mayer 1991). These findings provide further evidence for the involvement of CCK in opioid tolerance, support the participation of spinal CCK circuitry in tolerance (see next section), and are consistent with a tolerance mechanism in which the level of compensatory CCK response elicited is proportional to the degree of activation of opioid pathways. However, unexpected dose–response relationships were observed in this study. Coadministration of morphine with doses of CCK antagonists (7 ng of lorglumide, 20 ng of proglumide) which acutely enhanced 1 μg of morphine (Figs 43.1 and 43.2) did not attenuate

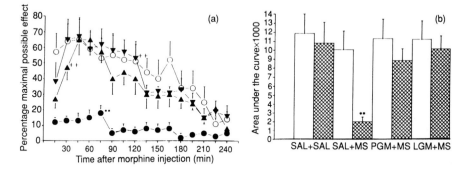

Fig. 43.4. Prevention of opioid tolerance by intrathecal coadministration of 1280 ng of proglumide (PGM) or 1400 ng of lorglumide (LGM) with a high (10 μg) dose of morphine (MS) from day 2 to day 7. (a) Analgesia induced by 1 μg of MS on day 8 in animals receiving saline (SAL) (○), SAL + MS (•), PGM + MS (▲), or LGM + MS (▼) from day 2 to day 7: $\star\star$ $p < 0.01$ compared with SAL: + + $p < 0.01$ compared with SAL + MS using ANOVA of area under the curve values. (b) Within-group comparisons of antinociception induced by 1 μg of MS on day 1 (open bars) to day 8 (hatched bars): $\star\star$ $p < 0.01$ compared with day 1 using ANOVA. Error bars indicate SEM (n = 7–8).

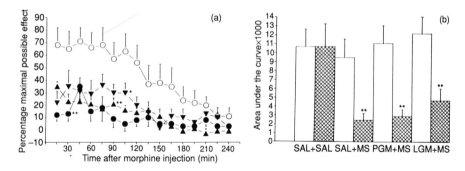

Fig. 43.5. Lack of effect on opioid tolerance of intrathecal coadministration of
64 ng of proglumide (PGM) or 70 ng of lorglumide (LGM) with a high (10 μg) dose
of morphine (MS) from day 2 to day 7. (a) Analgesia induced by 1 μg of MS on
day 8 in animals receiving saline (SAL) (○), SAL + MS (•), PGM + MS (▲), or
LGM + MS (▼) from day 2 to day 7. * p < 0.05, ** p < 0.01 compared with SAL
using ANOVA of area under the curve values. (b) Within-group comparisons of
antinociception induced by 1 μg of MS on day (open bars) to day 8 (hatched
bars): ** p < 0.01 compared with day 1 using ANOVA. Error bars indicate SEM
(n = 7–8).

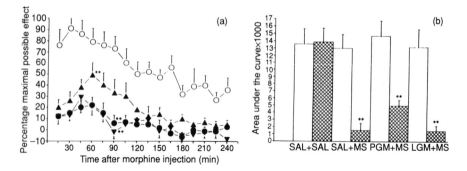

Fig. 43.6. Lack of effect on opioid tolerance of intrathecal coadministration of
20 ng of proglumide (PGM) or 7 ng of lorglumide (LGM) with a low (1 μg) dose of
morphine (MS) from day 2 to day 7. (a) Analgesia induced by 1 μg of MS on day 8
in animals receiving saline (SAL) (○), SAL + MS (•), PGM + MS (▲), or LGM +
MS (▼) from day 2 to day 7: ** p < 0.01 compared with SAL using ANOVA of area
under the curve values. (b) Within-group comparisons of antinociception induced
by 1 μg of MS on day 1 (open bars) to day 8 (hatched bars): ** p < 0.01 compared
with day 1 using ANOVA. Error bars indicate SEM (n = 7–8).

tolerance which developed to this dose (Fig. 43.6). Conversely, coinjection of
morphine with doses of proglumide (64 ng) or lorglumide (70 ng), which do not
acutely alter analgesia induced by 1 μg of morphine (Kellstein and Mayer,
unpublished observations), prevented tolerance to this dose of opioid

(Fig. 43.3). Therefore, although these findings indicate a potential therapeutic value for CCK antagonists in preventing opioid tolerance, the complex dose relationships suggest that further study is required before clinical application becomes practical.

Possible site(s) and mechanism(s) of CCK–opioid interaction

It is apparent from the previously discussed behavioural studies using intrathecal administration of CCK (Faris *et al*. 1983) or CCK antagonists (Watkins *et al*. 1984, 1985*a, b*; Kellstein and Mayer 1990) that the spinal cord is an important site for CCK antagonism of opioid analgesia. In addition, the observations that CCK is released from the spinal cord by intrathecal morphine (Tang *et al*. 1984) and inhibits the depressive effect of spinal morphine on the nociceptive flexion reflex (Wiesenfeld-Hallin and Duranti 1987) suggests that CCK is released in response to activation of spinal opioid antinociceptive pathways and functions to return pain responsiveness to normal.

In an attempt to elucidate the opioid receptor subtype(s) interacting with spinal CCK circuitry, Suberg and Watkins (1987) used a variety of nociceptive assays to demonstrate that intrathecal proglumide enhanced analgesia elicited by spinal injection of selective μ and δ opioid agonists or mixed μ–δ agonists but not that elicited by a κ agonist. In contrast, it was recently reported that exogenous CCK antagonized spinal analgesia elicited by μ or κ, but not δ, opioid agonists (Wang *et al*. 1990). Thus, although these two studies differ in their support for the involvement of δ or κ sites (see also Chapter 46 of this volume), they are both consistent with a role for μ receptors; they also agree with the observation that CCK antagonizes, whereas CCK blockers enhance, the antinociceptive effects of morphine, the prototypical μ agonist.

Electrophysiological studies in the spinal cord have demonstrated an opioid antagonistic effect of CCK at the neuronal level. Suberg *et al*. (1985*a, b, c*) found that CCK diminished morphine-induced inhibition of dorsal horn neuronal firing in response to painful stimuli, whereas CCK antibodies or proglumide enhanced it.

The neurophysiological mechanism by which CCK antagonizes the spinal antinociceptive effect of opioids is not known. Two studies reported that CCK increased the activity (firing) of dorsal horn neurons. However, these effects were not specific to nociceptive cells (either non-nociceptive neurons were also excited (Jeftinija *et al*. 1981) or the type of neuron was not determined (Willetts *et al*. 1985)); further, the data were not statistically analysed. In a recent study (Kellstein *et al*. 1991) we determined the effect of spinally superfused CCK (6.4 pmol–20 nmol) or lorglumide (145 fmol–145 pmol) on (i) the firing of dorsal horn nociceptive neurons and (ii) morphine-induced (4–100 nmol) inhibition of such activity. Neither CCK nor its antagonist lorglumide significantly altered spontaneous, A-fibre-evoked (not shown), or C-fibre evoked activity (Fig. 43.7) (see also Chapter 46 of this volume). However, pretreatment with

CCK and opioid analgesia

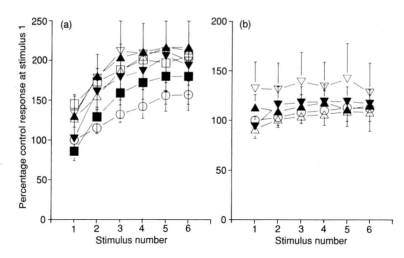

Fig. 43.7. Lack of effect of CCK or lorglumide on C-fibre evoked discharges of dorsal horn nociceptive neurons. Repetitive stimulation (six pulses at 0.5 Hz, 3 ma, 2 ms duration) was applied to the ipsilateral hind paw before (○) and after (a) CCK at doses of 6.4 pmol (△), 32 pmol (▲), 160 pmol (▽), 800 pmol (▼), 4 nmol (□), and 20 nmol (■) (n = 16), or (b) lorglumide at doses of 145 fmol (△), 1.45 pmol (▲), 14.5 pmol (▽), and 145 pmol (▼) (n = 8). Error bars indicate SEM.

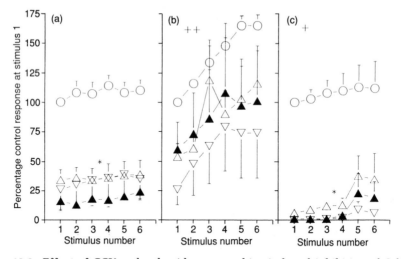

Fig. 43.8. Effect of CCK or lorglumide on morphine-induced inhibition of C-fibre evoked discharges of dorsal horn nociceptive neurons. Repetitive stimulation (six pulses at 0.5 Hz, 3 ma, 2 ms duration) was applied to the ipsilateral hind paw before (○) and after morphine at doses of 4 nmol (△), 20 nmol (▲), and 100 nmol (▽). Morphine was superfused following (a) saline (n = 12), (b) CCK (20 nmol, n = 12), or (c) lorglumide (145 pmol, n = 8). ∗ p < 0.0001 for the effect of morphine; + p < 0.05, ++ p < 0.005 for comparison of the morphine effect with that in (a) using ANOVA. Error bars indicate SEM.

CCK inhibited morphine inhibition of C-fibre evoked firing whereas pretreatment with lorglumide enhanced it (Fig. 43.8). The latter findings agree with those of Suberg *et al.* (1985*a, b, c*) and those reported in Chapter 46 of this volume. In addition, the lack of effect of CCK or lorglumide alone on neuronal activity agrees with the behavioural studies cited above which have consistently shown that neither CCK (in low and probably physiological doses) nor its antagonists alter nociceptive thresholds. Thus CCK apparently does not function as a neurotransmitter or neuromodulator of nociception, but rather as an endogenous antagonist of opioid analgesia.

Whether CCK inhibits opioid analgesia via interaction with opioid receptors is not clear. Stengaard-Pedersen and Larsson (1981) originally reported that CCK in concentrations of up to 1 μM did not displace radio-labelled naloxone or met-enkephalinamide binding to synaptosomes from rat brain homogenates, indicating that CCK does not bind to opioid receptors. However, in a recent study, also using rat brain synaptosomes, it was reported that CCK (1 pM–1 μM) decreased the affinity and maximal number of binding sites of radio-labelled etorphine binding to its high affinity sites (Wang *et al.* 1989). Further, this effect was reversed by proglumide (1 μM), suggesting that attachment of CCK to its receptor inhibits opioid ligand–receptor binding, possibly via a conformational change in the opioid site and/or steric hindrance (see Chapter 45 of this volume). Such a mechanism is consistent with the observation that intrathecal CCK prevented morphine-induced inhibition of the spinal nociceptive flexion reflex if administered before the opioid, but had no effect when superfused after morphine (Wiesenfeld-Hallin and Duranti 1987).

To summarize, the available evidence supports a possible mechanism in which application of exogenous opioids to, or release of endogenous opioids in, the spinal cord stimulates the compensatory release of CCK, which in turn binds to CCK receptors adjacent to opioid binding sites on pain transmission neurons in the spinal cord dorsal horn. Although binding of CCK to its receptor site does not excite the pain transmission neuron, it interferes with the attachment of the opioid to its receptor, thereby blocking the usual inhibitory effect of opioid on nociceptive transmission. In the case of opioid tolerance, prolonged drug application causes a progressive upregulation of spinal CCK activity, such that the normal antinociceptive effect of opioid is increasingly antagonized and progressively higher doses are required to attain initial levels of analgesia.

Although this mechanism is being suggested at the spinal level, it does not exclude the participation of supraspinal loci, e.g. the periaqueductal grey. Indeed, several studies support involvement of this site in CCK–opioid interactions. Endogenous opioids and CCK (Stengaard-Pedersen and Larsson 1981; Beinfeld and Palkovits 1982) as well as their respective receptors (Saito *et al.* 1980; Van Dijk *et al.* 1984) have been found in the periaqueductal grey. Furthermore, coinjection of proglumide with morphine into this area enhances opioid analgesia (Watkins *et al.* 1984, 1985*a*) whereas injection of CCK into the periaqueductal grey blocks morphine analgesia (Li and Han 1989). However, additional studies are required to determine if supraspinal CCK circuitry in the periaqueductal grey or other supraspinal sites antagonizes opioid analgesia and mediates opioid tolerance.

Conclusion

Since the initial observations approximately 8 years ago that CCK inhibited analgesia induced by both exogenous and endogenous opioids, considerable evidence has accumulated which indicates that CCK functions as a selective endogenous antagonist of opioid analgesia which may be involved in the development of opioid tolerance. Since opioids are powerful therapeutic tools in the clinical management of pain, and tolerance is a major problem which accompanies their chronic use, elucidation of CCK–opioid interactions and the underlying mechanisms are clearly important goals in the quest for improved pain treatment. The subsequent discoveries that (i) acute pretreatment with CCK antagonists enhanced opioid analgesia and (ii) coadministration of CCK blockers with opioids attenuated development of tolerance are exciting findings which may contribute directly to improved health care for millions of pain sufferers. However, the bell-shaped dose–response curves for CCK antagonist enhancement of opioid analgesia (Watkins *et al.* 1984, 1985*a*; Dourish *et al.* 1988*b*, 1990), the compensatory re-regulation of CCK and/or opioid systems which apparently accompanies chronic CCK blockade (Kellstein and Mayer 1990), and the unpredictable dose relationships for CCK antagonist attenuation of opioid tolerance (Figs 43.3–43.6) indicate that additional elucidation of the complex interactions between CCK and opioids is required before the clinical use of CCK antagonists becomes practical.

Acknowledgement

Portions of this work were supported by PHS Grants DA 00576 and NS 24009.

References

Baber, N. S., Dourish, C. T., and Hill, D. R. (1989). The role of CCK, caerulein, and CCK antagonists in nociception. *Pain* **39**, 307–28.
Barbaz, B. S., Autry, W. L., Ambrose, F. G., Hall, N. R., and Leibman, J. M. (1986). Antinociceptive profile of sulfated CCK: comparison with CCK-4, unsulfated CCK and other neuropeptides. *Neuropharmacology* **25**, 823–9.
Beinfeld, M. C. (1983). Cholecystokinin in the central nervous system: a minireview. *Neuropeptides* **3**, 411–27.
Beinfeld, M. C. and Palkovits, M. (1982). Distribution of cholecystokinin (CCK) in the rat lower brain stem nuclei. *Brain Res.* **238**, 260–5.
Chiodo, L. A. and Bunney, B. S. (1983). Proglumide: selective antagonism of excitatory effects of cholecystokinin in central nervous system. *Science* **219**, 1449–51.

Crawley, J. N. (1985). Comparative distribution of cholecystokinin and other neuropeptides. *Ann. NY Acad. Sci.* **448**, 1–9.

Crawley, J. N., Stivers, J. A., Hommer, D., Skirboll, L., and Paul, S. M. (1986). Antagonists of central and peripheral behavioral actions of cholecystokinin octapeptide. *J. Pharmacol. Exp. Ther.* **236**, 320–30.

Della-Fera, M. A. and Baile, C. A. (1979). Cholecystokinin octapeptide: continuous picomole injections into the cerebral ventricles of sheep suppress feeding. *Science* **206**, 471.

Dourish, C. T., Clark, M. L., and Iversen, S. D. (1988*a*). Analgesia induced by restraint stress is attenuated by CCK and enhanced by the CCK antagonists MK-329, L-365-031 and CR 1409. *Soc. Neurosci. Abstr.* **14**, 290.

Dourish, C. T., Hawley, D., and Iversen, S. D. (1988*b*). Enhancement of morphine analgesia and prevention of morphine tolerance in the rat by the cholecystokinin antagonist L-364,718. *Eur. J. Pharmacol.* **147**, 469–72.

Dourish, C. T., O'Neill, M. F., Coughlan, J., Kitchener, S. J., Hawley, D., and Iversen, S. D. (1990). The selective CCK-B receptor antagonist L-365,260 enhances morphine analgesia and prevents morphine tolerance in the rat. *Eur. J. Pharmacol.* **176**, 35–44.

Faris, P. L., Komisaruk, B., Watkins, L. R., and Mayer, D. J. (1983). Evidence for the neuropeptide cholecystokinin as an antagonist of opiate analgesia. *Science* **219**, 310–12.

Faris, P. L., McLaughlin, C. R., Baile, C. A., Olney, J. W., and Komisaruk, B. R. (1984). Morphine analgesia potentiated but tolerance not affected by active immunization against cholecystokinin. *Science* **226**, 1215–17.

Gall, C., Lauterhorn, J., Burks, D., and Seroogy, K. (1987). Co-localization of enkephalin and cholecystokinin in discrete areas of rat brain. *Brain Res.* **403**, 403–8.

Gillis, R. A., Quest, J. A., Pagani, F. D., Dias-Souza, J., Traveira DaSilva, A. M., Jensen, R. T., Garvey, T. Q., and Hamosh, P. (1983). Activation of central nervous system cholecystokinin receptors stimulates respiration in the cat. *J. Pharmacol. Exp. Ther.* **224**, 408–14.

Gillis, R. A., Quest, J. A., Pagani, F. D., Dias-Souza, J., Traveira DaSilva, A. M., Crawley, J. N., Stivers, J. A., Hommer, D., Skirboll, L., and Paul, S. M. (1986). Antagonists of central and peripheral behavioral actions of cholecystokinin. *J. Pharmacol. Exp. Ther.* **236**, 320–30.

Han, J., Ding, X. Z., and Fan, S. G. (1986). Cholecystokinin octapeptide (CCK): antagonism to electroacpuncture analgesia and a possible role in electroacupuncture tolerance. *Pain* **27**, 101–15.

Hill, D. R., Shaw, T. M., and Woodruff, G. N. (1987*a*). Species differences in the localization of 'peripheral' type cholecystokinin receptors in rodent brain. *Neurosci. Lett.* **79**, 286–9.

Hill, D. R., Shaw, T. M., and Woodruff, G. N. (1988). Binding sites for ^{125}I-cholecystokinin in primate spinal cord are of the CCK-A subclass. *Neurosci. Lett.* **89**, 133–9.

Hill, R. G., Hughes, J., and Pittaway, K. M. (1987*b*). Antinociceptive action of cholecystokinin octapeptide (CCK) and related peptides in rats and mice: effects of naloxone and peptidase inhibitors. *Neuropharmacology* **26**, 289–300.

Itoh, S., Katsuura, G., and Maeda, Y. (1982). Caerulein and cholecystokinin suppress beta-endorphin-induced analgesia in the rat. *Eur. J. Pharmacol.* **80**, 421–5.

Jeftinija, S., Miletic, V., and Randic, M. (1981). Cholecystokinin octapeptide excites dorsal horn neurons both *in vivo* and *in vitro*. *Brain Res.* **213**, 231–6.

Jensen, T. Z., Garvey, T., and Hamosh, P. (1983). Activation of central nervous system cholecystokinin receptors stimulates respiration in the cat. *J. Pharmacol. Exp. Ther.* **224**, 408–14.

Jurna, I. and Zetler, G. (1981). Antinociceptive effect of centrally administered caerulein and cholecystokinin (CCK). *Eur. J. Pharmacol.* **73**, 323–31.

Kapas, L., Benedek, G., and Penke, B. (1989). Cholecystokinin interferes with the thermoregulatory effect of exogenous and endogenous opioids. *Neuropeptides* **14**, 85–92.

Kellstein, D. E. and Mayer, D. J. (1990). Chronic administration of cholecystokinin antagonists reverses the enhancement of spinal morphine analgesia induced by acute pretreatment. *Brain Res.* **516**, 263–70.

Kellstein, D. E. and Mayer, D. J. (1991). Spinal co-administration of cholecystokinin antagonists with morphine prevents the development of opioid tolerance. *Pain*, in press.

Kellstein, D. E., Price, D. D., and Mayer, D. J. (1991). Cholecystokinin and its antagonist lorglumide respectively attenuate and facilitate morphine-induced inhibition of C-fiber evoked discharges of dorsal horn nociceptive neurons. *Brain Res.* **540**, 302–6.

Kinscheck, I. B., Watkins, L. R., Kaufman, E., and Mayer, D. J. (1983). Evidence for a cholecystokinin (CCK)-like endogenous opiate antagonist. *Soc. Neurosci. Abstr.* **9**, 792.

Larssen, L. J. and Rehfeld, J. (1979). Localization and molecular heterogeneity of cholecystokinin in the central and peripheral nervous system. *Brain Res.* **165**, 201–18.

Lavigne, G., Hargreaves, K. M., Schmidt, E. S., and Dionne, R. A. (1989). Proglumide potentiates morphine analgesia for acute postsurgical pain. *Clin. Pharmacol. Ther.* **45**, 666–73.

Li, Y. and Han, J. (1989). Cholecystokinin octapeptide antagonizes morphine analgesia in periaqueductal gray of the rat. *Brain Res.* **480**, 105–10.

Makovec, F., Bani, M., Chiste, R., Revel, L., Rovati, L. C., and Rovati, L. A. (1985). Differentiation of central and peripheral cholecystokinin receptors by new glutaramic acid derivatives with cholecystokinin-antagonistic activity. *Drug Res.* **36**, 98–102.

Moran, T. H., Robinson, P. H., Goldrich, M. S., and McHugh, P. R. (1986). Two brain cholecystokinin receptors: implications for behavioral actions. *Brain Res.* **362**, 175–9.

Morley, J. E. (1982). The ascent of cholecystokinin from gut to brain. *Life Sci.* **30**, 479–93.

Morley, J. E. and Levine, A. S. (1983). Opioid modulation of appetite. *Neurosci. Biobehav. Rev.* **7**, 281.

Mutt, V. (1982). Chemistry of the gastrointestinal hormones and hormone-like peptides and a sketch of their physiology and pharmacology. *Vitam. Horm.* **39**, 231–427.

O'Neill, M. F., Dourish, C. T., and Iversen, S. D. (1989). Morphine-induced analgesia in the rat paw pressure test is blocked by CCK and enhanced by the CCK antagonist MK-329. *Neuropharmacology* **28**, 243–8.

Panerai, A. E., Rovati, L. C., Cocco, E., Sacerdote, P., and Mantegazza, P. (1987). Dissociation of tolerance and dependence to morphine: a possible role for cholecystokinin. *Brain Res.* **40**, 52–60.

Pittaway, K. M., Rodriguez, R. E., Hughes, J., and Hill, R. G. (1987). CCK analgesia and hyperalgesia after intrathecal administration in the rat: comparison with CCK-related peptides. *Neuropeptides* **10**, 87–108.

Price, D. D., Vondergruen, A., Miller, J., Rafii, A., and Price, C. (1985). Potentiation of systemic morphine analgesia in humans by proglumide, a cholecystokinin antagonist. *Anesth. Analg.* **64**, 801-6.

Rehfeld, J. F. (1978). Immunochemical studies on cholecystokinin. II. Distribution and molecular heterogeneity in the central nervous system and small intestine of man and hog. *J. Bio. Chem.* **253**, 4022-30.

Saito, A., Sankaran, H., Goldfine, J. D., and Williams, J. A. (1980). Cholecystokinin receptors in the brain: characterization and distribution. *Science* **208**, 1155-6.

Schnur, P., Raigoza, V. P., Sanchez, M. R., and Kulkovsky, P. J. (1986). Cholecystokinin antagonizes morphine induced hypoactivity and hyperactivity in hamsters. *Pharmacol. Biochem. Behav.* **25**, 1067-70.

Stengaard-Pedersen, K., and Larsson, L.-I. (1981). Localization and opiate receptor binding of enkephalin, CCK and ACTH/beta-endorphin in the rat central nervous system. *Peptides* **2** (Suppl. 1), 3-19.

Suberg, S. N. and Watkins, L. R. (1987). Interaction of cholecystokinin and opioids in pain modulation. In *Pain and headache*, Vol. 9, pp. 247-65. Karger, Basel.

Suberg, S. N., Culhane, E. S., Carstens, E., and Watkins, L. R. (1985*a*). Behavioral and electrophysiological investigations of opiate/cholecystokinin interactions. *Adv. Pain Res. Ther.* **9**, 541-53.

Suberg, S. N., Culhane, E. S., Rosenquist, G., Carstens, E., and Watkins, L. R. (1985*b*). Effect of anti-cholecystokinin antibody (AB) on morphine (MOR) induced suppression of spinal nociceptive transmission. *Soc. Neurosci. Abstr.* **11**, 284.

Suberg, S. N., Culhane, E. S., Carstens, E., and Watkins, L. R. (1985*c*). The potentiation of morphine-induced inhibition of spinal transmission by proglumide, a putative cholecystokinin antagonist. *Ann. NY Acad. Sci.* **448**, 660-2.

Tang, J., Chou, J., Iadarola, M., Yang, H. Y. T., and Costa, E. (1984). Proglumide prevents and curtails acute tolerance to morphine in rats. *Neuropharmacology* **23**, 715-18.

Van Dijk, A., Richards, J. G., Trzeciak, A., Gillessen, D., and Mohler, H. (1984). Cholecystokinin receptors: biochemical demonstrations and autoradiographical localization in rat brain and pancreas using ^{3}H-cholecystokinin as radioligand. *J. Neurosci.* **4**, 1021-33.

Vigna, S. R., Szecowka, J., and Williams, J. A. (1985). Do antagonists of pancreatic cholecystokinin receptors interact with central nervous system cholecystokinin receptors? *Brain Res.* **343**, 394-7.

Wang, X. J., Fan, S. G., Ren, M. F., and Han, J. S. (1989). Cholecystokinin-8 suppressed ^{3}H-etorphine binding to rat brain opiate receptors. *Life Sci.* **45**, 117-23.

Wang, X.-J., Wang, X.-H., and Han, J.-S. (1990). Cholecystokinin octapeptide antagonized opioid analgesia mediated by mu- and kappa- but not delta-receptors in the spinal cord of the rat. *Brain Res.* **523**, 5-10.

Watkins, L. R., Cobelli, D. A., Faris, P., Aceto, M. D., and Mayer, D. J. (1982). Opiate vs nonopiate foot shock-induced analgesia (FSIA). The body region shocked is a critical factor. *Brain Res.* **242**, 299-308.

Watkins, L. R., and Mayer, D. J. (1982). The neural organisation of endogenous opiate and nonopiate pain control systems. *Science* **216**, 1185-92.

Watkins, L. R., Kinscheck, I. B., and Mayer, D. J. (1984). Potentiation of opiate analgesia and apparent reversal of morphine tolerance by proglumide. *Science* **224**, 395-6.

Watkins, L. R., Kinscheck, I. B., and Mayer, D. J. (1985*a*). Potentiation of morphine analgesia by the cholecystokinin antagonist proglumide. *Brain Res.* **327**, 169-80.

Watkins, L. R., Kinscheck, I. B., Kaufman, E. F. S., Miller, J., Frenk, H., and Mayer,

D. J. (1985*b*). Cholecystokinin antagonists selectively potentiate analgesia induced by endogenous opiates. *Brain Res.* **327**, 181–90.

Watkins, L. R., Kinscheck, I. B., Rosenquist, G., Miller, J., Wimberg, S., Frenk, H., Kaufman, E., Coghill, R., Suberg, S. N., and Mayer, D. J. (1985*c*). The enhancement of opiate analgesia and the possible reversal of morphine tolerance by proglumide. *Ann. NY Acad. Sci.* **448**, 676–7.

Wiesenfeld-Hallin, Z. and Duranti, R. (1987). Intrathecal cholecystokinin interacts with morphine but not substance P in modulating the nociceptive flexion reflex in the rat. *Peptides* **8**, 153–8.

Willetts, J., Urban, L., Murase, K., and Randic, M. (1985). Actions of cholecystokinin octapeptide on rat spinal dorsal horn neurons. *Ann. NY Acad. Sci.* **448**, 385–402.

Williams, J. A., Gryson, K. A., and McChesney, D. J. (1986). Brain CCK receptors: species differences in regional distribution and selectivity. *Peptides* **7**, 293–6.

Zarbin, M. A., Innis, R. B., Wamsley, J. K., Snyder, S. H., and Kuhar, M. J. (1983). Autoradiographic localization of cholecystokinin receptors in rodent brain. *J. Neurosci.* **3**, 877–906.

Zetler, G. (1979). Antagonism of cholecystokinin-like peptides by opioid peptides, morphine, and by tetrodotoxin. *Eur. J. Pharmacol.* **60**, 67–77.

Zetler, G. (1980). Analgesia and ptosis caused by caerulein and cholecystokinin octapeptide (CCK). *Neuropharmacology* **19**, 415–22.

44. The role of CCK_A and CCK_B receptors in mediating the inhibitory effect of CCK on opioid analgesia

Colin T. Dourish

Introduction

Since the publication of the seminal finding that CCK octapeptide sulphated (CCK-8S) (in this chapter the abbreviation CCK refers to this fragment unless stated otherwise) can block opioid analgesia (Itoh *et al.* 1982; Faris *et al.* 1983) there has been considerable interest in the possibility that it may act as an endogenous opioid antagonist (Faris 1985; Han *et al.* 1985). There was already substantial evidence prior to this discovery that the physiological effects of CCK and opioids were often the antithesis of each other. For example, opioids increase food intake whereas CCK decreases feeding (see Chapter 17 of this volume), and CCK decreases locomotion whereas morphine increases motor activity (Fog 1970; Chapter 22 of this volume). Nevertheless, the discovery of a possible interaction between the two peptides in the control of nociception was of particular significance given the pre-eminence of opioid drugs as painkillers. Indeed, the widespread use of opioid analgesics suggests that examination of CCK–opioid interactions may be of considerable clinical relevance. The potential for CCK–opioid interactions is emphasized by the parallel distribution of the two peptide families and their receptors in the brain and spinal cord. Thus there are high concentrations of CCK and opioids and of CCK and opioid receptors in periaqueductal grey (PAG), ventromedial thalamus, and spinal dorsal horn, all of which are areas that play an important role in the processing of nociceptive impulses (Baber *et al.* 1989). In addition, there is evidence that CCK and opioids are colocalized in nociceptive neurons in the thalamus and brainstem (Gall *et al.* 1987). This chapter reviews the preclinical and limited clinical evidence for CCK–opioid interactions in nociception. Particular emphasis is placed on the results of recent studies with potent and selective CCK antagonists which provide strong evidence for a role of endogenous CCK in mediating opioid analgesia in both rodents and primates. In addition, the respective roles of CCK_A and CCK_B receptor sub-types (Moran *et al.* 1986; Dourish and Hill 1987; Chapter 5 of this volume) in mediating CCK–opioid interactions are assessed. Finally, a model is proposed which may explain the mechanism of CCK–opioid interactions in nociception.

Blockade of opioid analgesia by CCK agonists

The first studies of the effects of CCK on pain thresholds were by Zetler and colleagues who injected large doses of CCK (minimum effective doses of 0.03 mg/kg in the hot-plate test and 0.75 mg/kg in the writhing test) and the CCK-related peptide caerulein subcutaneously (s.c.) and observed that both peptides induced naloxone-sensitive analgesia (Zetler 1980; Jurna and Zetler 1981). Subsequent studies have suggested that the peptide doses used in these experiments are outside the physiological range and that CCK-induced analgesia may be a pharmacological rather than a physiological phenomenon (see Baber et al. (1989) and Chapter 43 of this volume for discussion). This interpretation was reinforced by the first observations that small doses of CCK, which had no intrinsic effect on pain thresholds, could block opioid analgesia.

The first study to report the opioid antagonistic action of CCK was that of Itoh et al. (1982). They injected 1.7–3.3 μg of CCK and 0.3–2.3 μg of caerulein intracerebroventricularly (i.c.v.) and observed a blockade of analgesia induced by i.c.v. injection of β-endorphin in the hot-plate test. The doses used in this study were relatively large, and it was not until the following year, when it was demonstrated by Faris et al. (1983) that a dose of CCK as small as 1.5 μg/kg injected intraperitoneally (i.p.) could block opioid analgesia, that a physiological action of CCK was clearly implicated. Faris et al. (1983) showed that CCK injected i.p. at doses of 1.5–6.0 μg/kg or intrathecally (i.t.h.) at doses of 3.6 and 360 ng blocked analgesia induced by morphine in the rat tail-flick test. The effect was specific to CCK-8S as i.t.h. injection of equivalent doses of the unsulphated octapeptide (CCK-8US) were ineffective in blocking opioid analgesia. The antagonism of morphine analgesia by CCK was prevented by pretreatment with the CCK antagonist proglumide, suggesting that the interaction was mediated by CCK receptors (Tang et al. 1984; Suh and Tseng 1990), but see below for discussion regarding the specificity of proglumide. CCK also blocked analgesia induced by front-paw shock or classically conditioned to front-paw shock (Faris et al. 1983). Both these responses are reversed by the opioid antagonist naloxone (Watkins et al. 1982), thus indicating that CCK modulates analgesia mediated by endogenous as well as exogenous opioids. Further evidence for CCK modulation of analgesia mediated by endogenous opioids has emerged from studies by Han and colleagues on acupuncture-induced analgesia. Thus, analgesia induced by electroacupuncture in rats was blocked by CCK-8S but unaffected by CCK-8US (Han et al. 1986) (see Chapter 45 of this volume for a detailed discussion of the role of CCK in acupuncture-induced analgesia).

Evidence for the specificity of the effect for opioid analgesia was provided by the lack of effect of CCK on analgesia induced by hind-paw shock (Faris *et al.* 1983) which is non-opioid in nature (Watkins *et al.* 1982). Similarly, CCK had no effect on antinociception induced by i.t.h. injection of serotonin or noradrenaline (Han *et al.* 1986). Interestingly, it has also been reported that i.t.h. injection of CCK has no effect on analgesia induced by i.c.v. injection of morphine in mice (Suh and Tseng 1990). This was interpreted by the authors as further evidence for opioid specificity of the CCK effect as analgesia induced by i.c.v. injection of morphine is thought to be mediated by release of noradrenalin and serotonin from the spinal cord and is not thought to involve spinal opioid mechanisms (Suh and Tseng 1990). The blockade of opioid analgesia by CCK has subsequently been replicated in a number of studies using both the tail-flick test and the paw pressure test in rats (see Table 44.1 for a summary of results). In addition, it has been shown in a recent study by Li and Han (1989) that injection of CCK into the PAG blocks opiate analgesia. Han and colleagues have compared the potency of CCK injected i.t.h., i.c.v., and into the PAG in blocking morphine analgesia. The rank order of potency was PAG > i.c.v. > i.t.h., suggesting that the PAG may be an important site for mediating the anti-opioid effects of CCK (see below and Chapter 45 of this volume for further discussion).

An intriguing facet of the blockade of opiate analgesia by systemically injected CCK is the time course of the effect. The half-life of CCK is less than

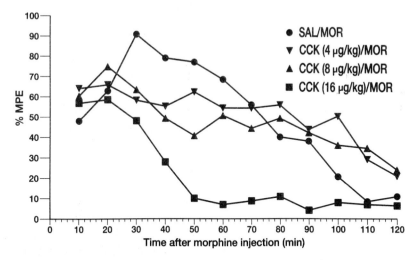

Fig. 44.1. Blockade of morphine analgesia by CCK in the rat tail-flick test. CCK was injected i.p. 10 min prior to 8 mg/kg i.p. of morphine (n = 8 rats per group). Significant blockade of morphine analgesia by CCK was observed 40–80 min after morphine injection. The percentage of maximum possible effect (% MPE) is a measure of the degree of analgesia and is calculated as % MPE = [(TL − BL)/ (10 − BL)] × 100, where TL is the test latency, BL is the baseline latency, and the maximum duration of the thermal stimulus is 10 s. (Reproduced with permission from Dourish *et al.* 1988*a*.)

Table 44.1. Blockade of opioid analgesia by CCK agonists in the rat

CCK Agonist	Dose/route	Opioid agonist	Dose/route	Species	Test	Reference
CCK-8S	5 μg/kg i.p.	Morphine	10 mg/kg i.p.	Rat	Tail-flick	Faris et al. 1983
CCK-8S	4–16 μg/kg i.p.	Morphine	8 mg/kg i.p.	Rat	Tail-flick	Dourish et al. 1988a
CCK-8S	1.5–6.0 μg/kg i.p.	Front-paw shock	1.6 mA	Rat	Tail-flick	Faris et al. 1983
CCK-8S	3.6–360 ng i.t.h.	Front-paw shock	1.6 mA	Rat	Tail-flick	Faris et al. 1983
CCK-8S	1–4 ng i.t.h.	Morphine	5 mg/kg s.c.	Rat	Tail-flick	Han et al. 1985
CCK-8S	0.25–0.5 ng i.t.h.	β-endorphin	1 μg i.c.v.	Mouse	Tail-flick	Suh and Tseng 1990
CCK-8S	1.7–3.3 μg i.c.v.	β-endorphin	2.5 μg i.c.v.	Rat	Hot-plate	Itoh et al. 1982
Caerulein	0.3–2.3 μg i.c.v.	β-endorphin	2.5 μg i.c.v.	Rat	Hot-plate	Itoh et al. 1982
CCK-8S	4–16 μg/kg i.p.	Morphine	8 mg/kg i.p.	Rat	Paw pressure	O'Neill et al. 1989
CCK-8S	1–4 μg i.c.v.	Morphine	5 mg/kg s.c.	Rat	Tail-flick	Han et al. 1985
CCK-8S	0.25–1.0 PAG	Morphine	5 mg/kg s.c.	Rat	Tail-flick	Li and Han 1989

i.p. intraperitoneal; s.c. subcutaneous; i.t.h. intrathecal; i.c.v. intracerebroventricular; PAG intra-periaqueductual grey

30 min and most of the effects of the peptide on behaviour are evident between 5 and 30 min after systemic injection (see, for example, the effects on feeding discussed in Chapter 17 of this volume). In contrast, the blockade of opioid analgesia by i.p. CCK is not evident until 40 min after injection and the peak effect is apparent between 60 and 120 min after injection. This distinctive time course was evident in the first report of the effect by Faris *et al.* (1983) and has been consistently observed in subsequent studies (Dourish *et al.* 1988*a*) (Fig. 44.1). One possible explanation for the long latency to onset of the CCK blockade of opioid analgesia could be that the initial site of action of the peptide is a peripheral receptor in the gastro-intestinal tract and activation of this receptor initiates a signal which is transmitted along afferent fibres to the spinal cord and/or brain. These afferent fibres may be located in the vagus nerve or in the visceral branches of the dorsal root ganglion neurons that project to the spinal cord. Such a complex mechanism might be slow to elicit a functional effect. Evidence in support of this hypothesis is the poor penetration of the blood–brain barrier by CCK (Zhu *et al.* 1986). In addition, a preliminary report has suggested that vagotomy can prevent the blockade of opioid analgesia by CCK (Steinman *et al.* 1986), although the interpretation of this result is complicated by the fact that vagotomy alone attenuated morphine analgesia in this study. Alternatively, small amounts of systemically injected CCK may slowly penetrate the brain in sufficient concentrations to cause a physiological effect. A further possibility which cannot be discounted is that a breakdown product of CCK-8S (e.g. CCK-5, CCK-4) may be responsible for the anti-opioid effect. To date, however, there is no evidence of an opioid antagonist action in anti-nociceptive tests of any fragment other than CCK-8S (Baber *et al.* 1989; Chapter 45 of this volume). Further studies are clearly warranted to probe this puzzling phenomenon.

Enhancement of opioid analgesia by CCK antagonists

The initial reports of blockade of opioid analgesia by CCK were quickly supported by the publication of evidence that the CCK antagonists proglumide and benzotript enhance morphine analgesia. Watkins *et al.* (1984) showed that proglumide injected systemically, i.t.h., or into the PAG enhanced the analgesic effect of a submaximal dose of morphine. Similarly, i.t.h. injection of proglumide enhanced opioid analgesia induced by front-paw shock, and i.t.h. injection of proglumide or benzotript enhanced analgesia induced by i.t.h. injection of the enkephalin analogue (D-Ala^2)-methionine enkephalinamide (DALA) (Watkins *et al.* 1984, 1985). These effects were not due to intrinsic analgesic actions of proglumide or benzotript as the drugs had no effect on pain thresholds when given alone (Watkins *et al.* 1984, 1985). The proglumide-induced enhancement of analgesia induced by DALA was reversed by the opioid antagonist naltrexone, indicating that the interaction was dependent on opioid receptors. Furthermore, the enhancement was selective for opioid mechanisms as proglumide attenuated or had no effect on various forms of non-opioid analgesia (Watkins *et al.* 1985). At approximately the same time, Faris *et al.* (1984) reported that rats immunized against CCK exhibited an enhanced analgesic response to morphine. When considered together, these results provided strong support for the hypothesis that endogenous CCK modulates opioid analgesia. Nevertheless, there were some puzzling discrepancies, not least the fact that the CNS concentrations of proglumide (a weak non-selective CCK antagonist) achieved in studies reporting enhancement of opioid analgesia (approximately micromolar) were of the order of 1000-fold below its affinity (millimolar) for CCK receptors (see Chapter 43 of this volume). Furthermore, it has been reported that proglumide possesses δ opioid agonist properties (Rezvani *et al.* 1987) and has a higher affinity for both μ and κ opioid receptors than for CCK receptors (Gaudreau *et al.* 1990). Similarly, lorglumide (a more potent analogue of proglumide) and benzotript have high affinity for both opioid and CCK receptors (Gaudreau *et al.* 1990). Such actions would clearly compromise the interpretation of opioid interactions of these drugs in terms of CCK receptor blockade. Therefore definitive evidence for CCK receptor involvement in the modulation of opioid analgesia was not obtained until potent selective CCK antagonists were developed. The first such compound to be developed and characterized was devazepide (Chang and Lotti 1986; Evans *et al.* 1986; Chapter 2 of this volume). Initially, the effects of devazepide were examined on morphine analgesia in the rat tail-flick test (Dourish *et al.* 1988*b*). Devazepide enhanced the peak analgesic effect of morphine and prolonged the duration of the opioid response. The dose–response curve was bell-shaped with the effective doses being 0.5, 1.0, and 2.0 mg/kg; doses above and below this range were ineffective (Dourish *et al.* 1988*b*). A bell-shaped dose–response curve was also noted in previous studies with proglumide (Watkins *et al.* 1984) and has been reported in all subsequent studies (with one exception, see below) in which enhancement of opioid analgesia by a CCK antagonist has been observed (see Chapter 43 of this volume). The explanation for the bell-shaped dose–response curve of CCK antagonists is unclear at present. However, it is not confined to analgesia studies, and has also been observed in feeding models (Dourish *et al.* 1989; Chapter 21 of this volume), anxiety models (Ravard and Dourish 1990; Chapter 12 of this volume), and dopamine models (Chapters 36 and 39 of this volume). It is possible that high doses of CCK antagonists activate an additional mechanism(s) which opposes the functional effect of CCK receptor blockade. The only apparent exception to this general rule is the recently disclosed compound PD 134308. This drug has been reported to enhance morphine analgesia in the rat hot-plate test over a 30-fold dose range and was claimed not to show a bell-shaped dose–response curve (Wiesenfeld-Hallin *et al.* 1990). However, this exception may be more apparent than real as devazepide was active in enhancing opiate analgesia in the squirrel monkey tail-withdrawal test over a 30-fold dose range (3–100 μg/kg), but increasing the dose of the drug still further revealed the presence of a bell-shaped

Table 44.2. Enhancement of opioid analgesia by CCK antagonists

Antagonist	Dose/route	Opioid	Dose/route	Species	Test	Reference
Proglumide	2–20 μg/kg i.p.	Morphine	3 mg/kg i.p.	Rat	Tail-flick	Watkins et al. 1984
Proglumide	0.001–0.01 μg i.t.h.	Morphine	1 μg i.t.h.	Rat	Tail-flick	Watkins et al. 1984
Proglumide	0.001–0.01 μg i.t.h.	DALA	3 μg i.t.h.	Rat	Tail-flick	Watkins et al. 1984
Proglumide	0.001–0.01 μg i.t.h.	FPS	1.6 mA	Rat	Tail-flick	Watkins et al. 1984
Proglumide	0.01 μg PAG	Morphine	3 μg PAG	Rat	Tail-flick	Watkins et al. 1984
Proglumide	2.5–10.0 μg i.t.h.	β-endorphin	0.25 μg i.c.v.	Mouse	Tail-flick	Suh and Tseng 1990
Proglumide	10 mg/kg s.c.	DAGOL	2.5 ng i.c.v.	Mouse	Tail-flick	Bodnar et al. 1990
Benzotript	0.01 μg i.t.h.	DALA	3 μg i.t.h.	Rat	Tail-flick	Watkins et al. 1985
Lorglumide	7 ng i.t.h.	Morphine	1 μg i.c.v.	Rat	Tail-flick	Kellstein and Mayer 1990
Devazepide	0.5–2.0 mg/kg s.c.	Morphine	4 mg/kg i.p.	Rat	Tail-flick	Dourish et al. 1988b
Devazepide	0.2 mg/kg i.p.	Morphine	30–300 μg/kg s.c.	Rat	Paw pressure	Rattray et al. 1988
Devazepide	1.0–2.0 mg/kg s.c.	Morphine	8 mg/kg i.p.	Rat	Paw pressure	O'Neill et al. 1989
Devazepide	3–100 μg/kg i.p.	Morphine	0.1 mg/kg s.c.	Squirrel monkey	Tail-withdrawal	Dourish et al. 1990b
Devazepide	10 μg/kg i.v.	Alfentanyl	0.25 μg/kg i.v.	Baboon	Dental dolorimetry	Chapter 50
Devazepide	0.05–0.1 mg/kg i.p.	Morphine	5 mg/kg i.p.	Mouse	Tail-flick	Hendrie et al. 1989
Devazepide	10–200 ng i.t.h.	Morphine	4 mg/kg s.c.	Rat	Tail-flick	Chapter 45
L-365,031	4–16 mg/kg s.c.	Morphine	4 mg/kg i.p.	Rat	Tail-flick	Dourish et al. 1990a
L-365,260	0.1–0.5 mg/kg s.c.	Morphine	4 mg/kg i.p.	Rat	Tail-flick	Dourish et al. 1990a
L-365,260	0.05–0.1 mg/kg s.c.	Morphine	8 mg/kg i.p.	Rat	Paw pressure	Dourish et al. 1990a
L-365,260	1–20 ng i.t.h.	Morphine	4 mg/kg s.c.	Rat	Tail-flick	Chapter 45.
PD 134308	0.1–1.0 mg/kg i.p.	Morphine	1 mg/kg i.p.	Rat	Hot-plate	Wiesenfeld-Hallin et al. 1990

DALA, (D-Ala^2)-methionine enkephalinamide; DAGOL (D-ala^2, MePhe^4, Gly(ol)^5) enkephalin; FPS, front-paw shock. See Table 44.1 for additional abbreviations.

dose–response curve (Dourish *et al.* 1990*b*). Therefore it is possible that an examination of somewhat higher doses of PD 134308 may unmask a similar bell-shaped curve.

Systemic injection of devazepide has now been shown to enhance opioid analgesia in rat, mouse, and monkey in tests that have used thermal, mechanical, and electrical stimuli (Rattray *et al.* 1988; Hendrie *et al.* 1989; O'Neill *et al.* 1989; Dourish *et al.* 1990*b*; Chapter 50 of this volume (see Table 44.2 for a summary)). Similarly, a number of other potent CCK antagonists from different chemical classes (L-365,031, L-365,260 (benzodiazepines), lorglumide (amino acid derivative), PD 134308 (peptoid)) have been shown to enhance opioid analgesia in thermal antinociceptive tests (see Table 44.2). I.t.h. injection of devazepide and L-365,260 has also been shown to enhance morphine analgesia in the rat tail-flick test (Chapter 45 of this volume). Furthermore, devazepide, L-365,031, and lorglumide were found to enhance analgesia induced by restraint stress, a response that is naloxone-reversible and presumably mediated by endogenous opioids (Dourish *et al.* 1988*c*). Surprisingly, however, devazepide was found to block rather than enhance the opioid-mediated analgesia induced by social conflict in the mouse (Hendrie *et al.* 1989). It is possible that the explanation for this apparent discrepancy may be the recently discovered anxiolytic effects of the drug (Hendrie and Dourish 1990; Ravard and Dourish 1990; Chapter 12 of this volume) as analgesia induced by social conflict in mice has been shown to be primarily under psychological rather than physiological control (Rodgers and Hendrie 1983).

Effects of CCK antagonists on opioid tolerance and dependence

It has been proposed that, in addition to enhancing the acute effects of morphine, blockade of CCK receptors or treatment with CCK antisera can prevent opioid tolerance (Tang *et al.* 1984; Watkins *et al.* 1984; Ding *et al.* 1986). It is debatable whether the effects observed in such studies reflect the prevention of tolerance or an enhancement of opioid analgesia in opioid-tolerant rats. Whichever interpretation proves to be valid, it is clear that rats which are tolerant to opioids show an analgesic response to morphine when pretreated with a CCK antagonist. The first studies to report this effect used proglumide and were published in 1984 (Tang *et al.* 1984; Watkins *et al.* 1984). The procedure used by Watkins *et al.* (1984) will be described in some detail as it has been used in a number of subsequent studies with more potent CCK antagonists. Tolerance was produced by injecting incremental doses of morphine i.p. twice daily for 6 days, beginning with a dose of 8 mg/kg on day 1 and ending with a dose of 48 mg/kg on day 6. Groups of animals were subjected to this treatment regime with or without cotreatment with proglumide at doses of 0.6 or 1.8 mg/kg i.p. In addition, other groups of animals were given appropriate control treatments. When challenged with a dose of 4 mg/kg of morphine on day 6, only a small analgesic effect was observed in animals that had been injected chronically with the drug. This tolerance to morphine analgesia was partially attenuated in the proglumide-treated groups (Watkins *et al.* 1984). Similarly, Tang *et al.* (1984) reported that proglumide prevented acute tolerance to morphine in rats induced by repeated hourly injections of the opioid. Results that are in broad agreement with the above findings were reported by Panerai *et al.* (1987), who showed an inhibition of morphine tolerance by both proglumide and benzotript when the CCK antagonists were given chronically in the drinking water of the animals. Furthermore, it was noted in this study that the CCK antagonists had no effect on opioid dependence (see below for further discussion). These three studies appear to provide evidence for the proposal that CCK receptors are involved in the development of opioid tolerance (Watkins *et al.* 1984). However, it is problematical to generalize from data obtained in studies using weak and non-selective compounds, such as proglumide, lorglumide, and benzotript, as reference CCK antagonists (as discussed above). Therefore a critical test of this hypothesis was an examination of the effects of potent and selective CCK antagonists on the response to chronic morphine treatment. The effects of three such drugs (devazepide, L-365,031, and L-365,260) have been examined to date (Dourish *et al.* 1988*b*, 1990*a*). In each case the procedure used was similar to that used by Watkins *et al.* (1984) (see above). Thus groups of rats were treated twice daily for 6 days with incremental doses of morphine with or without one of the CCK antagonists. Animals treated only with morphine became tolerant and exhibited a small analgesic response to a morphine challenge on day 6. In contrast, animals treated chronically with devazepide, L-365,031, or L-365,260 did not become tolerant and when challenged with morphine showed an analgesic response that was similar to that of rats given acute morphine (see Fig. 44.2 and Dourish *et al.* 1990*a*). To assess dependence the method of naloxone-precipitated withdrawal was used on the day after analgesia testing. Animals were given no morphine on this day, but instead were injected with 1 mg/kg of naloxone and the appearance of various characteristic withdrawal signs was noted. Neither devazepide nor L-365,260 had any effect on the frequency or intensity of the naloxone-precipitated withdrawal symptoms (Dourish *et al.* 1988*b*, 1990*a*).

Two conclusions can be drawn from the results of these studies. First, opioid tolerance may be due at least in part to a progressive compensatory increase in the activity of CCK neurons in response to chronic opioid administration. At present, however, there is no direct neurochemical evidence for this hypothesis. Second, CCK receptors do not appear to be involved in the development of opioid dependence and this conclusion is consistent with the suggestion that independent mechanisms mediate opioid tolerance and dependence (Ling *et al.* 1984). Nevertheless, a note of caution is warranted in this regard. In these studies opioid dependence was assessed using the opioid antagonist naloxone to precipitate withdrawal. It is conceivable that CCK receptors may influence opioid dependence but, in order to do so, the integrity of opioid receptors is a requirement (see below for further discussion). As opioid receptors were

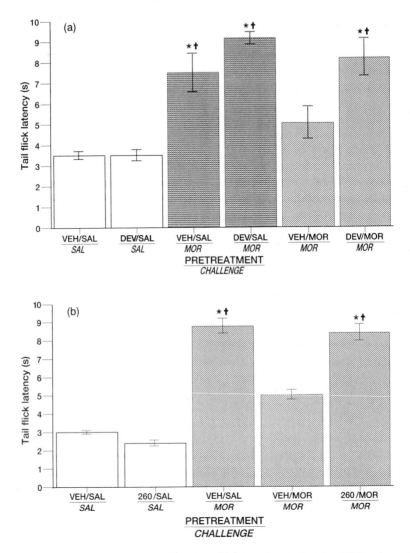

Fig. 44.2. Prevention of tolerance to morphine analgesia by the CCK antagonists (a) devazepide and (b) L-365,260 in the rat tail-flick test. Tolerance was produced over a 6 day period by twice daily injections of incremental doses of morphine starting at 8 mg/kg on day 1 and ending with 48 mg/kg on day 6. The data were recorded on the afternoon of day 6 during a 2 h period after injection of a challenge dose of 8 mg/kg of morphine. Symbols refer to statistically significant differences between the following groups: * significantly different from vehicle/saline/saline challenge; + significantly different from vehicle/morphine/morphine challenge. (Data from Dourish et al. 1988b, 1990a.)

blocked by naloxone in these studies, further experiments to examine the effects of CCK antagonists on spontaneous opioid withdrawal are necessary to test this possibility.

The role of CCK_A and CCK_B receptors in mediating CCK–opioid interactions

The existence of two subtypes of receptor for CCK was first proposed by Innis and Snyder (1980) on the basis of the differential potencies of various CCK fragments in displacing radio-labelled CCK in ligand binding studies. It was found that pentagastrin, CCK-8US, and CCK-4, the C-terminal tetrapeptide of CCK, had similar affinities to that of CCK-8S in displacing CCK binding to brain membranes, but that CCK-8S was three orders of magnitude more potent than the other fragments in displacing binding to peripheral tissues. However, although differences between brain and peripheral tissues were observed, this study did not provide evidence for CCK receptor heterogeneity in the brain. The first evidence of multiple CCK receptors in the brain was provided by the auto-radiographical studies of Moran *et al.* (1986) who discovered a limited number of brain binding sites with low affinity for CCK-8US, CCK-4, and pentagastrin that resembled those found in the periphery. These peripheral sites were termed CCK_A (alimentary) to distinguish them from the classical brain (CCK_B) receptor (Moran *et al.* 1986). The findings of Moran and colleagues and subsequent studies using the potent and selective CCK antagonists devazepide and L-365,260 as displacing agents indicated that CCK_A receptors in rodents were confined to a few regions of the brainstem (Moran *et al.* 1986; Hill *et al.* 1987; Chapter 5 of this volume). More recently, however, functional evidence has indicated that CCK_A receptors are also present in the forebrain of the rat (Vickroy and Bianchi 1989). Nevertheless, the majority of CCK receptors in the CNS of rodents are of the CCK_B subtype and these receptors are distributed ubiquitously throughout the brain and spinal cord (Moran *et al.* 1986; Hill *et al.* 1987; Chapter 5 of this volume).

The results of studies that have examined the effects of CCK agonists on opioid analgesia indicate that the opioid antagonist action of CCK may involve CCK_A receptors. Thus CCK-8S (which has high affinity for both CCK_A and CCK_B receptors) potently blocks opioid analgesia, whereas CCK-8US (which is a selective CCK_B agonist) is weak or ineffective (Faris *et al.* 1983; Faris 1985; Han *et al.* 1986; Chapter 45 of this volume). In contrast, the results of studies with CCK antagonists suggest the opposite conclusion, i.e. mediation of CCK–opioid interactions by CCK_B receptors (Dourish *et al.* 1990a). L-365,260 is a selective CCK_B antagonist (Bock *et al.* 1989; Lotti and Chang 1989; Kemp *et al.* 1989; Chapter 7 of this volume), whereas devazepide and L-365,031 are selective CCK_A antagonists (Chang and Lotti 1986; Evans *et al.* 1986, 1988). L-365,260 was 40 times more potent than L-365,031 and five times more potent than devazepide in enhancing morphine analgesia in the rat

Table 44.3. Relative potency of CCK antagonists in biochemical and pharmacological models: correlation of morphine enhancement with antagonism at CCK$_B$ receptors

Assay	Compound		
	L-365,260	Devazepide	L-365,031
CCK$_A$ binding affinity (Rat pancreas)*	6.3	9.5	9.0
CCK$_B$ binding affinity (Rat spinal cord)*	7.9	7.1	5.7
CCK$_B$ receptor blockade (Rat ventromedial hypothalamus)[†]	7.5	6.6	NT
Morphine enhancement (Rat tail-flick)[‡]	0.1	0.5	4.0
Morphine enhancement (Rat paw pressure)[‡]	0.05	1.0	NT
Prevention of morphine tolerance (Rat tail-flick)[‡]	0.2	1.0	8.0

*pIC$_{50}$ values; [†]pA$_2$ values; [‡]minimum effective dose (mg/kg s.c.); NT, not tested.

tail-flick test (Table 44.3). Similarly, L-365,260 was 20 times more potent than devazepide in enhancing morphine analgesia in the rat paw pressure test. This rank order of potency correlates with the affinities of these three CCK antagonists for CCK receptors in the rat spinal cord (Hill *et al.* 1988), but contrasts with their affinities for CCK$_A$ receptors in peripheral tissues where the rank order of potency is devazepide > L-365,031 >>> L-365,260 (Table 44.3). In addition, the opioid-enhancing properties of L-365,260 and devazepide are in accord with their relative potency to antagonize the excitatory action of CCK on CCK$_B$ receptors in slices of the rat ventromedial hypothalamus (Kemp *et al.* 1989; Chapter 50 this volume) (Table 44.3). Therefore these data indicate that CCK antagonists enhance opioid analgesia by blocking CCK$_B$ receptors. Recent studies by Wiesenfeld-Hallin *et al.* (1990) and Han (Chapter 46 of this volume) are in agreement with this conclusion. Thus the selective CCK$_B$ antagonist PD 134308 potentiated the analgesic effect of morphine in the rat hot-plate test and the depressive action of morphine on the spinal nociceptive flexor reflex (Wiesenfeld-Hallin *et al.* 1990). As PD 134308 has a very low affinity for CCK$_A$ receptors (see Chapter 5 of this volume), these data indicate CCK$_B$ receptor mediation of the response. Similarly, L-365,260 was 40 times more potent than devazepide in enhancing morphine analgesia in rats when both drugs were injected i.t.h. (Chapter 45 of this volume). Furthermore, L-365,260 was more effective than devazepide in enhancing analgesia induced by electro-acupuncture (Chapter 45 of this volume). The same conclusion has also been reached on the basis of results from electrophysiological studies. Thus the inhibition of firing of nociceptive neurons in the rat spinal cord by the μ opioid agonist (D-Ala^2, MePhe^4, Gly(ol)^5) enkephalin (DAGOL) was enhanced by the non-selective CCK$_A$ and CCK$_B$ antagonist IM 9/11 but not by the selective CCK$_A$ antagonist devazepide (Chapter 46 of this volume).

At present, then, there is a clear contradiction between the results of agonist studies (suggesting CCK$_A$ receptor involvement) and antagonist studies (implicating CCK$_B$ receptors). How might this issue be resolved? One possibility may be to examine the effects on opioid analgesia of a wider range of CCK$_A$ and CCK$_B$ agonists and a wider range of doses. To date, the only selective CCK$_B$ agonist to be tested as an opioid antagonist is CCK-8US (Faris *et al.* 1983; Faris 1985; Han *et al.* 1986; Chapter 45 of this volume). To the best of my knowledge, the effects of other selective CCK$_B$ agonists such as CCK-4, pentagastrin, and BC 264 (Chapter 6 of this volume) or the effects of selective CCK$_A$ agonists such as A-71623 (Shiosaki *et al.* 1990) have not been examined. In addition, it appears that the effects of only a narrow range of doses of CCK-8US have been probed to date (Faris *et al.* 1983; Faris 1985; Han *et al.* 1986). Given the prevalence of bell-shaped dose–response curves in CCK–opioid interactions (see above), future investigations of the effects of a wide range of doses of selective CCK$_B$ agonists on opioid analgesia may prove instructive.

Enhancement of opioid analgesia by CCK antagonists in monkeys and humans

Autoradiographic studies of the distribution and density of CCK receptor subtypes in rodents and primates have revealed profound species differences (see Chapter 5 of this volume for detailed discussion). One of the most striking differences was found in the spinal cord. Thus, CCK receptors in the spinal cord of rodents have been shown to be of the CCK$_B$ subtype, whereas CCK receptors in the spinal cord of monkeys and humans are of the CCK$_A$ subtype (Hill *et al.* 1988). It appears possible that CCK–opioid interactions rely, at least in part, on spinal nociceptive transmission (see below). If this is indeed the case, then it is uncertain whether the results from analgesia studies in rodents can be generalized to primates and humans. Consequently, the effects of CCK antagonists on opioid analgesia in primates assume increased importance. To date, the effects of the CCK$_A$ antagonist devazepide on opioid analgesia have been examined in two primate models, the squirrel monkey tail-withdrawal test (Dourish *et al.* 1990*b*) and the baboon dental dolorimetry model (Chapter 50 of this volume). In the tail-withdrawal test, squirrel monkeys were restrained in chairs and pain thresholds were measured as the latency to remove the tail from warm (55 °C) water. Morphine produced a dose-related increase in withdrawal latencies, with the minimum effective dose being 0.3 mg/kg s.c. When a subthreshold dose of morphine (0.1 mg/kg s.c.) was injected after treatment with devazepide (3–100 µg/kg i.p.), a significant analgesic response was observed (Fig. 44.3). Higher and lower doses of devazepide were ineffective, indicating the presence of a bell-shaped dose–response curve as observed in numerous previous studies of rodents (see above). In the dental dolorimetry model, baboons were restrained in chairs and pain thresholds measured as the voltage required to elicit a lick/chew response following electrical stimulation of the tooth pulp. Devazepide (10 μg/kg i.v.) enhanced both the peak analgesic effect and the duration of analgesia induced by a dose of 0.25 μg/kg i.v. of the short-acting opioid agonist alfentanyl (Chapter 50 of this volume). In both

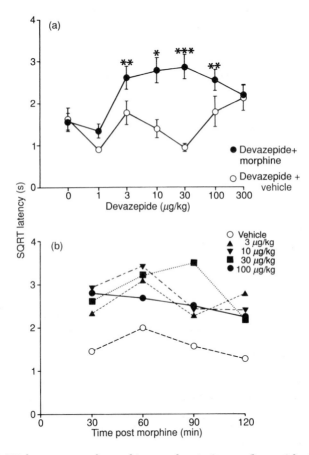

Fig. 44.3. (a) Enhancement of morphine analgesia (0.1 mg/kg s.c.) by i.p. injection of devazepide in the squirrel monkey tail-withdrawal test. Data are the mean (square root) latencies for five monkeys during a 2 h test. Vehicle-treated animals are represented by open symbols and devazepide-treated animals by solid symbols. Significant differences between devazepide/morphine and vehicle/morphine treatments are shown as follows: $\star$ $p < 0.1$, $\star\star$ $p < 0.05$, $\star\star\star$ $p < 0.01$.

(b) Time course of devazepide enhancement of morphine analgesia in the squirrel monkey tail-withdrawal test. (Reproduced from Dourish *et al.* 1990*b*.)

these models devazepide had no intrinsic effect on pain thresholds (Dourish *et al.* 1990*b*; Chapter 50 of this volume). In contrast, the selective CCK_B antagonist L-365,260 has recently been reported to possess intrinsic analgesic properties in the squirrel monkey tail-withdrawal test (O'Neill *et al.* 1990; Chapter 51 of this volume). It is unclear at present whether the intrinsic analgesia induced by L-365,260 is opioid-mediated.

It is clear from the preceding discussion that enhancement of opioid analgesia by devazepide does generalize from rodent to primate models. However, as yet there are no clinical data available with potent and selective CCK antagonists. Nevertheless, the results of two out of three clinical trials carried out with proglumide suggest that this is a promising area that is worthy of further investigation. In the first study, five groups of volunteers were injected i.v. with 0.04 or 0.06 mg/kg of morphine, with or without pretreatment with 10, 50, or 100 μg of proglumide, and their response to a thermal pain stimulus was measured using a visual analogue scale (Price *et al.* 1985). Proglumide enhanced morphine analgesia and the dose–response curve appeared to be bell-shaped, as observed in rodents (Watkins *et al.* 1984). Thus 100 μg of proglumide enhanced analgesia induced by 0.04 mg/kg of morphine but failed to enhance the analgesic action of 0.06 mg/kg of morphine (Price *et al.* 1985). However, 50 μg of proglumide plus 0.06 mg/kg of morphine was more effective than 100 μg of proglumide plus 0.04 mg/kg of morphine (Price *et al.* 1985). In the second study Lavigne *et al.* (1989) examined the effects of proglumide on morphine analgesia in subjects undergoing tooth extraction. The subjects were given 4 mg of morphine alone or in combination with 0.05, 0.5, or 5.0 mg of proglumide. A bell-shaped dose–response curve was apparent, with 0.05 mg of proglumide being observed to increase the magnitude and duration of morphine analgesia whereas the higher doses were ineffective (Lavigne *et al.* 1989). It is interesting to note that a 50 μg dose of proglumide was effective in enhancing morphine analgesia in both these studies. In contrast, the third study of Lehmann *et al.* (1989) using patient-controlled analgesia failed to find an enhancement of morphine analgesia by doses of 50 μg, 100 μg, or 50 mg of proglumide in elderly patients who had undergone major surgery.

It is difficult to draw any meaningful conclusions from the results of such a small number of studies, particularly when the data are contradictory. Furthermore, the problems inherent in using proglumide as a reference drug have already been discussed at length in this chapter. Therefore studies with potent selective CCK antagonists in human pain models, that are likely to take place during the next few years, will be required to permit a definitive test of the clinical relevance of the CCK–opioid modulation hypothesis.

The evidence from the studies discussed above suggests that CCK may play a physiological role as an opioid antagonist and that this action may be mediated by CCK_B receptors. Thus CCK attenuates analgesia induced by exogenous opioid injection or by endogenous opioid release. Conversely, blockade of CCK receptors enhances opioid analgesia and appears to prevent the development of opioid tolerance. At least three important questions are posed by these findings. First, what is the identity of the opioid receptor(s) involved in the CCK–opioid interaction? Second, how does the interaction between CCK and opioids occur at the neuronal level? Third, what is the anatomical location of this interaction? These three questions are addressed in turn below.

Involvement of multiple opioid receptors in mediating CCK–opioid interactions

The identity of the opioid receptor(s) involved in CCK–opioid interactions has been the subject of some debate. Both behavioural and electrophysiological studies of proglumide-induced enhancement of opioid analgesia by Watkins and colleagues (Watkins *et al.* 1984, 1985; Suberg and Watkins 1987) suggested the involvement of μ and/or δ receptors but not κ receptors. Similarly, Barbaz *et al.* (1989) reported that CCK attenuates analgesia induced by μ receptor agonists but has no effect on that induced by a κ agonist. In contrast, recent behavioural and receptor binding studies by Han and colleagues (Wang and Han 1990; Chapter 45 of this volume) suggest that CCK blocks analgesia induced by selective μ and κ agonists but not that induced by a δ agonist and, similarly, inhibits binding to μ and κ but not δ opioid receptors. The electrophysiological studies of Dickenson and colleagues (Chapter 46 of this volume) support the involvement of μ but not δ receptors, though the possible role of κ receptors was not addressed. At present the general consensus appears to support the involvement of μ opioid receptors in mediating CCK–opioid interactions but clearly further studies (particularly using potent and selective CCK antagonists) are needed to assess the respective roles of the other opioid receptor subtypes.

Neuronal mechanisms underlying CCK–opioid interactions: a model

In our initial conception of the mechanism of CCK–opioid interactions (Baber *et al.* 1989) we proposed the following scheme. A noxious thermal stimulus activates Aδ- and C-fibre afferents which in turn increase the firing of nociceptive neurons in the dorsal horn of the spinal cord. The Aδ- and C-fibre afferents also synapse with opioid and CCK neurons in the dorsal horn of the spinal cord. The Aδ- and C-fibre activation induced by the noxious stimulus may also increase the firing of these CCK and opioid neurons. Opioid receptors are located on nociceptive neurons and on the presynaptic terminals of Aδ and C fibres, and can be stimulated by endogenous or exogenous opioids. As a consequence of opioid receptor activation, the firing of the dorsal horn neuron is inhibited and analgesia is produced. CCK receptors are also thought to be present on nociceptive neurons and on presynaptic terminals, and we proposed that these receptors (of the CCK_B subtype in rodents) can be activated by endogenous CCK (released in response to pain, stress, or opioid injection) or exogenous CCK injection. The consequence of CCK receptor activation would be an increase in the firing of the dorsal horn neuron and an increase in pain transmission. It is proposed that under normal circumstances a dynamic balance exists, but that in response to acute pain the effects of opioid receptor activation overcome the effects of CCK receptor activation and analgesia ensues. However, injection of CCK may shift this balance towards dorsal horn neuronal excitation and opioid analgesia is blocked. Conversely, the addition of a CCK antagonist shifts the balance in the opposite direction and opioid analgesia is enhanced. There is electrophysiological data to support such a hypothesis. Thus i.t.h. application of CCK increases the firing of nociceptive neurons in the rat spinal cord in response to C-fibre stimulation (Chapter 46 of this volume). Furthermore, CCK prevented the inhibition of C-fibre-evoked neuronal firing induced by application of morphine or the selective μ opioid agonist DAGOL (Chapters 43 and 46 of this volume). In contrast, CCK had no effect on the inhibition of neuronal firing induced by the δ opioid agonist DSTBULET, suggesting that the blockade of opioid transmission in the spinal cord is specific for μ opioid receptors (Chapter 46 of this volume) (see above). The converse effect is observed on dorsal horn neurons when a CCK antagonist is applied together with an opioid. The first report of this phenomenon was by Suberg *et al.* (1985) who observed that proglumide enhanced the morphine-induced suppression of firing of dorsal horn neurons in response to noxious radiant heat. Similar findings have been reported recently with lorglumide (Chapter 43 of this volume) and IM 9/11 (Chapter 46 of this volume).

One prediction of the Baber *et al.* (1989) model was that CCK increases the firing of the dorsal horn neuron and, in the absence of opioids, hyperalgesia would be observed. However, as pointed out by Han (Chapter 45 of this volume) hyperalgesia is not observed after CCK injection (see above). This suggests that the functional output of the CCK receptor may depend upon a direct or indirect interaction with the opioid receptor. Nevertheless, until recently, there was no evidence for modulation of opioid receptor function by CCK. Studies by Han and colleagues (Wang *et al.* 1989; Wang and Han 1990) (see Chapter 45 of this volume for discussion) now suggest that CCK may inhibit binding to μ and κ but not δ opioid receptors. This action has been claimed to be mediated by CCK receptors, as it was prevented by proglumide. However, as proglumide has an equally high affinity for μ and κ opioid receptors as for CCK receptors (Gaudreau *et al.* 1990), further studies using selective CCK antagonists are necessary to determine whether this effect of CCK is due to a direct action of proglumide on opioid receptors or to an allosteric modulation of opioid receptors by CCK receptors. If the latter hypothesis were to be supported by sufficient additional evidence from studies with selective CCK antagonists, then a revision of the Baber *et al.* (1989) model would be necessary. A revised version of this model might suggest that activation of the CCK receptor by exogenous or endogenous CCK blocks opioid analgesia by a receptor–receptor interaction and that the CCK receptor and the opioid receptor are located on the same receptor protein (see Chapter 45 of this volume). However, this hypothesis implies that the two receptors would be linked to the same ion channel or second messenger system. This does not appear to be the case as there is evidence that stimulation of μ opioid receptors activates outward potassium conductance by a mechanism that is sensitive to pertussis toxin (North *et al.* 1987), whereas CCK receptor stimulation inhibits outward

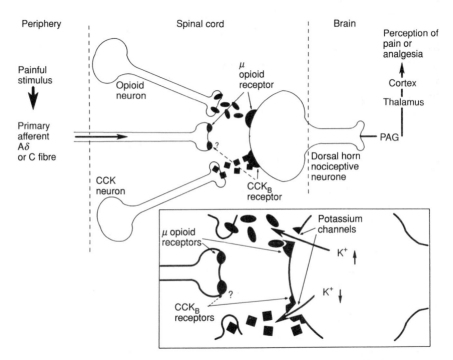

Fig. 44.4. Schematic illustration of the neuronal mechanisms underlying CCK–opioid interactions in the production of pain and analgesia. A noxious stimulus activates Aδ- and C-fibre afferents which, in turn, increase the firing of nociceptive neurons and CCK and opioid interneurons in the dorsal horn of the spinal cord. Opioid and CCK$_B$ receptors are present on nociceptive neurons and presynaptic terminals. It is proposed that the outcome of CCK–opioid interactions in nociception is determined by receptor-mediated alterations in potassium conductance of the membranes of the nociceptive neurons. It is hypothesized that CCK$_B$ receptor activation blocks opioid analgesia by opposing opioid activation of outward potassium conductance (induced by μ receptor stimulation) via the inhibition of outward potassium conductance (induced by CCK$_B$ receptor stimulation). It is proposed that the CCK$_B$ receptor and the μ opioid receptor are linked to separate potassium channels located on the same nociceptive neuron and thereby maintain a mutually interdependent dynamic balance in the control of pain and analgesia (see text for full details).

potassium conductance (Hill and Boden 1989) by a mechanism that is insensitive to pertussis toxin (Chapter 7 of this volume). A unifying hypothesis which accommodates all these data is as follows: CCK may block opioid analgesia by opposing the activating action of opioids on outward potassium conductance via the inhibition of outward potassium conductance at a different ion channel located on the same neuron. This model is illustrated schematically in Fig. 44.4.

Anatomical location of CCK–opioid interactions

The anatomical location of the interaction between CCK and opioids has also been debated. One proposal is that the interaction occurs in the spinal cord as illustrated in Fig. 44.4. However, there is also convincing evidence for an involvement of supraspinal mechanisms, particularly in the PAG and thalamus. The early studies by Faris and colleagues pointed towards a spinal site of action as CCK was effective in blocking opioid analgesia in rats when injected i.t.h. Similarly, recent studies in mice by Suh and Tseng (1990) implicate spinal mechanisms, as i.t.h. injection of CCK and proglumide attenuated and enhanced respectively analgesia induced by i.c.v. injection of β-endorphin in the tail-flick test, which is thought to be spinally mediated, but had no effect on responses to the opioid in the hot-plate test, which is thought to be supraspinally mediated. In contrast, it has been reported that i.c.v. injection of CCK is effective in blocking analgesia induced by β-endorphin in the rat hot-plate test (Itoh *et al.* 1982). This is in accordance with recent studies by Han and colleagues (Li and Han 1989; Chapter 45 of this volume) which suggest that CCK is significantly more potent in blocking opioid analgesia when injected i.c.v. or into the PAG than when injected i.t.h. Similarly, proglumide potentiates morphine analgesia when injected i.t.h. or into the PAG (Watkins *et al.* 1984, 1985). Furthermore, systemic injection of proglumide enhances the analgesia induced by i.c.v. injection of the μ receptor agonist DAGOL, but has no effect on the analgesia induced by i.t.h. infusion of the drug (Bodnar *et al.* 1990). Additional evidence for supraspinal involvement in CCK–opioid interactions was obtained in primate analgesia studies. Devazepide was approximately 16-fold more potent in enhancing morphine analgesia in the squirrel monkey (minimum effective dose 3 μg/kg) than in the rodent (minimum effective dose 50 μg/kg) when injected in the same vehicle (Hendrie *et al.* 1989; Dourish, unpublished results). This relatively small difference in potency of the drug between primate and rodent argues against an important role of spinal CCK receptors in mediating CCK–opioid interactions. Thus, if devazepide-induced enhancement of opioid analgesia were mediated by spinal CCK receptors, it was predicted that the drug would be at least 300-fold more potent in the primate as it is 300–500-fold more potent in displacing CCK binding to the spinal cord of monkeys and humans than to that of rodents (Hill *et al.* 1988).

Electrophysiological and neurochemical evidence has also implicated both spinal and supraspinal mechanisms in mediating CCK–opioid interactions. The depressive effect of i.t.h. morphine injection on the spinal nociceptive flexor reflex was blocked by CCK and potentiated by the CCK antagonist PD 134308 (Wiesenfeld-Hallin and Duranti 1987; Wiesenfeld-Hallin *et al.* 1990). This finding, together with the observation of a similar result with the antagonist in spinalized rats (Wiesenfeld-Hallin *et al.* 1990), argues strongly for a direct spinal interaction. Similarly, a spinal site of action is implicated by the results of Dickenson *et al.* (Chapter 46 of this volume) who show that i.t.h. application of CCK prevents the inhibition of dorsal horn neurons by the μ receptor agonist DAGOL. Furthermore, there is evidence that opioids can increase (Tang *et al.* 1984; Chapter 47 of this volume) or decrease (Rodriguez and Sacristan 1989; Chapter 47 of this volume) the release of CCK in the spinal cord. Suh and Tseng (1990) have proposed that i.t.h. injected CCK antagonizes the analgesic effect of i.c.v. injected β-endorphin by reducing the release of (Met^5)enkephalin from the spinal cord. Thus it was observed that the increased release of (Met^5)enkephalin induced by β-endorphin was attenuated when CCK was added to the spinal perfusate (unpublished observations cited by Suh and Tseng (1990)). Finally, and in contrast, there is evidence that CCK can block the effects of morphine or electroacupuncture on nociceptive neurons in the thalamus (see Chapter 45 of this volume for discussion).

Thus there is no clear consensus regarding a single site of action of CCK–opioid interactions. Indeed, the evidence appears to suggest that there may be multiple sites of interaction, spinally and supraspinally. This is not unexpected, given that the analgesic action of opioids is thought to involve both spinal and supraspinal components. Thus when morphine is injected simultaneously i.t.h. and into the brainstem a much lower total dose of the drug is required for analgesia indicating that there is a multiplicative interaction (Yeung and Rudy 1980). Future studies using combined i.t.h. and PAG injections of CCK with opioids may help to define the interaction. Furthermore, examination of the possible role of the thalamus and of cortical regions (e.g. somatosensory cortex, frontal cortex) in mediating CCK–opioid interactions in nociception has been neglected to date. There are high densities of both opioid and CCK receptors and high concentrations of opioid peptides and CCK in cortical regions, and it is possible that these sites may play an important role in the nociceptive interactions observed.

Clinical potential of CCK–opioid interactions

At present the clinical potential for CCK antagonists as analgesics or as adjuncts to opioid analgesics remains speculative. However, there are a number of possibilities. First, the use of a CCK antagonist may allow lower doses of opioids to be administered to relieve pain, possibly reducing the risk of side-effects. This potential use is supported by the finding that devazepide potentiates morphine analgesia but does not potentiate two of the major dose-limiting side-effects of the drug, respiratory depression and constipation (Dourish *et al.* 1990*b*). Second, it is possible that a CCK antagonist could be used to facilitate analgesia mediated by endogenous opioids such as that induced by post-operative stress or acupuncture. The combination of a CCK antagonist with acupuncture may be of particular interest as it has recently been shown that animals that do not respond to acupuncture alone show pronounced analgesia when it is combined with devazepide or L-365,260 (Chapter 46 of this volume). These data also suggest the possibility that patients who are opioid non-responders may become responsive to opioids when treated with a CCK antagonist. Finally, it is possible that CCK antagonists may be of value in the prevention of opioid tolerance in patients with intractable pain (e.g. cancer pain), although it should be noted that this use may be compromised by the compensatory changes in CCK–opioid interactions that have been observed following chronic i.t.h. infusion of the non-selective CCK antagonist lorglumide (see Chapter 43 of this volume). Further studies in animals and humans with potent selective CCK antagonists such as L-365,260 are likely to resolve this issue during the next few years.

Acknowledgements

I am grateful to Dr J. A. Kemp and Dr R. E. Rodriguez for valuable discussions regarding the model of CCK–opioid interactions proposed in this chapter and to A. Butler for the artwork in the illustration of the model shown in Fig. 44.4.

References

Baber, N. S., Dourish, C. T., and Hill, D. R. (1989). The role of CCK, caerulein and CCK antagonists in nociception. *Pain* **39**, 307-28.

Barbaz, B. S., Hall, N. R., and Liebman, J. M. (1989). Antagonism of morphine analgesia by CCK-8s does not extend to all assays nor to all opiate analgesics. *Peptides* **9**, 1295-1300.

Bock, M. G., DiPardo, R. M., Evans, B. E., Rittle, K. E., Whitter, W. L., Veber, D. F., Anderson, P. S., and Freidinger, R. M. (1989). Benzodiazepine gastrin and cholecystokinin receptor ligands: L-365,260. *J. Med. Chem.* **32**, 13-16.

Bodnar, R. J., Paul, D., and Pasternak, G. W. (1990). Proglumide selectively potentiates supraspinal μ_1 opioid analgesia in mice. *Neuropharmacology* **29**, 507-10.

Chang, R. S. L. and Lotti, V. J. (1986). Biochemical and pharmacological characterization of an extremely potent and selective non-peptide cholecystokinin antagonist. *Proc. Natl Acad. Sci. USA* **83**, 4923-6.

Ding, X. Z., Fan, S. G., Zhou, Z. F., and Han, J. S. (1986). Reversal of morphine tolerance but no potentiation of morphine analgesia by antiserum against cholecystokinin octapeptide. *Neuropharmacology* **25**, 1155-60.

Dourish, C. T. and Hill, D. R. (1987). Classification and function of CCK receptors. *Trends Pharmacol. Sci.* **8**, 207-8.

Dourish, C. T., Coughlan, J., Hawley, D., Clark, M. L., and Iversen, S. D. (1988*a*).

Blockade of CCK-induced hypophagia and prevention of morphine tolerance by the CCK antagonist L-364,718. In *Neurology and neurobiology*, Vol. 47, *Cholecystokinin antagonists* (ed. R. Y. Wang and R. Schoenfeld), pp. 307–25. Alan R. Liss, New York.

Dourish, C. T., Hawley, D., and Iversen, S. D. (1988*b*). Enhancement of morphine analgesia and prevention of morphine tolerance in the rat by the cholecystokinin antagonist L-364,718. *Eur. J. Pharmacol.* **147**, 469–72.

Dourish, C. T., Clark, M. L., and Iversen, S. D. (1988*c*). Analgesia induced by restraint stress is attenuated by CCK and enhanced by CCK antagonists. *Soc. Neurosci. Abstr.* **14**, 290.

Dourish, C. T., Rycroft, W., and Iversen, S. D. (1988). Postponement of satiety by blockade of brain cholecystokinin (CCK-B) receptors. *Science* **245**, 1509–11.

Dourish, C. T., O'Neill, M. F., Coughlan, J., Kitchener, S., Hawley, D., and Iversen, S. D. (1990*a*). The selective CCK-B receptor antagonist L-365,260 enhances morphine analgesia and prevents morphine tolerance in the rat. *Eur. J. Pharmacol.* **176**, 35–44.

Dourish, C. T., O'Neill, M. F., Schaffer, L. W., Siegl, P. K. S., and Iversen, S. D. (1990*b*). The cholecystokinin receptor antagonist devazepide enhances morphine-induced analgesia but not morphine-induced respiratory depression in the squirrel monkey. *J. Pharmacol. Exp. Ther.* **255**, 1158–65.

Evans, B. E., Bock, M. G., Rittle, K. E., DiPardo, R. M., Whitter, W. L., Veber, D. F., Anderson, P. S., and Freidinger, R. F. (1986). Design of potent, orally effective, non-peptidal antagonists of the peptide hormone cholecystokinin. *Proc. Natl Acad. Sci. USA* **83**, 4918–22.

Evans, B. E., Rittle, K. E., Bock, M. G., DiPardo, R. M., Freidinger, R. F., Whitter, W. L., Lundell, G. F., Veber, D. F., Anderson, P. S., Chang, R. S. L., Lotti, V. J., Cerino, D. J., Chen, T. B., Kling, P. J., Kunkel, K. A., Springer, J. P., and Hirschfield, J. (1988). Methods for drug discovery, development of potent, selective, orally effective cholecystokinin antagonists. *J. Med. Chem.* **31**, 2235–46.

Faris, P. L. (1985). Opiate antagonistic function of cholecystokinin in analgesia and energy balance systems. *Ann. NY Acad. Sci.* **448**, 437–47.

Faris, P. L., Komisaruk, B. R., Watkins, L. R., and Mayer, D. J. (1983). Evidence for the neuropeptide cholecystokinin as an antagonist of opiate analgesia. *Science* **219**, 310–12.

Faris, P. L., McLaughlin, C. R., Baile, C. A., Olney, J. W., and Komisaruk, B. R. (1984). Morphine analgesia potentiated but tolerance not affected by active immunization against cholecystokinin. *Science* **226**, 1215–17.

Fog, R. (1970). Behavioural effects in rats of morphine and amphetamine and a combination of the two drugs. *Psychopharmacologia* **16**, 305–12.

Gall, C., Lauterhorn, J., Burks, D., and Seroogy, K. (1987). Colocalization of enkephalin and cholecystokinin in discrete areas of rat brain. *Brain Res.* **403**, 403–8.

Gaudreau, P., Lavigne, G. J., and Quirion, R. (1990). Cholecystokinin antagonists proglumide, lorglumide and benzotript, but not L-364,718, interact with brain opioid binding sites. *Neuropeptides* **16**, 51–5.

Han, J. S., Ding, X. Z., and Fan, S. G. (1985). Is cholecystokinin octapeptide (CCK-8) a candidate for endogenous anti-opioid substrate? *Neuropeptides* **5**, 399–402.

Han, J. S., Ding, X. Z., and Fan, S. G. (1986). Cholecystokinin octapeptide (CCK-8): antagonism to electroacupuncture analgesia and a possible role in electroacupuncture tolerance. *Pain* **27**, 101–15.

Hendrie, C. A. and Dourish, C. T. (1990). Anxiolytic profile of the cholecystokinin antagonist devazepide in mice. *Br. J. Pharmacol.* **99**, 138P.

Hendrie, C. A., Shepherd, J. K., and Rodgers, R. J. (1989). Differential effects of the CCK antagonist devazepide on analgesia induced by morphine, social conflict (opioid) and defeat experience (non-opioid) in male mice. *Neuropharmacology* **28**, 1025–32.

Hill, R. G. and Boden, P. (1989). Electrophysiological actions of CCK in the mammalian central nervous system. In *The neuropeptide cholecystokinin (CCK)* (ed. J. Hughes, G. Dockray, and G. Woodruff) pp. 186–94. Ellis Horwood, Chichester.

Hill, D. R., Shaw, T. M., and Woodruff, G. N. (1988). Binding sites for ^{125}I-cholecystokinin in primate spinal cord are of the CCK-A subclass. *Neurosci. Lett.* **89**, 133–9.

Innis, R. B. and Snyder, S. H. (1980). Distinct cholecystokinin receptors in brain and pancreas. *Proc. Natl Acad. Sci. USA* **77**, 6917–21.

Itoh, S., Katsuura, G., and Maeda, Y. (1982). Caerulein and cholecystokinin suppress β-endorphin-induced analgesia in the rat. *Eur. J. Pharmacol.* **80**, 421–5.

Jurna, I. and Zetler, G. (1981). Antinociceptive effect of centrally administered caerulein and cholecystokinin (CCK). *Eur. J. Pharmacol.* **73**, 323–31.

Kellstein, D. E. and Mayer, D. J. (1990). Chronic administration of cholecystokinin antagonists reverses the enhancement of spinal morphine analgesia induced by acute pretreatment. *Brain Res.* **516**, 263–70.

Kemp, J. A., Marshall, G. R., and Woodruff, G. N. (1989). Antagonism of cholecystokinin-induced excitation of ventromedial hypothalamic neurones by a new selective CCK-B receptor antagonist. *Br. J. Pharmacol.* **98**, 630P.

Lavigne, G. J., Hargreaves, K. M., Schmidt, R. N., and Dionne, R. A. (1989). Proglumide potentiates morphine analgesia for acute postsurgical pain. *Clin. Pharmacol. Ther.* **45**, 666–73.

Lehmann, K. A., Schlusener, M., and Arabatsis, P. (1989). Failure of proglumide, a cholecystokinin antagonist, to potentiate clinical morphine analgesia. *Anesth. Analg.* **68**, 51–6.

Li, Y. and Han, J. S. (1989). Cholecystokinin octapeptide antagonizes morphine analgesia in periaqueductal gray of the rat. *Brain Res.* **480**, 105–10.

Ling, G. S. F., MacLeod, J. M., Lee, S., Lockhart, S. H., and Pasternak, G. W. (1984). Separation of morphine analgesia from physical dependence. *Science* **226**, 462–4.

Lotti, V. J. and Chang, R. S. L. (1989). A new potent and selective non-peptide gastrin antagonist and brain cholecystokinin receptor (CCK-B) ligand: L-365,260. *Eur. J. Pharmacol.* **163**, 273–80.

Moran, T. H., Robinson, P. H., Goldrich, M. S., and McHugh, P. R. (1986). Two brain cholecystokinin receptors: implications for behavioural actions. *Brain Res.* **362**, 175–9.

North, R. A., Williams, J. T., Suprenant, A.-M., and Christie, M. J. (1987). μ and δ receptors belong to a family of receptors that are coupled to potassium channels. *Proc. Natl Acad. Sci. USA* **84**, 5487–91.

O'Neill, M. F., Dourish, C. T., and Iversen, S. D. (1989). Morphine-induced analgesia in the rat paw pressure test is attenuated by CCK and enhanced by the CCK antagonist MK-329. *Neuropharmacology* **28**, 243–7.

O'Neill, M. F., Dourish, C. T., Tye, S., and Iversen, S. D. (1990). Blockade of CCK-B receptors by L-365,260 induces analgesia in the squirrel monkey. *Brain Res.* **534**, 287–90.

Panerai, A. E., Rovati, L. C., Cocco, E., Sacerdote, P., and Mantegazza, P. (1987).

Dissociation of tolerance and dependence to morphine: a possible role for cholecystokinin. *Brain Res.* **410**, 52–60.

Price, D. D., Vondergruen, A., Miller, J., Rafii, A., and Price, C. (1985). Potentiation of systemic morphine analgesia in humans by proglumide, a cholecystokinin antagonist. *Anesth. Analg.* **64**, 801–6.

Rattray, M., Jordan, C. C., and DeBelleroche, J. (1988). The novel CCK antagonist L-364,718 abolishes caerulein, but potentiates morphine-induced antinociception. *Eur. J. Pharmacol.* **152**, 163–6.

Ravard, S. and Dourish, C. T. (1990). Cholecystokinin and anxiety. *Trends Pharmacol. Sci.* **11**, 271–3.

Rezvani, A., Stokes, K. B., Rhoads, D. L., and Way, E. L. (1987). Proglumide exhibits delta opioid agonist properties. *Alcohol Drug Res.* **7**, 135.

Rodgers, R. J. and Hendrie, C. A. (1983). Social conflict activates status-dependent analgesic and hyperalgesic mechanisms in male mice: effects of naloxone on nociception and behaviour. *Physiol. Behav.* **30**, 775–80.

Rodriguez, R. E. and Sacristan, M. P. (1989). *In vivo* release of CCK-8 from the dorsal horn of the rat: inhibition by DAGOL. *FEBS Lett.* **250**, 215–17.

Shiosaki, K., Lin, C. W., Kopecka, H., Craig, R., Wagenaar, Bianchi, B., Miller, T., Witte, D., and Nazdan, A. M. (1990). Development of CCK-tetrapeptide analogues as potent and selective CCK-A agonists. *J. Med. Chem.* **33**, 2950–2.

Steinman, J. L., Faris, P. L., and Olney, J. W. (1986). Further evidence for vagal involvement in the peripheral peptidergic modulation of nociception. *Soc. Neurosci. Abstr.* **12**, 1490.

Suberg, S. N. and Watkins, L. R. (1987). Interaction of cholecystokinin and opioids in pain modulation. In *Pain and headache*, Vol. 9, pp. 247–65. Karger, Basel.

Suberg, S. N., Culhane, E. S., Carstens, E., and Watkins, L. R. (1985). The potentiation of morphine-induced inhibition of spinal transmission by proglumide, a putative cholecystokinin antagonist. *Ann. NY Acad. Sci.* **448**, 660–2.

Suh, H. H. and Tseng, L. F. (1990). Differential effects of sulfated cholecystokinin octapeptide and proglumide injected intrathecally on antinociception induced by β-endorphin and morphine administered intracerebroventricularly in mice. *Eur. J. Pharmacol.* **179**, 329–38.

Tang, J., Chou, J., Iadarola, M., Yang, H. Y. T., and Costa, E. (1984). Proglumide prevents and curtails acute tolerance to morphine in rats. *Neuropharmacology* **23**, 715–18.

Vickroy, T. W. and Bianchi, B. R. (1989). Pharmacological and mechanistic studies of cholecystokinin-facilitated [^{3}H]dopamine efflux from rat nucleus accumbens. *Neuropeptides* **13**, 43–50.

Wang, X. J. and Han, J. S. (1990). Modification by cholecystokinin octapeptide of the binding of μ, δ and k-opioid receptors. *J. Neurochem.* **55**, 1379–82.

Wang, X. J., Fan, S.-G., Ren, M. F. and Han, J. S. (1989). Cholecystokinin octapeptide suppressed [^{3}H]etorphine binding to rat brain opiate receptor. *Life Sci.* **45**, 117–123.

Watkins, L. R., Cobelli, D. A., Faris, P. L., Aceto, M. D., and Mayer, D. J. (1982). Opiate vs. non-opiate footshock-induced analgesia (FSIA): the body region shocked is a critical factor. *Brain Res.* **242**, 299–308.

Watkins, L. R., Kinscheck, I. B., and Mayer, D. J. (1984). Potentiation of opiate analgesia and apparent reversal of morphine tolerance by proglumide. *Science* **224**, 395–6.

Watkins, L. R., Kinscheck, I. B., Kaufman, E. F. S., Miller, J., Frank, H., and Mayer,

D. J. (1985). Cholecystokinin antagonists selectively potentiate analgesia induced by endogenous opiates. *Brain Res.* **327**, 181–90.

Wiesenfeld-Hallin, Z. and Duranti, R. (1987). Intrathecal cholecystokinin interacts with morphine but not substance P in modulating the nociceptive flexion reflex in the rat. *Peptides* **8**, 153–8.

Wiesenfeld-Hallin, Z., Xu, X.-J., Hughes, J., Horwell, D. C., and Hokfelt, T. (1990). PD 134308, a selective antagonist of cholecystokinin type B receptor, enhances the analgesic effect of morphine and synergistically interacts with intrathecal galanin to depress spinal nociceptive reflexes. *Proc. Natl Acad. Sci. USA* **87**, 7105–9.

Yeung, J. C. and Rudy, T. A. (1980). Multiplicative interaction between narcotic agonisms expressed at spinal and supraspinal sites of antinociceptive action as revealed by concurrent intrathecal and intracerebroventricular injections of morphine. *J. Pharmacol. Exp. Ther.* **215**, 633–42.

Zetler, G. (1980). Analgesia and ptosis caused by caerulein and cholecystokinin octapeptide (CCK). *Neuropharmacology* **19**, 415–22.

Zhu, X.-G., Greeley, G. H., Jr., Lewis, B. G., Lilja, P., and Thompson, J. C. (1986). Blood–CSF barrier to CCK and effect of centrally administered bombesin on release of brain CCK. *J. Neurosci. Res.* **15**, 393–403.

45. The role of CCK in electro-acupuncture analgesia and electro-acupuncture tolerance

J. S. Han

Introduction

From the functional point of view, one of the most important characteristics of neuropeptides compared with classical neurotransmitters appears to be the extreme complexity and plasticity of their role in signal transmission. A bell-shaped dose–response curve is very often encountered, yet the physiological implication of the paradoxical effects induced by the same peptide at different doses may not be clear at first sight.

With regard to the physiological effect of CCK* on nociception, opposite results have been reported: it is antagonistic to opioid analgesia in low doses (Itoh *et al*. 1982; Faris *et al*. 1983; Han *et al*. 1985), and analgesic when administered alone at high doses (Jurna and Zetler 1981; Hill *et al*. 1987*b*; Hong and Takemori 1989). Common sense suggests that the effect induced by threshold doses is more likely to be physiological in nature, and that of submaximal doses to be pharmacological. However, in order to document the physiological function of a peptide clearly, it is necessary to observe not only changes accompanying its accelerated release, but also (and even more importantly) changes resulting from the blockage of its function by a receptor antagonist or by antibodies against it (Han 1987).

Electro-acupuncture (EA) has been regarded as one of the physiological approaches to the release of endogenous opioids in the CNS (Han and Terenius 1982). If CCK antagonizes the analgesia elicited by exogenous morphine, it should also be expected to antagonize the analgesia elicited by endogenous opioids as a result of EA stimulation. Moreover, the possibility that endogenous CCK is involved as an anti-opioid to suppress EA analgesia (Han *et al*. 1980) has also been seriously considered. The main aim of the present research was to study the interaction between endogenously released opioids and endogenously released CCK. The interaction between exogenously administered opioids and CCK were observed in parallel for the purpose of comparison.

---

* Unless otherwise indicated, in this chapter the abbreviation CCK denotes the sulphated octapeptide (CCK-8S) and the abbreviation CCK-US denotes the unsulphated octapeptide (CCK-8US).

Antagonistic effect of CCK on morphine analgesia

In order to establish clearly the central effect of CCK on pain and analgesia it is necessary to establish a reliable dose-response relationship, either analgesic or anti-analgesic. Some of the results obtained in this laboratory using rats and rabbits are listed in Table 45.1.

It is obvious from Table 45.1 and Fig. 45.1(b) that the centrally administered CCK dose causing a 50 per cent suppression of morphine analgesia (ID_{50}) is in the nanogram range: 1.7 ng for i.c.v. injection, and 3.1 ng for i.t.h. injection. The i.c.v. route seems to be more effective than the i.t.h. route. For intra-periaqueductul grey (intra-PAG) injection, the dose needed (0.22 ng) was only 13 per cent of that used for i.c.v. injection, suggesting that the PAG is one of the principal sites for CCK to suppress opioid analgesia. In rabbit PAG the ID_{50} (2.47 ng) was about 10 times that of the rat (0.22 ng), which is roughly proportional to the ratio of the brain weight of rat (2 g) to that of rabbit (20 g).

It is interesting to note that CCK itself at doses equal to or higher than the ID_{50} mentioned above (rat, 4 ng i.c.v., 1 ng PAG, 4 ng–1 μg i.t.h.; rabbit, 3 ng PAG) produced no substantial changes in tail-flick latency (TFL) in rats (Han *et al.* 1986; Li and Han 1989; Wang *et al.* 1990) or escape response latency

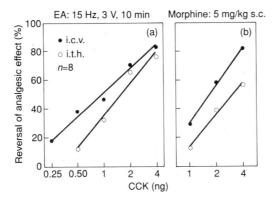

Fig. 45.1. Reversal of (a) EA analgesia and (b) morphine analgesia by i.c.v. or i.t.h. injection of CCK.

(a) Immediately before the EA stimulation, half the rats were injected with CCK and the other half were injected with CCK-US as control. The rats were changed over after 3–4 days. The percentage reversal of EA analgesia by CCK could be calculated from the ratio between the effect of EA analgesia (measured by changes in tail-flick latency) in the CCK and CCK-US groups.

(b) Morphine was injected subcutaneously (s.c.) at 5 mg/kg and CCK (or CCK-US as control) was injected i.c.v. or i.t.h. 25 min after the morphine administration. The effect of morphine analgesia measured 5 min before and 5 min after the CCK administration was used to calculate the percentage reversal of the morphine effect. (Reproduced with permission from Han *et al.* 1985.)

Table 45.1 CCK dose for 50 per cent inhibition of morphine analgesia

Administration route	Animal species	Dose range (ng)	ID$_{50}$ (ng)	Morphine dose (mg/kg)	Nociceptive test	Reference
I.c.v.	Rat	1.0–4.0	1.7	5 (s.c.)	Tail flick[a]	Han et al. 1985
I.t.h.	Rat	1.0–4.0	3.1	5 (s.c.)	Tail flick	Han et al. 1985
PAG	Rat	0.25–1.0	0.22	5 (s.c.)	Tail flick	Li and Han 1989
PAG	Rabbit	1.5–6.0	2.47	4 (i.v.)	Head jerk[b]	Cao et al. 1989

The i.c.v. injection was performed through an indwelling cannula with an injection volume of 10 μl. The i.t.h. injection was performed according to Yaksh and Rudy (1976) with an injection volume of 10 μl. [a] Tail-flick response was induced by the application of noxious heat to the lower third of the tail with a basal latency of 4–6 s and a cut-off limit of +150 per cent over the baseline level. [b] The nociceptive test used in rabbit was the escape response (head jerk) evoked by application of heat to the snout with a baseline latency of 4–6 s and a cut-off limit of +200 per cent over the baseline level.

(ERL) in rabbits (Cao *et al.* 1989). This implies that the antagonistic effect of CCK on opioid analgesia is not the result of a balance between opioid-induced hypo-algesia and CCK-induced hyperalgesia (see Fig. 45.5 below). It also implies that the tonic release of opioid peptides is very low, so that neither naloxone (Stacher *et al.* 1988) nor CCK causes substantial hyperalgesia.

The CCK antagonist proglumide was used as a pharmacological tool for three purposes (Li and Han 1989).

1. The anti-opioid effect of CCK (0.5 ng in PAG) was completely reversed by proglumide (0.2 μg in PAG), suggesting that the effect of CCK is mediated by a CCK receptor rather than acting directly on an opioid receptor.

2. Compared with the high efficacy in reversing the effect of exogenous CCK, the same dose of proglumide (0.2 μg PAG) produced little effect on baseline TFL, suggesting that the tonic release of CCK might be very low.

3. Proglumide (0.2 μg PAG) caused moderate (40–50 per cent) potentiation of morphine (2 mg/kg s.c.) analgesia, indicating that acute morphine produced a moderate increase in CCK release in rat PAG.

These results are in line with those reported by Watkins *et al.* (1985) who injected both morphine and proglumide into the same site in the PAG.

In summary, nanogram doses of CCK administered centrally do not affect the basal nociceptive threshold, yet have a powerful antagonistic effect on morphine-induced analgesia in rat and rabbit.

Reversal of morphine tolerance by CCK antiserum

In addition to the receptor antagonists, CCK antiserum (CCK-AS) has been used as a powerful tool to block the action of CCK, i.e. to sequester the endogenously released CCK and prevent it from acting on the receptors. CCK-AS injected into rabbit PAG produced only a mild and transient (10 min) antinociception in the naïve animal (Cao *et al.* 1989), suggesting that the baseline release of CCK is very low. However, the same dose of CCK-AS produced a marked potentiation of the analgesic effect induced by morphine (2 mg/kg i.v.), suggesting that exogenous morphine accelerated the release of CCK in the PAG to counteract the opioid effect.

The most profound release of CCK seems to occur in animals receiving repeated morphine administration. After repeated administration (5–30 mg/kg 3 times per day) for 6 days, the challenge dose of morphine (5 mg/kg s.c.) was no longer effective in raising the TFL of the rat. A dramatic (more than 50 per cent) restoration of the analgesic effect of morphine occurred after the i.c.v. injection of 2 μl of CCK-AS. The same dose of CCK-AS when injected i.t.h. caused a 25 per cent restoration of the morphine effect (Ding *et al.* 1986). Moreover, rats rendered tolerant to morphine showed a reduced response to EA. This cross-tolerance to EA analgesia was also reversed by CCK-AS administered i.c.v. or i.t.h. Again, i.c.v. injection produced a much stronger effect (more than 75 per cent reversal) than i.t.h. injection (40 per cent reversal) (Ding *et al.* 1986).

At first sight, these results seem to conflict with those of Faris *et al.* (1984) who found that morphine tolerance was not affected by active immunization against CCK. However, both results are reasonable if the difficulties that the antibodies encounter in crossing from the blood stream to the CNS are taken into account.

Although CCK-AS given i.c.v. was very effective in restoring morphine analgesia in morphine-tolerant rats, it was completely ineffective in potentiating morphine analgesia in non-tolerant rats (Ding *et al.* 1986), suggesting that CCK in the CNS was released only after repeated morphine exposure, but not after a single injection. However, this result contrasts with the finding that morphine analgesia can be potentiated by CCK-AS injected into the PAG of rat (Li and Han 1989) or rabbit (Cao *et al.* 1989). Although no satisfactory explanation for this can be provided at present, it nevertheless stresses the importance of PAG for the modulatory effect of CCK on opioid analgesia.

In summary, ample evidence supports the idea that the decrease in morphine's antinociceptive effect after prolonged administration is due, at least in part, to the overproduction and release of CCK in CNS. Whether central CCK can be activated by a single injection of morphine remains controversial.

Antagonism of electro-acupuncture analgesia by CCK

Rats were given EA stimulation via two pairs of stainless steel pins inserted into the acupoints ST 36 (located lateral and distal to the anterior tubercle of the tibia) and SP 6 (located posterior to the upper border of the medial malleolus of the tibia) on both legs. The pulse width of the electric impulse was fixed to 0.3 ms, the frequency could be adjusted in three steps (2, 15, and 100 Hz), and the intensity was increased automatically starting at 1 mA with an increment of 1 mA every 10 min up to a maximum of 3 mA over a total time period of 30 min. TFL measured before EA was taken as the baseline level and was usually in the range 4–6 s. The measurement was continued at 10 min intervals after the start of EA for a total of 60 min. The changes induced by EA were expressed as percentage changes with + 150 per cent as the cut-off limit to avoid unnecessary skin damage.

To test the effect of CCK on EA analgesia, rats were injected i.c.v. or i.t.h. with 0.25–4.0 ng CCK or CCK-US in a volume of 10 μl before the implementation of EA stimulation. I.c.v. or i.t.h. injection of CCK or CCK-US (4 ng) alone produced no significant changes in TFL. However, the antinociceptive effect induced by EA was almost totally abolished by CCK either i.c.v. or i.t.h. CCK-US was ineffective in this regard. Reducing the dose of CCK from 4 ng to 2, 1, 0.5, and 0.25 ng produced a linear recovery of the EA effect (Fig. 45.1(a)). The ID_{50} was calculated to be 1.0 ng and 1.6 ng via the i.c.v. and i.t.h. routes

Table 45.2. CCK dose for 50 per cent inhibition of electro-acupuncture analgesia

Administration route	Animal species	Dose range (ng)	ID_{50} (ng)	EA (15 Hz) parameter	Nociceptive test	Reference
I.c.v.	Rat	0.25–4.0	1.0	1–3 mA, 30′	Tail-flick	Han et al. 1986
I.t.h.	Rat	0.50–4.0	1.6	1–3 mA, 30′	Tail-flick	Han et al. 1986
PAG	Rabbit	1.50–6.0	2.38	2 mA, 10′	Head-jerk	Cao et al. 1989

respectively (Table 45.2). Compared with the corresponding values in Table 45.1, the CCK doses needed to suppress 50 per cent of EA analgesia were only 59 and 52 per cent of that required to suppress the effect of morphine. This may imply that CCK is more effective in antagonizing endogenously released opioids than exogenously administered morphine. Alternatively, it may simply reflect the fact that the amount of opioids released by EA was only 50–60 per cent of the morphine equivalent (5 mg/kg). In contrast with the results obtained in rats, the CCK dose required to antagonize EA analgesia in rabbits (2.38 ng) was almost the same as that required to antagonize morphine analgesia (2.47 ng).

To assess the time course of the antagonistic effect of CCK on EA analgesia, brief episodes of EA (100 Hz, 3 V, 10 min) were given 0.5, 1, 2, 4, 8, and 16 h after i.c.v. injection of 4 ng of CCK or CCK-US. CCK but not CCK-US caused a dramatic (more than 85 per cent) suppression of EA analgesia, an effect which lasted for at least 4 h. Recovery took place in 8 h (Han *et al.* 1986).

CCK was found to be more effective in antagonizing analgesia induced by low frequency EA, with a rank order of 2 Hz > 15 Hz > 100 Hz ($p < 0.01$) (Han *et al.* 1986); the ID_{50} values were 0.8 ng, 1.0 ng, and 1.15 ng respectively. The ratio between the three doses was 100: 125: 144.

Finally, CCK blocks opioid analgesia but does not affect analgesia induced by 5-hydroxytryptamine (5-HT) (200 μg i.t.h.) or norepinephrine (NE) (2 μg i.t.h.) (Han *et al.* 1986). This result indicates that CCK is probably a specific opioid antagonist, rather than a universal antagonist for analgesics.

Reversal of electro-acupuncture tolerance by CCK antiserum

Taking CCK as an endogenous anti-opioid substrate (AOS), we naturally need to know the conditions under which the CCK system is activated to play a role in antagonizing opioid effects.

Since the analgesic effect elicited by EA in naive rats could not be potentiated by CCK-AS (Han *et al.* 1986), the amount of opioids released by short-term EA stimulation appears to be insufficient to trigger the CCK system. However, after 6 h of continuous EA stimulation, when the analgesic effect disappeared (implying the development of EA tolerance), i.c.v. or i.t.h. injection of CCK-AS restored the EA effect to more than 80 per cent of the original value (Han *et al.* 1985) (Fig. 44.2).

To test whether EA tolerance can be prevented by CCK-AS, the antiserum was injected prior to EA administration. In this case EA remained fully effective in the first 2–3 h of continuous stimulation, whereas in the control group there was a significant decrease starting from the second hour. Nevertheless, the analgesic effect decreased sharply after 4–6 h of stimulation, i.e. CCK-AS could only postpone, but not prevent, the development of EA tolerance (Han *et al.* 1986). In fact, AOSs other than CCK have been characterized, one of which is angiotensin II (Kaneko *et al.* 1985). Evidence has been presented showing the

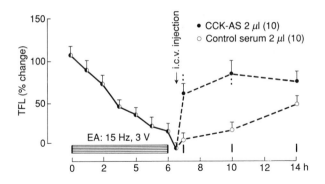

Fig. 45.2. Reversal of EA tolerance by i.c.v. injection of CCK-AS. Rats were given continuous EA stimulation for 6 h. The TFL was measured hourly to monitor the changes in the effect of EA analgesia. Tolerance to EA developed after 6 h. The rats were then given an i.c.v. injection of CCK-AS or normal rabbit serum as control. The recovery of the EA effect was monitored by brief episodes of EA stimulation as indicated by the vertical bars. (Reproduced with permission from Han *et al.* 1985.)

involvement of brain angiotensin II in the development of both morphine tolerance (Wang and Han 1989*a*) and EA tolerance (Wang and Han 1989*b*).

The results mentioned above indicate that the prolonged EA stimulation caused activation of the CCK system. However, we still do not know whether the CCK neurons were activated by EA itself or via the excessive amount of opioid released by prolonged EA stimulation. The latter interpretation is supported by the fact that animals rendered tolerant to morphine showed cross-tolerance to EA and vice versa (Han *et al.* 1981).

In summary, i.c.v. injection of CCK-AS not only reverses EA tolerance (the EA effect is restored to more than 80 per cent of the original value), but also reverses the cross-tolerance to morphine with equal efficacy. This may serve as unequivocal evidence showing the importance of CCK as a basic substrate mediating opioid tolerance.

Electrophysiological studies of CCK antagonism of opioid analgesia

Most of the data presented above were obtained in rats using noxious heat-induced tail flick as the nociceptive test. Since the end-point of nociception depends on the occurrence of a motor response, the results may be subject to misinterpretation (analgesia versus motor paralysis). To overcome this problem, efforts have been made to study the phenomenon on pain-related neurons in the CNS using single-neuron recording techniques.

Pain-excited and pain-inhibited neurons in the nucleus parafascicularis of the rat thalamus

Recent studies have revealed two kinds of pain-related neurons in CNS, one which responds to noxious stimulation with an increase in discharge (pain-excited neuron (PEN)), and the other which responds with a decrease in firing rate (pain-inhibited neuron (PIN)). Among the brain nuclei involved in nociception, the intralaminar nuclei of the thalamus, especially the nucleus parafascicularis (NPF), have received special attention (Chang 1973). The intralaminar nuclei of the thalamus are also known to be innervated by nerve fibres containing enkephalin and CCK (Sugimoto *et al.* 1985; Wahle and Albus 1985; Covenas *et al.* 1986; Gall *et al.* 1987). Techniques have been developed to record the two types of neurons in the NPF of urethane-anaesthetized rats on each side of the thalamus simultaneously (Sun *et al.* 1980; Li *et al.* 1984). This is a unique approach which enables the selective effect of a drug to be observed or certain physiological manipulations to be performed on the two types of neurons differentially and simultaneously. An antinociceptive agent would be expected to inhibit the PENs and simultaneously to excite the PINs. Substances with generalized excitatory or inhibitory effects would fail to do this.

Suppression by CCK of the effect of morphine on pain-inhibited and pain-excited neurons

Intraperitoneal (i.p.) injection of 10 mg/kg of morphine suppressed the PEN and excited the PIN (Table 45.3). This behaviour was completely reversed by the i.c.v. injection of 15 ng of CCK, the effect of which lasted for 10 min. CCK-US had little effect in this regard (Xu *et al.* 1987). It should be noted that, whereas CCK had a powerful anti-opioid action, it had no effect on either the PEN or the PIN in baseline conditions (Xu *et al.* 1988). Therefore the effect

Table 45.3.   Criteria for identifying pain-excited and pain-inhibited neurons in the nucleus parafascicularis of the rat thalamus and the effect of opioid analgesics on their discharge patterns

	Noxious stimulation of sciatic nerve	Opioid analgesics or EA stimulation
PEN	Increase in firing, or positive NED	Decrease in NED Prolongation of the latency of NED
PIN	Decrease in firing, or negative NED Appearance of period of complete silence	Increase of NED Reduction of period of complete silence

To quantify the changes in neuronal discharge after noxious stimulation, the discharge rates averaged over a period of 2 s immediately before and 2 s after the noxious stimulation were recorded. The difference between the pre- and post-stimulation period was taken as the noxiously evoked discharge (NED). This is a positive value in pain-excited neurones (PEN), and a negative value in pain-inhibited neurones (PIN).

of CCK is most likely to block the action of morphine, whether it is excitatory (to PIN) or inhibitory (to PEN).

Repeated i.p. administration of morphine (5 mg/kg) (six doses at intervals of 2 h) resulted in a complete disappearance of the opioid effect in both the PEN and the PIN. In other words, a challenge dose of morphine at 5 or 10 mg/kg could neither inhibit the PEN nor excite the PIN. However, the morphine effect was restored when the dose was increased to 20 mg/kg. Thus a characteristic morphine tolerance phenomenon can be reproduced in the pain-related neurons in the NPF of the thalamus (Xu *et al.* 1990).

Suppression by CCK of electro-acupuncture analgesia in pain-inhibited and pain-excited neurons

The effect of EA on nociception was assessed in both the PEN and the PIN of the rat thalamus NPF. In the control period, noxious stimuli (a train of five electrical impulses (0.4 ms, 5 mA, 200 Hz) applied directly to the sciatic nerve) evoked a net increase of 15.2 ± 1.0 discharges/s (noxiously evoked discharge (NED)) in the PEN. This value was reduced to 7.2 ± 0.7 discharges/s immediately after the cessation of EA, i.e. the noxious reaction was suppressed by 53 per cent. In addition to the decrease in NED, there was a 94 per cent prolongation of the latency of NED (from 158 ± 14 ms in the pre-EA period to 307 ± 48 ms in the immediately post-EA ($p < 0.01$)). These EA effects lasted for at least 10 min. I.c.v. injection of 15 ng of CCK resulted in an immediate increase in NED and a decrease in the latency to a level approaching the control value (Fig. 45.3(a)) (Bian *et al.* 1991*a*).

In the PIN, the noxious stimulus caused a decrease in the ongoing neuronal discharges, which resulted in a period of complete silence lasting for 1.25 ± 0.10 s. Immediately after the EA, there was an increase in NED and a reduction of the complete silence to 0.55 ± 0.13 s, i.e. a 56 per cent reduction of the noxious reaction ($p < 0.01$). This EA effect was abolished by i.c.v. injection of CCK (15 ng). A complete reversal of the EA effect was observed 6 min after the i.c.v. injection (Fig. 45.3(b)).

Most of the results discussed above were obtained in animals with the microelectrode recording one PEN or one PIN at a time. However, there were 12 experiments in which one electrode recorded a PEN and the other recorded a PIN on the opposite side of the thalamus. In all cases the PEN and the PIN showed co-ordinated activity, i.e. the reduction of NED in the PEN was always accompanied by an increase of NED in the PIN, and vice versa.

The electrophysiological study (Bian *et al.* 1991*a*) reproduced almost every aspect observed in behavioural studies. Hence it can be concluded that the antinociceptive effect of morphine or EA can be antagonized by CCK at nanogram doses. The advantages of the electrophysiological study are that the following problems are avoided: (a) interference from the motor system; (b) the possibility that EA analgesia might simply be stress-induced analgesia or a learning process (conditioning), (in the present study the animal was in an anaesthetized condition); (c) the anti-opioid effect of CCK might be due to a

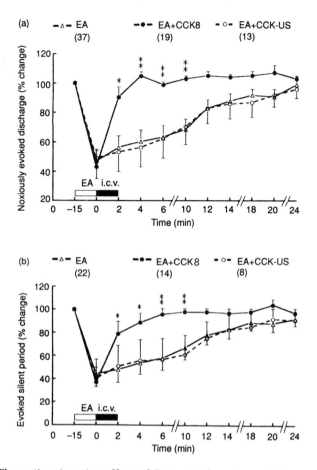

Fig. 45.3. The antinociceptive effect of EA in (a) the pain-excited neurones (PEN) and (b) the pain-inhibited neurones (PIN) in the nucleus parafascicularis (NPF) of the rat thalamus, and its reversal by CCK but not by CCK-US. The box indicates time of EA stimulation (6 Hz, 3 V, 15 min). The horizontal bar represents i.c.v. injection of 15 ng of CCK or CCK-US. Vertical bars indicate the SEM. $\star p < 0.05$, $\star\star p < 0.01$ compared with the corresponding values in the EA group.

generalized excitation of the neuronal elements rather than a specific antagonistic action on opioids.

While most of the results obtained with the two types of experiment are in good agreement, there is one major difference. It is apparent from Fig. 45.1 that, in the behavioural studies, the CCK dose needed for complete reversal of morphine analgesia was only 4 ng, whereas in electrophysiological studies doses as high as 15 ng were needed (Fig. 45.3). Moreover, in the behavioural studies the anti-opioid effect produced by one injection of CCK (4 ng) lasted as long as 4 h (Han *et al.* 1986), whereas in electrophysiological studies of anaesthetized rats the effect of CCK (15 ng) lasted for only 10 min (Fig. 45.3). However, these differences may be due to differences in technique. In order to minimize the influence of respiration and stabilize the microelectrode tip in the brain, the ventricular fluid was constantly drained from the cisterna magna. Therefore in the electrophysiological studies the peptide administered to the lateral ventricle would leave the ventricular system at a much faster rate than in the behavioural studies.

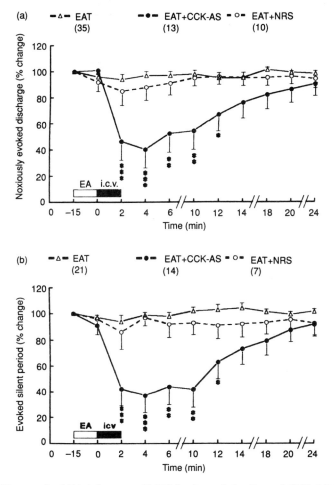

Fig. 45.4. Reversal of EA tolerance (EAT) by i.c.v. injection of CCK-AS in (a) the PEN and (b) the PIN in the NPF of the rat thalamus. In the EA tolerant rat, EA stimulation could no longer suppress the noxiously evoked discharge in PEN or shorten the noxiously evoked silent period in PIN. I.c.v. injection of 10 μl of CCK-AS restored the EA effect. I.c.v. injection of NRS was without effect. A significant difference was found between the EA tolerant group and the EA tolerant plus CCK-AS group: $\star$ $p < 0.05$; $\star\star$ $p < 0.01$; $\star\star\star$ $p < 0.001$.

Reversal of electro-acupuncture tolerance by CCK antiserum

In naive rats, EA stimulation suppressed the noxious reaction of both the PEN and the PIN by more than 50 per cent (Fig. 45.3). However, this effect was completely abolished after six sessions of EA stimulation (each session consisted of 30 min stimulation followed by 30 min rest), implying the development of EA tolerance (Fig. 45.4). To test whether central CCK is involved in the mechanisms of EA tolerance, CCK-AS (2 μl) was injected i.c.v. in an attempt to bind the CCK and prevent it from activating receptors. A complete restoration of the EA effect was observed in both PEN and PIN (Fig. 45.4), with the noxious reaction in the PIN being even stronger than that in the naive rat (compare Figs. 45.3(b) and 45.4(b)). Injection of rabbit serum obtained from non-immunized animals (normal rabbit serum (NRS)) proved to be ineffective.

In some of the animals one PEN and one PIN were recorded simultaneously (six pairs in the EA-tolerant rats, seven pairs in the EA-tolerant rats receiving CCK-AS, and five pairs in the EA-tolerant rats receiving NRS). In all cases there was a co-ordinated change, i.e. the suppression of PEN was always accompanied by excitation of the PIN and vice versa (Bian *et al.* 1991*b*).

In conclusion, noxious stimulation induces excitation of the PEN and inhibition of the PIN. These changes can be reversed by opioids, and the opioid effect can in turn be antagonized by CCK.

Possible mechanisms of the anti-opioid effect of CCK

The fact that opioid analgesia can be antagonized by a very small amount of CCK suggests that there is a strong biochemical mechanism underlying this phenomenon. The interaction between opioids and CCK may occur at the receptor level (direct occupation of the opioid receptor, or indirect action such as allosteric receptor–receptor cross-talk) or at post-receptor level (influence on second messengers, ion channels, etc.).

CCK suppression of high affinity opioid binding via CCK receptors

It has been reported that neither CCK nor sulphated CCK-7 are ligands for opioid receptors, although unsulphated CCK-7 has been shown to be such a ligand (Schiller *et al.* 1978; Stengaard-Pedersen and Larsson 1981). Therefore we investigated whether CCK could compete with the binding of [³H] etorphine (a highly potent opioid agonist which binds to all three types of opioid receptors) to opioid receptors in a rat brain synaptosomal preparation. In competition experiments, CCK produced a mild suppression (up to 31 per cent) of [³H]etorphine binding. This effect of CCK was completely reversed by the CCK receptor antagonist proglumide, indicating that the effect was mediated via a CCK receptor. Saturation experiments analysed by the Rosenthal plot revealed two populations of binding sites (high versus low affinity). CCK affected high affinity binding with an increase in K_d (+ 235 per cent,

$p < 0.001$) and a decrease in B_{max} (-80 per cent, $p < 0.01$) which could be reversed by administration of proglumide. CCK-US produced only a slight increase in K_d without affecting B_{max} (Wang *et al.* 1989*a*).

Two points deserve brief discussion. First, although the high affinity binding sites constituted only a small fraction of the total specific binding (Simon *et al.* 1975), they may be more important in physiological conditions compared with low affinity sites which might play a greater role in pharmacological conditions. Second CCK-US showed a much weaker effect than CCK on opioid binding, which is in line with the results obtained in behavioural studies that nanogram doses of CCK antagonized opioid analgesia whereas CCK-US had no effect (Han *et al.* 1986).

Modification by CCK of μ, δ, and κ opioid binding

To differentiate the effect of CCK on the three types of opioid receptors, we chose [^{3}H] (D-Ala^2, MePhe^4, Gly(ol)^5) enkephalin ([^{3}H]DAGOL), [^{3}H]DPDPE, and [^{3}H]U69593 as the specific μ, δ, and κ opioid ligands for the binding assays (Wang and Han 1990). In competition experiments, CCK suppressed the binding of [^{3}H]DAGOL and [^{3}H]U69593, but not that of [^{3}H]DPDPE, to the respective opioid receptors. These effects were blocked by proglumide. In the saturation experiments, CCK decreased the B_{max} of μ binding without affecting its K_d; in contrast, CCK increased the K_d of κ binding without affecting the B_{max}.

The fact that CCK does not affect δ binding was supported by recent work showing that there were no significant changes in the B_{max} and K_d values of δ binding in the presence or absence of CCK ($1\,\mu$M) in a mouse brain synaptosomal preparation (Hong and Takemori 1989).

It is interesting to note that the results of the *in vitro* study presented above are in good agreement with those obtained in behavioural studies (Wang *et al.* 1990) in that (a) CCK was very effective in antagonizing analgesia induced by the specific μ agonist PL017 (Chang *et al.* 1983) and the specific κ agonist 66A-078 (Tachibana *et al.* 1988), but did not antagonize that induced by the specific δ agonist DPDPE (even with a CCK dose as high as 40 ng), and (b) the dose-response curve of analgesia elicited by the μ receptor agonist PL017 was non-competitively antagonized by CCK whereas that of analgesia elicited by the κ agonist 66A-078 was competitively antagonized.

It should be noted that the results of the *in vitro* study appear to conflict with another *in vivo* study (Han *et al.* 1986) showing that CCK was almost equally effective in antagonizing analgesia induced by low and high frequency EA. Since it has been shown (Fei *et al.* 1987) that analgesia induced by low frequency (2 Hz) EA is mediated by met-enkephalin, which is a δ and μ agonist, and analgesia induced by high frequency (100 Hz) EA is mediated by dynorphin, which is a κ agonist, we would expect the enkephalin-mediated low frequency EA effect to resist CCK antagonism, which is not the case. One possibility is that the analgesia induced by enkephalin is mediated by a combined effect of

μ and δ receptors, or even by a μ–δ opioid receptor complex (Porreca 1990; Schoffelmeer *et al.* 1990), and that blockade of the μ receptor would abolish the whole effect.

Effect of CCK on calcium uptake

Since the prevention of calcium uptake into the nerve terminals has been considered as one of the mechanisms for morphine analgesia (Guerrero-Munoz *et al.* 1979; Chapman and Way 1980), it has been speculated that CCK might augment calcium uptake and thereby counteract the effect of morphine (physiological antagonism). However, the outcome of an experiment to investigate this was unexpected (Wang *et al.* 1989*b*). In the rat brain synaptosomal preparation, morphine at a concentration of 0.01–1 μM produced a dose-dependent suppression of ^{45}Ca uptake with a maximal effect of $-$ 41 per cent ($p < 0.001$). This effect was naloxone-reversible. Instead of augmenting calcium uptake, CCK at concentrations of 0.01–1 μM also caused a decrease in ^{45}Ca uptake, with a maximal effect of $-$ 40 per cent at 1 μM, which was reversed by the CCK antagonist proglumide. In other words, morphine and CCK work in the same direction via their respective receptors. However, when CCK (0.01–0.1 μM) was added to the synaptosomal preparation together with morphine (1 μM), the blockade of calcium flux was diminished. This anti-opiate effect of CCK was proglumide-reversible and was dose-dependent at lower concentrations of CCK (0.01–0.1 μM), but was ineffective at higher concentrations (1–10 μM). The results suggest that CCK at low concentrations could antagonize the calcium blocking effect of morphine via activation of CCK receptors. This effect may constitute one of the mechanisms for CCK (at small doses) to exert an anti-opioid effect.

Effect of opiates and CCK on spinal cAMP

It has been shown that opioids suppress the activity of adenylate cyclase and lower the content of cyclic adenosine monophosphato (cAMP) in the caudate and neuroblastoma glioma hybrid cell NG 108–15 (Law *et al.* 1981, 1982), whereas CCK increased the cAMP level in rat pancreatic acinar cells (Willems *et al.* 1987). Therefore it appeared worthwhile to determine whether opioids and CCK would induce opposite changes in cAMP content. The results of an experimental study using rats (Sun *et al.* 1991) showed that the spinal content of cAMP was significantly decreased by i.t.h. injection of μ and δ agonists, but not by a κ agonist. These changes were not affected by the i.t.h. injection of CCK, nor was the cAMP content affected by CCK alone. Therefore the opioid–CCK interaction does not appear to involve cAMP.

Type of CCK receptors involved in the anti-opioid effect

The long discussion of the classification and function of CCK receptors has led to the understanding that 'peripheral' type CCK receptors which are located in peripheral tissues and discrete brain regions should be referred to as CCK_A receptors, and that 'brain' type CCK receptors which are abundant throughout the CNS should be referred to as CCK_B receptors (Moran *et al.* 1986; Dourish and Hill 1987).

CCK receptor types have been identified according to their affinity for the sulphated (CCK) and unsulphated (CCK-US) forms. CCK_A receptors have high affinity for CCK but a low affinity for CCK-US, whereas CCK_B receptors discriminate poorly between CCK and CCK-US (Innis and Snyder 1980; Baber *et al.* 1989). Since *in vivo* studies showed that only CCK had an anti-opioid effect, while CCK-US was totally ineffective, we might speculate that the CCK receptor in question belongs to type A.

The development of the potent non-peptide CCK_A antagonists devazepide and L-365,031 (Chang and Lotti 1986; Evans *et al.* 1986; Chapter 2 of this volume) was a breakthrough in CCK research which provided a firmer basis to differentiate types of CCK receptor. The synthesis of the CCK_B antagonist L-365,260 provided the possibility of comparing the relative potency of different antagonists for a given physiological or pharmacological event. For example, evidence was presented that decreased feeding induced by systemic injection of CCK is mediated by CCK_A receptors (Dourish *et al.* 1989*a*), whereas satiety caused by endogenous CCK involves CCK_B receptors in the brain (Dourish *et al.* 1989*b*). With regard to pain control, L-365,260 administered parenterally was reported to be 40 times more potent than L-365,031 and five times more potent than devazepide in enhancing morphine analgesia in the rat tail-flick test, and 20 times more potent than devazepide in enhancing morphine analgesia in the rat paw pressure test (Dourish *et al.* 1990), suggesting that the effect of CCK in antagonizing morphine analgesia is mediated by CCK_B receptors (see Chapter 43 of this volume).

In order to clarify the type of CCK receptor involved in antagonizing endogenously released opioids for pain modulation, studies were performed to test the effects of devazepide and L-365,260 on EA analgesia and EA tolerance. The influence of the antagonists on morphine analgesia was also assessed for comparison (Zhou *et al.*, unpublished results).

Effect of devazepide and L-365,260 on morphine analgesia

Rats were given an s.c. injection of 4 mg/kg of morphine to produce a moderate antinociceptive effect (an increase of about 80 per cent in TFL). The CCK antagonists were injected i.t.h. in dose ranges of 10–200 ng for devazepide and 1–20 ng for L-365,260. Bell-shaped dose–response curves were obtained. The optimal dose which potentiated opioid analgesia was 100 ng for devazepide and 2.5 ng for L-365,260, i.e. L-365,260 was 40 times more potent than devazepide. Thus the results obtained by i.t.h. injection confirm those obtained by s.c. injection that the effect of CCK in antagonizing opioid analgesia is mediated by CCK_B receptors (Dourish *et al.* 1988, 1990).

Effect of devazepide and L-365,260 on electro-acupuncture analgesia and tolerance

In clinical practice, acupuncture treatment does not have a positive therapeutic effect on all patients. The effective rate for treating chronic pain ranged from 55 to 85 per cent according to different authors (Pomeranz 1989). We obtained similar results with our animal (rat and rabbit) models. About 70–90 per cent of the animals show an antinociceptive effect, as measured using tail-flick or head-jerk tests, and the remainder (10–30 per cent) are non-responders.

Both the responder and non-responder rats were given devazepide to determine whether the effect of EA analgesia was potentiated by the CCK antagonist. The results are shown in Table 45.4. In no case was EA analgesia potentiated by devazepide in responder rats. Rather, it suppressed the EA effect at higher doses. In contrast, non-responder rats showed marked potentiation of EA analgesia after the i.p. injection of devazepide at a dose of 100 μg/kg, i.e. non-responders were transformed into responders by the CCK antagonist (Zhou *et al.*, unpublished results). This result confirmed the previous finding that non-responder rats became responders after i.t.h. injection of CCK-AS (Han *et al.* 1985).

To test whether EA tolerance could be prevented by devazepide, the responder rats were given an injection of devazepide (1 mg/kg i.p., 1.25 μg i.c.v., 0.25 μg i.t.h.) or normal saline prior to the EA stimulation which lasted for 6 h. While both groups showed a similar degree of analgesia in response to EA stimulation in the first hour (an increase of TFL of more than 90 per cent over the baseline level), they showed a significant difference after 3–4 h of stimulation, i.e. EA remained effective in animals injected with devazepide whereas in the control group its effect was substantially reduced. These results are consistent with the finding that EA tolerance could be postponed by the central administration of CCK-AS (Han *et al.* 1986).

Taken together, these results could be interpreted to mean that a high level

Table 45.4.   Effect of devazepide and L-365,260 on electro-acupuncture analgesia and tolerance

Administration route	Effect on EA analgesia			Reversal of EA tolerance
	Increased	Not affected	Decreased	
Devazepide				
I.p. (mg/kg)		0.003–0.3	1, 3	1
I.c.v. (μg)		1.25, 2.5	5, 10	1.25
I.t.h. (μg)		0.1–1.0	2	0.25
L-365,260				
I.t.h. (ng)	2.5, 5	10, 25	50	NT

NT, not tested.

of central CCK activity may constitute one of the mechanisms for the animal's being an EA non-responder, and that prolonged EA may cause CCK release to change a responder into a non-responder, which could explain the development of EA tolerance. These 'non-responders' can in turn be transformed into responders by treatment with a CCK antagonist.

It has been shown that, although devazepide has a higher affinity for CCK_A receptors, it can also interact with CCK_B receptors at high concentrations (Hill *et al.* 1987*a*). We therefore used the CCK_B antagonist L-365,260 in a similar experimental design for the purpose of comparison. A bell-shaped dose–response curve was obtained: EA analgesia was potentiated by i.t.h. injection of L-365,260 at 2.5 and 5 ng, and significantly attenuated at 50 ng. Thus the L-365,260 dose which produces a substantial reversal of EA tolerance is only a fiftieth of the devazepide dose ($0.25\ \mu g$) required to produce the same effect (Table 44.4). These results support the proposal that the antagonistic effect of CCK on EA analgesia is mediated by CCK_B receptors.

How can these results be reconciled with those showing that CCK-US was ineffective in antagonizing opioid analgesia when an equal dose (4 ng by the i.c.v. or i.t.h. route, 1 ng for intra-PAG injection) to that of CCK which was effective was used. One possibility is that, while CCK_B receptors differentiate poorly between CCK and CCK-US (26-fold difference in favour of CCK) compared with CCK_A receptors (800-fold difference) (Baber *et al.* 1989), there is still a difference. It would be possible for CCK-US to antagonize opioid analgesia if the dose were increased 26-fold (to 100 ng for i.c.v. or i.t.h. injection), whereas this possibility would not exist if the effect were mediated by CCK_A receptors.

General discussion and conclusions

In contrast with the number of classical neurotransmitters, which does not exceed 10, the number of newly characterized neuropeptides is increasing rapidly and may soon reach three figures. The complex interactions among the neuropeptides and neurotransmitters at the receptor and post-receptor levels are a great challenge that neuroscientists have to face in the coming decade. The opioid–CCK interaction, or, in a broader sense, the opioid–anti-opioid interaction, can be taken as a model system in this regard.

On the large scale, prolonged opioid exposure may trigger the release of CCK (Tang *et al.* 1984) to counteract the opioid effect, and a large dose of CCK may accelerate opioid release (Cesselin *et al.* 1984; Hong and Takemori 1989). These interactions may occur in a single neuron in which opioids and CCK coexist (Gall *et al.* 1987; Baber *et al.* 1989), or may involve complex neuronal circuitry.

In a smaller system, the opioid–CCK interaction may occur between two receptors located adjacent to each other on the cell membrane. In this case at least two models can be proposed (Fig. 45.5). In one the pain transmission neuron (e.g. the dorsal horn neuron or a neuron in the PAG) is inhibited by

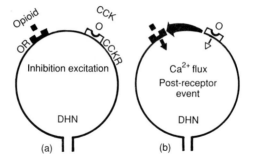

Fig. 45.5. Diagram showing the possible modes of interaction between opioids and CCK.

(a) Opioids cause inhibition of the dorsal horn neuron (DHN), which is an example of the pain-related neurons in the CNS, and CCK causes excitation of the DHN. The firing rate of the neuron, and hence the degree of nociception, depends on the balance between opioids and CCK.

(b) CCK activates the CCK receptor (CCKR) which in turn affects the opioid receptor (OR) by lowering its affinity and reducing the number of binding sites. Alternatively, opposite effects may be produced on calcium flux and other post-receptor events.

opioids and excited by CCK to maintain a dynamic balance, as proposed by Baber *et al.* (1989). According to this model, injection of CCK increases firing of the neuron and, as a result, induces hyperalgesia, which is not the case. An alternative model is that CCK itself does not affect the firing of pain transmission neurons (as was seen in the PEN and PIN of the thalamus NPF), but interferes with opioid receptor binding, calcium flux, and many other post-receptor events, thereby antagonizing the effect of opioids. This hypothesis is supported by electrophysiological data obtained in pain-related neurons in the medial thalamus showing that, regardless of whether the opioid effect is excitatory or inhibitory in nature, these effects can be reversed by CCK. The CCK receptor involved is probably of the CCK_B type.

In order to verify the hypothesis outlined above more studies are required to provide (a) morphological and electrophysiological evidence that opioid receptors and CCK receptors are located in close proximity on the cell membranes of pain-related neurons, and (b) biochemical evidence showing how CCK interferes with opioid-induced signal transduction at different levels of the pathways.

The use of non-peptide CCK_B antagonists clinically as an adjunct for the potentiation of EA analgesia (changing non-responders into responders and thus broadening the applicability of EA therapy) and postponement of EA tolerance is strongly supported.

Acknowledgements

This work was supported in part by the National Natural Science Foundation of China and grant DA 03983 from the National Institute of Drug Abuse, USA. The author wishes to thank Dr M. C. Beinfeld of St. Louis University and Dr J. S. Hong of the National Institute of Environmental Health Sciences, Research Triangle, NC, for the generous gifts of CCK antiserum, Squibb and Sons Inc. for the donation of CCK and CCK-US, and Dr R. F. Freidinger of Merck Sharp & Dohme for the supply of devazepide and L-365,260 used in this study.

References

Baber, N. S., Dourish, C. T., and Hill, D. R. (1989). The role of CCK, caerulein, and CCK antagonists in nociception. *Pain* **39**, 307–28.

Bian, J. T., Sun, M. Z., Xu, M. Y., and Han, J. S. (1991*a*). Antagonism by CCK-8 of the antinociceptive effect of electroacupuncture on pain-related neurons in nucleus parafascicularis of the rat, *Neuropeptides*, submitted for publication.

Bian, J. T., Sun, M. Z., and Han, J. S. (1991*b*). Reversal of electroacupuncture tolerance by CCK-8 antiserum. An electrophysiology study on pain-related neurons in nucleus parafascicularis of the rat. *Pain*, submitted for publication.

Cao, W., Zhou, Z. F., and Han, J. S. (1989). Antagonism of morphine analgesia and electroacupuncture analgesia by cholecystokinin octapeptide administered in periaqueductal gray of the rabbit. *Acta Physiol. Sin.* **41**, 388–94.

Chang, H.-T. (1973). Integrative action of thalamus in the process of acupuncture for analgesia, *Sci. Sin.* **16**, 25–60.

Chang, R. S. L., and Lotti, V. J. (1986). Biochemical and pharmacological characterization of an extremely potent and selective non-peptide cholecystokinin antagonist. *Proc. Natl Acad. Sci. USA* **83**, 4923–6.

Chang, K. J., Wei, E. T., Killian, A., and Chang, J. K. (1983). Potent morphiceptin analogues: structure activity relationships and morphine like activities. *J. Pharmacol. Exp. Ther.* **227**, 403–8.

Chapman, D. B., and Way, E. L. (1980). Metal ion interaction with opiates. *Ann. Rev. Pharmacol. Toxicol.* **20** (1980) 535–57.

Cesselin, F., Bourgoin, S., Artaud, F., and Hamon, M. (1984). Basic and regulatory mechanisms of *in vitro* release of metenkephalin from the dorsal zone of the rat spinal cord. *J. Neurochem.* **43**, 763–74.

Covenas, R., Romo, R., Cheramy, A., Cesselin, F., and Conrath, M. (1986). Immunocytochemical study of enkephalin-like cell bodies in the thalamus of the cat. *Brain Res.* **377**, 355–61.

Ding, X. Z., Fan, S. G., Zhou, Z. F., and Han, J. S. (1986). Reversal of morphine tolerance but no potentiation of morphine analgesia by antiserum against cholecystokinin octapeptide. *Neuropharmacology* **25**, 1155–60.

Dourish, C. T. and Hill, D. R. (1987). Classification and function of CCK receptors. *Trends Pharmacol. Sci.* **8**, 207–8.

Dourish, C. T., Hawley, D., and Iversen, S. D. (1988). Enhancement of morphine analgesia and prevention of morphine tolerance in the rat by the cholecystokinin antagonist L-364,718. *Eur. J. Pharmacol.* **147**, 469–72.

500   *Role of CCK in electro-acupuncture analgesia and tolerance*

Dourish, C. T., Ruckert, A. C., Tattersall, F. D., and Iversen, S. D. (1989*a*). Evidence that decreased feeding induced by systemic injection of cholecystokinin is mediated by CCK-A receptors. *Eur. J. Pharmacol.* **173**, 233–4.

Dourish, C. T., Rycroft, W., and Iversen, S. D. (1989*b*). Postponement of satiety by blockade of brain cholecystokinin (CCK-B) receptors. *Science* **245**, 1509–11.

Dourish, C. T., O'Neill, M. F., Coughlan, J., Kitchener, S. J., Hawley, D., and Iversen, S. D. (1990). The selective CCK-B receptor antagonist L-365,260 enhances morphine analgesia and prevents morphine tolerance in the rat, *Eur. J. Pharmacol.* **176**, 35–44.

Evans, B. E., Bock, M. G., Rittle, K. E., DiPardo, R. M., Freidinger, R. F., Whitter, W. L., Lundell, G. F, Veber, D. F., Anderson, P. S., Chang, R. S. L., Lotti, V. J., Cerino, D. J., Chen, T. B., Kling, P. J., Kunkel, K. A., Springer, J. P., and Hirschfeld, J. (1986). Design of potent, orally effective, non-peptide antagonists of the peptide hormone cholecystokinin. *Proc. Natl Acad. Sci. USA* **83**, 4918–22.

Faris, S., Komisaruk, B. R., Watkins, L. R., and Mayer, D. J. (1983). Evidence for the neuropeptide cholecystokinin as an antagonist of opioid analgesia. *Science* **219**, 310–12.

Faris, P. L., MacLaughlin, C. L., Baile, C. A., Loney, J. W., and Komisaruk, B. R. (1984). Morphine analgesia potentiated but tolerance not affected by active immunization against cholecystokinin. *Science* **226**, 1215–17.

Fei, H., Xie, G. X., and Han, J. S. (1987). Low and high frequency electroacupuncture release met[5]-enkephalin and dynorphin A in rat spinal cord. *Kexue Tongbao (Chin. Sci. Bull.)* **32**, 1496–1501.

Gall, C., Lauterborn, J., Burks, D., and Seroogy, K. (1987). Colocalization of enkephalin and cholecystokinin in discrete areas of rat brain. *Brain Res.* **403**, 403–8.

Guerrero-Munoz, F., Cerreta, K. V., Guerrero, M. L., and Way, E. L. (1979). Effect of morphine on synaptosomal calcium uptake. *J. Pharmacol. Exp. Ther.* **209**, 132–6.

Han, J. S. (1987). Antibody microinjection: a new approach for studying the function of neuropeptides. *Chin. Med. J.* **100**, 459–64.

Han, J. S., and Terenius, L. (1982). Neurochemical basis of acupuncture analgesia. *Ann. Rev. Pharmacol. Toxicol.* **22**, 193–220.

Han, J. S., Tang, J., Huang, B. S., Liang, X. N., and Zhang, N. H. (1980). Acupuncture tolerance in rats. Anti-opiate substrates implicated. In *Endogenous and exogenous opiate agonists and antagonists* (ed. E. L. Way), pp. 395–8. Pergamon, New York.

Han, J. S., Li, S. J., and Tang, J. (1981). Tolerance to electro-acupuncture and its cross tolerance to morphine. *Neuropharmacology* **20**, 593–6.

Han, J. S., Ding, X. Z., and Fan, S. G. (1985). Is cholecystokinin octapeptide (CCK-8) a candidate for endogenous anti-opioid substrate? *Neuropeptides* **5**, 399–402.

Han, J. S., Ding, X. Z., and Fan, S. G. (1986). Cholecystokinin octapeptide (CCK-8): antagonism to electroacupuncture analgesia and a possible role in electroacupuncture tolerance. *Pain* **27**, 101–15.

Hill, D. R., Campbell, N. J., Shaw, T. M., and Woodruff, G. N (1987*a*). Autoradiographic localization and biochemical characterization of peripheral type CCK receptors in rat CNS using highly selective non-peptide CCK antagonists. *J. Neurosci.* **7**, 2967–76.

Hill, R. G., Hughes, J., and Pittaway, K. M. (1987*b*). Antinociceptive action of cholecystokinin octapeptide (CCK-8) and related peptides in rats and mice: effects of naloxone and peptidase inhibitors. *Neuropharmacology* **26**, 289–300.

Hong, E. K.,and Takemori, A. E. (1989). Indirect involvement of delta opioid receptors in cholecystokinin octapeptide-induced analgesia in mice, *J. Pharmacol. Exp. Ther.* **251**, 594–8.

Innis, R. B.,and Snyder, S. H. (1980). Distinct cholecystokinin receptors in brain and pancreas. *Proc. Natl Acad. Sci. USA* **77**, 6917–21.

Itoh, S., Katssura, G., and Maeda, Y. (1982). Caerulein and CCK suppress β-endorphin-induced analgesia in the rat. *Eur. J. Pharmacol.* **80**, 270–89.

Jurna, I.,and Zetler, G. (1981). Antinociceptive effect of centrally administered caerulein and cholecystokinin octapeptide (CCK-8). *Eur. J. Pharmacol.* **73**, 323–31.

Kaneko, S., Tamura, S., and Takagi, H. (1985). Purification and identification of endogenous anti-opioid substances from bovine brain. *Biochem. Biophys. Res. Commun.* **126**, 587–93.

Law, P. Y., Wu, J., Koehler, J. E., and Loh, H. H. (1981). Demonstration and characterization of opiate inhibition of the striatal adenylate cyclase. *J. Neurochem.* **36**, 1834–46.

Law, P. Y., Hom, D. S., and Loh, H. H. (1982). Opiate regulation of adenosine 3′, 5′-cyclic monophosphate level in glioma neuroblastoma NG 108-15 hybrid cell. *Mol. Pharmacol.* **23**, 26–35.

Li, Y., and Han, J. S. (1989). Cholecystokinin-octapeptide antagonizes morphine analgesia in periaqueductal gray of the rat. *Brain Res.* **480**, 105–10.

Li, Y. R., Sun, M. Z., Zhang, J. Y., and Xu, T. (1984). The simultaneous discharges of two neurons in N. Pf and RF and the influence on them by stimulating central gray matter, *Sci. Sin. B* **27**, 1273–81.

Moran, T. H., Robinson, M. S., Goldrich, M.S., and McHugh, P. R. (1986). Two brain cholecystokinin receptors: implication for behavioral actions. *Brain Res.* **362**, 175–9.

Pomeranz, B. (1989). Acupuncture analgesia for chronic pain: brief survey of clinical trials. In *Scientific basis of acupuncture* (ed. B. Pomeranz and G. Stux), pp. 197–8. Springer-Verlag, Heidelberg.

Porreca, F. (1990). Functional interactions between opioid mu and delta receptors *in vivo*. In *New leads in opioid research* (ed. J. M. Van-Ree *et al.*), pp. 280–2. Excerpta Medica, Amsterdam.

Schiller, P. W., Lipton, A., Horrobin, D. F., and Bodansky, M. (1978). Unsulfated C-terminal 7-peptide of cholecystokinin. A new ligand of the opiate receptor. *Biochem. Biophys. Res. Commun.* **85**, 1332–8.

Schoffelmeer, A. N. M., Yao, Y. H., Hiller, J. M., Gioannini, T. L., *et al.* (1990). Beta-endorphin binds to a 80 kD glycoprotein with interacting mu and delta sites in rat striatum. In *New leads in opioid research* (ed. J. M. Van-Ree *et al.*), pp. 284–6. Excerpta Medica, Amsterdam.

Simon, E. J., Hiller, J. M., Groth, J., and Edelman, I. (1975). Further properties of stereospecific opiate binding sites in rat brain: on the nature of sodium effect. *J. Pharmacol. Exp. Ther.* **192**, 531–7.

Stacher, G., Abatzi, Y. A., Schulte, F., Schneider, C., Atacher-Janotta, G., Gaupmann, G., Mittelbach, G., and Steiringer, H. (1988). Naloxone does not alter the perception of pain induced by electrical and thermal stimulation of the skin in healthy humans. *Pain* **34**, 271–6.

Stengaard-Pedersen, K. and Larsson, L. I. (1981). Localization and opiate receptor binding of enkephalin, CCK and ACTH/β-endorphin in the rat central nervous system. *Peptides* **2**, 3–19.

Sugimoto, T., Itoh, K., Yasui, Y., Kaneko, T., and Mizuno, N. (1985). Coexistence of neuropeptides in projection neurons of the thalamus in the cat. *Brain Res.* **347**, 381–4.

Sun, M. Z., Chen, L. S., Gu, H. L., Cheng, J., and Yue, L. S. (1980). Effect of acupuncture on unit discharge in nucleus parafascicularis of rat hypothalamus, *Acta Physiol. Sin.* **32**, 207–13.

Sun, S. J., Zhang, L. J., and Han, J. S. (1991). The anti-opioid effect of CCK-8 is not mediated by cAMP in rat spinal cord. *Chin. Sci. Bull.*, in the press.

Tachibana, S., Yoshino, H., Arakawa, Y., Nakazawa, T., Kaneko, T., Yamatsu, K., and Miyagawa, H. (1988). Design and synthesis of metabolically stable analogues of dynorphin-A and their analgesia characteristics. In *Biowarning system in the brain* (ed. H. Takagi, Y. Oomura, M. Ito, and M. Otsuka), pp. 101–9. Tokyo University Press, Tokyo.

Tang, J., Chiu, J., Iadarola, M., Yang, H. Y. T., and Costa, E. (1984). Proglumide prevents and curtails acute tolerance to morphine in rats. *Neuropharmacology* **23**, 715–18.

Wahle, P., and Albus, K. (1985). Cholecystokinin octapeptide-like immunoreactive material in neurons of the intralaminar nuclei of the cat's thalamus, *Brain Res.* **327**, 348–53.

Wang, K. W., and Han, J. S. (1989*a*). Evidence for brain angiotensin II being involved in morphine tolerance in the rat. *Chin. J. Pharmacol. Toxicol.* **3**, 7–11.

Wang, K. W., and Han, J. S. (1989*b*). Evidence for involvement of brain angiotensin II in tolerance to electroacupuncture analgesia in rats. *Chin. J. Appl. Physiol.* **5**, 32–6.

Wang, X. J., and Han, J. S. (1990). Modification by cholecystokinin octapeptide of the binding of μ-, δ-, and κ-opioid receptors. *J. Neurochem.* **55**, 1379–82.

Wang, X. J., Fan, S. G., Ren, M. F., and Han, J. S. (1989*a*). Cholecystokinin-8 suppressed ^{3}H-etorphine binding to rat brain opiate receptors. *Life Sci.* **45**, 117–23.

Wang, X. J., Wang, X. M., and Han, J. S. (1989*b*). Antagonism of the suppressive effect of morphine on rat brain synaptosomal 45Ca uptake by cholecystokinin octapeptide, *Chin. J. Pharmacol. Toxicol.* **3**, 241–6.

Wang, X. J., Wang, X. H., and Han, J. S. (1990). Cholecystokinin octapeptide antagonized opioid analgesia mediated by μ- and κ- but not δ-receptors in the spinal cord of the rat, *Brain Res.*, **523**, 5–10.

Watkins, L. R., Kinscheck, I. B., and Mayer, D. J. (1985). Potentiation of morphine analgesia by the CCK antagonist proglumide. *Brain Res.* **27**, 169–80.

Willems, P. H., Tilly, R. H. J., and de Pont, J. J. (1987). Pertussis toxin stimulates cholecystokinin-induced cyclic AMP formation but is without effect on secretagofue-induced calcium mobilization, *Biochim. Biophys. Acta* **928**, 179–85.

Xu, M. Y., Sun, M. Z., Zhang, L. M., Wang, B. M., and Han, J. S. (1987). Antagonism to the effect of morphine on electric discharges of pain-related neurons in nucleus parafascicularis of thalamus by cholecystokinin octapeptide. *Acta Physiol. Sin.* **39**, 317–25.

Xu, M. Y., Sun, M. Z., Zhang, L. M., and Han, L. S. (1988). Effects of intracerebroventricularly injected CCK-8 on the electric activities of pain-related neurons in nucleus parafascicularis of thalamus in rats. *Acupuncture Res.* **14**, 305–9.

Xu, M. Y., Sun, M. Z., Yang, L. Z., Zhang, L. M., and Han, J. S. (1990). Simultaneous electric activities of pain-excitation and pain-inhibition neurons in nucleus parafascicularis of thalamus in rats during acute morphine tolerance, *Acta Pharmacol. Sin.* **11**, 200–3.

Yaksh, T. L., and Rudy, T. A. (1976). Chronic catheterization of the spinal subarachnoid space. *Physiol. Behav.* **17**, 1031–6.

46. CCK and opioid interactions in the spinal cord

Anthony H. Dickenson, Ann F. Sullivan, and David S. Magnuson

CCK is present in widespread areas of the CNS including the spinal cord, where it is found in high levels, and released from those areas of the dorsal horn involved in the transmission and modulation of sensory messages. (Yaksh *et al.* 1982; Yaksh and Noueihed 1985; Williams *et al.* 1987). There is evidence that CCK not only produces neuronal excitation in the dorsal born (Jeftinija *et al.* 1981; Dickenson and Magnuson 1989) but can also act to reduce the antinociception produced by morphine and yet produce analgesia in other circumstances (Baber *et al.* 1989). The ability of CCK to interfere with morphine analgesia may be relevant to the clinical states where morphine responsiveness is reduced (Arner and Meyerson 1988). We have used recordings of nociceptive neurons in the rat spinal cord to gauge the effects of a variety of opioid receptor agonists on the transmission of peripheral sensory responses in the anaesthetized animal (Dickenson *et al.* 1987). Clear but independent inhibitions of the noxious evoked activities of these cells follows the activation of μ and δ opioid receptors (Dickenson *et al.* 1987; Sullivan *et al.* 1989). Here we have studied the ability of CCK to interfere with μ and δ opioid actions in the spinal cord and by the use of antagonists attempted to identify the CCK receptor subtypes involved. The recording of sensory neurons allows the actions of these agents to be studied without the complication of sedative, motor, or motivational effects influencing the findings.

In the first set of experiments the effect of sulphated CCK octapeptide (CCK-8S) was gauged on the responses of the neurons to A β- and C-fibre stimulation and on the post-discharges of the cells resulting from the C-fibre stimuli. Intrathecal (i.t.h.) CCK (at a dose of 1 μg which produced maximal changes) elicited a modest but significant elevation of the C-fibre responses of the cells and a marked increase in the post-discharges. The effects lasted about 15–20 min. The responses of the cells to A β-fibre stimulation were not influenced by the peptide. Interestingly, this combination of effects is the exact opposite of those produced by *N*-methyl-D-aspartate (NMDA) antagonists on these cells; *in vitro* studies of a longitudinal slice of the spinal cord reveal that the excitatory effects of pressure-ejected L-glutamate and CCK-8 are very similar in onset and degree, but that CCK excited a subpopulation of the glutamate-responsive cells (Dickenson and Magnuson 1989).

If the application of CCK was followed by the administration of the selective μ opioid agonist (D-Ala^2, MePhe^4, Gly(ol)^5) enkephalin (DAGOL) at a dose which in control experiments caused a profound (70 per cent) inhibition of the neurons (Dickenson *et al.* 1987), the opioid inhibitions were abolished (Fig. 46.1). This effect is remarkably similar to that seen in studies on the flexion reflex where CCK blocked morphine-induced depressions of the motor response (Weisenfeld-Hallin and Duranti 1987). However, when equi-effective doses of the δ opioid agonist Tyr–D–Ser(Otbu)–Gly–Phe–Leu–Thr (DSTBULET) (Sullivan *et al.* 1989) were applied instead of DAGOL, the inhibitions were unaltered by the presence of CCK (Fig. 46.1). Thus the ability of CCK to reduce opioid antinociception in the rat spinal cord is specific for that elicited by μ opioid receptor activation. This is important since it strongly suggests that the action of CCK in producing this effect is not due to generalized changes in neuronal activity produced either by CCK itself or via other spinal transmitter systems (Magnuson *et al.* 1990). A similar finding that CCK attenuated the spinal inhibitions of nociceptive neurons produced by morphine has recently been reported (Kellstein *et al.* 1991).

In a second series of studies we found identical actions of the peptide FLFQPQRF-NH2 (Allard *et al.* 1989) on μ and δ opioid neuronal inhibitions, so that μ but not δ effects were prevented by this mammalian equivalent of FMRF amide, which is also found in the superficial levels of the spinal cord.

We then examined the ability of CCK antagonists to enhance the inhibitory effects of DAGOL on the premise that any endogenous tonic or evoked CCK activity could reduce μ opioid effects. The study used two antagonists: the non-selective CCK_A and CCK_B receptor antagonist IM9/11 (Boc–Tyr (SO$_3$H)–Nle–Gly–Trp–Orn(z)–Asp–Phe–NH$_2$), (see Chapter 6 of this volume) and the selective CCK_A antagonist devazepide (Chang and Lotti 1986). In these experiments it was first established that neither antagonist caused any change in either A- or C-fibre activity when applied at $10\ \mu$M doses by the i.t.h. route. This indicates at most a minor role of CCK in the responses of the cells to peripheral stimulation in these conditions. However, the responses of these cells to both acute and prolonged noxious peripheral stimulation can be reduced by antagonism at the NMDA receptor channel complex and abolished by non-NMDA antagonists (Dickenson and Sullivan, 1990; Haley *et al.* 1990). Thus the excitatory input on to the cells and any subsequent amplification of input are likely to be mediated by amino acids rather than by peptides, although this does not preclude important roles for peptides in the priming of these responses and in other facets of the complex systems activated by peripheral stimuli.

The application of the non-selective antagonist IM9/11 (Marseigne *et al.* 1988) by the i.t.h. route prior to the administration of spinal DAGOL caused a potentiation of the inhibitory effects of the selective μ opioid. Thus in control experiments the dose of 0.05 μg of DAGOL produced a mean maximal 4 ± 7 per cent inhibition of the C-fibre responses of the cells which became a 32 ± 5

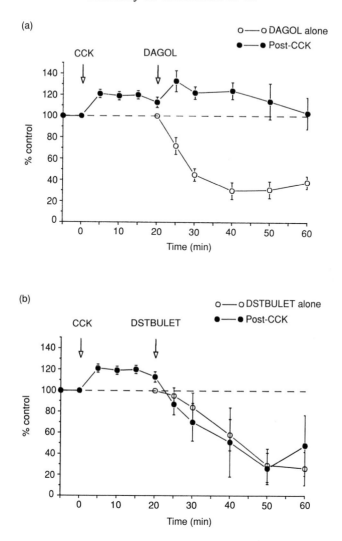

Fig. 46.1. (a) The i.t.h. application of CCK8S (1 μg) prevents the inhibition of dorsal horn neurons in the rat spinal cord by the selective μ opioid agonist DAGOL (0.5 μg). The inhibitory effects of DAGOL alone on the C-fibre-evoked responses of the cells (○) are abolished by pretreatment with CCK 20 min prior to the administration of DAGOL (●).

(b) In contrast, the δ opioid agonist DSTBULET (100 μg) produces inhibition of the same magnitude which is not influenced in any way by CCK (lower panel).

per cent inhibition in the presence of IM9/11 (Fig. 46.2). In contrast, 10 μM of devazepide did not alter the inhibitory effects of DAGOL in either direction. This supports the idea that the CCK_B receptor subtype is important in the modulatory effects of CCK on opioid analgesia in the rat as suggested by behavioural studies (Dourish *et al.* 1990; Chapter 44 of this volume). The present studies indicate that an important site of action of CCK is the spinal cord. In addition, the potentiating dose of the antagonist IM9/11 reduced the excitatory effects of CCK at the dose of the agonist (1 μg) that blocked the actions of DAGOL.

There is a large body of behavioural studies by Watkins and colleagues on i.t.h. and supraspinal morphine analgesia and the opioid analgesia produced by front-paw stimulation which is mediated by endogenous opioid systems (Watkins *et al.* 1982, 1984, 1985). The administration of proglumide potentiated the analgesia, and the routes of administration of the opioids in the various studies clearly demonstrate that this interaction occurs at both spinal and supraspinal levels (Baber *et al.* 1989; Chapter 43 of this volume). However, proglumide is a relatively weak and non-selective antagonist of CCK and has been shown to interact with opioid receptors (Rezvani *et al.* 1987). This latter point obviously causes problems in the interpretation of studies where proglumide potentiates opioid actions. The use of the recently developed potent and selective antagonists avoids this problem, and overall there can be little doubt that CCK antagonism augments opioid analgesia (Chapters 43, 44, and 45 of this volume).

The site of this interaction has not yet been determined, and problems are caused by the distribution of opioid receptors in the dorsal horn of the spinal

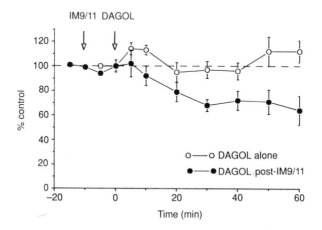

Fig. 46.2. In this experiment the non-selective CCK_A and CCK_B receptor antagonist IM9/11 (10 μM) potentiated the effects of a low dose of DAGOL (0.05 μg) when applied 10 min prior to the opioid (n = 7 in each group). The antagonist had no effect alone but prevented the excitation of cells produced by CCK alone.

cord. The receptors are predominantly located on presynaptic terminals of the C-fibre afferents, but are also located postsynaptically on the soma of the output cells, their dendrites, and on interneurons, or most probably a combination of these sites (Zajac *et al.* 1989). In recordings from an *in vitro* longitudinal slice of the dorsal horn we used the pressure ejection of glutamate to excite neurons directly and so bypassed the afferent terminals. Under these conditions we could see no interaction between bath applied morphine and CCK, in contrast with the *in vivo* experiments described earlier (Dickenson and Magnuson 1989). Since the presynaptic opioid receptors can modulate only in the latter situation and only postsynaptic receptors are present in the *in vitro* studies, the site of action of CCK and the μ opioid receptor may well be the presynaptic terminals of the C fibres. In keeping with the idea that presynaptic opioid receptors are of importance in the modulating effect of CCK, we have observed that μ opioids at low doses facilitate the C-fibre responses of these nociceptive neurons prior to inhibiting them as the dose increases. This has also been observed in recordings of dorsal root ganglion cells, which provide a model for presynaptic opioid receptors (Crain and Shen 1990). CCK not only reduced the inhibitory effects of opioids but also enhanced the facilitation indicative of a presynaptic locus. It would be of considerable interest to study the distribution of CCK receptor sites before and after rhizotomy in order to gauge whether the receptors are located pre- or postsynaptically or are a combination of both.

Thus all these electrophysiological studies on sensory dorsal horn neurons suggest that CCK plays an important role acting via the CCK_B receptor in the spinal analgesia produced by μ opioids in the rat. This is in keeping with the receptor subtype in the spinal cord of this species, in contrast with humans and primates where the spinal receptor is the CCK_A type (Hill *et al.* 1988). The studies where only blockade of both the CCK_A and the CCK_B receptors, and not the CCK_A receptor alone, potentiated the effects of the μ opioid DAGOL show that there is either an evoked or a tonic release of CCK which reduces μ opioid effectiveness via the CCK_B receptor. The experiments with the μ and δ opioid agonists in the spinal cord pretreated with CCK show that the action of CCK is specific for the μ receptor since δ opioid antinociception was unaltered by CCK. This is in agreement with the behavioural results described in Chapter 45 of this volume, where morphine analgesia is attenuated by CCK, and the supraspinal studies described in Chapter 45. However, there are some studies showing that δ opioid effects are influenced by CCK, but these have used either poorly selective ligands or supraspinal administration where the role of δ agonists is controversial (Heyman *et al.* 1988). According to some studies κ opioid analgesia is not influenced by CCK (Barbaz *et al.* 1989), but interactions between CCK and κ opioid analgesia have recently been described by other workers (Chapter 45 of this volume).

However, because of the selectivity of CCK for μ but not δ antinociception in the rat spinal cord, the action of CCK on μ opioid analgesia is unlikely to be due to increased neuronal responses caused either directly or via changes in descending or intrinsic circuits. Nor can it be due to any general change in other transmitters or effector mechanisms, since the same neurons can be inhibited by activation of μ and δ opioid receptors and any general effect on the cell would then alter both opioid modulations. Evidence suggests that CCK does not directly bind to opioid receptors, yet CCK can suppress μ and κ, but not δ, agonist binding to the appropriate opioid receptor, effects that are blocked by proglumide (Wang and Han 1990). Thus a probable explanation is that CCK activates a CCK receptor which in turn influences opioid receptors. Since the electrophysiological effects described here may be due to CCK_B receptor activation, the simplest explanation would be some form of CCK_B–μ opioid receptor interaction, perhaps involving allosteric or post-receptor changes which lead to reduced μ opioid actions (Chapters 44 and 45). Thus CCK_B receptor antagonists could be used as adjuncts to low doses of morphine to reduce the side-effects of morphine but with the same analgesia as high doses. They may also be useful in determining the contribution of CCK to those clinical states where morphine has reduced effectiveness, many of which involve pathology of the peripheral nerve (Arner and Meyerson 1988). This peripheral disruption could conceivably cause enhanced release of CCK, or a loss of control of the release of CCK from the intrinsic spinal cord cells, and the elevated levels of CCK could in turn reduce morphine analgesia. It will be important to determine whether enhanced morphine analgesia is accompanied by increased side-effects. (see Chapter 44 of this volume) This point is important, since the synthesis of morphine analogues over the years has not led to any compound which has combined analgesia with an absence of side-effects. However, the ability of CCK antagonists to enhance morphine analgesia but with differential effects on dependence and tolerance suggests that the various facets of morphine actions may be separated (Dourish *et al.* 1990; Chapter 44 of this volume).

The use of morphine to relieve pain in man can be limited by the unwanted effects of the drug, tolerance, and the existence of pain states with reduced sensitivity to the opioid. The use of CCK antagonists as adjuvants to morphine could then be of value in allowing lower doses of morphine to be used, thereby reducing side-effects and tolerance. In addition, they will be valuable tools in determining whether alterations in endogenous CCK contribute to morphine-insensitive pains and if this was the case, CCK antagonists would be valuable therapeutic agents.

References

Allard, M., Geoffre, S., Legendre, P., Vincent, J. D., and Simonnet, G., (1989). Characterization of rat spinal cord receptors to FLFQPQRF amide, a mammalian morphine modulating peptide: a binding study. *Brain Res.* **500**, 169–76.

Arner, S. and Meyerson, B. A. (1988). Lack of analgesic effect of opioids on neuropathic and idiopathic forms of pain. *Pain* **33**, 11–24.

Baber, N. S., Dourish, C. T., and Hill, D. R. (1989). The role of CCK, caerulein and CCK antagonists in nociception. *Pain* **39**, 307–28.

Barbaz, B. S., Hall, N. R., and Liebman, J. M. (1989). Antagonism of morphine analgesia by CCK-8-S does not extend to all assays nor to all opiate analgesics. *Peptides* **9**, 1295–1300.

Chang, R. S. L. and Lotti, V. J. (1986). Biochemical and pharmacological characterization of an extremely potent and selective non-peptide cholecystokinin antagonist. *Proc. Natl Acad. Sci. USA* **83**, 4923–6.

Crain, S. M. and Shen, K.-F. (1990). Opioids can evoke direct receptor-mediated excitatory effects on sensory neurons. *Trends Pharmacol. Sci.* **11**, 77–81.

Dickenson, A. H. and Magnuson, D. S. K. (1989). The actions of cholecystokinin, morphine and DSTBULET in the rat spinal cord *in vitro*. *J. Physiol.* **415**, 52P.

Dickenson, A. H. and Sullivan, A. F. (1990). Differential effects of excitatory amino-acid antagonists on dorsal horn nociceptive neurons in the rat. *Brain Res.* **506**, 31–9.

Dickenson, A. H., Sullivan, A. F., Knox, R. J., Zajac, Z. M., and Roques, B. P. (1987). Opioid receptor sub-types in the rat spinal cord: electrophysiological studies with μ and δ opioid receptor agonists in the control of nociception. *Brain Res.* **413**, 36–44.

Dourish, C. T., O'Neill, M. F., Coughlan, J., Kitchener, S. J., Hawley, D., and Iversen, S. D. (1990). The selective CCK-B antagonist L-365,260 enhances morphine analgesia and prevents morphine tolerance in the rat. *Eur. J. Pharmacol.* **176**, 35–44.

Haley, J. E., Sullivan, A. F., and Dickenson, A. H. (1990). Evidence for spinal N-methyl-D-aspartate receptor involvement in prolonged chemical nociception in the rat. *Brain Res.* **518**, 218–26.

Heyman, J. S., Vaught, J. L., Raffa, R. B., and Porreca, F. (1988). Can supraspinal δ opioid receptors mediate antinociception? *Trends Pharmacol. Sci.* **9**, 134–8.

Hill, D. R., Shaw, T. M., and Woodruff, G. N. (1988). Binding sites for ^{125}I-cholecystokinin in primate spinal cord are of the CCK-A subclass. *Neurosci. Lett.* **89**, 133–9.

Jeftinija, S., Miletic, V., and Randic, M. (1981). Cholecystokinin octapeptide excites dorsal horn neurons both *in vivo* and *in vitro*. *Brain Res.* **213**, 231–6.

Kellstein, D. E., Price, D. D., and Mayer, D. J. (1991). Cholecystokinin and its antagonist lorglumide respectively attenuate and facilitate morphine induced inhibition of C-fibre evoked discharges of dorsal horn neurons. *Brain Res.* **540**, 302–6.

Magnuson, D. S. K., Sullivan, A. F., Simonnet, G., Roques, B. P., and Dickenson, A. H. (1990). Differential interactions of cholecystokinin and FLFQRF-amide with μ and δ antinociception in the rat spinal cord. *Neuropeptides* **16**, 213–18.

Marseigne, I., Dor, A., Begue, D., Reibaud, M., Zundel., Blanchard, J. C., Peleprat, D., and Roques, B. P. (1988). Synthesis and biological activity of CCK$_{26-33}$ related analogues modified in position 31. *J. Med. Chem.* **31**, 966–70.

Rezvani, A., Stokes, B., Rhoads, D. L., and Way, E. L. (1987). Proglumide exhibits delta-opioid agonist properties. *Alcohol Drug Res.* **7**, 135–46.

Sullivan, A. F., Dickenson, A. H., and Roques, B. P. (1989). δ-opioid inhibition of acute and prolonged noxious evoked responses in rat dorsal horn neurons. *Br. J. Pharmacol.* **98**, 1039–49.

Wang, X.-J. and Han, J.-S. (1990). Modification by cholecystokinin octapeptide of the binding of μ-, δ- and κ-opioid receptors. *J. Neurochem.* **55**, 1379–82.

Watkins, L. R., Cobelli, D. A., Faris, Aceto, M. D., and Mayer, D. J. (1982). Opiate vs. non-opiate footshock-induced analgesia. (FSIA): the body region shocked is a critical factor. *Brain Res.* **242**, 299–308.

Watkins, L. R., Kinsheck, I. B., and Mayer, D. J. (1984). Potentiation of morphine analgesia and apparant reversal of morphine tolerance by proglumide. *Science* **224**, 395-6.

Watkins, L. R., Kinsheck, I. B., Kaufman, E. F. S., Miller, J., Frenk, H., and Mayer, D. J. (1985). Cholecystokinin antagonists selectively potentiate analgesia induced by endogenous opiates. *Brain Res.* **327**, 181-90.

Weisenfeld-Hallin, Z. and Duranti, R. (1987). Intrathecal cholecystokinin interacts with morphine but not substance P in modulating the nociceptive flexion reflex in the rat. *Peptides.* **8**, 153-8.

Williams, R. G., Dimaline, R., Varo, A., Isetta, A. M., Trizio, D., and Dockray, G. J. (1987). Cholecystokinin octapeptide in the rat central nervous system: immunohisto-chemical studies using a monoclonal antibody that does not react with CGRP. *Neurochem. Int.* **11**, 433-42.

Yaksh, T. L. and Noueihed, R. (1985). The physiology and pharmacology of spinal opiates. *Ann. Rev. Pharmacol. Toxicol.* **25**, 433-62.

Yaksh, T. L., Eustaquio, O. A., and Go, V. L. W. (1982). Studies on the location of release of cholecystokinin and vasoactive intestinal polypeptide in rat and cat spinal cord. *Brain Res.* **242**, 279-90.

Zajac, J.-M., Lombard, M.-C., Peschanski, M., Besson, J.-M., and Roques, B. P. (1989). Autoradiographic study of μ and δ opioid binding sites and neutral endopep-tidase in rat after dorsal root rhizotomy. *Brain Res.* **500**, 169-76.

47. Opioid control of CCK release from the rat spinal cord—interaction between μ, δ, and κ receptors

J. J. Benoliel, S. Bourgoin, A. Mauborgne, M. Hamon, and F. Cesselin

Numerous data in the literature suggest that CCK interacts with opioids in pain mechanisms. Thus large doses of CCK have been shown to induce a naloxone-reversible analgesia, whereas small doses of this peptide can prevent the anti-nociceptive action of opioids (Chapters 43 and 44 of this volume). The latter pharmacological observation is in line with a hypothetical opioid antagonist property of CCK under physiological conditions. Indeed, the blockade of CCK receptors by proglumide or selective CCK_A and CCK_B antagonists results in a marked enhancement in morphine- or β-endorphin-induced analgesia (Chapters 43 and 44 of this volume). Furthermore, tolerance to the analgesic effect of morphine is markedly delayed when CCK receptors are blocked, which suggests that chronic stimulation of opioid receptors by morphine may trigger a progressive compensatory increase in the activity of CCK-containing neurons in the CNS (Baber *et al.* 1989; Dourish *et al.* 1990; Chapters 43 and 44 of this volume).

Although this brief overview of the relevant literature suggests the hypo-thesis of a stimulatory effect of opioid receptor activation on central CCKergic systems, there have been few direct investigations of this possibility (see below). Therefore we examined the effect of μ, δ, and κ opioid agonists on the release of CCK-like material (CCKLM) from the dorsal horn of the rat spinal cord, a region which plays a critical role in the transfer and control of nociceptive messages. As some studies suggest that κ agonists can modulate the effects of agonists at other types of opioid receptors (e.g. Dickenson and Knox 1987), we also investigated the possible changes in CCLKM release from tissues exposed to both a κ agonist and either a μ or a δ agonist.

Slices of the dorsal zone of the lumbar enlargement of adult male rats were superfused at a flow rate of 1 ml/4 min with artificial cerebrospinal fluid (aCSF) in thermostatic (37 °C) chambers (see Mauborgne *et al.* (1987) for details). After washing for 20 min, superfusate fractions (15 × 1 ml) were collected at 0 °C and stored at −30 °C prior to measurement of their CCKLM content. Tissue depolarization was achieved by increasing the KCl concentration from 5.4 to 30 mM in the superfusing fluid during collection of fractions 3 and 4 (K_1) and fractions 12 and 13 (K_2). The ratio K_2/K_1 of CCKLM overflow was calculated as described by Mauborgne *et al.* (1987). Compounds to be tested were added from the beginning of the eighth fraction until the end of the experiment so that any change in K_2/K_1 in the presence of a given substance could be ascribed to the effect of this substance on K^+-induced CCKLM overflow. CCKLM in tissues and superfusates was radio-immunoassayed as described previously (Studler *et al.* 1982).

The spontaneous CCKLM outflow from dorsal lumbar cord slices was stable throughout the experiments (2.23 ± 0.15 pg/ml, i.e. 0.56 ± 0.04 pg/min) and corresponded to 0.012 per cent (fractional rate constant) of CCKLM tissue content released per minute. K^+-induced depolarization produced a marked enhancement of CCKLM release, in as much as a fivefold overflow was observed for the K_1 pulse. The CCKLM outflow returned to baseline levels 4 min after the K^+-enriched medium was changed to the normal medium. The second exposure (K_2) to 30 mM K^+ also induced a significant, but less

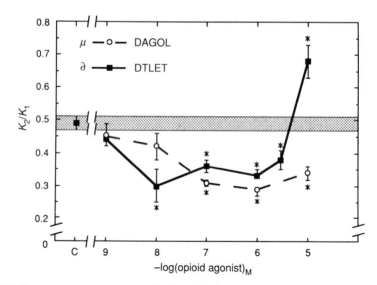

Fig. 47.1 Dose-response curves of the effects of DTLET and DAGOL on the K^+-evoked release of CCKLM from the dorsal part of the lumbar enlargement. Slices of the dorsal half of the lumbar enlargement were superfused with aCSF and depolarized twice (K_1, K_2) by 30 mM K^+ as described in the text. CCKLM was measured in each superfusate fraction (1 ml/4 min), and the ratio K_2/K_1 of K^+-evoked CCKLM overflow due to the second K^+ pulse to that due to the first K^+ pulse was calculated. Each point is the mean ± SEM of data obtained in at least 10 independent experiments. * $p < 0.001$ when compared with K_2/K_1 in superfusion experiments where tissues were depolarized in the absence of drugs (Point C on the abscissa).

pronounced, increase in CCKLM release, leading to a K_2/K_1 ratio of 0.49 ± 0.02.

Under control conditions, none of the drugs tested affected the spontaneous outflow of CCKLM from spinal cord slices. In contrast, significant modifications of the K^+-induced CCKLM overflow were observed when opioids were added to the superfusing fluid.

Thus morphine (10 μM) induced a significant increase in K^+-evoked CCKLM release ($K_2/K_1 = 0.67 \pm 0.07$, $p < 0.05$). Although inactive at concentrations below 10 nM, the selective μ agonist (D-Ala^2, MePhe^4, Gly(ol)^5) enkephalin (DAGOL) (0.1–10 μM) significantly reduced (−40 per cent) the K^+-evoked CCKLM overflow (as shown by the decrease in K_2/K_1 (Fig. 47.1)). Similarly, the selective δ opioid receptor agonist Tyr–D–Thr–Gly–Phe–Leu–Thr (DTLET) (0.01–3 μM) significantly decreased (−30 per cent) the K^+-evoked peptide overflow (Fig. 47.1). In contrast, a marked increase in K^+-evoked CCKLM overflow (+40 per cent) was induced by a higher concentration (10 μM) of DTLET (Fig. 47.1).

Although neither the selective μ antagonist naloxone (1 μM) nor the selective δ antagonist ICI 154129 (50 μM) affected the K^+-evoked CCKLM overflow, these drugs prevented the inhibitory effect of DAGOL (10 μM) and

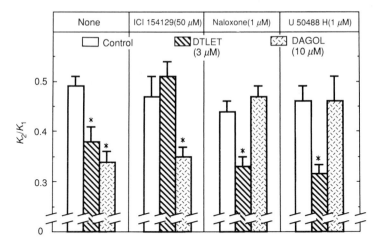

Fig. 47.2 Effects of the opioid antagonists (naloxone and ICI 154129) and the κ opioid agonist (U 50488 H) on the inhibition by DAGOL, and DTLET of the K^+-evoked release of CCKLM from the dorsal zone of the lumbar enlargement. The experimental protocol was the same as that described in Fig. 47.1. Drug-induced changes in the K^+-evoked CCKLM release were assessed from the ratio K_2/K_1. Each bar is the mean ± SEM of values calculated from at least 10 independent experiments. $\star$ $p < 0.001$ when compared with K_2/K_1 in superfusion experiments where the second K^+ pulse was applied without drugs ('none', control) or in the presence of an opioid antagonist (1 μM naloxone or 50 μM ICI 154129) or the κ opioid agonist U 50488 H (1 μM) alone.

DTLET (3 μM) respectively (Fig. 47.2). The selective κ opioid agonist U 50488 H (1 μM) did not modify the K^+-induced release of CCKLM. However, it prevented the inhibitory effect of 10 μM of DAGOL without affecting that of 3 μM of DTLET (Fig. 47.2).

These data show that the selective μ opioid agonist DAGOL, at concentrations above 10 nM, exerted an inhibitory effect on CCKLM release from slices of the dorsal horn of the rat spinal cord. This effect probably involved μ receptors since it was reversed by 1 μM of naloxone (a concentration which preferentially blocks μ opioid receptors) but not by 50 μM of ICI 154129 (a selective δ opioid antagonist). This result is consistent with the findings of Rodriguez and Sacristan (1989), who showed that CCK release from the rat spinal cord *in vivo* can be reduced by DAGOL. Previous *in vitro* studies also demonstrated that the stimulation of μ opioid receptors by morphine (Micevych *et al.* 1985) and by DAGOL and PL 017 (Benoliel *et al.* 1989) exerts an inhibitory influence on the Ca^{2+}-dependent release of CCKLM from slices of rat hypothalamus and substantia nigra.

The modulation of CCKLM release by δ opioid agonists seems to be more complex than that induced by the simulation of μ opioid receptors. Indeed, it has been reported that a low concentration (10 nM) of Tyr–D–Ala–Gly–Phe–D–Leu (DADLE) inhibits CCKLM release from slices of rat hypothalamus but not from slices of rat frontal cortex (Micevych *et al.* 1985). In contrast, 3 μM of DTLET enhanced release of CCKLM from slices of rat substantia nigra (Benoliel *et al.* 1989). In the present study, only 10 μM of DTLET increased CCKLM release from rat spinal cord slices, whereas low to moderate concentrations (from 10 nM to 3 μM) of the δ opioid agonist were inhibitory. The opposite effects of 3 μM of DTLET on CCKLM release from slices of substantia nigra or spinal cord are probably mediated by δ receptors since both can be prevented by ICI 154129 but not naloxone (1 μM). Furthermore, the possible involvement of μ opioid receptors in the enhancing effect of 10 μM of DTLET on CCKLM release from spinal cord slices can be ruled out because the stimulation of these receptors by DAGOL, even at a concentration of 10 μM, produced a decrease in the peptide release. Therefore differences seem to exist regarding the δ opioid receptor control of CCK-containing neurons in different regions of the CNS. These regional differences may be partly explained by subtypes of δ receptors, as recently suggested by Vaughn *et al.* (1990).

Although it would be necessary to explore a larger range of concentrations of various opioid agonists before drawing any firm conclusions, the present data suggest that activation of CCKergic systems by opioids (Baber *et al.* 1989) probably involves δ rather than μ opioid receptors. However, it must be emphasized that our experiments may be relevant to the acute effects of opioids on CCKLM release, whereas the compensatory increase in the activity of central CCKergic systems has been proposed in the case of chronic treatment with these drugs (see above). The possible modifications of CCKLM release after chronic morphine administration in rats are currently being investigated in our laboratory.

In contrast with DAGOL and DTLET, the selective κ agonist U 50488 H, at least at the concentration used, did not modify the release of CCKLM from spinal cord slices. However, this compound prevented the effect of DAGOL, but not that of DTLET, on peptide release. These data are in line with numerous studies showing that activation of κ receptors can reduce or abolish the effects of μ agonists, notably the DAGOL-induced inhibition of nociceptive neurons in the rat spinal cord (Dickenson and Knox 1987).

Owing to the lack of selectivity of morphine towards the various types of opioid receptors, not only μ but also δ and κ receptors are probably stimulated by high doses of this drug. Accordingly, its inhibitory effect through the stimulation of μ receptors might be masked by that of κ receptors, leading to a net increase in CCKLM release through the activation of δ receptors. The increased release of CCKLM that we observed in the presence of 10 μM of morphine is compatible with such a hypothesis.

References

Baber, N. S., Dourish, C. T., and Hill, D. R. (1989). The role of CCK, caerulein, and CCK antagonists in nociception. *Pain* **39**, 307–28.

Benoliel, J. J., Mauborgne, A., Bourgoin, S., Hamon, M., and Cesselin, F. (1989). Modulations of the *in vitro* release of CCK8-like material from the rat substantia nigra by μ, δ and κ opioid receptor agonists. In *The neuropeptide cholecystokinin (CCK)* (ed. J. Hughes, G. Dockray, and G. Woodruff), pp. 96–100. Wiley, Chichester.

Dickenson, A. H. and Knox, R. J. (1987). Antagonism of μ-opioid receptor-mediated inhibitions of nociceptive neurones by U 50488 H and dynorphin A_{1-13} in the rat dorsal horn. *Neurosci. Lett.* **75**, 229–34.

Dourish, C. T., O'Neill, M. F., Coughlan, J., Kitchener, S. J., Hawley, D., and Iversen, S. D. (1990). The selective CCK-B receptor antagonist L-365, 260 enhances morphine analgesia and prevents morphine tolerance in the rat. *Eur. J. Pharmacol.* **176**, 35–44.

Mauborgne, A., Bourgoin, S., Benoliel, J. J., Hirsch, M., Berthier, J. L., Hamon, M., and Cesselin, F. (1987). Enkephalinase is involved in the degradation of endogenous substance P released from slices of rat substantia nigra. *J. Pharmacol. Exp. Ther.* **243**, 674–80.

Micevych, P. E., Yaksh, T. L., Go, V. L. W., and Finkelstein, J. A. (1985). Effect of opiates on the release of cholecystokinin from *in vitro* hypothalamus and frontal cortex of Zucker lean (Fa/-) and obese (fa/fa) rats. *Brain Res.* **337**, 382–5.

Rodriguez, R. E. and Sacristan, M. P. (1989). *In vivo* release of CCK-8 from the dorsal horn of the rat: inhibition by DAGOL. *FEBS Lett.* **250**, 215–17.

Studler, J., Javoy-Agid, F., Cesselin, F., Legrand, J. C., and Agid, Y. (1982). CCK8 immunoreactivity distribution in human brain: selective decrease in the substantia nigra from parkinsonian patients. *Brain. Res.* **243**, 715–18.

Vaughn, L. K., Wire, W. S., Davis, P., Shimohigashi, Y., Toth, G., Knapp, R. J., Hruby, V. J., Burks, T. F., and Yamamura, H. I. (1990). Differentiation between rat brain and mouse vas deferens δ opioid receptors. *Eur. J. Pharmacol.* **177**, 99–101.

48. *In vivo* release of CCK-8 from the dorsal horn of the rat: modulation by selective opioid agonists

Raquel E. Rodriguez, and Maria P. Sacristan

Introduction

The octapeptide form of CCK (CCK-8) predominates in the CNS of mammalian species, including man (Beinfeld 1983). Many of the physiological roles of CCK-8 in the CNS are unknown, and the physiological actions attributed to this peptide are as diverse as its distribution. CCK-8 shows an excitatory action in the gastro-intestinal tract, and it is a potent activator of digestion, pancreatic secretion, and gall bladder contraction. In the CNS, the physiological actions attributed to CCK-8 include regulation of satiation, modulation of catecolaminergic activity, and regulation of hypothalamic peptides. A variety of effects have been observed after central administration of CCK-8. These include hypothermia (Katsuura *et al*. 1981), hyperglycemia (Morley and Levine 1980), and sometimes analgesia (Pittaway *et al*. 1987). Also, electrophysiological data have shown both excitatory and inhibitory actions of CCK-8 in the CNS (Salt and Hill 1982).

There is supporting evidence that CCK-8 may interact with opioid systems. CCK-8 increases respiration physiologically (Gillis *et al*. 1983), whereas a well-known side-effect of morphine is respiratory depression. Probably the most important point is that many of the CCK-8-containing loci are areas observed to contain high concentrations of endogenous opioids, many of which are implicated in pain modulation (Baber *et al*. 1989). Also, the distributions of enkephalins and CCK are very similar, and a functional relationship may exist between them. It should be noted, however, that sulphated CCK-8 does not bind to opioid receptors in the brain (Stengaard-Pedersen *et al*. 1981). In antinociceptive tests CCK has been shown to block opioid analgesia, whereas CCK antagonists produce the opposite effect (Baber *et al*. 1989; Chapters 43 and 44 of this volume).

Quantification of the peptide content in subcellular fractions derived from brain tissue has shown that CCK-8 is largely recovered in the synaptosomal fraction (Dodd *et al*. 1980; Malesci *et al*. 1980). There is evidence that the CCK-8 localized within nerve terminals is part of a releasable fraction (Rehfeld 1978). In addition, depolarization releases CCK-8 immunoreactivity from *in vitro* brain preparations in a calcium-dependent fashion (Pinget *et al.* 1978; Dodd *et al.* 1980).

In vivo release of CCK-8 from the dorsal horn of the rat

We have used a spinal superfusion system to examine factors governing the movement of CCK-8 from the spinal cord into the extracellular space and its dependence upon selective opioid agonists. To perfuse the spinal subarachnoid space an exposure of the cisterna magna was made by a mid-line incision and blunt dissection, followed by incision of the dura and arachnoid membranes. A polythene tube 10 cm long (0.28 mm i.d., 0.61 mm o.d.) was advanced down the subarachnoid space to the lumbar area. The tip of this tube served as an inflow cannula and artificial cerebrospinal fluid (aCSF) was pumped in at a rate of 110 μl/min. The aCSF was withdrawn at a rate of 100 μl/min using another polythene tube positioned at the opening of the cisterna magna. Before entering the animal's body, the perfusion fluid was warmed (37 °C) and continuously oxygenated (95 per cent O_2, CO_2). Experiments were conducted by first flushing the subarachnoid space with aCSF for 1 h to allow stabilization of the preparation, and then collecting 5 min (500 μl) fractions. The composition of the aCSF used and the radio-immunoassay (RIA) procedure have been published elsewhere (Rodriguez and Sacristan 1989). We have used synthetic CCK rabbit antiserum as our CCK antibody source. The sensitivity of the assay is 0.2 fmol/tube with 6 per cent intra-assay and 10 per cent inter-assay variation.

Stimulation procedure

Potassium stimulation was obtained by raising the concentration of KCl to 50 mM, while a corresponding amount of NaCl was removed from the aCSF solution to minimize undesirable osmotic effects. To stimulate somatosensory input into the lumbar spinal cord of the rat, the sciatic nerve was exposed and prepared for stimulation and recording of the compound action potential. Stimulation of the nerve was performed using rectangular pulses: 3–4 V, 0.05 ms, 50 Hz for activation of A$\alpha\beta$ fibres, and 40–50 V, 0.05 ms, 50 Hz for recruitment of Aδ and C fibres.

CCK-8-like immunoreactivity was found to be present in the resting superfusate of the spinal cord of the anaesthetized rat. The mean resting levels in these samples were above the absolute sensitivity of the RIA used (0.2 ± 0.08 fmol/ml per 5 min) ($n = 20$), which is in agreement with other authors (Yaksh *et al.* 1982). The addition of potassium (50 mM excess) to the infused aCSF resulted in a 160 per cent increase in CCK-8 above baseline in the spinal superfusate ($n = 5$; $p < 0.05$, Student's t test) (Fig. 48.1). Upon substitution of normal aCSF, the CCK-8 levels fell within the limits of baseline variability.

To examine whether CCK-8 is released from the spinal cord following

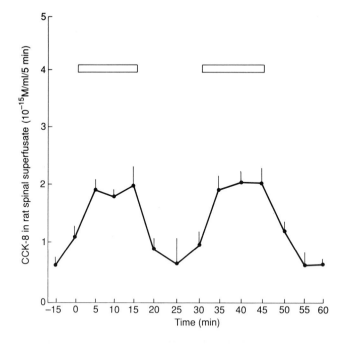

Fig. 48.1. Release of CCK-8 from superfused rat spinal cord in response to high potassium levels. The figure shows the amount (in fmol/ml per 5 min fractions) of CCK-8 released in the perfusing medium. The baseline as observed 15 min prior to the administration of 50 mM K^+ is shown to indicate the baseline levels of CCK-8 release in the perfusate. The results shown are the mean from five animals.

activation of specific afferent fibre populations, we stimulated the rat sciatic nerve bilaterally, which resulted in a 282 per cent increase in CCK-8 release (after stimulation of $A\delta/c$ fibres). This provides evidence that CCK-8 is released from rat spinal cord *in vivo* by activating nociceptive primary afferents.

The addition of capsaicin (3×10^{-4} M) to the superfusion medium failed to have any significant effect on CCK-8 levels. The activity of capsaicin in the spinal cord seems to be restricted to nociceptive primary sensory neurons (Jessel *et al.* 1978). In our experiments (Fig. 48.2) direct superfusion of capsaicin into the spinal cord failed to increase the release of CCK-8 from basal levels. On the other hand, chemical stimulation with potassium (50 mM) or electrical stimulation of the sciatic nerve produced a significant release of CCK-8. This suggests that we are measuring intrinsic CCK-8 rather than CCK-8 from primary afferents.

We have also investigated whether spinal application of the μ selective opioid agonist (D-Ala[2], MePhe[4], Gly (ol)[5]) enkephalin (DAGOL), at concentrations known to produce analgesia (1 nM), can affect CCK-8 release from rat spinal cord *in vivo*. Following superfusion with 1 nM of DAGOL, the resting release

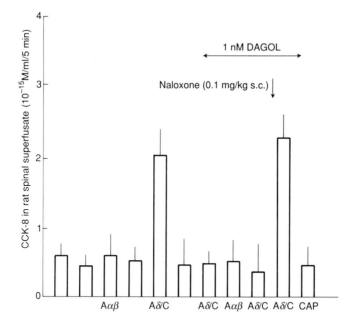

Fig. 48.2 Release of CCK-8 from superfused rat spinal cord in response to sciatic nerve stimulation and stimulation with capsaicin (CAP). The effects of the selective μ opioid agonist DAGOL and naloxone are also shown. Each value is the mean ± SE from five experiments. The statistical significance was assessed using Student's t test ($p \leq 0.05$). To determine whether CCK-8 is released from the spinal cord following activation of specific afferent fibre populations, we stimulated the rat sciatic nerve bilaterally while superfusing segments of the spinal cord receiving sensory input from the sciatic nerve. During stimulation compound action potential was monitored to determine the stimulus intensities required to activate Aαβ fibres alone and to recruit Aδ/C fibres.

of CCK-8 was substantially decreased (Fig. 48.2). Furthermore, bilateral stimulation of the sciatic nerve at intensities that clearly cause the release of CCK-8 before the addition of DAGOL now failed to increase CCK-8 release. Following subcutaneous (s.c.) injection of naloxone hydrochloride (0.1 mg/kg), stimulation of the sciatic nerve at the same intensity in the continued presence of intrathecal DAGOL fully restored the evoked release of CCK-8, suggesting that the effect observed is mediated through opioids.

In order to test the involvement of opioid selectivity in the *in vivo* release of CCK-8, we have also examined the effect produced by the δ selective agonist (D-Pen[2], D-Pen[5])enkephalin (DPDPE) and by the κ selective agonist U 69595. As shown in Fig. 48.3, the *in vivo* release of CCK-8 was partially attenuated by the δ selective agonist DPDPE in a naloxone-reversible (1 mg/kg) fashion. However, perfusion with the κ selective agonist U 69595 did not affect *in vivo*

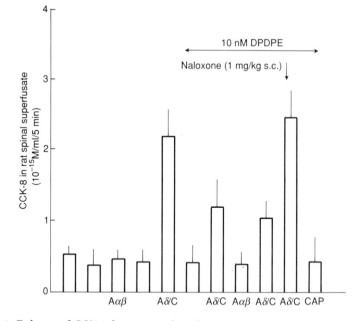

Fig. 48.3 Release of CCK-8 from superfused rat spinal cord in response to sciatic nerve stimulation and stimulation with capsaicin (CAP). The effects of the selective δ opioid agonist DPDPE and naloxone are shown. Each value is the mean ± SE from five experiments. Statistical significance was assessed using Student's t test ($p \leq 0.05$).

release of CCK-8 (Fig. 48.4). These results suggest a μ–δ opioid modulation of the releasable CCK-8 in the spinal cord of the rat (see Chapters 44 and 46 for further discussion).

References

Baber, N. S., Dourish, C. T., and Hill, D. R., (1989). The role of CCK, caerulein and CCK antagonists in nociception. *Pain* **39**, 307–28.

Beinfeld, M. C. (1983). Cholecystokinin in the central nervous system. *Neuropeptides* **3**, 411–21.

Dodd, P. R., Edwardson, J. A., and Dockray, G. J. (1980). The depolarization-induced release of cholecystokinin C-terminal octapeptide (CCK-8) from rat synaptosomes and brain slices. *Regul. Pept.* **1**, 17–29.

Gillis, R. A., Quest, J. A., Pagani, F. D., Souza, J. D., Taveira da Silva, A. M., Jensen, R. T. Garvey, T. Q., and Hamosh, P. (1983). Activation of central nervous system cholecystokinin receptors stimulate respiration in the cat. *J. Pharmacol Exp. Ther.* **224**, 408–14.

Jessell, T. M., Iversen, L. L., and Cuello, A. C. (1978). Capsaicin induced depletion of subtance P from primary sensory neurons. *Brain Res.* **152**, 183–8.

Katsuura, G., Hirota, R., and Itoh, S. (1981) Cholecystokinin-induced hypothermia in the rat. *Experientia* **37**, 60–6.

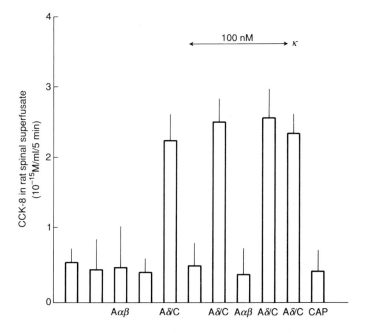

Fig. 48.4 Release of CCK-8 from superfused rat spinal cord in response to sciatic nerve stimulation and stimulation with capsaicin (CAP). The lack of effect of the selective κ agonist U 69595 is shown.

Malesci, A., Straus, E., and Yalow, R. S. (1980). Cholecystokinin converting enzymes in brain. *Proc. Natl Acad. Sci. USA* **77**, 597–9.

Morley, J. E. and Levine, A. S. (1980). Intraventricular CCK-8 produces hyperglycemia and hypothermia. *Clin. Res.* **28**, 712A.

Pinget, M., Straus, E., and Yalow, R. S. (1978). Release of cholecystokinin from a synaptosome-enriched fraction of rat cerebral cortex. *Life Sci.* **25**, 339–42.

Pittaway, K. M., Rodriguez, R. E., Hughes, J., and Hill, R. G. (1987). CCK-8 analgesia and hyperalgesia after intrathecal administration in the rat: comparison with CCK-related peptides. *Neoropeptides* **10**, 87–108.

Rehfeld, J. F. (1978). Immunochemical studies on cholecystokinin. *J. Biol. Chem.* **253**, 4016–21.

Rodriguez, R. E. and Sacristan, M. P. (1989). *In vivo* release of CCK-8 from the dorsal horn of the rat: inhibition by DAGOL. *FEBS Lett.* **250**, 251–17.

Salt, T. E. and Hill, R. G. (1982). The effects of C-terminal fragments of cholecystokinin on the firing of single neurons in the caudal trigeminal nucleus of the rat, *Neuropeptides* **2**, 301–6.

Stengaard-Pedersen, K. and Larson, L. I. (1981). Localization and opiate receptor binding of enkephalin, CCK and ACTH/beta endorphin in the rat central nervous system. *Peptides* **2**, 3–19.

Yaksh, T. L., Abay, E. O., and Go, V. L. W. (1982). Studies on location and release of cholecystokinin and vasoactive intestinal peptide in rat and cat spinal cord. *Brain Res.* **242**, 279–90.

49. Seeking a role for CCK in control of spinal reflexes

Robert W. Clarke, John Harris, and Timothy W. Ford

Introduction

Endogenous opioid peptides are inhibitory modulators of spinal reflex function. Work from many laboratories, including our own, has demonstrated that opioid antagonists such as naloxone increase reflex responses in dog (McClane and Martin 1967), cat (Goldfarb and Hu 1975; Bell and Martin 1977; Duggan *et al.* 1984), man (Boureau *et al.* 1978), and rabbit (Catley *et al.* 1983; Clarke and Ford 1987; Clarke *et al.* 1989*a*). These findings indicate that there is a tonic release of opioid peptides in the spinal cord which results in suppression of reflex responses. Further, intense electrical stimulation of the high threshold afferents of peripheral nerves has been shown to evoke prolonged (15–40 min) naloxone-reversible inhibition of reflexes in cat (Chung *et al.* 1983) and rabbit (Catley *et al.* 1984; Clarke *et al.* 1989*b*), suggesting that intense afferent barrage augments activity in spinal opioidergic neurons.

Our search for a role for CCK in controlling spinal reflexes was stimulated by reports which suggested that this peptide might be an excitatory transmitter or modulator acting in opposition to endogenous opioids, which led to CCK being described as a 'physiological' opioid antagonist. The octapeptide form of CCK has been shown to reduce suppression of the tail-flick reflex of rats induced by morphine or by front-paw shock, a form of inhibition which is probably mediated by endogenous opioids (Faris *et al.* 1983). Morphine-induced suppression of limb-withdrawal reflexes in the rat is also reduced by CCK (Wiesenfeld-Hallin and Duranti 1987). Furthermore, the non-selective CCK receptor antagonist proglumide can enhance the effects of morphine or paw shock, indicating that endogenous CCK actively opposes opioid-mediated inhibition (Watkins *et al.* 1984, 1985; Chapter 43 of this volume). Proglumide also potentiates the effects of opioids on dorsal horn neurons in rat but has no effect when given alone (Suberg *et al.* 1985).

These findings prompted us to examine the effects of proglumide on the sural–gastrocnemius medialis (GM) reflex of the spinalized rabbit, which is tonically inhibited by opioid peptides (Catley *et al.* 1983) and is depressed by a naloxone-reversible mechanism after intense stimulation of the common peroneal (CP) nerve (Catley *et al.* 1984). Proglumide, given intravenously (i.v.)

in doses which were reported to be effective in rat, had no effect on the sural–GM reflex *per se* or on CP-evoked inhibition of the reflex. We concluded that endogenous CCK was probably not involved in modulating this reflex in spinalized animals (Clarke *et al.* 1988*b*). However, doubts about the actions of proglumide, and the development of new, potent, and selective antagonists for CCK receptors (see Baber *et al.* (1989) and Chapter 44 of this volume for a review) led us to reinvestigate the possible roles of endogenous CCK in controlling reflexes. In this study we have used the selective CCK$_A$ receptor antagonist devazepide (O'Neill *et al.* 1989) to probe for possible tonic and stimulus-evoked effects of endogenous CCK. Experiments were performed in rabbits decerebrated under halothane–nitrous oxide anaesthesia. Reflexes were evoked by electrical stimulation of the sural nerve at Aβ (in spinalized animals) or Aδ (in non-spinalized animals) intensity, and were recorded from the GM nerve to be averaged and integrated by computer. Devazepide was dissolved in a 25 per cent dimethyl sulphoxide–75 per cent propylene glycol mixture and given i.v. in volumes of 25–400 μl.

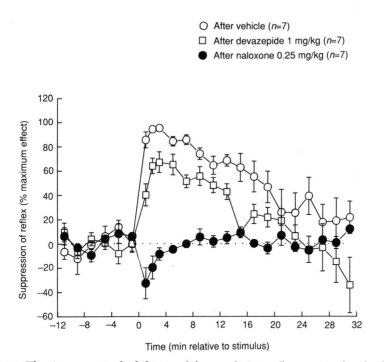

Fig. 49.1 The time course of inhibition of the sural–GM reflex in spinalized rabbits after stimulation of the common peroneal nerve with 50 pulses of 20 V, 1 ms, and the effects upon it of devazepide (open squares) and naloxone (full circles). The effect of the stimulus (applied around time zero) is expressed as a percentage of the maximum suppression of the reflex seen in the vehicle-treated control state, and each point is the mean ± SEM.

The effects of devazepide in spinalized rabbits

In rabbits spinalized at the T12-L1 level, i.v. injection of vehicle or devazepide in doses of 1 mg/kg ($n = 13$) or 2 mg/kg ($n = 5$) had no significant effects on the size or latency of the GM reflex response. The vehicle alone (400 μl) caused mean arterial blood pressure to increase by an average of 9.4 ± 1.7 mmHg, with a concomitant fall in heart rate of 12 ± 2.5 beats/min. Devazepide had no cardiovascular actions over and above those of the vehicle. The effects of repetitive stimulation of the CP nerve were also examined. In the first group of experiments the conditioning stimulus applied to the CP nerve was a train of 100 shocks of 20 V, 1 ms given at 5 Hz, which has been shown to evoke maximal levels of naloxone-reversible inhibition (Taylor *et al.* 1990). The suppression produced by such a stimulus was not affected by vehicle or devazepide at 1 mg/kg (seven experiments) or 2 mg/kg (five experiments). We then went on to look at the effects of conditioning stimuli previously shown to produce submaximal inhibition of the sural–GM reflex, i.e. 50 shocks applied to the CP nerve at 5 Hz. When given at 1 mg/kg, devazepide significantly *reduced* the inhibition evoked by the CP stimulus (Fig. 49.1) and naloxone (0.25 mg/kg i.v.) completely blocked stimulus-evoked inhibition of the reflex as reported previously (Clarke *et al.* 1989b).

The effects of devazepide in non-spinalized rabbits

In decerebrated non-spinalized rabbits, the sural–GM reflex is tonically inhibited by opioid peptides and by noradrenalin released from descending fibres (Clarke *et al.* 1988a), and inhibition evoked from the CP nerve is blocked only by coadministration of naloxone with the selective α_2 adrenoceptor antagonist idazoxan (Clarke *et al.* 1989b). Devazepide was administered to six decerebrated non-spinalized rabbits in volumes of 25–50μl. This volume of vehicle had no effect on cardiovascular parameters. In these preparations, the sural–GM reflex increased after devazepide (0.2 and 0.8 mg/kg i.v.) (Fig. 49.2). This effect was observed in five of six animals, was slow in onset, and showed little sign of recovery over 1 h; in fact the reflex sometimes continued to increase over this period. Stimulation of the CP nerve with 50 pulses of 20 V, 1 ms at 5 Hz resulted in profound suppression of the reflex (Fig. 49.3). Vehicle had no effect on the inhibition generated by this stimulus, but after a 0.2 mg/kg dose of devazepide stimulus-evoked suppression was significantly increased compared with the vehicle-treated control. The inhibition obtained from CP stimulation

Fig. 49.3 The time course of inhibition of the sural–GM reflex in non-spinalized rabbits after stimulation of the common nerve with 50 pulses at 20 V, 1 ms, and the effects upon it of devazepide at doses of 0.2 mg/kg (open squares) and 0.8 mg/kg (full circles), and i.t.h. naloxone + idazoxan (full squares). Details as for Fig. 49.1.

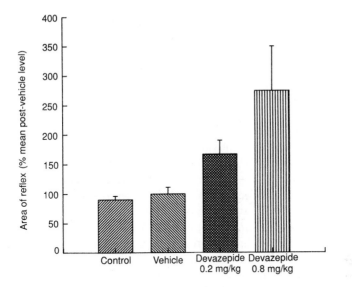

Fig. 49.2 The effects of vehicle (25 per cent dimethyl sulphoxide–75 per cent propylene glycol) and devazepide on the sural–GM reflex in six non-spinalized rabbits. The reflex response is expressed as a percentage of the mean post-vehicle level.

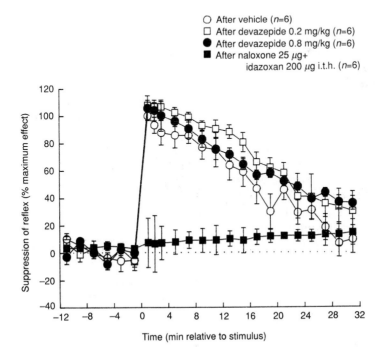

after the 0.8 mg/kg dose of the CCK antagonist was not significantly different from control or from that seen after the lower dose of devazepide. Stimulus-evoked inhibition was almost entirely blocked by intrathecal (i.t.h.) co-administration of naloxone (25 μg) and idazoxan (200 μg).

Conclusions

We were probably premature in dismissing CCK as a modulator of spinal reflexes, but it is not at all obvious from these studies what the role of CCK might be. The most striking action of devazepide was an increase in reflex responses in non-spinalized rabbits, which is the opposite of what had been anticipated. The size of the sural–GM reflex in non-spinalized animals is determined mainly by the balance of activity in descending facilitatory and inhibitory systems (Harris *et al.* 1990), and devazepide seemed to tip this balance in favour of facilitation in a self-reinforcing manner. There are some reports of the presence of CCK-like immunoreactivity in neurons which project to the spinal cord from the regions of the brain associated with descending inhibition, such as the reticular core of the brain stem (Mantyth and Hunt 1984) and the periaqueductal grey matter (Skirboll *et al.* 1983), so that CCK could be a mediator of inhibition from the brain. An interesting observation is that CCK-like immunoreactivity is present in neurons which project rostrally from the spinal cord to the brainstem reticular formation (Nahin 1987; Leah *et al.* 1988), and it is possible that CCK might contribute to driving descending inhibition from neurons in that region (Le Bars and Villanueva 1988).

The effects of devazepide on CP-evoked inhibition in spinalized rabbits indicate that, rather than acting to oppose the actions of opioids, CCK might be one factor triggering their release; indeed, high intensity electrical stimulation of peripheral nerves in rat (Rodriguez and Sacristan 1989; Chapter 47 of this volume) and cat (Yaksh *et al.* 1982) results in the release of CCK-like immunoreactivity from the surface of the spinal cord. The insensitivity of supramaximal CP stimuli to devazepide might result from parallel non-CCK opioid-activating systems being sufficient to drive the release of opioid peptides, notwithstanding the blockade of CCK receptors. In non-spinalized rabbits it appeared that low, but not high, doses of devazepide enhanced the effects of CP stimulation. This is consistent with previous reports of bell-shaped dose–response curves for the CCK antagonist (Dourish *et al.* 1988, O'Neil *et al.* 1989; Chapter 44 of this volume), and suggests that endogenous CCK may after all act counter to the effects of opioids in this situation. It seems that CCK has more than one role in modulating reflexes, and further experiments, using a wider range of doses of devazepide in both spinalized and non-spinalized preparations, are required to clarify the involvement of this peptide in spinal reflex function.

Acknowledgements

This research was supported by the Agricultural and Food Research Council. We wish to thank Caroline Northway for expert technical assistance and Dr C. T. Dourish of Merck Sharp & Dohme for the supply of devazepide.

References

Baber, N. S., Dourish, C. T., and Hill, D. R. (1989). The role of CCK, caerulein and CCK antagonists in nociception. *Pain* **39**, 307–28.

Bell, J. A. and Martin, W. R. (1977). The effect of the narcotic antagonists naloxone, naltrexone and nalorphine on spinal C-fiber reflexes evoked by electrical stimulation or radiant heat. *Eur. J. Pharmacol.* **42**, 147–54.

Boureau, F., Willer, J.-C., and Dauthier, C. (1978). Study of naloxone in normal awake man: effects on spinal reflexes. *Neuropharmacology* **17**, 565–8.

Catley, D. M., Clarke, R. W., and Pascoe, J. E. (1983). Naloxone enhancement of spinal reflexes in the rabbit. *J. Physiol. (Lond.)* **339**, 61–73.

Catley, D. M., Clarke, R. W., and Pascoe, J. E. (1984). Post-tetanic depression of spinal reflexes in the rabbit and the possible involvement of opioid peptides. *J. Physiol. (Lond.)* **352**, 483–93.

Chung, J. M., Fang, Z. R., Cargill, C. L., and Willis, W. D. (1983). Prolonged, naloxone-reversible inhibition of the flexion reflex in the cat. *Pain* **15**, 35–53.

Clarke, R. W. and Ford, T. W. (1987). The contributions of μ, δ and κ opioid receptors to the actions of endogenous opioids on spinal reflexes in the rabbit. *Br. J. Pharmacol.* **91**, 579–89.

Clarke, R. W., Ford, T. W., Harris, S. M., and Taylor, J. S. (1988a). Thyrotropin releasing hormone, cholecystokinin and endogenous opioids in the modulation of spinal reflexes in the rabbit. *Neuropharmacology* **27**, 1279–84.

Clarke, R. W., Ford, T. W., and Taylor, J. S. (1988b). Adrenergic and opioidergic modulation of a spinal reflex in the decerebrated rabbit. *J. Physiol. (Lond.)* **404**, 407–17.

Clarke, R. W., Ford, T. W., Galloway, F. J., and Taylor J. S. (1989a). Naloxone enhancement of flexor and extensor reflexes in the decerebrated and spinalized rabbit. *J. Physiol. (Lond.)* **416**, 18P.

Clarke, R. W., Ford, T. W., and Taylor, J. S. (1989b). Activation by high intensity peripheral nerve stimulation of adrenergic and opioidergic suppression of a spinal reflex in the decerebrated rabbit. *Brain Res.* **505**, 1–6.

Dourish, C. T., Hawley, D., and Iversen, S. D. (1988). Enhancement of morphine analgesia and prevention of morphine tolerance in the rat by the cholecystokinin antagonist L-364,718. *Eur. J. Pharmacol.* **147**, 469–72.

Duggan, A. W., Morton, C. R., Johnson, S. M., and Zhao, Z. Q. (1984). Opioid antagonists and spinal reflexes in the anaesthetized cat. *Brain Res.* **297**, 33–40.

Faris, P. L., Komisaruk, B. R., Watkins, L. R., and Mayer, D. J. (1983). Evidence for the neuropeptide cholecystokinin as an antagonist of opiate analgesia. *Science* **219**, 310–12.

Goldfarb, J. and Hu, J. W. (1976). Enhancement of reflexes by naloxone in spinal cats. *Neuropharmacology* **15**, 785–92.

Harris, J., Ford, T. W., and Clarke, R. W. (1990). The effects of intrathecal application of α-adrenoceptor antagonists on the sural-gastrocnemius reflex of the decerebrated rabbit. *J. Physiol. (Lond.)* **429**, 35P.

Le Bars, D. and Villanueva, L. (1988). Electrophysiological evidence for the activation of descending controls by nociceptive afferent pathways. *Prog. Brain Res.* **77**, 275–99.

Leah, J., Menétrey, D., and de Pommery, J. (1988). Neuropeptides in long ascending spinal tract cells in the rat: evidence for parallel processing of ascending information. *Neuroscience* **24**, 195–207.

Mantyth, P. W. and Hunt, S. P. (1984). Evidence for cholecystokinin-like immunoreactive neurons in the rat medulla oblongata which project to the spinal cord. *Brain Res.* **291**, 49–54.

McClane, T. K. and Martin, W. R. (1967). The effects of morphine, naloxone, cyclazocine on the flexor reflex. *Int. J. Neuropharmacol.* **6**, 89–98.

Nahin, R. L. (1987). Immunocytochemical identification of long ascending peptidergic neurons contributing to the spinoreticular tract in the rat. *Neuroscience* **23**, 859–69.

O'Neill, M. F., Dourish, C. T., and Iversen, S. D. (1989). Morphine-induced analgesia in the rat paw-pressure test is blocked by CCK and enhanced by the CCK antagonist MK-329. *Neuropharmacology* **28**, 243–7.

Rodriguez, R. E. and Sacristan, M. P. (1989). *In vivo* release of CCK-8 from the dorsal horn of the rat: inhibition by DAGOL. *FEBS Lett.* **250**, 215–17.

Skirboll, L., Hökfelt, T., Dockray, G., Rehfeld, J., Brownstein, M., and Cuello, A. C. (1983). Evidence for peri-aqueductal cholecystokinin-substance P neurons projecting to the spinal cord. *J. Neurosci.* **3**, 1151–7.

Suberg, S. N., Culhane, E. S., Carstens, E., and Watkins, L. R. (1985). The potentiation of morphine-induced inhibition of spinal transmission by proglumide, a putative cholecystokinin antagonist. *Ann. NY Acad. Sci.* **448**, 660–2.

Taylor, J. S., Pettit, J. S., Harris, J., Ford, T. W., and Clarke, R. W. (1990). Noxious stimulation of the toes evokes naloxone-reversible suppression of the sural-gastrocnemius reflex of the rabbit. *Brain Res.*, **531**, 263–8.

Watkins, L. R., Kinschek, I. B., and Mayer, D. J. (1984). Potentiation of opiate analgesia and apparent reversal of morphine tolerance by proglumide. *Science* **224**, 395–6.

Watkins, L. R., Kinschek, I. B., Kaufman, E. F. S., Miller, J., Frenk, H., and Mayer, D. J. (1985). Cholecystokinin antagonists selectively potentiate analgesia induced by endogenous opiates. *Brain Res.* **327**, 181–90.

Weisenfeld-Hallin, Z. and Duranti, R. (1987). Intrathecal cholecystokinin interacts with morphine but not substance P in modulating the nociceptive flexion reflex in the rat. *Peptides* **8**, 153–8.

Yaksh, T. L., Abay, E. O., and Go, V. L. (1982). Studies on the location and release of cholecystokinin and vasoactive intestinal peptide in rat and cat spinal cord. *Brain Res.* **242**, 279–90.

50. Enhancement of opiate analgesia by devazepide in a baboon dolorimetry model

Hilton Klein, Robert Jackson, Gwendolyn McCormick, Tamara Montgomery, Dale Frankenfield, Walter Pouch, Keith Soper, and Kathy Murray

Electrical stimulation of tooth pulp provides a means of evaluating the effects of analgesics on pain without the complicating factors associated with peripheral stimulation. It is generally accepted that tooth pulp stimulation primarily elicits pain sensation (Scorr 1972; Anderson *et al.* 1973; Matthews and Searle 1976; Matthews 1979; Sessle 1979). Because the impulses from the dental nerve enter the brainstem and medulla via the trigeminal nerve, the afferent conduction pathway is short and less likely to be modulated by inhibitory neurons of reflex arcs. It has also been shown that all species have essentially the same dental sensory mechanisms (Byers *et al.* 1982). These facts make dental nerve stimulation an ideal means of evaluating the central modulation of analgesia as well as providing a procedure in which the basic mechanism of pain remains consistent between the different species. Dental dolorimetry is the measurement of pain responses by stimulation of the tooth pulp nerves and has been used to study pain in the rat (Iriki and Toda 1980*a*, *b*; Foong *et al.* 1982*a*), rabbit (Foong *et al.* 1982*b*; Wynn *et al.* 1984), cat (Mahon and Anderson 1970; Miyagawa *et al.* 1980), dog (Skingle and Tyers 1980), horse (Brunson and Majors 1987; Brunson *et al.* 1987), monkey (Van Hassle *et al.* 1972; Ha *et al.* 1978; Biedenbach *et al.* 1979; Chudler *et al.* 1986), and man (Matthews and Searle 1976). It provides an ideal means of studying pain perception because it is sensitive, reproducible, and allows for the quantitative evaluation of the response to pain with minimal interference from reflex responses. Dental dolorimetry also allows for repeated measurements of the same nerve and sensory pathway without tissue damage (Brunson *et al.* 1987). Its sensitivity is such that the pain-inducing stimulus is brief and can be discontinued as soon as the threshold is reached, thus minimizing discomfort to the subject and meeting the criteria for the evaluation of pain in animals (Vyklicky 1979; Covino *et al.* 1980). Previous studies have shown that CCK antagonists enhance opiate analgesia in rodents, primates, and humans (Chapters 43 and 44 of this volume). We have used dental dolorimetry to evaluate potentiation of alfentanil-induced analgesia by the CCK antagonist devazepide in the baboon. Alfentanil was chosen for these studies as it is a short-acting μ opioid receptor agonist and therefore it was possible to assess whether a CCK antagonist would increase the duration of opioid analgesia.

Baboons (*Papio anubis*) were used in these studies. All subjects were maintained in standard stainless steel caging in an American Association for Accreditation of Laboratory Animal Care (AAALAC) accredited facility and in accordance with the *Guide for the care and use of laboratory animals*. Each baboon was conditioned to chair restraint prior to the initiation of the investigative phases of the protocol. Conditioning consisted of lightly anaesthetizing the animal with ketamine (10 mg/kg intramuscularly (i.m.)) prior to placement in a restraint chair. The baboon was then allowed to recover in the restraint chair and maintained in restraint for 2–4 h. This process was repeated until the primate showed no observable signs of distress, i.e. struggling against the restraint, vocalization, facial expression.

Two to five days prior to the initiation of the investigative sessions, a monopolar electrode was implanted into the tooth pulp of an upper incisor at the gingival margin and anchored with silver amalgam or dental acrylic. Baboons were anesthetized i.m. with 8.0 mg/kg of ketamine plus 0.4 mg/kg of xylazine, intubated, and maintained on isoflurane and oxygen. An area extending from the nares to behind the vertical ramus of the mandible was clipped and prepared for surgery. The upper lip was elevated dorsally, the mouth flushed with water, and the gingiva swabbed with a povoiodine solution. A variable-speed dental drill and a round No. 4 dental burr were used to drill into the dental enamel at the gingival margin. Drilling was continued through the dentine layer which was identified by a pinkish grey coloration and softer texture.

A 30-gauge multistrand Teflon-coated stainless steel wire was used as an electrode. A 16-gauge over-the-needle catheter was used to pass the wire under the gingiva and subcutaneous tissue along the maxilla to a surgically created subcutaneous pouch located behind the vertical ramus of the mandible and below the ear. The electrode wire was then stripped of approximately 1 cm of the Teflon coating at the gingival end, knotted, and placed into the cavity produced in the incisor. Successful electrical stimulation of the tooth pulp was assured by extending the insulating coating of the electrode into the cavity to prevent short-circuits. The electrode was secured using silver amalgam or dental acrylic. Once secured, the distal end of the electrode wire was looped, sprayed thoroughly with a povoiodine solution, and placed into the subcutaneous pouch (Fig. 50.1). The pouch incision was closed using 2-0 non-absorbable suture in a simple interrupted pattern.

Multiple test sessions were carried out on all baboons. When successive sessions were conducted in the same animal, a minimum of 24 h was allowed between each session. Prior to a test session, each baboon was chemically restrained with ketamine (10 mg/kg i.m.), placed in a restraint chair, recovered, and allowed to sit for at least 2 h post-recovery to minimize the effect of

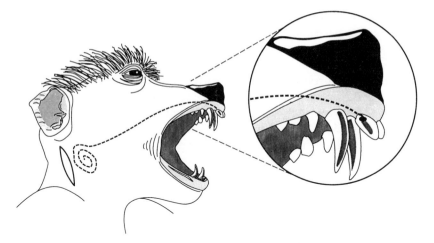

Fig. 50.1 Schematic diagram of wire electrode placement. Wire is run subgingivally/subcutaneously from the gingival margin to a subcutaneous pocket located beneath the ear. Care must be taken to ensure that the wire is not exposed within the cheek pouch.

ketamine on the dolorimetry studies. While the animal was still under the influence of the ketamine restraint, the sutures to the subcutaneous pouch were removed and the incision was reopened to exteriorize the free end of the electrode. The electrode tip was cleaned with a sterile gauze sponge and then connected to a Grass stimulator located behind the restraint chair. An earthed electrode was placed on a shaved area of the animal's right forearm.

Each session was conducted by a team of at least two investigators. One investigator operated the Grass stimulator to stimulate the tooth pulp electrically. Each stimulation sequence started at zero volts and was gradually increased at 0.1 V intervals until a lick/chew response was elicited. Each electrical pulse was given at a frequency of 0.5 pulses/s with a duration of 0.5 ms and a 0.5 ms delay. The other investigator served as a blind observer. At no time did the observer know what voltage was being used for stimulation. When a lick/chew response was noted by the observer, the operator was notified and recorded the voltage which stimulated the response. At this time the voltage was reduced to zero and the stimulation sequence was repeated.

Each session consisted of several 'runs' in which stimulation was provided at 1 pulse/s for 0.5 ms at gradually increasing voltages. The first run was conducted before any drug was administered in order to establish a minimum baseline voltage required to induce pain as evidenced by the lick/chew response. During this run, the stimulation procedure was repeated until three consecutive readings that were within 0.2 V of each other were attained. The second run was to establish that the vehicle (glycerol, PEG 4000, and sterile water) used to make the drug dilutions had no analgesic effect when dosed intravenously (i.v.). This run was used to determine a baseline with vehicle and served as a vehicle control. Subsequent runs were used to test compounds for minimal analgesic response.

A total of seven test sessions were carried out using four baboons. When successive sessions were conducted in the same animal, a minimum of 24 h was allowed between each session. For the 'alfentanil run', a single dose of alfentanil at (0.25 µg/kg i.v.) was administered. The stimulation voltage was then recorded at 2 min intervals to establish a dose-response curve over a 15 min period. When necessary, additional runs were made increasing the dosage of alfentanil by 0.25 µg/kg i.v. until an increase in the voltage required to elicit a lick/chew response was observed. This minimal analgesic dose was then used as the target dose of alfentanil. No baboon in this study required more then 0.5 µg/kg to achieve an analgesic effect. Thus alfentanil served as a positive control. A 10–15 min wash-out period was allowed after the alfentanil run, after which the baboon was tested to ensure that the lick/chew response could be induced at the baseline voltage.

Once the voltage required for stimulation returned to baseline voltage, a single dose of devazepide (10 µg/kg i.v.) was administered. Based on data in the squirrel monkey showing that devazepide had no significant analgesic effect when measured at 30 min intervals over 2 h (Dourish *et al.* 1990), a voltage response was measured at 0 min and 15 min to demonstrate that devazepide given alone had no immediate short-acting analgesic effect. At this time, the i.v. target dose of alfentanil was administered and the test session defined as the 'potentiation run' commenced. A voltage response was then measured every 2 min until the analgesic effect returned to baseline or below on two successive measurements.

In order to minimize the number of primates studied, multiple sessions were conducted on three of four animals. By using each test session as the unit of analysis the sample size was increased from four baboons to seven sessions. Each animals served as its own control, thus reducing variation between animals.

Four criteria for evaluating analgesic potentiation were defined. These criteria were time of onset, duration, maximal response, and area beneath the response curve. Each variable was assessed using the Wilcoxon matched-pair rank sum test (Hollander and Wolfe 1973). Since the area beneath the response curve combined maximal response and duration of response, it was selected as the principal measure. Once the effect of devazepide was established by demonstrating that the area was increased, time of onset, duration, and maximal response were used to indicate the nature of the increase.

In order to assess the variability in response among baboons, a second analysis was performed for the principle outcome variable area using the animal as the unit of analysis. Multiple sessions using the same animal were averaged so that each baboon would provide one area for the alfentanil run and one area for the potentiation run. Because of the small sample size (four), the data were analysed using a *t* test (Snedecor and Cochran 1967). Although the data have a natural pairing, an unpaired *t* test was used in order to increase the degrees of freedom for error from three to six. The use of the unpaired *t* test is conservative, provided that the responses within each animal are positively correlated, and it was felt that the increase in precision with six degrees of freedom for error would outweigh the reduction in variance from using paired data (Snedecor and Cochran 1967). Additionally, each area was normalized by the starting value for the run to reduce variation between animals.

No formal statistical tests were made to evaluate a trend in voltage response over time within a session because each successive run included a higher dosage of alfentanil or the addition of devazepide. However, an inspection of the data showed no marked trend in voltage response over time prior to the administration of compound for each successive run.

There was no pattern of analgesic effect with the administration of 10 μg/kg devazepide alone ($p = 0.812$). In every session, the voltage required to elicit the lick/chew response 15 min after the i.v. administration of devazepide was not significantly higher than the baseline response. Two of the early sessions included voltage measurements at 5 and 10 min after devazepide administration to ensure that the compound did not have a transitory analgesic effect. In each case there was no indication of analgesia based on voltage measurements.

Four criteria were used to evaluate the effect of devazepide on alfentanil analgesia. The first response evaluated was the difference between the area below the dose-response curve in the alfentanil run compared with the dose-response area in the potentiation run (Fig. 50.2). In order to assess the variability in response among the baboons, a *t* test for the dose–response area was computed using each animal as the unit of analysis. Because the variance in the potentiation runs appeared to be greater than that in the alfentanil runs, the variance was stabilized using a logarithmic transformation (Snedecor and Cochran 1967). For all seven sessions, the *t* test indicated that the combination of devazepide and alfentanil (potentiation-run) yielded a larger dose–response area than did the alfentanil run ($p = 0.044$).

Changes in the time of onset, duration of response, and maximal response of analgesia were also evaluated between the alfentanil and potentiation runs. In five of the seven sessions, there was no difference in the time of onset between the runs. One session showed a more rapid time of onset with devazepide, and one session showed a longer time of onset with devazepide. There was no evidence that devazepide alters the onset of alfentanil analgesia. In all seven sessions, the duration of analgesia was longer in the potentiation run than in the alfentanil run ($p = 0.008$). The maximal response was defined as the maximum increase over baseline in the voltage required to induce the lick/chew response. Five of the seven exhibited a larger maximal response for the potentiation run than for the alfentanil run. However, this trend did not achieve statistical significance.

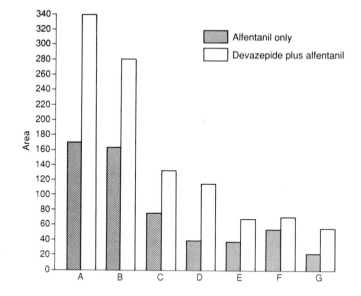

Fig. 50.2 Comparison of the areas under the curves for alfentanil alone and devazepide plus alfentanil: A–G, the seven test sessions run using four baboons. The area under the curve is a primary measure of anlagesic potency as determined by dental dolorimetry. It is a composite of the maximum effect and duration of analgesia induced by a drug. Devazepide caused significant potentiation of the analgesic effect produced by alfentanil ($p = 0.008$).

The decision to use a repeated measures design in this study was based on the three Rs of animal research (Russell and Burch 1959; Bennett *et al.* 1990): *reduction* of the number of animals used, *refinement* of techniques which reduced the pain and distress to which an animal is subjected, and *replacement* of animal models when appropriate. The design also allowed for minimal variation of the target dose of alfentanil required between individual baboons and different test sessions. Each session was then evaluated as to the results of the alfentanil run, the devazepide response, and the potentiation run (alfentanil plus devazepide). Alfentanil was used in this study because it is a potent μ opioid analgesic with an immediate onset of action. Additionally, alfentanil has a short half-life compared with other opioids. The short distribution half-life and rapid elimination of alfentanil permitted the use of the same baboon for both the alfentanil and potentiation runs within the same day, thus eliminating possible variables within the same subject on different days and resulting in more accurate comparisons between runs in an individual baboon.

The results of this study suggest that the CCK antagonist devazepide potentiates opioid analgesia and that this interaction does not involve neural pathways other than those related to pain. Although devazepide had no intrinsic analgesic effect, it enhanced alfentanil analgesia by increasing the duration of response and/or maximal effect when compared with alfentanil given alone. This is consistent with findings in a thermal pain model in the squirrel monkey (Dourish *et al.* 1990). The mechanism of potentiation appears to be pharmacologically based rather than a direct analgesic effect on opioid receptors (see Chapter 44 of this volume). Since devazepide enhances opioid analgesia without potentiating morphine-induced respiratory depression (Dourish *et al.* 1990), this compound may serve as a valuable adjuvant in the treatment of pain with opioid analgesics.

References

Anderson, S. A., Keller, O., and Vyklicky, L. (1973). Cortical activity evoked from tooth pulp afferents. *Brain Res.* **50**, 473–5.

Bennett, B. T., Brown, M. J. and Schofield, J. C. (1990). *Essentials for animal research. A primer for research personnel*. National Agriculture Library, Bethesda, MD.

Biedenbach, M. A., Van Hassel, H. J., and Brown, A. C. (1979). Tooth pulp-driven neurons in somatosensory cortex of primates: role of pain mechanisms including a review of the literature. *Pain* **7**, 31–50.

Brunson, D. B., and Majors, L. J. (1987). Comparative analgesia of xylazine, xylazine/morphine, xylazine/butorphanol, and xylazine/nalbupphine in the horse, using dental dolorimetry. *Am. J. Vet. Res.* **48**, 1087–91.

Brunson, D. B., Collier, M. A., Scott, E. A., and Majors, L. J. (1987). Dental dolorimetry for the evaluation of an analgesic agent in the horse. *Am. J. Vet. Res.* **48**, 1082–86.

Byers, M. R., Neuhaus, S. J., and Gehrig, J. D. (1982). Dental sensory receptor structure in human teeth. *Pain* **13**, 221–35.

Chudler, E. H., Dong, W. K., and Kawakami, Y. (1986). Cortical nociceptive responses and behavioral correlates in the monkey. *Brain Res.* **397**, 47–60.

Covino, B. G., Dubner, R., Gybels, J., Kosterlitz, H. W., Liebeskind, J. C., Sternbach, R. A., Vyklicky, L., Yamamura, H., and Zimmermann, H. (1990). Ethical standards for investigation of experimental pain in animals. *Pain* **9**, 141–43.

Dourish, C. T., O'Neill, M. F., Schaffer, L. W., Siegl, P. K. S., Iversen, S. D., (1990). The cholecystokinin receptor antagonist devazepide enhances morphine-induced analgesia but not morphine-induced respiratory depression in the squirrel monkey. *J. Pharmacol. Exp. Ther.* **255**, 1158–65.

Foong, F. W., Satoh, M., and Taskagi, H. (1982*a*). A newly devised reliable method for evaluating analgesic potencies of drugs on trigeminal pain. *J. Pharmacol. Methods* **7**, 271–8.

Foong, F. W., Satoh, M., and Takagi, H. (1982*b*). Sites of analgesic action of cyclazocine: a study using evoked potentials at various regions of the CNS by electrical stimulation of rabbit tooth pulp. *Arch. Int. Pharmacodyn. Ther.* **256**, 212–18.

Ha, H., Wu, R. S., Contreras, R. A., and Tan E. (1978). Measurement of pain threshold by electrical stimulation of tooth pulp afferents in the monkey. *Exp. Neurol.* **61**, 260–9.

Hollander, M. and Wolfe, D. A. (1973). *Nonparametric statistical methods*. Wiley, New York.

Iriki, A. and Toda, K. (1980*a*). Difference in the effects of electro-acupuncture and morphine on thalamic-evoked responses in ventrobasal complex and the posterior nuclear group after tooth pulp stimulation in the rat. *Arch. Oral Biol.* **25**, 697–99.

Iriki, A. and Toda, K. (1980*b*) Morphine and electroacupunture: comparison of the effects on the cortical-evoked responses after tooth pulp stimulation in rats. *Eur. J. Pharmacol.* **68**, 83–7.

Mahon, P. E. and Anderson, K. V. (1970). Activation of pain pathways in animals. *Am. J. Anat.* **128**, 235–8.

Matthews, B. (1979). Functions of tooth-pulp afferents. In *Advances in pain research theraphy*, Vol 3 (ed. J. J. Bonica, J. C. Liebeskind, and D. G. Albe-Fessard), pp. 261–4. Raven Press, New York.

Matthews, B. and Searle, B. N. (1976). Electrical stimulation to teeth. *Pain* **2**, 245–51.

Miyagawa, T., Sakurada, S., Shima, K., Ando, R., Takashi, N., and Kisara, K. (1980). Effects of morphine on evoked potentials recorded from the amygdala by tooth pulp stimulation in cats. *Jpn J. Pharmacol.* **30**, 463–9.

Russell, W. M. S. and Burch, R. L. (1959). *The principles of humane experimental technique*. Methuen, London.

Scorr, D. (1972). The arousal and suppression of pain in the tooth. *Int. Dent. J.* **22**, 20.

Sessle, B. J. (1979). Is the tooth pulp a 'pure' source of noxious input. In *Advances in pain research therapy*, Vol. 3 (ed. J. J. Bonica, J. C. Liebeskind, and D. G. Albe-Fessard), pp. 245–60. Raven Press, New York.

Skingle, M. and Tyers, M. B. (1980). Further studies on opiate receptors that mediate antinociception: tooth pulp stimulation in the dog. *Br. J. Pharmacol.* **70**, 323–7.

Snedecor, G. W. and Cochran, W. G. (1967). *Statistical methods*. Iowa State University Press, Ames, IA.

Van Hassle, H. J., Biedenbach, M. A., and Brown A. C. (1972). Cortical potentials evoked by tooth pulp stimulation in rhesus monkeys. *Arch. Oral Biol.* **17**, 1059–66.

Vyklicky, L. (1979). Techniques for the study of pain in animals. In *Advances in pain research and therapy*, Vol. 3 (ed. J. J. Bonica, J. C. Liebeskind, and D. G. Albe-Fessard), pp. 727–46. Raven Press, New York.

Wynn, R. L., ElBaghdady, Y. M., and Ford R. D. (1984). A rabbit tooth pulp assay to determine ED_{50} values and duration of action of analgesics. *J. Pharmacol. Methods* **11**, 109–17.

51. Blockade of CCK$_B$ receptors by L-365,260 induces analgesia in the squirrel monkey

Michael F. O'Neill, Spencer Tye, and Colin T. Dourish

CCK receptors are found in high concentrations in the CNS in areas that are associated with nociception such as the dorsal horn of the spinal cord and periaqueductal grey (PAG) (Baber *et al.* 1989). Two subtypes of CCK receptor have been identified, CCK$_A$ and CCK$_B$, which have greater preponderance in alimentary and brain tissues respectively (Moran *et al.* 1986). Recently a series of non-peptide CCK antagonists have been described which differ in their relative affinities for these subtypes of CCK receptor. L-365,031 and devazepide (Chang and Lotti 1986; Evans *et al.* 1986) have nanomolar affinity for CCK$_A$ receptors, whereas L-365,260 has nanomolar affinity for CCK$_B$ receptors (Bock *et al.* 1989; Lotti and Chang 1989). Thus the functional roles of CCK$_A$ and CCK$_B$ receptor subtypes in mediating the effects of CCK can be studied.

CCK exerts a bimodal influence on pain thresholds in rodents. Large doses ($> 50\,\mu g/kg$) cause sedation, ptosis, and analgesia in rodents (Zetler 1980). Lower doses ($2-16\,\mu g/kg$) have no intrinsic analgesic activity and, furthermore, block analgesia induced by endogenous or exogenous opiates (Faris *et al.* 1983). It has been suggested that CCK may act as an endogenous antagonist of opioid activity. Hence blockade of CCK receptors may enhance opioid analgesia by preventing inhibition of opioid action by endogenous CCK (Baber *et al.* 1989). Indeed, the weak non-selective CCK antagonist proglumide and the potent and selective antagonists devazepide, L-365,031, and L-365,260 have no intrinsic antinociceptive actions but enhance opioid analgesia in a variety of rodent pain models (Baber *et al.* 1989; Dourish *et al.* 1990*a*). Similarly, devazepide had no intrinsic effect on pain thresholds in the squirrel monkey but potentiated morphine analgesia (Dourish *et al.* 1990*b*). Enhancement of morphine analgesia by CCK antagonists in rodents appears to be mediated by CCK$_B$ receptors (Dourish *et al.* 1990*a*; Chapters 44 and 45 of this volume). However, there are profound interspecies differences in the distribution of CCK receptor subtypes (Hill *et al.* 1988) and hence results obtained in rodent models cannot necessarily be generalized to primates and humans. Therefore in the present study we examined the role of CCK$_B$ receptors in the mediation of nociceptive responses in the primate using the selective CCK_B antagonist L-365,260. In contrast with previous findings with other CCK antagonists, we discovered that L-365,260 possesses intrinsic antinociceptive properties in the squirrel monkey.

The method used was essentially the same as that described by Dourish *et al.* (1990*b*). Six adult male squirrel monkeys (*Saimiri sciureus*) were group housed under standard laboratory conditions. They were adapted to sitting to Perspex testing chairs for 3 h daily for 1 week prior to testing. During this time the animals were also habituated to leaving their tails immersed in cool water (35 °C). On drug test days the animals had to satisfy two criteria. First, each monkey had to leave its tail in the cool water (35 °C) for 20 s. Second, each monkey had to display an appropriate response to noxious stimuli by withdrawing its tail from warm water (55 °C) within 4 s. If any animal failed to meet either of these two criteria it was excluded from that day's testing. Testing was carried out twice weekly. Dose was assigned on the basis of a pseudo-random latin square design such that each animal received each treatment without confounding dose-order effects.

L-365,260 (R(+)-*N*-(2,3-dihydro-1-methyl-2-oxo-5-phenyl-1*H*-1, 4-benzodia-zepin-3-yl-)-*N*-(3-methyl-phenyl)urea) was suspended in 0.5 per cent methylcellulose and injected intraperitoneally (i.p.) in a volume of 0.1 ml/kg. Immediately following drug administration, the latency to tail withdrawal for each animal was tested at 30 min intervals for 120 min. If the monkey failed to remove its tail from the warm water within 20 s, the tail was removed by the experimenter and a latency of 20 s was recorded. The cut-off of 20 s was used to prevent any possible tissue damage that might result from prolonged exposure to warm water.

Data were square root transformed to normalize the variance and analysed by ANOVA with planned contrasts. L-365,260 significantly increased tail-withdrawal latency in the squirrel monkey (L-365,260 main effect $F(7, 35) = 3.97$, $p < 0.001$). The dose–response curve for L-365,260-induced analgesia was bell-shaped (Fig. 51.1). Thus doses from 100 ng/kg to 100 µg/kg significantly raised withdrawal latencies, while doses above and below this range were without significant effect. ANOVA revealed that there was no time effect nor was there a significant interaction between L-365,260 and time (time main effect $F(3, 15) = 0.97$, $p < 0.5$; interaction $F(21, 105) = 0.66$, $p < 0.7$), indicating that there was no change in the analgesic effect of L-365,260 over the 2 h test period. For example, tail-withdrawal latencies were consistently elevated by a 1 µg/kg dose of L-365,260 for the entire 2 h test period (Fig. 51.2).

Previously, L-365,260 has been shown to enhance the analgesic effect of morphine in rodents. This study reports for the first time that blockade of the action of endogenous CCK produces analgesia in the primate. The dose-response curve was bell-shaped which appears to be characteristic of CCK antagonist activity in analgesic tests (see Chapter 44 of this volume). L-365,260 significantly raised withdrawal thresholds over a 1000-fold dose range (100 ng/kg–100 µg/kg). This is a broader range of doses than that reported in previous

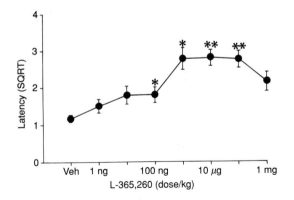

Fig. 51.1 Effect of i.p. injection of L-365,260 on withdrawal latencies in the squirrel monkey tail-withdrawal test. Data are mean (square root) latencies (± SE) for six monkeys during a 120 min test period. Significant differences were determined by planned contrasts following a significant ANOVA: $*p < 0.05$; $**p < 0.01$.

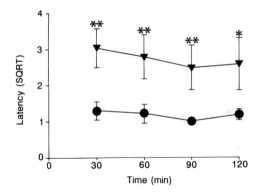

Fig. 51.2 Time course of analgesic action of i.p. L-365,260 (1 μg/kg) versus vehicle control in the squirrel monkey tail-withdrawal test. Data are mean (square root) latency at each time point. The x-axis shows the time after injection and the y-axis shows latency (square root): $*p < 0.1$; $**p < 0.05$.

studies in rodents using CCK antagonists to potentiate opioid analgesia (Baber *et al.* 1989; Dourish *et al.* 1990a; Chapter 44 of this volume).

The neuroanatomical site of action for the antinociceptive effect of L-365,260 is unclear. In the primate, the spinal cord receptors are of the CCK_A subtype and L-365,260 has only micromolar affinity for this subtype (Hill *et al.* 1988). In the squirrel monkey tail-withdrawal test L-365,260 was active at doses as low as 100 ng/kg, and at this dose the drug is unlikely to bind to CCK_A receptors (Lotti and Chang 1989; see Chapter 9 of this volume). Furthermore, devazepide has subnanomolar affinity for CCK_A receptors but does not induce analgesia in the primate (Dourish *et al.* 1990*b*). L-365,260 has nanomolar affinity for CCK_B/gastrin receptors (Lotti and Chang 1989). CCK_B receptors are absent in primate spinal cord but are widely distributed throughout primate brain and are observed in high concentrations in the cortex, hypothalamus, and PAG (Hill *et al.* 1988, 1990). The PAG is an area known to be crucially involved in the mediation of nociception (Baber *et al.* 1989). Infusion of CCK into rat PAG blocks morphine analgesia, whereas infusion of the non-selective CCK antagonist proglumide into this site enhances morphine analgesia (Chapters 43 and 45 of this volume). Thus endogenous CCK may tonically inhibit opioid neurons in the PAG, and blockade of CCK receptors by L-365,260 may cause disinhibition of these neurons and thereby induce analgesia.

References

Baber, N. S., Dourish, C. T., and Hill, D. R. (1989). The role of CCK, caerulein and CCK antagonists in nociception. *Pain* **39**, 307–28.

Bock, M. G., DiPardo, R. M., Evans, B. E., Rittle, K. E., Whitter, W. L., Veber, D. F., Anderson, P. S., and Freidinger, R. M. (1989). Benzodiazepine gastrin and cholecystokinin receptor ligands: L-365,260. *J. Med. Chem.* **32**, 13–16.

Chang, R. S. L. and Lotti, V. J. (1986). Biochemical and pharmacological characterization of an extremely potent and selective nonpeptide cholecystokinin antagonist. *Proc. Natl Acad. Sci. USA* **83**, 4923–6.

Dourish, C. T., O'Neill, M. F., Coughlan, J., Kitchener, S. J., Hawley, D., and Iversen, S. D. (1990*a*). The selective CCK_B receptor antagonist L-365,260 enhances morphine analgesia and prevents morphine tolerance in the rat. *Eur. J. Pharmacol.* **176**, 35–44.

Dourish, C. T., O'Neill, M. F., Schaeffer, L. W., Siegl, P. K. S., and Iversen, S. D. (1990*b*). The cholecystokinin receptor antagonist devazepide enhances morphine-induced analgesia but not morphine-induced respiratory depression in the squirrel monkey. *J. Pharmacol. Exp. Ther.* **255**, 1158–65.

Evans, B. E., Bock, M. G., Rittle, K. E., DiPardo, R. M., Whitter, W. L., Veber, D. F., Anderson, P. S., and Freidinger, R. M. (1986). Design of potent, orally effective, nonpeptidal antagonists of the peptide hormone cholecystokinin. *Proc. Natl Acad. Sci. USA* **83**, 4918–22.

Faris, P. L., Komisaruk, B. R., Watkins, L. R., and Mayer, D. J. (1983). Evidence for the neuropeptide cholecystokinin as an antagonist of opiate analgesia. *Science* **219**, 310–12.

Hill, D. R., Shaw, T. M., and Woodruff, G. N. (1988). Binding sites for ^{125}I-cholecystokinin in primate spinal cord are of the CCK_A subclass. *Neurosci. Lett.* **89**, 133–9.

Hill, D. R., Shaw, T. M., Graham, W., and Woodruff, G. N. (1990). Autoradiographical detection of cholecystokinin (CCK_A) receptors in primate brain using ^{125}I-Bolton Hunter CCK-8 and [^{3}H]-MK-329. *J. Neurosci.* **10**, 1070–81.

Lotti, V. J. and Chang, R. S. L. (1989). A new potent and selective non-peptide gastrin

antagonist and brain cholecystokinin receptor (CCK_B) ligand: L-365,260. *Eur. J. Pharmacol.* **163**, 273–80.

Moran, T. H., Robinson, P. H., Goldrich, M. S., and McHugh, P. R. (1986). Two brain cholecystokinin receptors: implications for behavioural actions. *Brain Res.* **362**, 175–9.

Zetler, G. (1980). Analgesia and ptosis caused by caerulein and cholecystokinin octapeptide (CCK-8). *Neuropharmacology* **19**, 415–22.

Index

A-57696, in ventromedial hypothalamic nucleus neurons, effects 91, 93
A-65186, hypolocomotion and effects of, CCK-induced 257
A 71623 80, 81
abdominal organs, feeding behaviour and receptors in 170-2; *see also specific organs*
Aβ-fibres, opioid–CCK interactions and 503
acid secretion, gastric 98
  agonist effects 99-100
acupuncture, electro-, *see* electro-acupuncture
acylsemicarbazide derivatives 30-3
adenosine monophosphate, *see* AMP, cyclic
adenylate cyclase, dopamine-induced 370
Aδ-fibres, opioid–CCK interactions and 471-2
afferent fibres, gastric, feeding behaviour and 170, 199-200
A-fibres, opioid–CCK interactions and 471-2, 503, 504
agonists, CCK receptor 77-90, 91-4, 136-8, 142-7, 166-72, 456-9
  behavioural effects 77-90, 136-8, 142-7, 166-72
  binding characteristics in brain 81-3
  design 78-80
  in dyskinesia treatment 410, 411
  gastrointestinal effects 99-102
  opioid analgesia blockade by 456-9, 465
  in schizophrenia treatment 390-1
  in ventromedial hypothalamic nuclei, effects 91-4
  see also specific agonists and receptors
agonists, non-CCK, *see specific agonists and receptors*
agoraphobia, panic disorder with 122
  CCK-4-induced symptoms in patients with 126-7
A(h)-receptor 4
alprazolam, predator responses and effects of 138-40
A(l)-receptor 4
amino acid sequence, CCK family 3, 4
3-aminobenzodiazepine derivatives 15, 16
3-aminobenzolactams 17
γ-aminobutyric acid, CCK and, interactions 155-8
3-amino-5-phenyl-1,4-benzodiazepine derivatives 15

Ammon's horn, CCK mRNA distribution and 51
AMP, cyclic
  c-*fos* mRNA induction and 320
  spinal, opiate and CCK effects on 494
amygdala/amygdaloid nuclei
  CCK injected into, anxiogenicity 129
  CCK mRNA in 41, 42
analgesia
  electro-acupuncture, *see* electro-acupuncture
  non-opioid endogenous 132
  opioid-mediated endogenous 132, 431-541
animals, farm, satiety in 206-21; *see also specific animals*
anorexia
  aspartame-induced, post-ingestive, in humans 309-14
    mechanisms 310-12
  aspartame-induced, post-ingestive, in rats, relative absence 311
  D-fenfluramine-induced 266, 267, 287-8
anorexia nervosa 305
  cause/mechanisms 305
  treatment 274
antagonists (of CCK receptor family) 5, 6, 8-37, 57-76, 173-6, 214-16, 243-6, 280-5, 290-300, 315-18, 459-65, 467-70, 474-5, 495-7, 522-41
  apomorphine-induced hypotensive response and effects of 427
  behavioural effects (in general) of non-peptide 69-70, 71-2, 138-40
  biochemical studies using non-peptide 61-4
  blood–brain transfer of 107-14
  electrophysiological studies 28-37, 64-9
  feeding behaviour and effects of 173-6, 214-16, 243-6, 264, 265, 280-5, 290-300
  gastrointestinal effects 102-4
  hypolocomotion and effects of, CCK-induced 257
  medicinal chemistry 28-37
  in opioid analgesia 433-5, 436-7, 441-7, 459-65, 465, 467-70, 474-5, 495-7, 522-41
    clinical/therapeutic potential 436-7, 474-5
  oxytocin neuronal activation blocked by 315-18

Index